W9-AKD-867

DEMCO

STEVE JOBS

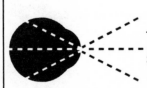

This Large Print Book carries the
Seal of Approval of N.A.V.H.

STEVE JOBS

WALTER ISAACSON

THORNDIKE PRESS
A part of Gale, Cengage Learning

GALE
CENGAGE Learning

Detroit • New York • San Francisco • New Haven, Conn • Waterville, Maine • London

GALE
CENGAGE Learning

Library of Congress CIP data on file. Cataloguing in Publication Data for this book is available from the Library of Congress.

ISBN-13: 978-1-4104-4522-3 (hardcover)
ISBN-10: 1-4104-4522-4 (hardcover)

Published in 2011 by arrangement with Simon & Schuster, Inc.

Printed in the United States of America
3 4 5 6 7 15 14 13 12 11

The people who are crazy enough
to think they can change
the world are the ones who do.
— Apple's "Think Different" commercial, 1997

CONTENTS

CHARACTERS

AL ALCORN. Chief engineer at Atari, who designed Pong and hired Jobs.

GIL AMELIO. Became CEO of Apple in 1996, bought NeXT, bringing Jobs back.

BILL ATKINSON. Early Apple employee, developed graphics for the Macintosh.

CHRISANN BRENNAN. Jobs's girlfriend at Homestead High, mother of his daughter Lisa.

LISA BRENNAN-JOBS. Daughter of Jobs and Chrisann Brennan, born in 1978; became a writer in New York City.

NOLAN BUSHNELL. Founder of Atari and entrepreneurial role model for Jobs.

BILL CAMPBELL. Apple marketing chief during Jobs's first stint at Apple and board member and confidant after Jobs's return in 1997.

EDWIN CATMULL. A cofounder of Pixar and later a Disney executive.

KOBUN CHINO. A Sōtō Zen master in California who became Jobs's spiritual teacher.

LEE CLOW. Advertising wizard who created Apple's "1984" ad and worked with Jobs for three decades.

DEBORAH "DEBI" COLEMAN. Early Mac team manager who took over Apple manufacturing.

11

TIM COOK. Steady, calm, chief operating officer hired by Jobs in 1998; replaced Jobs as Apple CEO in August 2011.

EDDY CUE. Chief of Internet services at Apple, Jobs's wingman in dealing with content companies.

ANDREA "ANDY" CUNNINGHAM. Publicist at Regis McKenna's firm who handled Apple in the early Macintosh years.

MICHAEL EISNER. Hard-driving Disney CEO who made the Pixar deal, then clashed with Jobs.

LARRY ELLISON. CEO of Oracle and personal friend of Jobs.

TONY FADELL. Punky engineer brought to Apple in 2001 to develop the iPod.

SCOTT FORSTALL. Chief of Apple's mobile device software.

ROBERT FRIEDLAND. Reed student, proprietor of an apple farm commune, and spiritual seeker who influenced Jobs, then went on to run a mining company.

JEAN-LOUIS GASSÉE. Apple's manager in France, took over the Macintosh division when Jobs was ousted in 1985.

BILL GATES. The other computer wunderkind born in 1955.

ANDY HERTZFELD. Playful, friendly software engineer and Jobs's pal on the original Mac team.

JOANNA HOFFMAN. Original Mac team member with the spirit to stand up to Jobs.

ELIZABETH HOLMES. Daniel Kottke's girlfriend at Reed and early Apple employee.

ROD HOLT. Chain-smoking Marxist hired by Jobs

in 1976 to be the electrical engineer on the Apple II.

ROBERT IGER. Succeeded Eisner as Disney CEO in 2005.

JONATHAN "JONY" IVE. Chief designer at Apple, became Jobs's partner and confidant.

ABDULFATTAH "JOHN" JANDALI. Syrian-born graduate student in Wisconsin who became biological father of Jobs and Mona Simpson, later a food and beverage manager at the Boomtown casino near Reno.

CLARA HAGOPIAN JOBS. Daughter of Armenian immigrants, married Paul Jobs in 1946; they adopted Steve soon after his birth in 1955.

ERIN JOBS. Middle child of Laurene Powell and Steve Jobs.

EVE JOBS. Youngest child of Laurene and Steve.

PATTY JOBS. Adopted by Paul and Clara Jobs two years after they adopted Steve.

PAUL REINHOLD JOBS. Wisconsin-born Coast Guard seaman who, with his wife, Clara, adopted Steve in 1955.

REED JOBS. Oldest child of Steve Jobs and Laurene Powell.

RON JOHNSON. Hired by Jobs in 2000 to develop Apple's stores.

JEFFREY KATZENBERG. Head of Disney Studios, clashed with Eisner and resigned in 1994 to cofound DreamWorks SKG.

DANIEL KOTTKE. Jobs's closest friend at Reed, fellow pilgrim to India, early Apple employee.

JOHN LASSETER. Cofounder and creative force at Pixar.

DAN'L LEWIN. Marketing exec with Jobs at Apple

13

and then NeXT.

MIKE MARKKULA. First big Apple investor and chairman, a father figure to Jobs.

REGIS MCKENNA. Publicity whiz who guided Jobs early on and remained a trusted advisor.

MIKE MURRAY. Early Macintosh marketing director.

PAUL OTELLINI. CEO of Intel who helped switch the Macintosh to Intel chips but did not get the iPhone business.

LAURENE POWELL. Savvy and good-humored Penn graduate, went to Goldman Sachs and then Stanford Business School, married Steve Jobs in 1991.

GEORGE RILEY. Jobs's Memphis-born friend and lawyer.

ARTHUR ROCK. Legendary tech investor, early Apple board member, Jobs's father figure.

JONATHAN "RUBY" RUBINSTEIN. Worked with Jobs at NeXT, became chief hardware engineer at Apple in 1997.

MIKE SCOTT. Brought in by Markkula to be Apple's president in 1977 to try to manage Jobs.

JOHN SCULLEY. Pepsi executive recruited by Jobs in 1983 to be Apple's CEO, clashed with and ousted Jobs in 1985.

JOANNE SCHIEBLE JANDALI SIMPSON. Wisconsin-born biological mother of Steve Jobs, whom she put up for adoption, and Mona Simpson, whom she raised.

MONA SIMPSON. Biological full sister of Jobs; they discovered their relationship in 1986 and became close. She wrote novels loosely based on her mother Joanne (*Anywhere but Here*), Jobs and

14

his daughter Lisa (*A Regular Guy*), and her father Abdulfattah Jandali (*The Lost Father*).

ALVY RAY SMITH. A cofounder of Pixar who clashed with Jobs.

BURRELL SMITH. Brilliant, troubled programmer on the original Mac team, afflicted with schizophrenia in the 1990s.

AVADIS "AVIE" TEVANIAN. Worked with Jobs and Rubinstein at NeXT, became chief software engineer at Apple in 1997.

JAMES VINCENT. A music-loving Brit, the younger partner with Lee Clow and Duncan Milner at the ad agency Apple hired.

RON WAYNE. Met Jobs at Atari, became first partner with Jobs and Wozniak at fledgling Apple, but unwisely decided to forgo his equity stake.

STEPHEN WOZNIAK. The star electronics geek at Homestead High; Jobs figured out how to package and market his amazing circuit boards and became his partner in founding Apple.

INTRODUCTION

HOW THIS BOOK CAME TO BE

In the early summer of 2004, I got a phone call from Steve Jobs. He had been scattershot friendly to me over the years, with occasional bursts of intensity, especially when he was launching a new product that he wanted on the cover of *Time* or featured on CNN, places where I'd worked. But now that I was no longer at either of those places, I hadn't heard from him much. We talked a bit about the Aspen Institute, which I had recently joined, and I invited him to speak at our summer campus in Colorado. He'd be happy to come, he said, but not to be onstage. He wanted instead to take a walk so that we could talk.

That seemed a bit odd. I didn't yet know that taking a long walk was his preferred way to have a serious conversation. It turned out that he wanted me to write a biography of him. I had recently published one on Benjamin Franklin and was writing one about Albert Einstein, and my initial reaction was to wonder, half jokingly, whether he saw himself as the natural successor in that sequence. Because I assumed that he was still in the middle of an oscillating career that had many more ups and downs left, I demurred. Not now, I said. Maybe in a decade or two, when you retire.

I had known him since 1984, when he came to Manhattan to have lunch with *Time*'s editors and extol his new Macintosh. He was petulant even then, attacking a *Time* correspondent for having wounded him with a story that was too revealing. But talking to him afterward, I found myself rather captivated, as so many others have been over the years, by his engaging intensity. We stayed in touch, even after he was ousted from Apple. When he had something to pitch, such as a NeXT computer or Pixar movie, the beam of his charm would suddenly refocus on me, and he would take me to a sushi restaurant in Lower Manhattan to tell me that whatever he was touting was the best thing he had ever produced. I liked him.

When he was restored to the throne at Apple, we put him on the cover of *Time,* and soon thereafter he began offering me his ideas for a series we were doing on the most influential people of the century. He had launched his "Think Different" campaign, featuring iconic photos of some of the same people we were considering, and he found the endeavor of assessing historic influence fascinating.

After I had deflected his suggestion that I write a biography of him, I heard from him every now and then. At one point I emailed to ask if it was true, as my daughter had told me, that the Apple logo was an homage to Alan Turing, the British computer pioneer who broke the German wartime codes and then committed suicide by biting into a cyanide-laced apple. He replied that he wished he had thought of that, but hadn't. That started an exchange about the early history of Apple, and I found myself gathering string on the subject, just in case I ever decided to do such a book. When my Einstein biography came out, he came to a book event in Palo Alto and pulled me aside to suggest, again, that he would make a good subject.

His persistence baffled me. He was known to guard his privacy, and I had no reason to believe he'd ever read any of my books. Maybe someday, I continued to say. But in 2009 his wife, Laurene Powell, said bluntly, "If you're ever going to do a book on Steve, you'd better do it now." He had just taken a second medical leave. I confessed to her that when he had first raised the idea, I hadn't known he was sick. Almost nobody knew, she said. He had called me right before he was going to be operated on for cancer, and he was still keeping it a secret, she explained.

I decided then to write this book. Jobs surprised me by readily acknowledging that he would have no control over it or even the right to see it in advance. "It's your book," he said. "I won't even read it." But later that fall he seemed to have second thoughts about cooperating and, though I didn't know it, was hit by another round of cancer complications. He stopped returning my calls, and I put the project aside for a while.

Then, unexpectedly, he phoned me late on the afternoon of New Year's Eve 2009. He was at home in Palo Alto with only his sister, the writer Mona Simpson. His wife and their three children had taken a quick trip to go skiing, but he was not healthy enough to join them. He was in a reflective mood, and we talked for more than an hour. He began by recalling that he had wanted to build a frequency counter when he was twelve, and he was able to look up Bill Hewlett, the founder of HP, in the phone book and call him to get parts. Jobs said that the past twelve years of his life, since his return to Apple, had been his most productive in terms of creating new products. But his more important goal, he said, was to do what Hewlett and his friend David Packard had done, which was create a company that was so imbued with innovative creativity that it would outlive them.

"I always thought of myself as a humanities person as a kid, but I liked electronics," he said. "Then I read something that one of my heroes, Edwin Land of Polaroid, said about the importance of people who could stand at the intersection of humanities and sciences, and I decided that's what I wanted to do." It was as if he were suggesting themes for his biography (and in this instance, at least, the theme turned out to be valid). The creativity that can occur when a feel for both the humanities and the sciences combine in one strong personality was the topic that most interested me in my biographies of Franklin and Einstein, and I believe that it will be a key to creating innovative economies in the twenty-first century.

I asked Jobs why he wanted me to be the one to write his biography. "I think you're good at getting people to talk," he replied. That was an unexpected answer. I knew that I would have to interview scores of people he had fired, abused, abandoned, or otherwise infuriated, and I feared he would not be comfortable with my getting them to talk. And indeed he did turn out to be skittish when word trickled back to him of people that I was interviewing. But after a couple of months, he

20

began encouraging people to talk to me, even foes and former girlfriends. Nor did he try to put anything off-limits. "I've done a lot of things I'm not proud of, such as getting my girlfriend pregnant when I was twenty-three and the way I handled that," he said. "But I don't have any skeletons in my closet that can't be allowed out." He didn't seek any control over what I wrote, or even ask to read it in advance. His only involvement came when my publisher was choosing the cover art. When he saw an early version of a proposed cover treatment, he disliked it so much that he asked to have input in designing a new version. I was both amused and willing, so I readily assented.

I ended up having more than forty interviews and conversations with him. Some were formal ones in his Palo Alto living room, others were done during long walks and drives or by telephone. During my two years of visits, he became increasingly intimate and revealing, though at times I witnessed what his veteran colleagues at Apple used to call his "reality distortion field." Sometimes it was the inadvertent misfiring of memory cells that happens to us all; at other times he was spinning his own version of reality both to me and to himself. To check and flesh out his story, I interviewed more than a hundred friends, relatives, competitors, adversaries, and colleagues.

His wife also did not request any restrictions or control, nor did she ask to see in advance what I would publish. In fact she strongly encouraged me to be honest about his failings as well as his strengths. She is one of the smartest and most grounded people I have ever met. "There are parts of his life and personality that are extremely messy, and that's the truth," she told me early on. "You shouldn't whitewash it. He's good at spin, but he also has a remarkable story, and I'd like to see that it's all told truthfully."

I leave it to the reader to assess whether I have succeeded in this mission. I'm sure there are players in this drama who will remember some of the events differently or think that I sometimes got trapped in Jobs's distortion field. As happened when I wrote a book about Henry Kissinger, which in some ways was good preparation for this project, I found that people had such strong positive and negative emotions about Jobs that the Rashomon effect was often evident. But I've done the best I can to balance conflicting accounts fairly and be transparent about the sources I used.

This is a book about the roller-coaster life and searingly intense personality of a creative entrepreneur whose passion for perfection and ferocious drive revolutionized six industries: personal computers, animated movies, music, phones, tablet computing, and digital publishing. You might even add a seventh, retail stores, which Jobs did not quite revolutionize but did reimagine. In addition, he opened the way for a new market for digital content based on apps rather than just websites. Along the way he produced not only transforming products but also, on his second try, a lasting company, endowed with his DNA, that is filled with creative designers and daredevil engineers who could carry forward his vision. In August 2011, right before he stepped down as CEO, the enterprise he started in his parents' garage became the world's most valuable company.

This is also, I hope, a book about innovation. At a time when the United States is seeking ways to sustain its innovative edge, and when societies around the world are trying to build creative digital-age economies, Jobs stands as the ultimate icon of inventiveness, imagination, and sustained innovation. He knew that the best way to create value in the twenty-first century was to connect creativity with technology, so he built a company where leaps of the imagination were combined with remarkable feats of engineering. He and his colleagues at Apple were able to think differently: They developed not merely modest product advances based on focus groups, but whole new devices and services that consumers did not yet know they needed.

He was not a model boss or human being, tidily packaged for emulation. Driven by demons, he could drive those around him to fury and despair. But his personality and passions and products were all interrelated, just as Apple's hardware and software tended to be, as if part of an integrated system. His tale is thus both instructive and cautionary, filled with lessons about innovation, character, leadership, and values.

Shakespeare's *Henry V* — the story of a willful and immature prince who becomes a passionate but sensitive, callous but sentimental, inspiring but flawed king — begins with the exhortation "O for a Muse of fire, that would ascend / The brightest heaven of invention." For Steve Jobs, the ascent to the brightest heaven of invention begins with a tale of two sets of parents, and of growing up in a valley that was just learning how to turn silicon into gold.

CHAPTER ONE: CHILDHOOD
ABANDONED AND CHOSEN

THE ADOPTION

When Paul Jobs was mustered out of the Coast Guard after World War II, he made a wager with his crewmates. They had arrived in San Francisco, where their ship was decommissioned, and Paul bet that he would find himself a wife within two weeks. He was a taut, tattooed engine mechanic, six feet tall, with a passing resemblance to James Dean. But it wasn't his looks that got him a date with Clara Hagopian, a sweet-humored daughter of Armenian immigrants. It was the fact that he and his friends had a car, unlike the group she had originally planned to go out with that evening. Ten days later, in March 1946, Paul got engaged to Clara and won his wager. It would turn out to be a happy marriage, one that lasted until death parted them more than forty years later.

Paul Reinhold Jobs had been raised on a dairy farm in Germantown, Wisconsin. Even though his father was an alcoholic and sometimes abusive, Paul ended up with a gentle and calm disposition under his leathery exterior. After dropping out of high school, he wandered through the Midwest picking up work as a mechanic until, at age nine-

25

teen, he joined the Coast Guard, even though he didn't know how to swim. He was deployed on the USS *General M. C. Meigs* and spent much of the war ferrying troops to Italy for General Patton. His talent as a machinist and fireman earned him commendations, but he occasionally found himself in minor trouble and never rose above the rank of seaman.

Clara was born in New Jersey, where her parents had landed after fleeing the Turks in Armenia, and they moved to the Mission District of San Francisco when she was a child. She had a secret that she rarely mentioned to anyone: She had been married before, but her husband had been killed in the war. So when she met Paul Jobs on that first date, she was primed to start a new life.

Like many who lived through the war, they had experienced enough excitement that, when it was over, they desired simply to settle down, raise a family, and lead a less eventful life. They had little money, so they moved to Wisconsin and lived with Paul's parents for a few years, then headed for Indiana, where he got a job as a machinist for International Harvester. His passion was tinkering with old cars, and he made money in his spare time buying, restoring, and selling them. Eventually he quit his day job to become a full-time used car salesman.

Clara, however, loved San Francisco, and in 1952 she convinced her husband to move back there. They got an apartment in the Sunset District facing the Pacific, just south of Golden Gate Park, and he took a job working for a finance company as a "repo man," picking the locks of cars whose owners hadn't paid their loans and repossessing them. He also bought, repaired, and sold some of the cars, making a decent enough living in the process.

There was, however, something missing in their lives. They wanted children, but Clara had suffered an ectopic pregnancy, in which the fertilized egg was implanted in a fallopian tube rather than the uterus, and she had been unable to have any. So by 1955, after nine years of marriage, they were looking to adopt a child.

Like Paul Jobs, Joanne Schieble was from a rural Wisconsin family of German heritage. Her father, Arthur Schieble, had immigrated to the outskirts of Green Bay, where he and his wife owned a mink farm and dabbled successfully in various other businesses, including real estate and photo-engraving. He was very strict, especially regarding his daughter's relationships, and he had strongly disapproved of her first love, an artist who was not a Catholic. Thus it was no surprise that he threatened to cut Joanne off completely when, as a graduate student at the University of Wisconsin, she fell in love with Abdulfattah "John" Jandali, a Muslim teaching assistant from Syria.

Jandali was the youngest of nine children in a prominent Syrian family. His father owned oil refineries and multiple other businesses, with large holdings in Damascus and Homs, and at one point pretty much controlled the price of wheat in the region. His mother, he later said, was a "traditional Muslim woman" who was a "conservative, obedient housewife." Like the Schieble family, the

Jandalis put a premium on education. Abdulfattah was sent to a Jesuit boarding school, even though he was Muslim, and he got an undergraduate degree at the American University in Beirut before entering the University of Wisconsin to pursue a doctoral degree in political science.

In the summer of 1954, Joanne went with Abdulfattah to Syria. They spent two months in Homs, where she learned from his family to cook Syrian dishes. When they returned to Wisconsin she discovered that she was pregnant. They were both twenty-three, but they decided not to get married. Her father was dying at the time, and he had threatened to disown her if she wed Abdulfattah. Nor was abortion an easy option in a small Catholic community. So in early 1955, Joanne traveled to San Francisco, where she was taken into the care of a kindly doctor who sheltered unwed mothers, delivered their babies, and quietly arranged closed adoptions.

Joanne had one requirement: Her child must be adopted by college graduates. So the doctor arranged for the baby to be placed with a lawyer and his wife. But when a boy was born — on February 24, 1955 — the designated couple decided that they wanted a girl and backed out. Thus it was that the boy became the son not of a lawyer but of a high school dropout with a passion for mechanics and his salt-of-the-earth wife who was working as a bookkeeper. Paul and Clara named their new baby Steven Paul Jobs.

When Joanne found out that her baby had been placed with a couple who had not even graduated from high school, she refused to sign the adoption papers. The standoff lasted weeks, even after the baby had settled into the Jobs household. Eventually Joanne relented, with the stipulation that the couple promise — indeed sign a pledge — to fund a savings account to pay for the boy's college education.

There was another reason that Joanne was balky about signing the adoption papers. Her father was about to die, and she planned to marry Jandali soon after. She held out hope, she would later tell family members, sometimes tearing up at the memory, that once they were married, she could get their baby boy back.

Arthur Schieble died in August 1955, after the adoption was finalized. Just after Christmas that year, Joanne and Abdulfattah were married in St. Philip the Apostle Catholic Church in Green Bay. He got his PhD in international politics the next year, and then they had another child, a girl named Mona. After she and Jandali divorced in 1962, Joanne embarked on a dreamy and peripatetic life that her daughter, who grew up to become the acclaimed novelist Mona Simpson, would capture in her book *Anywhere but Here*. Because Steve's adoption had been closed, it would be twenty years before they would all find each other.

Steve Jobs knew from an early age that he was adopted. "My parents were very open with me about that," he recalled. He had a vivid memory of sitting on the lawn of his house, when he was six or seven years old, telling the girl who lived across the street. "So does that mean your real parents didn't want you?" the girl asked. "Lightning bolts went

29

off in my head," according to Jobs. "I remember running into the house, crying. And my parents said, 'No, you have to understand.' They were very serious and looked me straight in the eye. They said, 'We specifically picked you out.' Both of my parents said that and repeated it slowly for me. And they put an emphasis on every word in that sentence."

Abandoned. Chosen. Special. Those concepts became part of who Jobs was and how he regarded himself. His closest friends think that the knowledge that he was given up at birth left some scars. "I think his desire for complete control of whatever he makes derives directly from his personality and the fact that he was abandoned at birth," said one longtime colleague, Del Yocam. "He wants to control his environment, and he sees the product as an extension of himself." Greg Calhoun, who became close to Jobs right after college, saw another effect. "Steve talked to me a lot about being abandoned and the pain that caused," he said. "It made him independent. He followed the beat of a different drummer, and that came from being in a different world than he was born into."

Later in life, when he was the same age his biological father had been when he abandoned him, Jobs would father and abandon a child of his own. (He eventually took responsibility for her.) Chrisann Brennan, the mother of that child, said that being put up for adoption left Jobs "full of broken glass," and it helps to explain some of his behavior. "He who is abandoned is an abandoner," she said. Andy Hertzfeld, who worked with Jobs at Apple in the early 1980s, is among the few who remained close to both Brennan and Jobs. "The key question about Steve is why he can't control himself at times from being so reflexively cruel and harmful to some people," he said. "That goes back to being abandoned at birth. The real underlying problem was the theme of abandonment in Steve's life."

Jobs dismissed this. "There's some notion that because I was abandoned, I worked very hard so I could do well and make my parents wish they had me back, or some such nonsense, but that's ridiculous," he insisted. "Knowing I was adopted may have made me feel more independent, but I have never felt abandoned. I've always felt special. My parents made me feel special." He would later bristle whenever anyone referred to Paul and Clara Jobs as his "adoptive" parents or implied that they were not his "real" parents. "They were my parents 1,000%," he said. When speaking about his biological parents, on the other hand, he was curt: "They were my sperm and egg bank. That's not harsh, it's just the way it was, a sperm bank thing, nothing more."

SILICON VALLEY

The childhood that Paul and Clara Jobs created for their new son was, in many ways, a stereotype of the late 1950s. When Steve was two they adopted a girl they named Patty, and three years later they moved to a tract house in the suburbs. The finance company where Paul worked as a repo man, CIT, had transferred him down to its Palo Alto office, but he could not afford to live there, so they landed in a subdivision in Mountain View, a less expensive town just to the south.

31

There Paul tried to pass along his love of mechanics and cars. "Steve, this is your workbench now," he said as he marked off a section of the table in their garage. Jobs remembered being impressed by his father's focus on craftsmanship. "I thought my dad's sense of design was pretty good," he said, "because he knew how to build anything. If we needed a cabinet, he would build it. When he built our fence, he gave me a hammer so I could work with him."

Fifty years later the fence still surrounds the back and side yards of the house in Mountain View. As Jobs showed it off to me, he caressed the stockade panels and recalled a lesson that his father implanted deeply in him. It was important, his father said, to craft the backs of cabinets and fences properly, even though they were hidden. "He loved doing things right. He even cared about the look of the parts you couldn't see."

His father continued to refurbish and resell used cars, and he festooned the garage with pictures of his favorites. He would point out the detailing of the design to his son: the lines, the vents, the chrome, the trim of the seats. After work each day, he would change into his dungarees and retreat to the garage, often with Steve tagging along. "I figured I could get him nailed down with a little mechanical ability, but he really wasn't interested in getting his hands dirty," Paul later recalled. "He never really cared too much about mechanical things."

"I wasn't that into fixing cars," Jobs admitted. "But I was eager to hang out with my dad." Even as he was growing more aware that he had been adopted, he was becoming more attached to his father. One day when he was about eight, he discovered a photograph of his father from his time in the Coast Guard. "He's in the engine room, and he's got his shirt off and looks like James Dean. It was one of those *Oh wow* moments for a kid. *Wow, oooh,* my parents were actually once very young and really good-looking."

Through cars, his father gave Steve his first exposure to electronics. "My dad did not have a deep understanding of electronics, but he'd encountered it a lot in automobiles and other things he would fix. He showed me the rudiments of electronics, and I got very interested in that." Even more interesting were the trips to scavenge for parts. "Every weekend, there'd be a junkyard trip. We'd be looking for a generator, a carburetor, all sorts of components." He remembered watching his father negotiate at the counter. "He was a good bargainer, because he knew better than the guys at the counter what the parts should cost." This helped fulfill the pledge his parents made when he was adopted. "My college fund came from my dad paying $50 for a Ford Falcon or some other beat-up car that didn't run, working on it for a few weeks, and selling it for $250 — and not telling the IRS."

The Jobses' house and the others in their neighborhood were built by the real estate developer Joseph Eichler, whose company spawned more than eleven thousand homes in various California subdivisions between 1950 and 1974. Inspired by Frank Lloyd Wright's vision of simple modern homes for the American "everyman," Eichler built inexpensive houses that featured floor-to-ceiling

33

glass walls, open floor plans, exposed post-and-beam construction, concrete slab floors, and lots of sliding glass doors. "Eichler did a great thing," Jobs said on one of our walks around the neighborhood. "His houses were smart and cheap and good. They brought clean design and simple taste to lower-income people. They had awesome little features, like radiant heating in the floors. You put carpet on them, and we had nice toasty floors when we were kids."

Jobs said that his appreciation for Eichler homes instilled in him a passion for making nicely designed products for the mass market. "I love it when you can bring really great design and simple capability to something that doesn't cost much," he said as he pointed out the clean elegance of the houses. "It was the original vision for Apple. That's what we tried to do with the first Mac. That's what we did with the iPod."

Across the street from the Jobs family lived a man who had become successful as a real estate agent. "He wasn't that bright," Jobs recalled, "but he seemed to be making a fortune. So my dad thought, 'I can do that.' He worked so hard, I remember. He took these night classes, passed the license test, and got into real estate. Then the bottom fell out of the market." As a result, the family found itself financially strapped for a year or so while Steve was in elementary school. His mother took a job as a bookkeeper for Varian Associates, a company that made scientific instruments, and they took out a second mortgage. One day his fourth-grade teacher asked him, "What is it you don't understand about the universe?" Jobs replied,

"I don't understand why all of a sudden my dad is so broke." He was proud that his father never adopted a servile attitude or slick style that may have made him a better salesman. "You had to suck up to people to sell real estate, and he wasn't good at that and it wasn't in his nature. I admired him for that." Paul Jobs went back to being a mechanic.

His father was calm and gentle, traits that his son later praised more than emulated. He was also resolute. Jobs described one example:

Nearby was an engineer who was working at Westinghouse. He was a single guy, beatnik type. He had a girlfriend. She would babysit me sometimes. Both my parents worked, so I would come here right after school for a couple of hours. He would get drunk and hit her a couple of times. She came over one night, scared out of her wits, and he came over drunk, and my dad stood him down — saying "She's here, but you're not coming in." He stood right there. We like to think everything was idyllic in the 1950s, but this guy was one of those engineers who had messed-up lives.

What made the neighborhood different from the thousands of other spindly-tree subdivisions across America was that even the ne'er-do-wells tended to be engineers. "When we moved here, there were apricot and plum orchards on all of these corners," Jobs recalled. "But it was beginning to boom because of military investment." He soaked up the history of the valley and developed a yearning to play his own role. Edwin Land of Polaroid later told him about being asked by Eisenhower to help build the U-2 spy plane cameras to see how real the Soviet threat was. The film was dropped in canisters and returned to the NASA Ames Research Center in Sunnyvale, not far from where Jobs lived. "The first computer terminal I ever saw was when my dad brought me to the Ames Center," he said. "I fell totally in love with it."

Other defense contractors sprouted nearby during the 1950s. The Lockheed Missiles and Space Division, which built submarine-launched ballistic missiles, was founded in 1956 next to the NASA Center; by the time Jobs moved to the area four years later, it employed twenty thousand people. A few hundred yards away, Westinghouse built facilities that produced tubes and electrical transformers for the missile systems. "You had all these military companies on the cutting edge," he recalled. "It was mysterious and high-tech and made living here very exciting."

In the wake of the defense industries there arose a booming economy based on technology. Its roots stretched back to 1938, when David Packard and his new wife moved into a house in Palo Alto that had a shed where his friend Bill Hewlett was soon ensconced. The house had a garage — an appendage that would prove both useful and iconic in the valley — in which they tinkered around until they had their first product, an audio oscillator. By the 1950s, Hewlett-Packard was a fast-growing company making technical instruments.

Fortunately there was a place nearby for entrepreneurs who had outgrown their garages. In a move that would help transform the area into the cradle of the tech revolution, Stanford University's

36

dean of engineering, Frederick Terman, created a seven-hundred-acre industrial park on university land for private companies that could commercialize the ideas of his students. Its first tenant was Varian Associates, where Clara Jobs worked. "Terman came up with this great idea that did more than anything to cause the tech industry to grow up here," Jobs said. By the time Jobs was ten, HP had nine thousand employees and was the blue-chip company where every engineer seeking financial stability wanted to work.

The most important technology for the region's growth was, of course, the semiconductor. William Shockley, who had been one of the inventors of the transistor at Bell Labs in New Jersey, moved out to Mountain View and, in 1956, started a company to build transistors using silicon rather than the more expensive germanium that was then commonly used. But Shockley became increasingly erratic and abandoned his silicon transistor project, which led eight of his engineers — most notably Robert Noyce and Gordon Moore — to break away to form Fairchild Semiconductor. That company grew to twelve thousand employees, but it fragmented in 1968, when Noyce lost a power struggle to become CEO. He took Gordon Moore and founded a company that they called Integrated Electronics Corporation, which they soon smartly abbreviated to Intel. Their third employee was Andrew Grove, who later would grow the company by shifting its focus from memory chips to microprocessors. Within a few years there would be more than fifty companies in the area making semiconductors.

The exponential growth of this industry was correlated with the phenomenon famously discovered by Moore, who in 1965 drew a graph of the speed of integrated circuits, based on the number of transistors that could be placed on a chip, and showed that it doubled about every two years, a trajectory that could be expected to continue. This was reaffirmed in 1971, when Intel was able to etch a complete central processing unit onto one chip, the Intel 4004, which was dubbed a "microprocessor." Moore's Law has held generally true to this day, and its reliable projection of performance to price allowed two generations of young entrepreneurs, including Steve Jobs and Bill Gates, to create cost projections for their forward-leaning products.

The chip industry gave the region a new name when Don Hoefler, a columnist for the weekly trade paper *Electronic News,* began a series in January 1971 entitled "Silicon Valley USA." The forty-mile Santa Clara Valley, which stretches from South San Francisco through Palo Alto to San Jose, has as its commercial backbone El Camino Real, the royal road that once connected California's twenty-one mission churches and is now a bustling avenue that connects companies and startups accounting for a third of the venture capital investment in the United States each year. "Growing up, I got inspired by the history of the place," Jobs said. "That made me want to be a part of it."

Like most kids, he became infused with the passions of the grown-ups around him. "Most of the dads in the neighborhood did really neat stuff, like photovoltaics and batteries and radar," Jobs recalled. "I grew up in awe of that stuff and asking people about it." The most important of these neighbors, Larry Lang, lived seven doors away. "He was my model of what an HP engineer was supposed to be: a big ham radio operator, hard-core electronics guy," Jobs recalled. "He would bring me stuff to play with." As we walked up to Lang's old house, Jobs pointed to the driveway. "He took a carbon microphone and a battery and a speaker, and he put it on this driveway. He had me talk into the carbon mike and it amplified out of the speaker." Jobs had been taught by his father that microphones always required an electronic amplifier. "So I raced home, and I told my dad that he was wrong."

"No, it needs an amplifier," his father assured him. When Steve protested otherwise, his father said he was crazy. "It can't work without an amplifier. There's some trick."

"I kept saying no to my dad, telling him he had to see it, and finally he actually walked down with me and saw it. And he said, 'Well I'll be a bat out of hell.' "

Jobs recalled the incident vividly because it was his first realization that his father did not know everything. Then a more disconcerting discovery began to dawn on him: He was smarter than his parents. He had always admired his father's competence and savvy. "He was not an educated man, but I had always thought he was pretty damn smart. He didn't read much, but he could do a lot. Almost everything mechanical, he could figure it out." Yet the carbon microphone incident, Jobs said, began a jarring process of realizing that he was in fact more clever and quick than his parents.

39

"It was a very big moment that's burned into my mind. When I realized that I was smarter than my parents, I felt tremendous shame for having thought that. I will never forget that moment." This discovery, he later told friends, along with the fact that he was adopted, made him feel apart — detached and separate — from both his family and the world.

Another layer of awareness occurred soon after. Not only did he discover that he was brighter than his parents, but he discovered that they knew this. Paul and Clara Jobs were loving parents, and they were willing to adapt their lives to suit a son who was very smart — and also willful. They would go to great lengths to accommodate him. And soon Steve discovered this fact as well. "Both my parents got me. They felt a lot of responsibility once they sensed that I was special. They found ways to keep feeding me stuff and putting me in better schools. They were willing to defer to my needs."

So he grew up not only with a sense of having once been abandoned, but also with a sense that he was special. In his own mind, that was more important in the formation of his personality.

School

Even before Jobs started elementary school, his mother had taught him how to read. This, however, led to some problems once he got to school. "I was kind of bored for the first few years, so I occupied myself by getting into trouble." It also soon became clear that Jobs, by both nature and nurture, was not disposed to accept authority. "I encountered authority of a different kind than I had ever encountered before, and I did not like it. And they really almost got me. They came close to really beating any curiosity out of me."

His school, Monta Loma Elementary, was a series of low-slung 1950s buildings four blocks from his house. He countered his boredom by playing pranks. "I had a good friend named Rick Ferrentino, and we'd get into all sorts of trouble," he recalled. "Like we made little posters announcing 'Bring Your Pet to School Day.' It was crazy, with dogs chasing cats all over, and the teachers were beside themselves." Another time they convinced some kids to tell them the combination numbers for their bike locks. "Then we went outside and switched all of the locks, and nobody could get their bikes. It took them until late that night to straighten things out." When he was in third grade, the pranks became a bit more dangerous. "One time we set off an explosive under the chair of our teacher, Mrs. Thurman. We gave her a nervous twitch."

Not surprisingly, he was sent home two or three times before he finished third grade. By then, however, his father had begun to treat him as special, and in his calm but firm manner he made it clear that he expected the school to do the same. "Look, it's not his fault," Paul Jobs told the teachers, his son recalled. "If you can't keep him interested, it's your fault." His parents never punished him for his transgressions at school. "My father's father was an alcoholic and whipped him with a belt, but I'm not sure if I ever got spanked." Both of his parents, he added, "knew the school was at fault for trying to make me memorize stupid stuff rather than stimulating me." He was already starting to show the admixture of sensitivity and insensitivity, bristliness and detachment, that would mark him for the rest of his life.

When it came time for him to go into fourth grade, the school decided it was best to put Jobs and Ferrentino into separate classes. The teacher for the advanced class was a spunky woman named Imogene Hill, known as "Teddy," and she became, Jobs said, "one of the saints of my life." After watching him for a couple of weeks, she figured that the best way to handle him was to bribe him. "After school one day, she gave me this workbook with math problems in it, and she said, 'I want you to take it home and do this.' And I thought, 'Are you nuts?' And then she pulled out one of these giant lollipops that seemed as big as the world. And she said, 'When you're done with it, if you get it mostly right, I will give you this and five dollars.' And I handed it back within two days." After a few months, he no longer required the bribes. "I just wanted to learn and to please her."

She reciprocated by getting him a hobby kit for grinding a lens and making a camera. "I learned more from her than any other teacher, and if it hadn't been for her I'm sure I would have gone to jail." It reinforced, once again, the idea that he was special. "In my class, it was just me she cared about. She saw something in me."

It was not merely intelligence that she saw. Years later she liked to show off a picture of that year's class on Hawaii Day. Jobs had shown up without the suggested Hawaiian shirt, but in the picture he is front and center wearing one. He had, literally, been able to talk the shirt off another kid's back.

Near the end of fourth grade, Mrs. Hill had Jobs tested. "I scored at the high school sophomore level," he recalled. Now that it was clear, not only to himself and his parents but also to his teachers, that he was intellectually special, the school made the remarkable proposal that he skip two grades and go right into seventh; it would be the easiest way to keep him challenged and stimulated. His parents decided, more sensibly, to have him skip only one grade.

The transition was wrenching. He was a socially awkward loner who found himself with kids a year older. Worse yet, the sixth grade was in a different school, Crittenden Middle. It was only eight blocks from Monta Loma Elementary, but in many ways it was a world apart, located in a neighborhood filled with ethnic gangs. "Fights were a daily occurrence; as were shakedowns in bathrooms," wrote the Silicon Valley journalist Michael S. Malone. "Knives were regularly brought to school as a show of macho." Around the time that Jobs arrived, a group of students were jailed for a gang rape, and the bus of a neighboring school was destroyed after its team beat Crittenden's in a wrestling match.

Jobs was often bullied, and in the middle of seventh grade he gave his parents an ultimatum. "I insisted they put me in a different school," he recalled. Financially this was a tough demand. His parents were barely making ends meet, but by this point there was little doubt that they would eventually bend to his will. "When they resisted,

I told them I would just quit going to school if I had to go back to Crittenden. So they researched where the best schools were and scraped together every dime and bought a house for $21,000 in a nicer district."

The move was only three miles to the south, to a former apricot orchard in Los Altos that had been turned into a subdivision of cookie-cutter tract homes. Their house, at 2066 Crist Drive, was one story with three bedrooms and an all-important attached garage with a roll-down door facing the street. There Paul Jobs could tinker with cars and his son with electronics.

Its other significant attribute was that it was just over the line inside what was then the Cupertino-Sunnyvale School District, one of the safest and best in the valley. "When I moved here, these corners were still orchards," Jobs pointed out as we walked in front of his old house. "The guy who lived right there taught me how to be a good organic gardener and to compost. He grew everything to perfection. I never had better food in my life. That's when I began to appreciate organic fruits and vegetables."

Even though they were not fervent about their faith, Jobs's parents wanted him to have a religious upbringing, so they took him to the Lutheran church most Sundays. That came to an end when he was thirteen. In July 1968 *Life* magazine published a shocking cover showing a pair of starving children in Biafra. Jobs took it to Sunday school and confronted the church's pastor. "If I raise my finger, will God know which one I'm going to raise even before I do it?"

The pastor answered, "Yes, God knows every-thing."

Jobs then pulled out the *Life* cover and asked, "Well, does God know about this and what's going to happen to those children?"

"Steve, I know you don't understand, but yes, God knows about that."

Jobs announced that he didn't want to have any-thing to do with worshipping such a God, and he never went back to church. He did, however, spend years studying and trying to practice the tenets of Zen Buddhism. Reflecting years later on his spiritual feelings, he said that religion was at its best when it emphasized spiritual experiences rather than received dogma. "The juice goes out of Christianity when it becomes too based on faith rather than on living like Jesus or seeing the world as Jesus saw it," he told me. "I think different reli-gions are different doors to the same house. Some-times I think the house exists, and sometimes I don't. It's the great mystery."

Paul Jobs was then working at Spectra-Physics, a company in nearby Santa Clara that made lasers for electronics and medical products. As a machin-ist, he crafted the prototypes of products that the engineers were devising. His son was fascinated by the need for perfection. "Lasers require preci-sion alignment," Jobs said. "The really sophisti-cated ones, for airborne applications or medical, had very precise features. They would tell my dad something like, 'This is what we want, and we want it out of one piece of metal so that the coefficients of expansion are all the same.' And he had to fig-ure out how to do it." Most pieces had to be made from scratch, which meant that Paul had to create custom tools and dies. His son was impressed, but he rarely went to the machine shop. "It would have been fun if he had gotten to teach me how to use a mill and lathe. But unfortunately I never went, because I was more interested in electronics."

One summer Paul took Steve to Wisconsin to visit the family's dairy farm. Rural life did not appeal to Steve, but one image stuck with him. He saw a calf being born, and he was amazed when the tiny animal struggled up within minutes and began to walk. "It was not something she had learned, but it was instead hardwired into her," he recalled. "A human baby couldn't do that. I found it remarkable, even though no one else did." He put it in hardware-software terms: "It was as if something in the animal's body and in its brain had been engineered to work together instantly rather than being learned."

In ninth grade Jobs went to Homestead High, which had a sprawling campus of two-story cinderblock buildings painted pink that served two thousand students. "It was designed by a famous prison architect," Jobs recalled. "They wanted to make it indestructible." He had developed a love of walking, and he walked the fifteen blocks to school by himself each day.

He had few friends his own age, but he got to know some seniors who were immersed in the counterculture of the late 1960s. It was a time when the geek and hippie worlds were beginning to show some overlap. "My friends were the really smart kids," he said. "I was interested in math and science and electronics. They were too, and also into LSD and the whole counterculture trip."

His pranks by then typically involved electronics. At one point he wired his house with speakers. But since speakers can also be used as microphones, he built a control room in his closet, where he could listen in on what was happening in other rooms. One night, when he had his headphones on and was listening in on his parents' bedroom, his father caught him and angrily demanded that he dismantle the system. He spent many evenings visiting the garage of Larry Lang, the engineer who lived down the street from his old house. Lang eventually gave Jobs the carbon microphone that had fascinated him, and he turned him on to Heathkits, those assemble-it-yourself kits for making ham radios and other electronic gear that were beloved by the soldering set back then. "Heathkits came with all the boards and parts color-coded, but the manual also explained the theory of how it operated," Jobs recalled. "It made you realize you could build and understand anything. Once you built a couple of radios, you'd see a TV in the catalogue and say, 'I can build that as well,' even if you didn't. I was very lucky, because when I was a kid both my dad and the Heathkits made me believe I could build anything."

Lang also got him into the Hewlett-Packard Explorers Club, a group of fifteen or so students who met in the company cafeteria on Tuesday nights. "They would get an engineer from one of the labs to come and talk about what he was working on," Jobs recalled. "My dad would drive me there. I was in heaven. HP was a pioneer of light-emitting diodes. So we talked about what to do with them."

47

Because his father now worked for a laser company, that topic particularly interested him. One night he cornered one of HP's laser engineers after a talk and got a tour of the holography lab. But the most lasting impression came from seeing the small computers the company was developing. "I saw my first desktop computer there. It was called the 9100A, and it was a glorified calculator but also really the first desktop computer. It was huge, maybe forty pounds, but it was a beauty of a thing. I fell in love with it."

The kids in the Explorers Club were encouraged to do projects, and Jobs decided to build a frequency counter, which measures the number of pulses per second in an electronic signal. He needed some parts that HP made, so he picked up the phone and called the CEO. "Back then, people didn't have unlisted numbers. So I looked up Bill Hewlett in Palo Alto and called him at home. And he answered and chatted with me for twenty minutes. He got me the parts, but he also got me a job in the plant where they made frequency counters." Jobs worked there the summer after his freshman year at Homestead High. "My dad would drive me in the morning and pick me up in the evening."

His work mainly consisted of "just putting nuts and bolts on things" on an assembly line. There was some resentment among his fellow line workers toward the pushy kid who had talked his way in by calling the CEO. "I remember telling one of the supervisors, 'I love this stuff, I love this stuff,' and then I asked him what he liked to do best. And he said, 'To fuck, to fuck.'" Jobs had an easier time ingratiating himself with the engineers who worked one floor above. "They served doughnuts and coffee every morning at ten. So I'd go upstairs and hang out with them."

Jobs liked to work. He also had a newspaper route — his father would drive him when it was raining — and during his sophomore year spent weekends and the summer as a stock clerk at a cavernous electronics store, Haltek. It was to electronics what his father's junkyards were to auto parts: a scavenger's paradise sprawling over an entire city block with new, used, salvaged, and surplus components crammed onto warrens of shelves, dumped unsorted into bins, and piled in an outdoor yard. "Out in the back, near the bay, they had a fenced-in area with things like Polaris submarine interiors that had been ripped and sold for salvage," he recalled. "All the controls and buttons were right there. The colors were military greens and grays, but they had these switches and bulb covers of amber and red. There were these big old lever switches that, when you flipped them, it was awesome, like you were blowing up Chicago."

At the wooden counters up front, laden with thick catalogues in tattered binders, people would haggle for switches, resistors, capacitors, and sometimes the latest memory chips. His father used to do that for auto parts, and he succeeded because he knew the value of each better than the clerks. Jobs followed suit. He developed a knowledge of electronic parts that was honed by his love of negotiating and turning a profit. He would go to electronic flea markets, such as the San Jose swap meet, haggle for a used circuit board that contained some valuable chips or components, and

49

then sell those to his manager at Haltek.

Jobs was able to get his first car, with his father's help, when he was fifteen. It was a two-tone Nash Metropolitan that his father had fitted out with an MG engine. Jobs didn't really like it, but he did not want to tell his father that, or miss out on the chance to have his own car. "In retrospect, a Nash Metropolitan might seem like the most wickedly cool car," he later said. "But at the time it was the most uncool car in the world. Still, it was a car, so that was great." Within a year he had saved up enough from his various jobs that he could trade up to a red Fiat 850 coupe with an Abarth engine. "My dad helped me buy and inspect it. The satisfaction of getting paid and saving up for something, that was very exciting."

That same summer, between his sophomore and junior years at Homestead, Jobs began smoking marijuana. "I got stoned for the first time that summer. I was fifteen, and then began using pot regularly." At one point his father found some dope in his son's Fiat. "What's this?" he asked. Jobs coolly replied, "That's marijuana." It was one of the few times in his life that he faced his father's anger. "That was the only real fight I ever got in with my dad," he said. But his father again bent to his will. "He wanted me to promise that I'd never use pot again, but I wouldn't promise." In fact by his senior year he was also dabbling in LSD and hash as well as exploring the mind-bending effects of sleep deprivation. "I was starting to get stoned a bit more. We would also drop acid occasionally, usually in fields or in cars."

He also flowered intellectually during his last

two years in high school and found himself at the intersection, as he had begun to see it, of those who were geekily immersed in electronics and those who were into literature and creative endeavors. "I started to listen to music a whole lot, and I started to read more outside of just science and technology — Shakespeare, Plato. I loved *King Lear*." His other favorites included *Moby-Dick* and the poems of Dylan Thomas. I asked him why he related to King Lear and Captain Ahab, two of the most willful and driven characters in literature, but he didn't respond to the connection I was making, so I let it drop. "When I was a senior I had this phenomenal AP English class. The teacher was this guy who looked like Ernest Hemingway. He took a bunch of us snowshoeing in Yosemite."

One course that Jobs took would become part of Silicon Valley lore: the electronics class taught by John McCollum, a former Navy pilot who had a showman's flair for exciting his students with such tricks as firing up a Tesla coil. His little stockroom, to which he would lend the key to pet students, was crammed with transistors and other components he had scored.

McCollum's classroom was in a shed-like building on the edge of the campus, next to the parking lot. "This is where it was," Jobs recalled as he peered in the window, "and here, next door, is where the auto shop class used to be." The juxtaposition highlighted the shift from the interests of his father's generation. "Mr. McCollum felt that electronics class was the new auto shop."

McCollum believed in military discipline and respect for authority. Jobs didn't. His aversion to authority was something he no longer tried to hide, and he affected an attitude that combined wiry and weird intensity with aloof rebelliousness. McCollum later said, "He was usually off in a corner doing something on his own and really didn't want to have much of anything to do with either me or the rest of the class." He never trusted Jobs with a key to the stockroom. One day Jobs needed a part that was not available, so he made a collect call to the manufacturer, Burroughs in Detroit, and said he was designing a new product and wanted to test out the part. It arrived by air freight a few days later. When McCollum asked how he had gotten it, Jobs described — with defiant pride — the collect call and the tale he had told. "I was furious," McCollum said. "That was not the way I wanted my students to behave." Jobs's response was, "I don't have the money for the phone call. They've got plenty of money."

Jobs took McCollum's class for only one year, rather than the three that it was offered. For one of his projects, he made a device with a photocell that would switch on a circuit when exposed to light, something any high school science student could have done. He was far more interested in playing with lasers, something he learned from his father. With a few friends, he created light shows for parties by bouncing lasers off mirrors that were attached to the speakers of his stereo system.

CHAPTER TWO: ODD COUPLE
THE TWO STEVES

Woz

While a student in McCollum's class, Jobs became friends with a graduate who was the teacher's all-time favorite and a school legend for his wizardry in the class. Stephen Wozniak, whose younger brother had been on a swim team with Jobs, was almost five years older than Jobs and far more knowledgeable about electronics. But emotionally and socially he was still a high school geek.

Like Jobs, Wozniak learned a lot at his father's knee. But their lessons were different. Paul Jobs was a high school dropout who, when fixing up cars, knew how to turn a tidy profit by striking the right deal on parts. Francis Wozniak, known as Jerry, was a brilliant engineering graduate from Cal Tech, where he had quarterbacked the football team, who became a rocket scientist at Lockheed. He exalted engineering and looked down on those in business, marketing, and sales. "I remember him telling me that engineering was the highest level of importance you could reach in the world," Steve Wozniak later recalled. "It takes society to a new level."

One of Steve Wozniak's first memories was going

to his father's workplace on a weekend and being shown electronic parts, with his dad "putting them on a table with me so I got to play with them." He watched with fascination as his father tried to get a waveform line on a video screen to stay flat so he could show that one of his circuit designs was working properly. "I could see that whatever my dad was doing, it was important and good." Woz, as he was known even then, would ask about the resistors and transistors lying around the house, and his father would pull out a blackboard to illustrate what they did. "He would explain what a resistor was by going all the way back to atoms and electrons. He explained how resistors worked when I was in second grade, not by equations but by having me picture it."

Woz's father taught him something else that became ingrained in his childlike, socially awkward personality: Never lie. "My dad believed in honesty. Extreme honesty. That's the biggest thing he taught me. I never lie, even to this day." (The only partial exception was in the service of a good practical joke.) In addition, he imbued his son with an aversion to extreme ambition, which set Woz apart from Jobs. At an Apple product launch event in 2010, forty years after they met, Woz reflected on their differences. "My father told me, 'You always want to be in the middle,' " he said. "I didn't want to be up with the high-level people like Steve. My dad was an engineer, and that's what I wanted to be. I was way too shy ever to be a business leader like Steve."

By fourth grade Wozniak became, as he put it, one of the "electronics kids." He had an easier time making eye contact with a transistor than with a girl, and he developed the chunky and stooped look of a guy who spends most of his time hunched over circuit boards. At the same age when Jobs was puzzling over a carbon microphone that his dad couldn't explain, Wozniak was using transistors to build an intercom system featuring amplifiers, relays, lights, and buzzers that connected the kids' bedrooms of six houses in the neighborhood. And at an age when Jobs was building Heathkits, Wozniak was assembling a transmitter and receiver from Hallicrafters, the most sophisticated radios available.

Woz spent a lot of time at home reading his father's electronics journals, and he became enthralled by stories about new computers, such as the powerful ENIAC. Because Boolean algebra came naturally to him, he marveled at how simple, rather than complex, the computers were. In eighth grade he built a calculator that included one hundred transistors, two hundred diodes, and two hundred resistors on ten circuit boards. It won top prize in a local contest run by the Air Force, even though the competitors included students through twelfth grade.

Woz became more of a loner when the boys his age began going out with girls and partying, endeavors that he found far more complex than designing circuits. "Where before I was popular and riding bikes and everything, suddenly I was socially shut out," he recalled. "It seemed like nobody spoke to me for the longest time." He found an outlet by playing juvenile pranks. In twelfth grade he built an electronic metronome — one of those tick-tick-tick devices that keep time in music class — and realized it sounded like a bomb. So he took the labels off some big batteries, taped them together, and put it in a school locker; he rigged it to start ticking faster when the locker opened. Later that day he got called to the principal's office. He thought it was because he had won, yet again, the school's top math prize. Instead he was confronted by the police. The principal had been summoned when the device was found, bravely ran onto the football field clutching it to his chest, and pulled the wires off. Woz tried and failed to suppress his laughter. He actually got sent to the juvenile detention center, where he spent the night. It was a memorable experience. He taught the other prisoners how to disconnect the wires leading to the ceiling fans and connect them to the bars so people got shocked when touching them.

Getting shocked was a badge of honor for Woz. He prided himself on being a hardware engineer, which meant that random shocks were routine. He once devised a roulette game where four people put their thumbs in a slot; when the ball landed, one would get shocked. "Hardware guys will play this game, but software guys are too chicken," he noted.

During his senior year he got a part-time job at Sylvania and had the chance to work on a computer for the first time. He learned FORTRAN from a book and read the manuals for most of the systems of the day, starting with the Digital Equipment PDP-8. Then he studied the specs for the latest microchips and tried to redesign the computers using these newer parts. The challenge he set himself was to replicate the design using the fewest components possible. Each night he would try to improve his drawing from the night before. By the end of his senior year, he had become a master. "I was now designing computers with half the number of chips the actual company had in their own design, but only on paper." He never told his friends. After all, most seventeen-year-olds were getting their kicks in other ways.

On Thanksgiving weekend of his senior year, Wozniak visited the University of Colorado. It was closed for the holiday, but he found an engineering student who took him on a tour of the labs. He begged his father to let him go there, even though the out-of-state tuition was more than the family could easily afford. They struck a deal: He would be allowed to go for one year, but then he would transfer to De Anza Community College back home. After arriving at Colorado in the fall of 1969, he spent so much time playing pranks (such as producing reams of printouts saying "Fuck Nixon") that he failed a couple of his courses and was put on probation. In addition, he created a program to calculate Fibonacci numbers that burned up so much computer time the university threatened to bill him for the cost. So he readily lived up to his bargain with his parents and transferred to De Anza.

After a pleasant year at De Anza, Wozniak took time off to make some money. He found work at a company that made computers for the California Motor Vehicle Department, and a coworker made him a wonderful offer: He would provide some spare chips so Wozniak could make one of the computers he had been sketching on paper. Wozniak decided to use as few chips as possible, both as a personal challenge and because he did not want to take advantage of his colleague's largesse.

Much of the work was done in the garage of a friend just around the corner, Bill Fernandez, who was still at Homestead High. To lubricate their efforts, they drank large amounts of Cragmont cream soda, riding their bikes to the Sunnyvale Safeway to return the bottles, collect the deposits, and buy more. "That's how we started referring to it as the Cream Soda Computer," Wozniak recalled. It was basically a calculator capable of multiplying numbers entered by a set of switches and displaying the results in binary code with little lights.

When it was finished, Fernandez told Wozniak there was someone at Homestead High he should meet. "His name is Steve. He likes to do pranks like you do, and he's also into building electronics like you are." It may have been the most significant meeting in a Silicon Valley garage since Hewlett went into Packard's thirty-two years earlier. "Steve and I just sat on the sidewalk in front of Bill's house for the longest time, just sharing stories — mostly about pranks we'd pulled, and also what kind of electronic designs we'd done," Wozniak recalled. "We had so much in common. Typically, it was really hard for me to explain to people what kind of design stuff I worked on, but Steve got it right away. And I liked him. He was kind of skinny and wiry and full of energy." Jobs was also impressed. "Woz was the first person I'd met who knew more electronics than I did," he once said, stretching his own expertise. "I liked him right away. I was a little more mature than my years, and he was a little less mature than his, so it evened out. Woz was very bright, but emotionally he was my age."

In addition to their interest in computers, they shared a passion for music. "It was an incredible time for music," Jobs recalled. "It was like living at a time when Beethoven and Mozart were alive. Really. People will look back on it that way. And Woz and I were deeply into it." In particular, Wozniak turned Jobs on to the glories of Bob Dylan. "We tracked down this guy in Santa Cruz who put out this newsletter on Dylan," Jobs said. "Dylan taped all of his concerts, and some of the people around him were not scrupulous, because soon there were tapes all around. Bootlegs of everything. And this guy had them all."

Hunting down Dylan tapes soon became a joint venture. "The two of us would go tramping through San Jose and Berkeley and ask about Dylan bootlegs and collect them," said Wozniak. "We'd buy brochures of Dylan lyrics and stay up late interpreting them. Dylan's words struck chords of creative thinking." Added Jobs, "I had more than a hundred hours, including every concert on the '65 and '66 tour," the one where Dylan went electric. Both of them bought high-end TEAC reel-to-reel tape decks. "I would use mine at a low speed to record many concerts on one tape," said Wozniak. Jobs matched his obsession: "Instead of big speakers I bought a pair of awesome headphones and would just lie in my bed and listen to that stuff for hours."

Jobs had formed a club at Homestead High to put on music-and-light shows and also play pranks. (They once glued a gold-painted toilet seat onto a flower planter.) It was called the Buck Fry Club, a play on the name of the principal. Even though they had already graduated, Wozniak and his friend Allen Baum joined forces with Jobs, at the end of his junior year, to produce a farewell gesture for the departing seniors. Showing off the Homestead campus four decades later, Jobs paused at the scene of the escapade and pointed. "See that balcony? That's where we did the banner prank that sealed our friendship." On a big bedsheet Baum had tie-dyed with the school's green and white colors, they painted a huge hand flipping the middle-finger salute. Baum's nice Jewish mother helped them draw it and showed them how to do the shading and shadows to make it look more real. "I know what that is," she snickered. They devised a system of ropes and pulleys so that it could be dramatically lowered as the graduating class marched past the balcony, and they signed it "SWAB JOB," the initials of Wozniak and Baum combined with part of Jobs's name. The prank became part of school lore — and got Jobs suspended one more time.

Another prank involved a pocket device Wozniak built that could emit TV signals. He would take it to a room where a group of people were watching TV, such as in a dorm, and secretly press the button so that the screen would get fuzzy with static. When someone got up and whacked the set, Wozniak would let go of the button and the picture would clear up. Once he had the unsuspecting viewers hopping up and down at his will, he would make things harder. He would keep the picture fuzzy until someone touched the antenna. Eventually he would make people think they had to hold the antenna while standing on one foot or touching the top of the set. Years later, at a keynote presentation where he was having his own trouble getting a video to work, Jobs broke from his script and recounted the fun they had with the device. "Woz would have it in his pocket and we'd go into a dorm . . . where a bunch of folks would be, like, watching *Star Trek,* and he'd screw up the TV, and someone would go up to fix it, and just as they had the foot off the ground he would turn it back on, and as they put their foot back on the ground he'd screw it up again." Contorting himself into a pretzel onstage, Jobs concluded to great laughter, "And within five minutes he would have someone like this."

THE BLUE BOX

The ultimate combination of pranks and electronics — and the escapade that helped to create Apple — was launched one Sunday afternoon when Wozniak read an article in *Esquire* that his mother had left for him on the kitchen table. It was September 1971, and he was about to drive off the next day to Berkeley, his third college. The story, Ron Rosenbaum's "Secrets of the Little Blue Box," described how hackers and phone phreakers had found ways to make long-distance calls for free by replicating the tones that routed signals on the AT&T network. "Halfway through the article, I had to call my best friend, Steve Jobs, and read parts of this long article to him," Wozniak recalled. He knew that Jobs, then beginning his senior year, was one of the few people who would share his excitement.

A hero of the piece was John Draper, a hacker known as Captain Crunch because he had discovered that the sound emitted by the toy whistle that came with the breakfast cereal was the same 2600 Hertz tone used by the phone network's call-routing switches. It could fool the system into allowing a long-distance call to go through without extra charges. The article revealed that other tones that served to route calls could be found in an issue of the *Bell System Technical Journal,* which AT&T immediately began asking libraries to pull from their shelves.

As soon as Jobs got the call from Wozniak that Sunday afternoon, he knew they would have to get their hands on the technical journal right away. "Woz picked me up a few minutes later, and we went to the library at SLAC [the Stanford Linear Accelerator Center] to see if we could find it," Jobs recounted. It was Sunday and the library was closed, but they knew how to get in through a door that was rarely locked. "I remember that we were furiously digging through the stacks, and it was Woz who finally found the journal with all the frequencies. It was like, holy shit, and we opened it and there it was. We kept saying to ourselves, 'It's real. Holy shit, it's real.' It was all laid out — the tones, the frequencies."

Wozniak went to Sunnyvale Electronics before it closed that evening and bought the parts to make an analog tone generator. Jobs had built a frequency counter when he was part of the HP Explorers Club, and they used it to calibrate the desired tones. With a dial, they could replicate and tape-record the sounds specified in the article. By midnight they were ready to test it. Unfortunately the oscillators they used were not quite stable enough to replicate the right chirps to fool the phone company. "We could see the instability using Steve's frequency counter," recalled Wozniak, "and we just couldn't make it work. I had to leave for Berkeley the next morning, so we decided I would work on building a digital version once I got there."

No one had ever created a digital version of a Blue Box, but Woz was made for the challenge. Using diodes and transistors from Radio Shack, and with the help of a music student in his dorm who had perfect pitch, he got it built before Thanksgiving. "I have never designed a circuit I was prouder of," he said. "I still think it was incredible."

One night Wozniak drove down from Berkeley to Jobs's house to try it. They attempted to call Wozniak's uncle in Los Angeles, but they got a wrong number. It didn't matter; their device had worked. "Hi! We're calling you for free! We're calling you for free!" Wozniak shouted. The person on the other end was confused and annoyed. Jobs chimed in, "We're calling from California! From California! With a Blue Box." This probably baffled the man even more, since he was also in California.

At first the Blue Box was used for fun and pranks. The most daring of these was when they called the Vatican and Wozniak pretended to be Henry Kissinger wanting to speak to the pope. "Ve are at de summit meeting in Moscow, and ve need to talk to de pope," Woz intoned. He was told that it was 5:30 a.m. and the pope was sleeping. When he called back, he got a bishop who was supposed to serve as the translator. But they never actually got the pope on the line. "They realized that Woz wasn't Henry Kissinger," Jobs recalled. "We were at a public phone booth."

It was then that they reached an important milestone, one that would establish a pattern in their partnerships: Jobs came up with the idea that the Blue Box could be more than merely a hobby; they could build and sell them. "I got together the rest of the components, like the casing and power supply and keypads, and figured out how we could price it," Jobs said, foreshadowing roles he would play when they founded Apple. The finished product was about the size of two decks of playing cards. The parts cost about $40, and Jobs decided they should sell it for $150.

Following the lead of other phone phreaks such as Captain Crunch, they gave themselves handles. Wozniak became "Berkeley Blue," Jobs was "Oaf Tobark." They took the device to college dorms and gave demonstrations by attaching it to a phone and speaker. While the potential customers watched, they would call the Ritz in London or a dial-a-joke service in Australia. "We made a hundred or so Blue Boxes and sold almost all of them," Jobs recalled.

The fun and profits came to an end at a Sunnyvale pizza parlor. Jobs and Wozniak were about to drive to Berkeley with a Blue Box they had just finished making. Jobs needed money and was eager to sell, so he pitched the device to some guys at the next table. They were interested, so Jobs went to a phone booth and demonstrated it with a call to

Chicago. The prospects said they had to go to their car for money. "So we walk over to the car, Woz and me, and I've got the Blue Box in my hand, and the guy gets in, reaches under the seat, and he pulls out a gun," Jobs recounted. He had never been that close to a gun, and he was terrified. "So he's pointing the gun right at my stomach, and he says, 'Hand it over, brother.' My mind raced. There was the car door here, and I thought maybe I could slam it on his legs and we could run, but there was this high probability that he would shoot me. So I slowly handed it to him, very carefully." It was a weird sort of robbery. The guy who took the Blue Box actually gave Jobs a phone number and said he would try to pay for it if it worked. When Jobs later called the number, the guy said he couldn't figure out how to use it. So Jobs, in his felicitous way, convinced the guy to meet him and Wozniak at a public place. But they ended up deciding not to have another encounter with the gunman, even on the off chance they could get their $150.

The partnership paved the way for what would be a bigger adventure together. "If it hadn't been for the Blue Boxes, there wouldn't have been an Apple," Jobs later reflected. "I'm 100% sure of that. Woz and I learned how to work together, and we gained the confidence that we could solve technical problems and actually put something into production." They had created a device with a little circuit board that could control billions of dollars' worth of infrastructure. "You cannot believe how much confidence that gave us." Woz came to the same conclusion: "It was probably a bad idea selling them, but it gave us a taste of

65

what we could do with my engineering skills and his vision." The Blue Box adventure established a template for a partnership that would soon be born. Wozniak would be the gentle wizard coming up with a neat invention that he would have been happy just to give away, and Jobs would figure out how to make it user-friendly, put it together in a package, market it, and make a few bucks.

CHAPTER THREE: THE DROPOUT

TURN ON, TUNE IN . . .

CHRISANN BRENNAN

Toward the end of his senior year at Homestead, in the spring of 1972, Jobs started going out with a girl named Chrisann Brennan, who was about his age but still a junior. With her light brown hair, green eyes, high cheekbones, and fragile aura, she was very attractive. She was also enduring the breakup of her parents' marriage, which made her vulnerable. "We worked together on an animated movie, then started going out, and she became my first real girlfriend," Jobs recalled. As Brennan later said, "Steve was kind of crazy. That's why I was attracted to him."

Jobs's craziness was of the cultivated sort. He had begun his lifelong experiments with compulsive diets, eating only fruits and vegetables, so he was as lean and tight as a whippet. He learned to stare at people without blinking, and he perfected long silences punctuated by staccato bursts of fast talking. This odd mix of intensity and aloofness, combined with his shoulder-length hair and scraggly beard, gave him the aura of a crazed shaman. He oscillated between charismatic and creepy. "He shuffled around and looked half-mad," recalled Brennan. "He had a lot of angst. It was like a big darkness around him."

Jobs had begun to drop acid by then, and he turned Brennan on to it as well, in a wheat field just outside Sunnyvale. "It was great," he recalled. "I had been listening to a lot of Bach. All of a sudden the wheat field was playing Bach. It was the most wonderful feeling of my life up to that point. I felt like the conductor of this symphony with Bach coming through the wheat."

That summer of 1972, after his graduation, he and Brennan moved to a cabin in the hills above Los Altos. "I'm going to go live in a cabin with Chrisann," he announced to his parents one day. His father was furious. "No you're not," he said. "Over my dead body." They had recently fought about marijuana, and once again the younger Jobs was willful. He just said good-bye and walked out.

Brennan spent a lot of her time that summer painting; she was talented, and she did a picture of a clown for Jobs that he kept on the wall. Jobs wrote poetry and played guitar. He could be brutally cold and rude to her at times, but he was also entrancing and able to impose his will. "He was an enlightened being who was cruel," she recalled. "That's a strange combination."

Midway through the summer, Jobs was almost killed when his red Fiat caught fire. He was driving on Skyline Boulevard in the Santa Cruz Mountains with a high school friend, Tim Brown, who looked back, saw flames coming from the engine, and casually said to Jobs, "Pull over, your car is on fire." Jobs did. His father, despite their arguments, drove out to the hills to tow the Fiat home.

In order to find a way to make money for a new car, Jobs got Wozniak to drive him to De Anza College to look on the help-wanted bulletin board. They discovered that the Westgate Shopping Center in San Jose was seeking college students who could dress up in costumes and amuse the kids. So for $3 an hour, Jobs, Wozniak, and Brennan donned heavy full-body costumes and headgear to play Alice in Wonderland, the Mad Hatter, and the White Rabbit. Wozniak, in his earnest and sweet way, found it fun. "I said, 'I want to do it, it's my chance, because I love children.' I think Steve looked at it as a lousy job, but I looked at it as a fun adventure." Jobs did indeed find it a pain. "It was hot, the costumes were heavy, and after a while I felt like I wanted to smack some of the kids." Patience was never one of his virtues.

REED COLLEGE

Seventeen years earlier, Jobs's parents had made a pledge when they adopted him: He would go to college. So they had worked hard and saved dutifully for his college fund, which was modest but adequate by the time he graduated. But Jobs, becoming ever more willful, did not make it easy. At first he toyed with not going to college at all. "I think I might have headed to New York if I didn't go to college," he recalled, musing on how different his world — and perhaps all of ours — might have been if he had chosen that path. When his parents pushed him to go to college, he responded in a passive-aggressive way. He did not consider state schools, such as Berkeley, where Woz then was, despite the fact that they were more affordable.

69

Nor did he look at Stanford, just up the road and likely to offer a scholarship. "The kids who went to Stanford, they already knew what they wanted to do," he said. "They weren't really artistic. I wanted something that was more artistic and interesting."

Instead he insisted on applying only to Reed College, a private liberal arts school in Portland, Oregon, that was one of the most expensive in the nation. He was visiting Woz at Berkeley when his father called to say an acceptance letter had arrived from Reed, and he tried to talk Steve out of going there. So did his mother. It was far more than they could afford, they said. But their son responded with an ultimatum: If he couldn't go to Reed, he wouldn't go anywhere. They relented, as usual.

Reed had only one thousand students, half the number at Homestead High. It was known for its free-spirited hippie lifestyle, which combined somewhat uneasily with its rigorous academic standards and core curriculum. Five years earlier Timothy Leary, the guru of psychedelic enlightenment, had sat cross-legged at the Reed College commons while on his League for Spiritual Discovery (LSD) college tour, during which he exhorted his listeners, "Like every great religion of the past we seek to find the divinity within. . . . These ancient goals we define in the metaphor of the present — turn on, tune in, drop out." Many of Reed's students took all three of those injunctions seriously; the dropout rate during the 1970s was more than one-third.

When it came time for Jobs to matriculate in the fall of 1972, his parents drove him up to Portland, but in another small act of rebellion he refused to

70

let them come on campus. In fact he refrained from even saying good-bye or thanks. He recounted the moment later with uncharacteristic regret:

It's one of the things in life I really feel ashamed about. I was not very sensitive, and I hurt their feelings. I shouldn't have. They had done so much to make sure I could go there, but I just didn't want them around. I didn't want anyone to know I had parents. I wanted to be like an orphan who had bummed around the country on trains and just arrived out of nowhere, with no roots, no connections, no background.

In late 1972, there was a fundamental shift happening in American campus life. The nation's involvement in the Vietnam War, and the draft that accompanied it, was winding down. Political activism at colleges receded and in many late-night dorm conversations was replaced by an interest in pathways to personal fulfillment. Jobs found himself deeply influenced by a variety of books on spirituality and enlightenment, most notably *Be Here Now,* a guide to meditation and the wonders of psychedelic drugs by Baba Ram Dass, born Richard Alpert. "It was profound," Jobs said. "It transformed me and many of my friends."

The closest of those friends was another wispy-bearded freshman named Daniel Kottke, who met Jobs a week after they arrived at Reed and shared his interest in Zen, Dylan, and acid. Kottke, from a wealthy New York suburb, was smart but low-octane, with a sweet flower-child demeanor made even mellower by his interest in Buddhism. That spiritual quest had caused him to eschew material possessions, but he was nonetheless impressed by Jobs's tape deck. "Steve had a TEAC reel-to-reel and massive quantities of Dylan bootlegs," Kottke recalled. "He was both really cool and high-tech."

Jobs started spending much of his time with Kottke and his girlfriend, Elizabeth Holmes, even after he insulted her at their first meeting by grilling her about how much money it would take to get her to have sex with another man. They hitchhiked to the coast together, engaged in the typical dorm raps about the meaning of life, attended the love festivals at the local Hare Krishna temple, and went to the Zen center for free vegetarian meals. "It was a lot of fun," said Kottke, "but also philosophical, and we took Zen very seriously."

Jobs began sharing with Kottke other books, including *Zen Mind, Beginner's Mind* by Shunryu Suzuki, *Autobiography of a Yogi* by Paramahansa Yogananda, and *Cutting Through Spiritual Materialism* by Chögyam Trungpa. They created a meditation room in the attic crawl space above Elizabeth Holmes's room and fixed it up with Indian prints, a dhurrie rug, candles, incense, and meditation cushions. "There was a hatch in the ceiling leading to an attic which had a huge amount of space," Jobs said. "We took psychedelic drugs there sometimes, but mainly we just meditated."

Jobs's engagement with Eastern spirituality, and especially Zen Buddhism, was not just some passing fancy or youthful dabbling. He embraced it with his typical intensity, and it became deeply ingrained in his personality. "Steve is very much Zen," said Kottke. "It was a deep influence. You

72

see it in his whole approach of stark, minimalist aesthetics, intense focus." Jobs also became deeply influenced by the emphasis that Buddhism places on intuition. "I began to realize that an intuitive understanding and consciousness was more significant than abstract thinking and intellectual logical analysis," he later said. His intensity, however, made it difficult for him to achieve inner peace; his Zen awareness was not accompanied by an excess of calm, peace of mind, or interpersonal mellowness.

He and Kottke enjoyed playing a nineteenth-century German variant of chess called Kriegspiel, in which the players sit back-to-back; each has his own board and pieces and cannot see those of his opponent. A moderator informs them if a move they want to make is legal or illegal, and they have to try to figure out where their opponent's pieces are. "The wildest game I played with them was during a lashing rainstorm sitting by the fireside," recalled Holmes, who served as moderator. "They were tripping on acid. They were moving so fast I could barely keep up with them."

Another book that deeply influenced Jobs during his freshman year was *Diet for a Small Planet* by Frances Moore Lappé, which extolled the personal and planetary benefits of vegetarianism. "That's when I swore off meat pretty much for good," he recalled. But the book also reinforced his tendency to embrace extreme diets, which included purges, fasts, or eating only one or two foods, such as carrots or apples, for weeks on end.

Jobs and Kottke became serious vegetarians during their freshman year. "Steve got into it even more than I did," said Kottke. "He was living off Roman Meal cereal." They would go shopping at a farmers' co-op, where Jobs would buy a box of cereal, which would last a week, and other bulk health food. "He would buy flats of dates and almonds and lots of carrots, and he got a Champion juicer and we'd make carrot juice and carrot salads. There is a story about Steve turning orange from eating so many carrots, and there is some truth to that." Friends remember him having, at times, a sunset-like orange hue.

Jobs's dietary habits became even more obsessive when he read *Mucusless Diet Healing System* by Arnold Ehret, an early twentieth-century German-born nutrition fanatic. He believed in eating nothing but fruits and starchless vegetables, which he said prevented the body from forming harmful mucus, and he advocated cleansing the body regularly through prolonged fasts. That meant the end of even Roman Meal cereal — or any bread, grains, or milk. Jobs began warning friends of the mucus dangers lurking in their bagels. "I got into it in my typical nutso way," he said. At one point he and Kottke went for an entire week eating only apples, and then Jobs began to try even purer fasts. He started with two-day fasts, and eventually tried to stretch them to a week or more, breaking them carefully with large amounts of water and leafy vegetables. "After a week you start to feel fantastic," he said. "You get a ton of vitality from not having to digest all this food. I was in great shape. I felt I could get up and walk to San Francisco anytime I wanted."

Vegetarianism and Zen Buddhism, meditation

and spirituality, acid and rock — Jobs rolled together, in an amped-up way, the multiple impulses that were hallmarks of the enlightenment-seeking campus subculture of the era. And even though he barely indulged it at Reed, there was still an undercurrent of electronic geekiness in his soul that would someday combine surprisingly well with the rest of the mix.

ROBERT FRIEDLAND

In order to raise some cash one day, Jobs decided to sell his IBM Selectric typewriter. He walked into the room of the student who had offered to buy it only to discover that he was having sex with his girlfriend. Jobs started to leave, but the student invited him to take a seat and wait while they finished. "I thought, 'This is kind of far out,' " Jobs later recalled. And thus began his relationship with Robert Friedland, one of the few people in Jobs's life who were able to mesmerize him. He adopted some of Friedland's charismatic traits and for a few years treated him almost like a guru — until he began to see him as a charlatan.

Friedland was four years older than Jobs, but still an undergraduate. The son of an Auschwitz survivor who became a prosperous Chicago architect, he had originally gone to Bowdoin, a liberal arts college in Maine. But while a sophomore, he was arrested for possession of 24,000 tablets of LSD worth $125,000. The local newspaper pictured him with shoulder-length wavy blond hair smiling at the photographers as he was led away. He was sentenced to two years at a federal prison in Virginia, from which he was paroled in 1972. That fall

75

he headed off to Reed, where he immediately ran for student body president, saying that he needed to clear his name from the "miscarriage of justice" he had suffered. He won.

Friedland had heard Baba Ram Dass, the author of *Be Here Now,* give a speech in Boston, and like Jobs and Kottke had gotten deeply into Eastern spirituality. During the summer of 1973, he traveled to India to meet Ram Dass's Hindu guru, Neem Karoli Baba, famously known to his many followers as Maharaj-ji. When he returned that fall, Friedland had taken a spiritual name and walked around in sandals and flowing Indian robes. He had a room off campus, above a garage, and Jobs would go there many afternoons to seek him out. He was entranced by the apparent intensity of Friedland's conviction that a state of enlightenment truly existed and could be attained. "He turned me on to a different level of consciousness," Jobs said.

Friedland found Jobs fascinating as well. "He was always walking around barefoot," he later told a reporter. "The thing that struck me was his intensity. Whatever he was interested in he would generally carry to an irrational extreme." Jobs had honed his trick of using stares and silences to master other people. "One of his numbers was to stare at the person he was talking to. He would stare into their fucking eyeballs, ask some question, and would want a response without the other person averting their eyes."

According to Kottke, some of Jobs's personality traits — including a few that lasted throughout his career — were borrowed from Friedland. "Friedland taught Steve the reality distortion field," said Kottke. "He was charismatic and a bit of a con man and could bend situations to his very strong will. He was mercurial, sure of himself, a little dictatorial. Steve admired that, and he became more like that after spending time with Robert."

Jobs also absorbed how Friedland made himself the center of attention. "Robert was very much an outgoing, charismatic guy, a real salesman," Kottke recalled. "When I first met Steve he was shy and self-effacing, a very private guy. I think Robert taught him a lot about selling, about coming out of his shell, of opening up and taking charge of a situation." Friedland projected a high-wattage aura. "He would walk into a room and you would instantly notice him. Steve was the absolute opposite when he came to Reed. After he spent time with Robert, some of it started to rub off."

On Sunday evenings Jobs and Friedland would go to the Hare Krishna temple on the western edge of Portland, often with Kottke and Holmes in tow. They would dance and sing songs at the top of their lungs. "We would work ourselves into an ecstatic frenzy," Holmes recalled. "Robert would go insane and dance like crazy. Steve was more subdued, as if he was embarrassed to let loose." Then they would be treated to paper plates piled high with vegetarian food.

Friedland had stewardship of a 220-acre apple farm, about forty miles southwest of Portland, that was owned by an eccentric millionaire uncle from Switzerland named Marcel Müller. After Friedland became involved with Eastern spirituality, he turned it into a commune called the All

One Farm, and Jobs would spend weekends there with Kottke, Holmes, and like-minded seekers of enlightenment. The farm had a main house, a large barn, and a garden shed, where Kottke and Holmes slept. Jobs took on the task of pruning the Gravenstein apple trees. "Steve ran the apple orchard," said Friedland. "We were in the organic cider business. Steve's job was to lead a crew of freaks to prune the orchard and whip it back into shape."

Monks and disciples from the Hare Krishna temple would come and prepare vegetarian feasts redolent of cumin, coriander, and turmeric. "Steve would be starving when he arrived, and he would stuff himself," Holmes recalled. "Then he would go and purge. For years I thought he was bulimic. It was very upsetting, because we had gone to all that trouble of creating these feasts, and he couldn't hold it down."

Jobs was also beginning to have a little trouble stomaching Friedland's cult leader style. "Perhaps he saw a little bit too much of Robert in himself," said Kottke. Although the commune was supposed to be a refuge from materialism, Friedland began operating it more as a business; his followers were told to chop and sell firewood, make apple presses and wood stoves, and engage in other commercial endeavors for which they were not paid. One night Jobs slept under the table in the kitchen and was amused to notice that people kept coming in and stealing each other's food from the refrigerator. Communal economics were not for him. "It started to get very materialistic," Jobs recalled. "Everybody got the idea they were working very hard for Robert's farm, and one by one they started to leave. I got pretty sick of it."

Many years later, after Friedland had become a billionaire copper and gold mining executive — working out of Vancouver, Singapore, and Mongolia — I met him for drinks in New York. That evening I emailed Jobs and mentioned my encounter. He telephoned me from California within an hour and warned me against listening to Friedland. He said that when Friedland was in trouble because of environmental abuses committed by some of his mines, he had tried to contact Jobs to intervene with Bill Clinton, but Jobs had not responded. "Robert always portrayed himself as a spiritual person, but he crossed the line from being charismatic to being a con man," Jobs said. "It was a strange thing to have one of the spiritual people in your young life turn out to be, symbolically and in reality, a gold miner."

. . . Drop Out

Jobs quickly became bored with college. He liked being at Reed, just not taking the required classes. In fact he was surprised when he found out that, for all of its hippie aura, there were strict course requirements. When Wozniak came to visit, Jobs waved his schedule at him and complained, "They are making me take all these courses." Woz replied, "Yes, that's what they do in college." Jobs refused to go to the classes he was assigned and instead went to the ones he wanted, such as a dance class where he could enjoy both the creativity and the chance to meet girls. "I would never have refused to take the courses you were supposed to, that's a difference in our personality," Wozniak marveled.

Jobs also began to feel guilty, he later said, about spending so much of his parents' money on an education that did not seem worthwhile. "All of my working-class parents' savings were being spent on my college tuition," he recounted in a famous commencement address at Stanford. "I had no idea what I wanted to do with my life and no idea how college was going to help me figure it out. And here I was spending all of the money my parents had saved their entire life. So I decided to drop out and trust that it would all work out okay."

He didn't actually want to leave Reed; he just wanted to quit paying tuition and taking classes that didn't interest him. Remarkably, Reed tolerated that. "He had a very inquiring mind that was enormously attractive," said the dean of students, Jack Dudman. "He refused to accept automatically received truths, and he wanted to examine everything himself." Dudman allowed Jobs to audit classes and stay with friends in the dorms even after he stopped paying tuition.

"The minute I dropped out I could stop taking the required classes that didn't interest me, and begin dropping in on the ones that looked interesting," he said. Among them was a calligraphy class that appealed to him after he saw posters on campus that were beautifully drawn. "I learned about serif and sans serif typefaces, about varying the amount of space between different letter combinations, about what makes great typography great. It was beautiful, historical, artistically subtle in a way that science can't capture, and I found it fascinating."

It was yet another example of Jobs consciously positioning himself at the intersection of the arts and technology. In all of his products, technology would be married to great design, elegance, human touches, and even romance. He would be in the fore of pushing friendly graphical user interfaces. The calligraphy course would become iconic in that regard. "If I had never dropped in on that single course in college, the Mac would have never had multiple typefaces or proportionally spaced fonts. And since Windows just copied the Mac, it's likely that no personal computer would have them."

In the meantime Jobs eked out a bohemian existence on the fringes of Reed. He went barefoot most of the time, wearing sandals when it snowed. Elizabeth Holmes made meals for him, trying to keep up with his obsessive diets. He returned soda bottles for spare change, continued his treks to the free Sunday dinners at the Hare Krishna temple, and wore a down jacket in the heatless garage apartment he rented for $20 a month. When he needed money, he found work at the psychology department lab maintaining the electronic equipment that was used for animal behavior experiments. Occasionally Chrisann Brennan would come to visit. Their relationship sputtered along erratically. But mostly he tended to the stirrings of his own soul and personal quest for enlightenment.

"I came of age at a magical time," he reflected later. "Our consciousness was raised by Zen, and also by LSD." Even later in life he would credit psychedelic drugs for making him more enlightened. "Taking LSD was a profound experience,

81

one of the most important things in my life. LSD shows you that there's another side to the coin, and you can't remember it when it wears off, but you know it. It reinforced my sense of what was important — creating great things instead of making money, putting things back into the stream of history and of human consciousness as much as I could."

Chapter Four:
Atari and India

ZEN AND THE ART OF GAME DESIGN

Atari

In February 1974, after eighteen months of hanging around Reed, Jobs decided to move back to his parents' home in Los Altos and look for a job. It was not a difficult search. At peak times during the 1970s, the classified section of the *San Jose Mercury* carried up to sixty pages of technology help-wanted ads. One of those caught Jobs's eye. "Have fun, make money," it said. That day Jobs walked into the lobby of the video game manufacturer Atari and told the personnel director, who was startled by his unkempt hair and attire, that he wouldn't leave until they gave him a job.

Atari's founder was a burly entrepreneur named Nolan Bushnell, who was a charismatic visionary with a nice touch of showmanship in him — in other words, another role model waiting to be emulated. After he became famous, he liked driving around in a Rolls, smoking dope, and holding staff meetings in a hot tub. As Friedland had done and as Jobs would learn to do, he was able to turn charm into a cunning force, to cajole and intimidate and distort reality with the power of his personality. His chief engineer was Al Alcorn, beefy and jovial and a bit more grounded, the house grown-up trying to implement the vision and curb the enthusiasms of Bushnell. Their big hit thus far was a video game called Pong, in which two players tried to volley a blip on a screen with two movable lines that acted as paddles. (If you're under thirty, ask your parents.)

When Jobs arrived in the Atari lobby wearing sandals and demanding a job, Alcorn was the one who was summoned. "I was told, 'We've got a hippie kid in the lobby. He says he's not going to leave until we hire him. Should we call the cops or let him in?' I said bring him on in!"

Jobs thus became one of the first fifty employees at Atari, working as a technician for $5 an hour. "In retrospect, it was weird to hire a dropout from Reed," Alcorn recalled. "But I saw something in him. He was very intelligent, enthusiastic, excited about tech." Alcorn assigned him to work with a straitlaced engineer named Don Lang. The next day Lang complained, "This guy's a goddamn hippie with b.o. Why did you do this to me? And he's impossible to deal with." Jobs clung to the belief that his fruit-heavy vegetarian diet would prevent not just mucus but also body odor, even if he didn't use deodorant or shower regularly. It was a flawed theory.

Lang and others wanted to let Jobs go, but Bushnell worked out a solution. "The smell and behavior wasn't an issue with me," he said. "Steve was prickly, but I kind of liked him. So I asked him to go on the night shift. It was a way to save him." Jobs would come in after Lang and others had left and work through most of the night. Even thus isolated, he became known for his brashness. On those occasions when he happened to interact with others, he was prone to informing them that they were "dumb shits." In retrospect, he stands by that judgment. "The only reason I shone was that everyone else was so bad," Jobs recalled.

Despite his arrogance (or perhaps because of it) he was able to charm Atari's boss. "He was more philosophical than the other people I worked with," Bushnell recalled. "We used to discuss free will versus determinism. I tended to believe that things were much more determined, that we were programmed. If we had perfect information, we could predict people's actions. Steve felt the opposite." That outlook accorded with his faith in the power of the will to bend reality.

Jobs helped improve some of the games by pushing the chips to produce fun designs, and Bushnell's inspiring willingness to play by his own rules rubbed off on him. In addition, he intuitively appreciated the simplicity of Atari's games. They came with no manual and needed to be uncomplicated enough that a stoned freshman could figure them out. The only instructions for Atari's *Star Trek* game were "1. Insert quarter. 2. Avoid Klingons."

Not all of his coworkers shunned Jobs. He became friends with Ron Wayne, a draftsman at Atari, who had earlier started a company that built slot machines. It subsequently failed, but Jobs became fascinated with the idea that it was possible to start your own company. "Ron was an amazing guy," said Jobs. "He started companies. I had never met anybody like that." He proposed to Wayne that they go into business together; Jobs said he could borrow $50,000, and they could design and market a slot machine. But Wayne had already been burned in business, so he declined. "I said that was the quickest way to lose $50,000," Wayne recalled, "but I admired the fact that he had a burning drive to start his own business."

One weekend Jobs was visiting Wayne at his apartment, engaging as they often did in philosophical discussions, when Wayne said that there was something he needed to tell him. "Yeah, I think I know what it is," Jobs replied. "I think you like men." Wayne said yes. "It was my first encounter with someone who I knew was gay," Jobs recalled. "He planted the right perspective of it for me." Jobs grilled him: "When you see a beautiful woman, what do you feel?" Wayne replied, "It's like when you look at a beautiful horse. You can appreciate it, but you don't want to sleep with it. You appreciate beauty for what it is." Wayne said that it is a testament to Jobs that he felt like revealing this to him. "Nobody at Atari knew, and I could count on my toes and fingers the number of people I told in my whole life. But I guess it just felt right to tell him, that he would understand, and it didn't have any effect on our relationship."

INDIA

One reason Jobs was eager to make some money in early 1974 was that Robert Friedland, who had gone to India the summer before, was urging him to take his own spiritual journey there. Friedland had studied in India with Neem Karoli Baba (Maharaj-ji), who had been the guru to much of the sixties hippie movement. Jobs decided he should do the same, and he recruited Daniel Kottke to go with him. Jobs was not motivated by mere adventure. "For me it was a serious search," he said. "I'd been turned on to the idea of enlightenment and trying to figure out who I was and how I fit into things." Kottke adds that Jobs's quest seemed driven partly by not knowing his birth parents. "There was a hole in him, and he was trying to fill it."

When Jobs told the folks at Atari that he was quitting to go search for a guru in India, the jovial Alcorn was amused. "He comes in and stares at me and declares, 'I'm going to find my guru,' and I say, 'No shit, that's super. Write me!' And he says he wants me to help pay, and I tell him, 'Bullshit!' " Then Alcorn had an idea. Atari was making kits and shipping them to Munich, where they were built into finished machines and distributed by a wholesaler in Turin. But there was a problem: Because the games were designed for the American rate of sixty frames per second, there were frustrating interference problems in Europe, where the rate was fifty frames per second. Alcorn sketched out a fix with Jobs and then offered to pay for him to go to Europe to implement it. "It's got to be cheaper to get to India from there," he said. Jobs agreed. So Alcorn sent him on his way with the exhortation, "Say hi to your guru for me."

Jobs spent a few days in Munich, where he solved the interference problem, but in the process he flummoxed the dark-suited German managers. They complained to Alcorn that he dressed and smelled like a bum and behaved rudely. "I said,

'Did he solve the problem?' And they said, 'Yeah.' I said, 'If you got any more problems, you just call me, I got more guys just like him!' They said, 'No, no we'll take care of it next time.' " For his part, Jobs was upset that the Germans kept trying to feed him meat and potatoes. "They don't even have a word for vegetarian," he complained (incorrectly) in a phone call to Alcorn.

He had a better time when he took the train to see the distributor in Turin, where the Italian pastas and his host's camaraderie were more simpatico. "I had a wonderful couple of weeks in Turin, which is this charged-up industrial town," he recalled. "The distributor took me every night to dinner at this place where there were only eight tables and no menu. You'd just tell them what you wanted, and they made it. One of the tables was on reserve for the chairman of Fiat. It was really super." He next went to Lugano, Switzerland, where he stayed with Friedland's uncle, and from there took a flight to India.

When he got off the plane in New Delhi, he felt waves of heat rising from the tarmac, even though it was only April. He had been given the name of a hotel, but it was full, so he went to one his taxi driver insisted was good. "I'm sure he was getting some baksheesh, because he took me to this complete dive." Jobs asked the owner whether the water was filtered and foolishly believed the answer. "I got dysentery pretty fast. I was sick, really sick, a really high fever. I dropped from 160 pounds to 120 in about a week."

Once he got healthy enough to move, he decided that he needed to get out of Delhi. So he headed to the town of Haridwar, in western India near the source of the Ganges, which was having a festival known as the Kumbh Mela. More than ten million people poured into a town that usually contained fewer than 100,000 residents. "There were holy men all around. Tents with this teacher and that teacher. There were people riding elephants, you name it. I was there for a few days, but I decided that I needed to get out of there too."

He went by train and bus to a village near Nainital in the foothills of the Himalayas. That was where Neem Karoli Baba lived, or had lived. By the time Jobs got there, he was no longer alive, at least in the same incarnation. Jobs rented a room with a mattress on the floor from a family who helped him recuperate by feeding him vegetarian meals. "There was a copy there of *Autobiography of a Yogi* in English that a previous traveler had left, and I read it several times because there was not a lot to do, and I walked around from village to village and recovered from my dysentery." Among those who were part of the community there was Larry Brilliant, an epidemiologist who was working to eradicate smallpox and who later ran Google's philanthropic arm and the Skoll Foundation. He became Jobs's lifelong friend.

At one point Jobs was told of a young Hindu holy man who was holding a gathering of his followers at the Himalayan estate of a wealthy business-man. "It was a chance to meet a spiritual being and hang out with his followers, but it was also a chance to have a good meal. I could smell the food as we got near, and I was very hungry." As Jobs was eating, the holy man — who was not much older than Jobs — picked him out of the crowd, pointed at him, and began laughing maniacally. "He came running over and grabbed me and made a tooting sound and said, 'You are just like a baby,'" recalled Jobs. "I was not relishing this attention." Taking Jobs by the hand, he led him out of the worshipful crowd and walked him up to a hill, where there was a well and a small pond. "We sit down and he pulls out this straight razor. I'm thinking he's a nutcase and begin to worry. Then he pulls out a bar of soap — I had long hair at the time — and he lathered up my hair and shaved my head. He told me that he was saving my health."

Daniel Kottke arrived in India at the beginning of the summer, and Jobs went back to New Delhi to meet him. They wandered, mainly by bus, rather aimlessly. By this point Jobs was no longer trying to find a guru who could impart wisdom, but instead was seeking enlightenment through ascetic experience, deprivation, and simplicity. He was not able to achieve inner calm. Kottke remembers him getting into a furious shouting match with a Hindu woman in a village marketplace who, Jobs alleged, had been watering down the milk she was selling them.

Yet Jobs could also be generous. When they got to the town of Manali, Kottke's sleeping bag was stolen with his traveler's checks in it. "Steve covered my food expenses and bus ticket back to Delhi," Kottke recalled. He also gave Kottke the rest of his own money, $100, to tide him over.

During his seven months in India, he had written to his parents only sporadically, getting mail at the American Express office in New Delhi when he passed through, and so they were somewhat surprised when they got a call from the Oakland airport asking them to pick him up. They immediately drove up from Los Altos. "My head had been shaved, I was wearing Indian cotton robes, and my skin had turned a deep, chocolate brown-red from the sun," he recalled. "So I'm sitting there and my parents walked past me about five times and finally my mother came up and said 'Steve?' and I said 'Hi!'"

They took him back home, where he continued trying to find himself. It was a pursuit with many paths toward enlightenment. In the mornings and evenings he would meditate and study Zen, and in between he would drop in to audit physics or engineering courses at Stanford.

THE SEARCH

Jobs's interest in Eastern spirituality, Hinduism, Zen Buddhism, and the search for enlightenment was not merely the passing phase of a nineteen-year-old. Throughout his life he would seek to follow many of the basic precepts of Eastern religions, such as the emphasis on experiential *prajña,* wisdom or cognitive understanding that is intuitively experienced through concentration of the mind. Years later, sitting in his Palo Alto garden, he reflected on the lasting influence of his trip to India:

Coming back to America was, for me, much more of a cultural shock than going to India. The people in the Indian countryside don't use their intellect like we do, they use their intuition instead, and

91

their intuition is far more developed than in the rest of the world. Intuition is a very powerful thing, more powerful than intellect, in my opinion. That's had a big impact on my work.

Western rational thought is not an innate human characteristic; it is learned and is the great achievement of Western civilization. In the villages of India, they never learned it. They learned something else, which is in some ways just as valuable but in other ways is not. That's the power of intuition and experiential wisdom.

Coming back after seven months in Indian villages, I saw the craziness of the Western world as well as its capacity for rational thought. If you just sit and observe, you will see how restless your mind is. If you try to calm it, it only makes it worse, but over time it does calm, and when it does, there's room to hear more subtle things — that's when your intuition starts to blossom and you start to see things more clearly and be in the present more. Your mind just slows down, and you see a tremendous expanse in the moment. You see so much more than you could see before. It's a discipline; you have to practice it.

Zen has been a deep influence in my life ever since. At one point I was thinking about going to Japan and trying to get into the Eihei-ji monastery, but my spiritual advisor urged me to stay here. He said there is nothing over there that isn't here, and he was correct. I learned the truth of the Zen saying that if you are willing to travel around the world to meet a teacher, one will appear next door.

Jobs did in fact find a teacher right in his own neighborhood. Shunryu Suzuki, who wrote *Zen Mind, Beginner's Mind* and ran the San Francisco Zen Center, used to come to Los Altos every Wednesday evening to lecture and meditate with a small group of followers. After a while he asked his assistant, Kobun Chino Otogawa, to open a full-time center there. Jobs became a faithful follower, along with his occasional girlfriend, Chrisann Brennan, and Daniel Kottke and Elizabeth Holmes. He also began to go by himself on retreats to the Tassajara Zen Center, a monastery near Carmel where Kobun also taught.

Kottke found Kobun amusing. "His English was atrocious," he recalled. "He would speak in a kind of haiku, with poetic, suggestive phrases. We would sit and listen to him, and half the time we had no idea what he was going on about. I took the whole thing as a kind of lighthearted interlude." Holmes was more into the scene. "We would go to Kobun's meditations, sit on zafu cushions, and he would sit on a dais," she said. "We learned how to tune out distractions. It was a magical thing. One evening we were meditating with Kobun when it was raining, and he taught us how to use ambient sounds to bring us back to focus on our meditation."

As for Jobs, his devotion was intense. "He became really serious and self-important and just generally unbearable," according to Kottke. He began meeting with Kobun almost daily, and every few months they went on retreats together to meditate. "I ended up spending as much time as I could with him," Jobs recalled. "He had a wife who was

93

a nurse at Stanford and two kids. She worked the night shift, so I would go over and hang out with him in the evenings. She would get home about midnight and shoo me away." They sometimes discussed whether Jobs should devote himself fully to spiritual pursuits, but Kobun counseled otherwise. He assured Jobs that he could keep in touch with his spiritual side while working in a business. The relationship turned out to be lasting and deep; seventeen years later Kobun would perform Jobs's wedding ceremony.

Jobs's compulsive search for self-awareness also led him to undergo primal scream therapy, which had recently been developed and popularized by a Los Angeles psychotherapist named Arthur Janov. It was based on the Freudian theory that psychological problems are caused by the repressed pains of childhood; Janov argued that they could be resolved by re-suffering these primal moments while fully expressing the pain — sometimes in screams. To Jobs, this seemed preferable to talk therapy because it involved intuitive feeling and emotional action rather than just rational analyzing. "This was not something to think about," he later said. "This was something to do: to close your eyes, hold your breath, jump in, and come out the other end more insightful."

A group of Janov's adherents ran a program called the Oregon Feeling Center in an old hotel in Eugene that was managed by Jobs's Reed College guru Robert Friedland, whose All One Farm commune was nearby. In late 1974, Jobs signed up for a twelve-week course of therapy there costing $1,000. "Steve and I were both into personal growth, so I wanted to go with him," Kottke recounted, "but I couldn't afford it."

Jobs confided to close friends that he was driven by the pain he was feeling about being put up for adoption and not knowing about his birth parents. "Steve had a very profound desire to know his physical parents so he could better know himself," Friedland later said. He had learned from Paul and Clara Jobs that his birth parents had both been graduate students at a university and that his father might be Syrian. He had even thought about hiring a private investigator, but he decided not to do so for the time being. "I didn't want to hurt my parents," he recalled, referring to Paul and Clara.

"He was struggling with the fact that he had been adopted," according to Elizabeth Holmes. "He felt that it was an issue that he needed to get hold of emotionally." Jobs admitted as much to her. "This is something that is bothering me, and I need to focus on it," he said. He was even more open with Greg Calhoun. "He was doing a lot of soul-searching about being adopted, and he talked about it with me a lot," Calhoun recalled. "The primal scream and the mucusless diets, he was trying to cleanse himself and get deeper into his frustration about his birth. He told me he was deeply angry about the fact that he had been given up."

John Lennon had undergone the same primal scream therapy in 1970, and in December of that year he released the song "Mother" with the Plastic Ono Band. It dealt with Lennon's own feelings about a father who had abandoned him and a mother who had been killed when he was a teenager. The refrain includes the haunting chant

95

"Mama don't go, Daddy come home." Jobs used to play the song often.

Jobs later said that Janov's teachings did not prove very useful. "He offered a ready-made, buttoned-down answer which turned out to be far too oversimplistic. It became obvious that it was not going to yield any great insight." But Holmes contended that it made him more confident: "After he did it, he was in a different place. He had a very abrasive personality, but there was a peace about him for a while. His confidence improved and his feelings of inadequacy were reduced."

Jobs came to believe that he could impart that feeling of confidence to others and thus push them to do things they hadn't thought possible. Holmes had broken up with Kottke and joined a religious cult in San Francisco that expected her to sever ties with all past friends. But Jobs rejected that injunction. He arrived at the cult house in his Ford Ranchero one day and announced that he was driving up to Friedland's apple farm and she was to come. Even more brazenly, he said she would have to drive part of the way, even though she didn't know how to use the stick shift. "Once we got on the open road, he made me get behind the wheel, and he shifted the car until we got up to 55 miles per hour," she recalled. "Then he puts on a tape of Dylan's *Blood on the Tracks,* lays his head in my lap, and goes to sleep. He had the attitude that he could do anything, and therefore so can you. He put his life in my hands. So that made me do something I didn't think I could do."

It was the brighter side of what would become known as his reality distortion field. "If you trust him, you can do things," Holmes said. "If he's decided that something should happen, then he's just going to make it happen."

BREAKOUT

One day in early 1975 Al Alcorn was sitting in his office at Atari when Ron Wayne burst in. "Hey, Stevie is back!" he shouted.

"Wow, bring him on in," Alcorn replied.

Jobs shuffled in barefoot, wearing a saffron robe and carrying a copy of *Be Here Now,* which he handed to Alcorn and insisted he read. "Can I have my job back?" he asked.

"He looked like a Hare Krishna guy, but it was great to see him," Alcorn recalled. "So I said, sure!"

Once again, for the sake of harmony, Jobs worked mostly at night. Wozniak, who was living in an apartment nearby and working at HP, would come by after dinner to hang out and play the video games. He had become addicted to Pong at a Sunnyvale bowling alley, and he was able to build a version that he hooked up to his home TV set.

One day in the late summer of 1975, Nolan Bushnell, defying the prevailing wisdom that paddle games were over, decided to develop a single-player version of Pong; instead of competing against an opponent, the player would volley the ball into a wall that lost a brick whenever it was hit. He called Jobs into his office, sketched it out on his little blackboard, and asked him to design it. There would be a bonus, Bushnell told him, for every chip fewer than fifty that he used. Bushnell knew that Jobs was not a great engineer, but he assumed, correctly, that he would recruit Wozniak, who was always hanging around. "I looked at it as a two-for-one thing," Bushnell recalled. "Woz was a better engineer."

Wozniak was thrilled when Jobs asked him to help and proposed splitting the fee. "This was the most wonderful offer in my life, to actually design a game that people would use," he recalled. Jobs said it had to be done in four days and with the fewest chips possible. What he hid from Wozniak was that the deadline was one that Jobs had imposed, because he needed to get to the All One Farm to help prepare for the apple harvest. He also didn't mention that there was a bonus tied to keeping down the number of chips.

"A game like this might take most engineers a few months," Wozniak recalled. "I thought that there was no way I could do it, but Steve made me sure that I could." So he stayed up four nights in a row and did it. During the day at HP, Wozniak would sketch out his design on paper. Then, after a fast-food meal, he would go right to Atari and stay all night. As Wozniak churned out the design, Jobs sat on a bench to his left implementing it by wire-wrapping the chips onto a breadboard. "While Steve was breadboarding, I spent time playing my favorite game ever, which was the auto racing game Gran Trak 10," Wozniak said.

Astonishingly, they were able to get the job done in four days, and Wozniak used only forty-five chips. Recollections differ, but by most accounts Jobs simply gave Wozniak half of the base fee and not the bonus Bushnell paid for saving five chips.

It would be another ten years before Wozniak discovered (by being shown the tale in a book on the history of Atari titled *Zap*) that Jobs had been paid this bonus. "I think that Steve needed the money, and he just didn't tell me the truth," Wozniak later said. When he talks about it now, there are long pauses, and he admits that it causes him pain. "I wish he had just been honest. If he had told me he needed the money, he should have known I would have just given it to him. He was a friend. You help your friends." To Wozniak, it showed a fundamental difference in their characters. "Ethics always mattered to me, and I still don't understand why he would've gotten paid one thing and told me he'd gotten paid another," he said. "But, you know, people are different."

When Jobs learned this story was published, he called Wozniak to deny it. "He told me that he didn't remember doing it, and that if he did something like that he would remember it, so he probably didn't do it," Wozniak recalled. When I asked Jobs directly, he became unusually quiet and hesitant. "I don't know where that allegation comes from," he said. "I gave him half the money I ever got. That's how I've always been with Woz. I mean, Woz stopped working in 1978. He never did one ounce of work after 1978. And yet he got exactly the same shares of Apple stock that I did."

Is it possible that memories are muddled and that Jobs did not, in fact, shortchange Wozniak? "There's a chance that my memory is all wrong and messed up," Wozniak told me, but after a pause he reconsidered. "But no. I remember the details of this one, the $350 check." He confirmed his memory with Nolan Bushnell and Al Alcorn. "I remember talking about the bonus money to Woz, and he was upset," Bushnell said. "I said yes, there was a bonus for each chip they saved, and he just shook his head and then clucked his tongue."

Whatever the truth, Wozniak later insisted that it was not worth rehashing. Jobs is a complex person, he said, and being manipulative is just the darker facet of the traits that make him successful. Wozniak would never have been that way, but as he points out, he also could never have built Apple. "I would rather let it pass," he said when I pressed the point. "It's not something I want to judge Steve by."

The Atari experience helped shape Jobs's approach to business and design. He appreciated the user-friendliness of Atari's insert-quarter-avoid-Klingons games. "That simplicity rubbed off on him and made him a very focused product person," said Ron Wayne. Jobs also absorbed some of Bushnell's take-no-prisoners attitude. "Nolan wouldn't take no for an answer," according to Alcorn, "and this was Steve's first impression of how things got done. Nolan was never abusive, like Steve sometimes is. But he had the same driven attitude. It made me cringe, but dammit, it got things done. In that way Nolan was a mentor for Jobs."

Bushnell agreed. "There is something indefinable in an entrepreneur, and I saw that in Steve," he said. "He was interested not just in engineering, but also the business aspects. I taught him that if you act like you can do something, then it will work. I told him, 'Pretend to be completely in control and people will assume that you are.' "

Chapter Five: The Apple I

TURN ON, BOOT UP, JACK IN . . .

Machines of Loving Grace

In San Francisco and the Santa Clara Valley during the late 1960s, various cultural currents flowed together. There was the technology revolution that began with the growth of military contractors and soon included electronics firms, microchip makers, video game designers, and computer companies. There was a hacker subculture — filled with wireheads, phreakers, cyberpunks, hobbyists, and just plain geeks — that included engineers who didn't conform to the HP mold and their kids who weren't attuned to the wavelengths of the subdivisions. There were quasi-academic groups doing studies on the effects of LSD; participants included Doug Engelbart of the Augmentation Research Center in Palo Alto, who later helped develop the computer mouse and graphical user interfaces, and Ken Kesey, who celebrated the drug with music-and-light shows featuring a house band that became the Grateful Dead. There was the hippie movement, born out of the Bay Area's beat generation, and the rebellious political activists, born out of the Free Speech Movement at Berkeley. Overlaid on it all were various self-fulfillment movements pursuing paths to personal enlightenment: Zen and Hinduism, meditation and yoga, primal scream and sensory deprivation, Esalen and est.

This fusion of flower power and processor power, enlightenment and technology, was embodied by Steve Jobs as he meditated in the mornings, audited physics classes at Stanford, worked nights at Atari, and dreamed of starting his own business. "There was just something going on here," he said, looking back at the time and place. "The best music came from here — the Grateful Dead, Jefferson Airplane, Joan Baez, Janis Joplin — and so did the integrated circuit, and things like the *Whole Earth Catalog*."

Initially the technologists and the hippies did not interface well. Many in the counterculture saw computers as ominous and Orwellian, the province of the Pentagon and the power structure. In *The Myth of the Machine*, the historian Lewis Mumford warned that computers were sucking away our freedom and destroying "life-enhancing values." An injunction on punch cards of the period — "Do not fold, spindle or mutilate" — became an ironic phrase of the antiwar Left.

But by the early 1970s a shift was under way. "Computing went from being dismissed as a tool of bureaucratic control to being embraced as a symbol of individual expression and liberation," John Markoff wrote in his study of the counterculture's convergence with the computer industry, *What the Dormouse Said*. It was an ethos lyrically expressed in Richard Brautigan's 1967 poem, "All Watched Over by Machines of Loving Grace," and the cyberdelic fusion was certified when Timothy Leary declared that personal computers had become the new LSD and years later revised his famous mantra to proclaim, "Turn on, boot up, jack in." The musician Bono, who later became a friend of Jobs, often discussed with him why those immersed in the rock-drugs-rebel counterculture of the Bay Area ended up helping to create the personal computer industry. "The people who invented the twenty-first century were pot-smoking, sandal-wearing hippies from the West Coast like Steve, because they saw differently," he said. "The hierarchical systems of the East Coast, England, Germany, and Japan do not encourage this different thinking. The sixties produced an anarchic mind-set that is great for imagining a world not yet in existence."

One person who encouraged the denizens of the counterculture to make common cause with the hackers was Stewart Brand. A puckish visionary who generated fun and ideas over many decades, Brand was a participant in one of the early sixties LSD studies in Palo Alto. He joined with his fellow subject Ken Kesey to produce the acid-celebrating Trips Festival, appeared in the opening scene of Tom Wolfe's *The Electric Kool-Aid Acid Test,* and worked with Doug Engelbart to create a seminal sound-and-light presentation of new technologies called the Mother of All Demos. "Most of our generation scorned computers as the embodiment of centralized control," Brand later noted. "But a tiny contingent — later called hackers — embraced computers and set about transforming them into tools of liberation. That turned out to be the true royal road to the future."

Brand ran the Whole Earth Truck Store, which began as a roving truck that sold useful tools and educational materials, and in 1968 he decided to extend its reach with the *Whole Earth Catalog*. On its first cover was the famous picture of Earth taken from space; its subtitle was "Access to Tools." The underlying philosophy was that technology could be our friend. Brand wrote on the first page of the first edition, "A realm of intimate, personal power is developing — power of the individual to conduct his own education, find his own inspiration, shape his own environment, and share his adventure with whoever is interested. Tools that aid this process are sought and promoted by the *Whole Earth Catalog*." Buckminster Fuller followed with a poem that began: "I see God in the instruments and mechanisms that work reliably."

Jobs became a *Whole Earth* fan. He was particularly taken by the final issue, which came out in 1971, when he was still in high school, and he brought it with him to college and then to the All One Farm. "On the back cover of their final issue" Jobs recalled, "was a photograph of an early morning country road, the kind you might find yourself hitchhiking on if you were so adventurous. Beneath it were the words: 'Stay Hungry. Stay Foolish.' " Brand sees Jobs as one of the purest embodiments of the cultural mix that the catalog sought to celebrate. "Steve is right at the nexus of the counterculture and technology," he said. "He got the notion of tools for human use."

Brand's catalog was published with the help of the Portola Institute, a foundation dedicated to the fledgling field of computer education. The foundation also helped launch the People's Computer Company, which was not a company at all but a newsletter and organization with the motto "Computer power to the people." There were occasional Wednesday-night potluck dinners, and two of the regulars, Gordon French and Fred Moore, decided to create a more formal club where news about personal electronics could be shared.

They were energized by the arrival of the January 1975 issue of *Popular Mechanics,* which had on its cover the first personal computer kit, the Altair. The Altair wasn't much — just a $495 pile of parts that had to be soldered to a board that would then do little — but for hobbyists and hackers it heralded the dawn of a new era. Bill Gates and Paul Allen read the magazine and started working on a version of BASIC, an easy-to-use programming language, for the Altair. It also caught the attention of Jobs and Wozniak. And when an Altair kit arrived at the People's Computer Company, it became the centerpiece for the first meeting of the club that French and Moore had decided to launch.

THE HOMEBREW COMPUTER CLUB

The group became known as the Homebrew Computer Club, and it encapsulated the *Whole Earth* fusion between the counterculture and technology. It would become to the personal computer era something akin to what the Turk's Head coffeehouse was to the age of Dr. Johnson, a place where ideas were exchanged and disseminated. Moore wrote the flyer for the first meeting, held on March 5,

1975, in French's Menlo Park garage: "Are you building your own computer? Terminal, TV, typewriter?" it asked. "If so, you might like to come to a gathering of people with like-minded interests."

Allen Baum spotted the flyer on the HP bulletin board and called Wozniak, who agreed to go with him. "That night turned out to be one of the most important nights of my life," Wozniak recalled. About thirty other people showed up, spilling out of French's open garage door, and they took turns describing their interests. Wozniak, who later admitted to being extremely nervous, said he liked "video games, pay movies for hotels, scientific calculator design, and TV terminal design," according to the minutes prepared by Moore. There was a demonstration of the new Altair, but more important to Wozniak was seeing the specification sheet for a microprocessor.

As he thought about the microprocessor — a chip that had an entire central processing unit on it — he had an insight. He had been designing a terminal, with a keyboard and monitor, that would connect to a distant minicomputer. Using a microprocessor, he could put some of the capacity of the minicomputer inside the terminal itself, so it could become a small stand-alone computer on a desktop. It was an enduring idea: keyboard, screen, and computer all in one integrated personal package. "This whole vision of a personal computer just popped into my head," he said. "That night, I started to sketch out on paper what would later become known as the Apple I."

At first he planned to use the same microprocessor that was in the Altair, an Intel 8080. But each of those "cost almost more than my monthly rent," so he looked for an alternative. He found one in the Motorola 6800, which a friend at HP was able to get for $40 apiece. Then he discovered a chip made by MOS Technologies that was electronically the same but cost only $20. It would make his machine affordable, but it would carry a long-term cost. Intel's chips ended up becoming the industry standard, which would haunt Apple when its computers were incompatible with it.

After work each day, Wozniak would go home for a TV dinner and then return to HP to moonlight on his computer. He spread out the parts in his cubicle, figured out their placement, and soldered them onto his motherboard. Then he began writing the software that would get the microprocessor to display images on the screen. Because he could not afford to pay for computer time, he wrote the code by hand. After a couple of months he was ready to test it. "I typed a few keys on the keyboard and I was shocked! The letters were displayed on the screen." It was Sunday, June 29, 1975, a milestone for the personal computer. "It was the first time in history," Wozniak later said, "anyone had typed a character on a keyboard and seen it show up on their own computer's screen right in front of them."

Jobs was impressed. He peppered Wozniak with questions: Could the computer ever be networked? Was it possible to add a disk for memory storage? He also began to help Woz get components. Particularly important were the dynamic random-access memory chips. Jobs made a few calls and was able to score some from Intel for free. "Steve is just that sort of person," said Wozniak. "I mean, he knew how to talk to a sales representative. I could never have done that. I'm too shy."

Jobs began to accompany Wozniak to Homebrew meetings, carrying the TV monitor and helping to set things up. The meetings now attracted more than one hundred enthusiasts and had been moved to the auditorium of the Stanford Linear Accelerator Center. Presiding with a pointer and a free-form manner was Lee Felsenstein, another embodiment of the merger between the world of computing and the counterculture. He was an engineering school dropout, a participant in the Free Speech Movement, and an antiwar activist. He had written for the alternative newspaper *Berkeley Barb* and then gone back to being a computer engineer.

Woz was usually too shy to talk in the meetings, but people would gather around his machine afterward, and he would proudly show off his progress. Moore had tried to instill in the Homebrew an ethos of swapping and sharing rather than commerce. "The theme of the club," Woz said, "was 'Give to help others.' " It was an expression of the hacker ethic that information should be free and all authority mistrusted. "I designed the Apple I because I wanted to give it away for free to other people," said Wozniak.

This was not an outlook that Bill Gates embraced. After he and Paul Allen had completed their BASIC interpreter for the Altair, Gates was appalled that members of the Homebrew were making copies of it and sharing it without paying him. So he wrote what would become a famous letter to the club: "As the majority of hobbyists must be aware, most of you steal your software. Is this fair? . . . One thing you do is prevent good software from being written. Who can afford to do professional work for nothing? . . . I would appreciate letters from anyone who wants to pay up."

Steve Jobs, similarly, did not embrace the notion that Wozniak's creations, be it a Blue Box or a computer, wanted to be free. So he convinced Wozniak to stop giving away copies of his schematics. Most people didn't have time to build it themselves anyway, Jobs argued. "Why don't we build and sell printed circuit boards to them?" It was an example of their symbiosis. "Every time I'd design something great, Steve would find a way to make money for us," said Wozniak. Wozniak admitted that he would have never thought of doing that on his own. "It never crossed my mind to sell computers. It was Steve who said, 'Let's hold them in the air and sell a few.' "

Jobs worked out a plan to pay a guy he knew at Atari to draw the circuit boards and then print up fifty or so. That would cost about $1,000, plus the fee to the designer. They could sell them for $40 apiece and perhaps clear a profit of $700. Wozniak was dubious that they could sell them all. "I didn't see how we would make our money back," he recalled. He was already in trouble with his landlord for bouncing checks and now had to pay each month in cash.

Jobs knew how to appeal to Wozniak. He didn't argue that they were sure to make money, but instead that they would have a fun adventure. "Even if we lose our money, we'll have a company," said Jobs as they were driving in his Volkswagen bus.

"For once in our lives, we'll have a company." This was enticing to Wozniak, even more than any prospect of getting rich. He recalled, "I was excited to think about us like that. To be two best friends starting a company. Wow. I knew right then that I'd do it. How could I not?"

In order to raise the money they needed, Wozniak sold his HP 65 calculator for $500, though the buyer ended up stiffing him for half of that. For his part, Jobs sold his Volkswagen bus for $1,500. But the person who bought it came to find him two weeks later and said the engine had broken down, and Jobs agreed to pay for half of the repairs. Despite these little setbacks, they now had, with their own small savings thrown in, about $1,300 in working capital, the design for a product, and a plan. They would start their own computer company.

APPLE IS BORN

Now that they had decided to start a business, they needed a name. Jobs had gone for another visit to the All One Farm, where he had been pruning the Gravenstein apple trees, and Wozniak picked him up at the airport. On the ride down to Los Altos, they bandied around options. They considered some typical tech words, such as Matrix, and some neologisms, such as Executek, and some straightforward boring names, like Personal Computers Inc. The deadline for deciding was the next day, when Jobs wanted to start filing the papers. Finally Jobs proposed Apple Computer. "I was on one of my fruitarian diets," he explained. "I had just come back from the apple farm. It sounded fun, spirited, and not intimidating. Apple took the edge off the word 'computer.' Plus, it would get us ahead of Atari in the phone book." He told Wozniak that if a better name did not hit them by the next afternoon, they would just stick with Apple. And they did.

Apple. It was a smart choice. The word instantly signaled friendliness and simplicity. It managed to be both slightly off-beat and as normal as a slice of pie. There was a whiff of counterculture, back-to-nature earthiness to it, yet nothing could be more American. And the two words together — Apple Computer — provided an amusing disjuncture. "It doesn't quite make sense," said Mike Markkula, who soon thereafter became the first chairman of the new company. "So it forces your brain to dwell on it. Apple and computers, that doesn't go together! So it helped us grow brand awareness."

Wozniak was not yet ready to commit full-time. He was an HP company man at heart, or so he thought, and he wanted to keep his day job there. Jobs realized he needed an ally to help corral Wozniak and adjudicate if there was a disagreement. So he enlisted his friend Ron Wayne, the middle-aged engineer at Atari who had once started a slot machine company.

Wayne knew that it would not be easy to make Wozniak quit HP, nor was it necessary right away. Instead the key was to convince him that his computer designs would be owned by the Apple partnership. "Woz had a parental attitude toward the circuits he developed, and he wanted to be able to use them in other applications or let HP use them,"

Wayne said. "Jobs and I realized that these circuits would be the core of Apple. We spent two hours in a roundtable discussion at my apartment, and I was able to get Woz to accept this." His argument was that a great engineer would be remembered only if he teamed with a great marketer, and this required him to commit his designs to the partnership. Jobs was so impressed and grateful that he offered Wayne a 10% stake in the new partnership, turning him into a tie-breaker if Jobs and Wozniak disagreed over an issue.

"They were very different, but they made a powerful team," said Wayne. Jobs at times seemed to be driven by demons, while Woz seemed a naïf who was toyed with by angels. Jobs had a bravado that helped him get things done, occasionally by manipulating people. He could be charismatic, even mesmerizing, but also cold and brutal. Wozniak, in contrast, was shy and socially awkward, which made him seem childishly sweet. "Woz is very bright in some areas, but he's almost like a savant, since he was so stunted when it came to dealing with people he didn't know," said Jobs. "We were a good pair." It helped that Jobs was awed by Wozniak's engineering wizardry, and Wozniak was awed by Jobs's business drive. "I never wanted to deal with people and step on toes, but Steve could call up people he didn't know and make them do things," Wozniak recalled. "He could be rough on people he didn't think were smart, but he never treated me rudely, even in later years when maybe I couldn't answer a question as well as he wanted."

Even after Wozniak became convinced that his new computer design should become the property of the Apple partnership, he felt that he had to offer it first to HP, since he was working there. "I believed it was my duty to tell HP about what I had designed while working for them. That was the right thing and the ethical thing." So he demonstrated it to his managers in the spring of 1976. The senior executive at the meeting was impressed, and seemed torn, but he finally said it was not something that HP could develop. It was a hobbyist product, at least for now, and didn't fit into the company's high-quality market segments. "I was disappointed," Wozniak recalled, "but now I was free to enter into the Apple partnership."

On April 1, 1976, Jobs and Wozniak went to Wayne's apartment in Mountain View to draw up the partnership agreement. Wayne said he had some experience "writing in legalese," so he composed the three-page document himself. His "legalese" got the better of him. Paragraphs began with various flourishes: "Be it noted herewith . . . Be it further noted herewith . . . Now the refore [sic], in consideration of the respective assignments of interests . . ." But the division of shares and profits was clear — 45%-45%-10% — and it was stipulated that any expenditures of more than $100 would require agreement of at least two of the partners. Also, the responsibilities were spelled out. "Wozniak shall assume both general and major responsibility for the conduct of Electrical Engineering; Jobs shall assume general responsibility for Electrical Engineering and Marketing, and Wayne shall assume major responsibility for Mechanical Engineering and Documentation." Jobs signed in lowercase script, Wozniak in careful

114

cursive, and Wayne in an illegible squiggle.

Wayne then got cold feet. As Jobs started planning to borrow and spend more money, he recalled the failure of his own company. He didn't want to go through that again. Jobs and Wozniak had no personal assets, but Wayne (who worried about a global financial Armageddon) kept gold coins hidden in his mattress. Because they had structured Apple as a simple partnership rather than a corporation, the partners would be personally liable for the debts, and Wayne was afraid potential creditors would go after him. So he returned to the Santa Clara County office just eleven days later with a "statement of withdrawal" and an amendment to the partnership agreement. "By virtue of a reassessment of understandings by and between all parties," it began, "Wayne shall hereinafter cease to function in the status of 'Partner.' " It noted that in payment for his 10% of the company, he received $800, and shortly afterward $1,500 more.

Had he stayed on and kept his 10% stake, at the end of 2010 it would have been worth approximately $2.6 billion. Instead he was then living alone in a small home in Pahrump, Nevada, where he played the penny slot machines and lived off his social security check. He later claimed he had no regrets. "I made the best decision for me at the time. Both of them were real whirlwinds, and I knew my stomach and it wasn't ready for such a ride."

Jobs and Wozniak took the stage together for a presentation to the Homebrew Computer Club shortly after they signed Apple into existence. Wozniak held up one of their newly produced circuit boards and described the microprocessor, the eight kilobytes of memory, and the version of BASIC he had written. He also emphasized what he called the main thing: "a human-typable keyboard instead of a stupid, cryptic front panel with a bunch of lights and switches." Then it was Jobs's turn. He pointed out that the Apple, unlike the Altair, had all the essential components built in. Then he challenged them with a question: How much would people be willing to pay for such a wonderful machine? He was trying to get them to see the amazing value of the Apple. It was a rhetorical flourish he would use at product presentations over the ensuing decades.

The audience was not very impressed. The Apple had a cut-rate microprocessor, not the Intel 8080. But one important person stayed behind to hear more. His name was Paul Terrell, and in 1975 he had opened a computer store, which he dubbed the Byte Shop, on Camino Real in Menlo Park. Now, a year later, he had three stores and visions of building a national chain. Jobs was thrilled to give him a private demo. "Take a look at this," he said. "You're going to like what you see." Terrell was impressed enough to hand Jobs and Woz his card. "Keep in touch," he said.

"I'm keeping in touch," Jobs announced the next day when he walked barefoot into the Byte Shop. He made the sale. Terrell agreed to order fifty computers. But there was a condition: He didn't want just $50 printed circuit boards, for which customers would then have to buy all the chips and do the assembly. That might appeal to a few hardcore hobbyists, but not to most customers. Instead

116

he wanted the boards to be fully assembled. For that he was willing to pay about $500 apiece, cash on delivery.

Jobs immediately called Wozniak at HP. "Are you sitting down?" he asked. Wozniak said he wasn't. Jobs nevertheless proceeded to give him the news. "I was shocked, just completely shocked," Wozniak recalled. "I will never forget that moment."

To fill the order, they needed about $15,000 worth of parts. Allen Baum, the third prankster from Homestead High, and his father agreed to loan them $5,000. Jobs tried to borrow more from a bank in Los Altos, but the manager looked at him and, not surprisingly, declined. He went to Haltek Supply and offered an equity stake in Apple in return for the parts, but the owner decided they were "a couple of young, scruffy-looking guys," and declined. Alcorn at Atari would sell them chips only if they paid cash up front. Finally, Jobs was able to convince the manager of Cramer Electronics to call Paul Terrell to confirm that he had really committed to a $25,000 order. Terrell was at a conference when he heard over a loudspeaker that he had an emergency call (Jobs had been persistent). The Cramer manager told him that two scruffy kids had just walked in waving an order from the Byte Shop. Was it real? Terrell confirmed that it was, and the store agreed to front Jobs the parts on thirty-day credit.

GARAGE BAND

The Jobs house in Los Altos became the assembly point for the fifty Apple I boards that had to be delivered to the Byte Shop within thirty days, when

117

the payment for the parts would come due. All available hands were enlisted: Jobs and Wozniak, plus Daniel Kottke, his ex-girlfriend Elizabeth Holmes (who had broken away from the cult she'd joined), and Jobs's pregnant sister, Patty. Her vacated bedroom as well as the kitchen table and garage were commandeered as work space. Holmes, who had taken jewelry classes, was given the task of soldering chips. "Most I did well, but I got flux on a few of them," she recalled. This didn't please Jobs. "We don't have a chip to spare," he railed, correctly. He shifted her to bookkeeping and paperwork at the kitchen table, and he did the soldering himself. When they completed a board, they would hand it off to Wozniak. "I would plug each assembled board into the TV and keyboard to test it to see if it worked," he said. "If it did, I put it in a box. If it didn't, I'd figure what pin hadn't gotten into the socket right."

Paul Jobs suspended his sideline of repairing old cars so that the Apple team could have the whole garage. He put in a long old workbench, hung a schematic of the computer on the new plasterboard wall he built, and set up rows of labeled drawers for the components. He also built a burn box bathed in heat lamps so the computer boards could be tested by running overnight at high temperatures. When there was the occasional eruption of temper, an occurrence not uncommon around his son, Paul would impart some of his calm. "What's the matter?" he would say. "You got a feather up your ass?" In return he occasionally asked to borrow back the TV set so he could watch the end of a football game. During some of these breaks, Jobs

118

and Kottke would go outside and play guitar on the lawn.

Clara Jobs didn't mind losing most of her house to piles of parts and houseguests, but she was frustrated by her son's increasingly quirky diets. "She would roll her eyes at his latest eating obsessions," recalled Holmes. "She just wanted him to be healthy, and he would be making weird pronouncements like, 'I'm a fruitarian and I will only eat leaves picked by virgins in the moonlight.' "

After a dozen assembled boards had been approved by Wozniak, Jobs drove them over to the Byte Shop. Terrell was a bit taken aback. There was no power supply, case, monitor, or keyboard. He had expected something more finished. But Jobs stared him down, and he agreed to take delivery and pay.

After thirty days Apple was on the verge of being profitable. "We were able to build the boards more cheaply than we thought, because I got a good deal on parts," Jobs recalled. "So the fifty we sold to the Byte Shop almost paid for all the material we needed to make a hundred boards." Now they could make a real profit by selling the remaining fifty to their friends and Homebrew compatriots.

Elizabeth Holmes officially became the part-time bookkeeper at $4 an hour, driving down from San Francisco once a week and figuring out how to port Jobs's checkbook into a ledger. In order to make Apple seem like a real company, Jobs hired an answering service, which would relay messages to his mother. Ron Wayne drew a logo, using the ornate line-drawing style of Victorian illustrated fiction, that featured Newton sitting under a tree framed by a quote from Wordsworth: "A mind forever voyaging through strange seas of thought, alone." It was a rather odd motto, one that fit Wayne's self-image more than Apple Computer. Perhaps a better Wordsworth line would have been the poet's description of those involved in the start of the French Revolution: "Bliss was it in that dawn to be alive / But to be young was very heaven!" As Wozniak later exulted, "We were participating in the biggest revolution that had ever happened, I thought. I was so happy to be a part of it."

Woz had already begun thinking about the next version of the machine, so they started calling their current model the Apple I. Jobs and Woz would drive up and down Camino Real trying to get the electronics stores to sell it. In addition to the fifty sold by the Byte Shop and almost fifty sold to friends, they were building another hundred for retail outlets. Not surprisingly, they had contradictory impulses: Wozniak wanted to sell them for about what it cost to build them, but Jobs wanted to make a serious profit. Jobs prevailed. He picked a retail price that was about three times what it cost to build the boards and a 33% markup over the $500 wholesale price that Terrell and other stores paid. The result was $666.66. "I was always into repeating digits," Wozniak said. "The phone number for my dial-a-joke service was 255-6666." Neither of them knew that in the Book of Revelation 666 symbolized the "number of the beast," but they soon were faced with complaints, especially after 666 was featured in that year's hit movie, *The Omen*. (In 2010 one of the original

Apple I computers was sold at auction by Christie's for $213,000.)

The first feature story on the new machine appeared in the July 1976 issue of *Interface,* a now-defunct hobbyist magazine. Jobs and friends were still making them by hand in his house, but the article referred to him as the director of marketing and "a former private consultant to Atari." It made Apple sound like a real company. "Steve communicates with many of the computer clubs to keep his finger on the heartbeat of this young industry," the article reported, and it quoted him explaining, "If we can rap about their needs, feelings and motivations, we can respond appropriately by giving them what they want."

By this time they had other competitors, in addition to the Altair, most notably the IMSAI 8080 and Processor Technology Corporation's SOL-20. The latter was designed by Lee Felsenstein and Gordon French of the Homebrew Computer Club. They all had the chance to go on display during Labor Day weekend of 1976, at the first annual Personal Computer Festival, held in a tired hotel on the decaying boardwalk of Atlantic City, New Jersey. Jobs and Wozniak took a TWA flight to Philadelphia, cradling one cigar box with the Apple I and another with the prototype for the successor that Woz was working on. Sitting in the row behind them was Felsenstein, who looked at the Apple I and pronounced it "thoroughly unimpressive." Wozniak was unnerved by the conversation in the row behind him. "We could hear them talking in advanced business talk," he recalled, "using businesslike acronyms we'd never heard before."

Wozniak spent most of his time in their hotel room, tweaking his new prototype. He was too shy to stand at the card table that Apple had been assigned near the back of the exhibition hall. Daniel Kottke had taken the train down from Manhattan, where he was now attending Columbia, and he manned the table while Jobs walked the floor to inspect the competition. What he saw did not impress him. Wozniak, he felt reassured, was the best circuit engineer, and the Apple I (and surely its successor) could beat the competition in terms of functionality. However, the SOL-20 was better looking. It had a sleek metal case, a keyboard, a power supply, and cables. It looked as if it had been produced by grown-ups. The Apple I, on the other hand, appeared as scruffy as its creators.

CHAPTER SIX: THE APPLE II

DAWN OF A NEW AGE

AN INTEGRATED PACKAGE

As Jobs walked the floor of the Personal Computer Festival, he came to the realization that Paul Terrell of the Byte Shop had been right: Personal computers should come in a complete package. The next Apple, he decided, needed to have a great case and a built-in keyboard, and be integrated end to end, from the power supply to the software. "My vision was to create the first fully packaged computer," he recalled. "We were no longer aiming for the handful of hobbyists who liked to assemble their own computers, who knew how to buy transformers and keyboards. For every one of them there were a thousand people who would want the machine to be ready to run."

In their hotel room on that Labor Day weekend of 1976, Wozniak tinkered with the prototype of the new machine, to be named the Apple II, that Jobs hoped would take them to this next level. They brought the prototype out only once, late at night, to test it on the color projection television in one of the conference rooms. Wozniak had come up with an ingenious way to goose the machine's chips into creating color, and he wanted to see if it would work on the type of television that uses a projector to display on a movie-like screen. "I figured a projector might have a different color circuitry that would choke on my color method," he recalled. "So I hooked up the Apple II to this projector and it worked perfectly." As he typed on his keyboard, colorful lines and swirls burst on the screen across the room. The only outsider who saw this first Apple II was the hotel's technician. He said he had looked at all the machines, and this was the one he would be buying.

To produce the fully packaged Apple II would require significant capital, so they considered selling the rights to a larger company. Jobs went to Al Alcorn and asked for the chance to pitch it to Atari's management. He set up a meeting with the company's president, Joe Keenan, who was a lot more conservative than Alcorn and Bushnell. "Steve goes in to pitch him, but Joe couldn't stand him," Alcorn recalled. "He didn't appreciate Steve's hygiene." Jobs was barefoot, and at one point put his feet up on a desk. "Not only are we not going to buy this thing," Keenan shouted, "but get your feet off my desk!" Alcorn recalled thinking, "Oh, well. There goes that possibility."

In September Chuck Peddle of the Commodore computer company came by the Jobs house to get a demo. "We'd opened Steve's garage to the sunlight, and he came in wearing a suit and a cowboy hat," Wozniak recalled. Peddle loved the Apple II, and he arranged a presentation for his top brass a few weeks later at Commodore headquarters. "You might want to buy us for a few hundred thousand dollars," Jobs said when they got there. Wozniak was stunned by this "ridiculous" suggestion, but Jobs persisted. The Commodore honchos called a few days later to say they had decided it would be cheaper to build their own machine. Jobs was not upset. He had checked out Commodore and decided that its leadership was "sleazy." Wozniak did not rue the lost money, but his engineering sensibilities were offended when the company came out with the Commodore PET nine months later. "It kind of sickened me. They made a real crappy product by doing it so quick. They could have had Apple."

The Commodore flirtation brought to the surface a potential conflict between Jobs and Wozniak: Were they truly equal in what they contributed to Apple and what they should get out of it? Jerry Wozniak, who exalted the value of engineers over mere entrepreneurs and marketers, thought most of the money should be going to his son. He confronted Jobs personally when he came by the Wozniak house. "You don't deserve shit," he told Jobs. "You haven't produced anything." Jobs began to cry, which was not unusual. He had never been, and would never be, adept at containing his emotions. He told Steve Wozniak that he was willing to call off the partnership. "If we're not fifty-fifty," he said to his friend, "you can have the whole thing." Wozniak, however, understood better than his father the symbiosis they had. If it had not been for Jobs, he might still be handing out schematics of his boards for free at the back of Homebrew meetings. It was Jobs who had turned his ingenious designs into a budding business, just as he had with the Blue Box. He agreed they should remain partners.

It was a smart call. To make the Apple II successful required more than just Wozniak's awesome circuit design. It would need to be packaged into a fully integrated consumer product, and that was Jobs's role.

He began by asking their erstwhile partner Ron Wayne to design a case. "I assumed they had no money, so I did one that didn't require any tooling and could be fabricated in a standard metal shop," he said. His design called for a Plexiglas cover attached by metal straps and a rolltop door that slid down over the keyboard.

Jobs didn't like it. He wanted a simple and elegant design, which he hoped would set Apple apart from the other machines, with their clunky gray metal cases. While haunting the appliance aisles at Macy's, he was struck by the Cuisinart food processors and decided that he wanted a sleek case made of light molded plastic. At a Homebrew meeting, he offered a local consultant, Jerry Manock, $1,500 to produce such a design. Manock, dubious about Jobs's appearance, asked for the money up front. Jobs refused, but Manock took the job anyway. Within weeks he had produced a simple foam-molded plastic case that was uncluttered and exuded friendliness. Jobs was thrilled.

Next came the power supply. Digital geeks like Wozniak paid little attention to something so analog and mundane, but Jobs decided it was a key component. In particular he wanted — as he would his entire career — to provide power in a way that avoided the need for a fan. Fans inside computers were not Zen-like; they distracted. He dropped by Atari to consult with Alcorn, who knew old-fashioned electrical engineering. "Al turned me on to this brilliant guy named Rod Holt, who was a chain-smoking Marxist who had been through many marriages and was an expert on everything," Jobs recalled. Like Manock and others meeting Jobs for the first time, Holt took a look at him and was skeptical. "I'm expensive," Holt said. Jobs sensed he was worth it and said that cost was no problem. "He just conned me into working," said Holt, who ended up joining Apple full-time.

Instead of a conventional linear power supply, Holt built one like those used in oscilloscopes. It switched the power on and off not sixty times per second, but thousands of times; this allowed it to store the power for far less time, and thus throw off less heat. "That switching power supply was as revolutionary as the Apple II logic board was," Jobs later said. "Rod doesn't get a lot of credit for this in the history books, but he should. Every computer now uses switching power supplies, and they all rip off Rod's design." For all of Wozniak's brilliance, this was not something he could have done. "I only knew vaguely what a switching power supply was," Woz admitted.

Jobs's father had once taught him that a drive for perfection meant caring about the craftsmanship even of the parts unseen. Jobs applied that to the layout of the circuit board inside the Apple II. He rejected the initial design because the lines were not straight enough.

This passion for perfection led him to indulge his instinct to control. Most hackers and hobbyists liked to customize, modify, and jack various things into their computers. To Jobs, this was a threat to a seamless end-to-end user experience. Wozniak, a hacker at heart, disagreed. He wanted to include eight slots on the Apple II for users to insert whatever smaller circuit boards and peripherals they might want. Jobs insisted there be only two, for a printer and a modem. "Usually I'm really easy to get along with, but this time I told him, 'If that's what you want, go get yourself another computer,' " Wozniak recalled. "I knew that people like me would eventually come up with things to add to any computer." Wozniak won the argument that time, but he could sense his power waning. "I was in a position to do that then. I wouldn't always be."

MIKE MARKKULA

All of this required money. "The tooling of this plastic case was going to cost, like, $100,000," Jobs said. "Just to get this whole thing into production was going to be, like, $200,000." He went back to Nolan Bushnell, this time to get him to put in some money and take a minority equity stake. "He asked me if I would put $50,000 in and he would give me a third of the company," said Bushnell. "I was so smart, I said no. It's kind of fun to think about that, when I'm not crying."

Bushnell suggested that Jobs try Don Valentine, a straight-shooting former marketing manager at National Semiconductor who had founded Sequoia Capital, a pioneering venture capital firm. Valentine arrived at the Jobses' garage in a Mercedes wearing a blue suit, button-down shirt, and rep tie. His first impression was that Jobs looked and smelled odd. "Steve was trying to be the embodiment of the counterculture. He had a wispy beard, was very thin, and looked like Ho Chi Minh."

Valentine, however, did not become a preeminent Silicon Valley investor by relying on surface appearances. What bothered him more was that Jobs knew nothing about marketing and seemed content to peddle his product to individual stores one by one. "If you want me to finance you," Valentine told him, "you need to have one person as a partner who understands marketing and distribution and can write a business plan." Jobs tended to be either bristly or solicitous when older people offered him advice. With Valentine he was the latter. "Send me three suggestions," he replied. Valentine did, Jobs met them, and he clicked with one of them, a man named Mike Markkula, who would end up playing a critical role at Apple for the next two decades.

Markkula was only thirty-three, but he had already retired after working at Fairchild and then Intel, where he made millions on his stock options when the chip maker went public. He was a cautious and shrewd man, with the precise moves of someone who had been a gymnast in high school, and he excelled at figuring out pricing strategies, distribution networks, marketing, and finance. Despite being slightly reserved, he had a flashy side when it came to enjoying his newly minted wealth. He built himself a house in Lake Tahoe and later an outsize mansion in the hills of Woodside. When he showed up for his first meeting at Jobs's garage, he was driving not a dark Mercedes like Valentine, but a highly polished gold Corvette convertible. "When I arrived at the garage, Woz was at the workbench and immediately began showing off the Apple II," Markkula recalled. "I looked past the fact that both guys needed a haircut and was amazed by what I saw on that workbench. You can always get a haircut."

Jobs immediately liked Markkula. "He was short and he had been passed over for the top marketing job at Intel, which I suspect made him want to prove himself." He also struck Jobs as decent and fair. "You could tell that if he could screw you, he wouldn't. He had a real moral sense to him." Wozniak was equally impressed. "I thought he was the nicest person ever," he recalled. "Better still, he actually liked what we had!"

Markkula proposed to Jobs that they write a business plan together. "If it comes out well, I'll invest," Markkula said, "and if not, you've got a few weeks of my time for free." Jobs began going to Markkula's house in the evenings, kicking around projections and talking through the night. "We made a lot of assumptions, such as about how many houses would have a personal computer, and there were nights we were up until 4 a.m.," Jobs recalled. Markkula ended up writing most of the plan. "Steve would say, 'I will bring you this section next time,' but he usually didn't deliver on time, so I ended up doing it."

Markkula's plan envisioned ways of getting beyond the hobbyist market. "He talked about introducing the computer to regular people in regular homes, doing things like keeping track of your favorite recipes or balancing your checkbook," Woz-

130

niak recalled. Markkula made a wild prediction: "We're going to be a Fortune 500 company in two years," he said. "This is the start of an industry. It happens once in a decade." It would take Apple seven years to break into the Fortune 500, but the spirit of Markkula's prediction turned out to be true.

Markkula offered to guarantee a line of credit of up to $250,000 in return for being made a one-third equity participant. Apple would incorporate, and he along with Jobs and Wozniak would each own 26% of the stock. The rest would be reserved to attract future investors. The three met in the cabana by Markkula's swimming pool and sealed the deal. "I thought it was unlikely that Mike would ever see that $250,000 again, and I was impressed that he was willing to risk it," Jobs recalled.

Now it was necessary to convince Wozniak to come on board full-time. "Why can't I keep doing this on the side and just have HP as my secure job for life?" he asked. Markkula said that wouldn't work, and he gave Wozniak a deadline of a few days to decide. "I felt very insecure in starting a company where I would be expected to push people around and control what they did," Wozniak recalled. "I'd decided long ago that I would never become someone authoritative." So he went to Markkula's cabana and announced that he was not leaving HP.

Markkula shrugged and said okay. But Jobs got very upset. He cajoled Wozniak; he got friends to try to convince him; he cried, yelled, and threw a couple of fits. He even went to Wozniak's parents' house, burst into tears, and asked Jerry for help. By this point Wozniak's father had realized there was real money to be made by capitalizing on the Apple II, and he joined forces on Jobs's behalf. "I started getting phone calls at work and home from my dad, my mom, my brother, and various friends," Wozniak recalled. "Every one of them told me I'd made the wrong decision." None of that worked. Then Allen Baum, their Buck Fry Club mate at Homestead High, called. "You really ought to go ahead and do it," he said. He argued that if he joined Apple full-time, he would not have to go into management or give up being an engineer. "That was exactly what I needed to hear," Wozniak later said. "I could stay at the bottom of the organization chart, as an engineer." He called Jobs and declared that he was now ready to come on board.

On January 3, 1977, the new corporation, the Apple Computer Co., was officially created, and it bought out the old partnership that had been formed by Jobs and Wozniak nine months earlier. Few people noticed. That month the Homebrew surveyed its members and found that, of the 181 who owned personal computers, only six owned an Apple. Jobs was convinced, however, that the Apple II would change that.

Markkula would become a father figure to Jobs. Like Jobs's adoptive father, he would indulge Jobs's strong will, and like his biological father, he would end up abandoning him. "Markkula was as much a father-son relationship as Steve ever had," said the venture capitalist Arthur Rock. He began to teach Jobs about marketing and sales. "Mike really

132

took me under his wing," Jobs recalled. "His values were much aligned with mine. He emphasized that you should never start a company with the goal of getting rich. Your goal should be making something you believe in and making a company that will last."

Markkula wrote his principles in a one-page paper titled "The Apple Marketing Philosophy" that stressed three points. The first was *empathy,* an intimate connection with the feelings of the customer: "We will truly understand their needs better than any other company." The second was *focus:* "In order to do a good job of those things that we decide to do, we must eliminate all of the unimportant opportunities." The third and equally important principle, awkwardly named, was *impute.* It emphasized that people form an opinion about a company or product based on the signals that it conveys. "People *DO* judge a book by its cover," he wrote. "We may have the best product, the highest quality, the most useful software etc.; if we present them in a slipshod manner, they will be perceived as slipshod; if we present them in a creative, professional manner, we will *impute* the desired qualities."

For the rest of his career, Jobs would understand the needs and desires of customers better than any other business leader, he would focus on a handful of core products, and he would care, sometimes obsessively, about marketing and image and even the details of packaging. "When you open the box of an iPhone or iPad, we want that tactile experience to set the tone for how you perceive the product," he said. "Mike taught me that."

Regis McKenna

The first step in this process was convincing the Valley's premier publicist, Regis McKenna, to take on Apple as a client. McKenna was from a large working-class Pittsburgh family, and bred into his bones was a steeliness that he cloaked with charm. A college dropout, he had worked for Fairchild and National Semiconductor before starting his own PR and advertising firm. His two specialties were doling out exclusive interviews with his clients to journalists he had cultivated and coming up with memorable ad campaigns that created brand awareness for products such as microchips. One of these was a series of colorful magazine ads for Intel that featured racing cars and poker chips rather than the usual dull performance charts. These caught Jobs's eye. He called Intel and asked who created them. "Regis McKenna," he was told. "I asked them what Regis McKenna was," Jobs recalled, "and they told me he was a person." When Jobs phoned, he couldn't get through to McKenna. Instead he was transferred to Frank Burge, an account executive, who tried to put him off. Jobs called back almost every day.

Burge finally agreed to drive out to the Jobs garage. "Holy Christ, this guy is going to be something else," he recalled thinking. "What's the least amount of time I can spend with this clown without being rude." Then, when he was confronted with the unwashed and shaggy Jobs, two things hit him: "First, he was an incredibly smart young man. Second, I didn't understand a fiftieth of what he was talking about."

So Jobs and Wozniak were invited to have a meeting with, as his impish business cards read, "Regis McKenna, himself." This time it was the normally shy Wozniak who became prickly. McKenna glanced at an article Wozniak was writing about Apple and suggested that it was too technical and needed to be livened up. "I don't want any PR man touching my copy," Wozniak snapped. McKenna suggested it was time for them to leave his office. "But Steve called me back right away and said he wanted to meet again," McKenna recalled. "This time he came without Woz, and we hit it off."

McKenna had his team get to work on brochures for the Apple II. The first thing they did was to replace Ron Wayne's ornate Victorian woodcut-style logo, which ran counter to McKenna's colorful and playful advertising style. So an art director, Rob Janoff, was assigned to create a new one. "Don't make it cute," Jobs ordered. Janoff came up with a simple apple shape in two versions, one whole and the other with a bite taken out of it. The first looked too much like a cherry, so Jobs chose the one with a bite. He also picked a version that was striped in six colors, with psychedelic hues sandwiched between whole-earth green and sky blue, even though that made printing the logo significantly more expensive. Atop the brochure McKenna put a maxim, often attributed to Leonardo da Vinci, that would become the defining precept of Jobs's design philosophy: "Simplicity is the ultimate sophistication."

THE FIRST LAUNCH EVENT

The introduction of the Apple II was scheduled to coincide with the first West Coast Computer

Faire, to be held in April 1977 in San Francisco, organized by a Homebrew stalwart, Jim Warren. Jobs signed Apple up for a booth as soon as he got the information packet. He wanted to secure a location right at the front of the hall as a dramatic way to launch the Apple II, and so he shocked Wozniak by paying $5,000 in advance. "Steve decided that this was our big launch," said Wozniak. "We would show the world we had a great machine and a great company."

It was an application of Markkula's admonition that it was important to "impute" your greatness by making a memorable impression on people, especially when launching a new product. That was reflected in the care that Jobs took with Apple's display area. Other exhibitors had card tables and poster board signs. Apple had a counter draped in black velvet and a large pane of backlit Plexiglas with Janoff's new logo. They put on display the only three Apple IIs that had been finished, but empty boxes were piled up to give the impression that there were many more on hand.

Jobs was furious that the computer cases had arrived with tiny blemishes on them, so he had his handful of employees sand and polish them. The imputing even extended to gussying up Jobs and Wozniak. Markkula sent them to a San Francisco tailor for three-piece suits, which looked faintly ridiculous on them, like tuxes on teenagers. "Markkula explained how we would all have to dress up nicely, how we should appear and look, how we should act," Wozniak recalled.

It was worth the effort. The Apple II looked solid yet friendly in its sleek beige case, unlike the intimidating metal-clad machines and naked boards on the other tables. Apple got three hundred orders at the show, and Jobs met a Japanese textile maker, Mizushima Satoshi, who became Apple's first dealer in Japan.

The fancy clothes and Markkula's injunctions could not, however, stop the irrepressible Wozniak from playing some practical jokes. One program that he displayed tried to guess people's nationality from their last name and then produced the relevant ethnic jokes. He also created and distributed a hoax brochure for a new computer called the "Zaltair," with all sorts of fake ad-copy superlatives like "Imagine a car with five wheels." Jobs briefly fell for the joke and even took pride that the Apple II stacked up well against the Zaltair in the comparison chart. He didn't realize who had pulled the prank until eight years later, when Woz gave him a framed copy of the brochure as a birthday gift.

MIKE SCOTT

Apple was now a real company, with a dozen employees, a line of credit, and the daily pressures that can come from customers and suppliers. It had even moved out of the Jobses' garage, finally, into a rented office on Stevens Creek Boulevard in Cupertino, about a mile from where Jobs and Wozniak went to high school.

Jobs did not wear his growing responsibilities gracefully. He had always been temperamental and bratty. At Atari his behavior had caused him to be banished to the night shift, but at Apple that was not possible. "He became increasingly

137

tyrannical and sharp in his criticism," according to Markkula. "He would tell people, 'That design looks like shit.' " He was particularly rough on Wozniak's young programmers, Randy Wigginton and Chris Espinosa. "Steve would come in, take a quick look at what I had done, and tell me it was shit without having any idea what it was or why I had done it," said Wigginton, who was just out of high school.

There was also the issue of his hygiene. He was still convinced, against all evidence, that his vegan diets meant that he didn't need to use a deodorant or take regular showers. "We would have to literally put him out the door and tell him to go take a shower," said Markkula. "At meetings we had to look at his dirty feet." Sometimes, to relieve stress, he would soak his feet in the toilet, a practice that was not as soothing for his colleagues.

Markkula was averse to confrontation, so he decided to bring in a president, Mike Scott, to keep a tighter rein on Jobs. Markkula and Scott had joined Fairchild on the same day in 1967, had adjoining offices, and shared the same birthday, which they celebrated together each year. At their birthday lunch in February 1977, when Scott was turning thirty-two, Markkula invited him to become Apple's new president.

On paper he looked like a great choice. He was running a manufacturing line for National Semiconductor, and he had the advantage of being a manager who fully understood engineering. In person, however, he had some quirks. He was overweight, afflicted with tics and health problems, and so tightly wound that he wandered the halls with clenched fists. He also could be argumentative. In dealing with Jobs, that could be good or bad.

Wozniak quickly embraced the idea of hiring Scott. Like Markkula, he hated dealing with the conflicts that Jobs engendered. Jobs, not surprisingly, had more conflicted emotions. "I was only twenty-two, and I knew I wasn't ready to run a real company," he said. "But Apple was my baby, and I didn't want to give it up." Relinquishing any control was agonizing to him. He wrestled with the issue over long lunches at Bob's Big Boy hamburgers (Woz's favorite place) and at the Good Earth restaurant (Jobs's). He finally acquiesced, reluctantly.

Mike Scott, called "Scotty" to distinguish him from Mike Markkula, had one primary duty: managing Jobs. This was usually accomplished by Jobs's preferred mode of meeting, which was taking a walk together. "My very first walk was to tell him to bathe more often," Scott recalled. "He said that in exchange I had to read his fruitarian diet book and consider it as a way to lose weight." Scott never adopted the diet or lost much weight, and Jobs made only minor modifications to his hygiene. "Steve was adamant that he bathed once a week, and that was adequate as long as he was eating a fruitarian diet."

Jobs's desire for control and disdain for authority was destined to be a problem with the man who was brought in to be his regent, especially when Jobs discovered that Scott was one of the only people he had yet encountered who would not bend to his will. "The question between Steve and me was who could be most stubborn, and I

139

was pretty good at that," Scott said. "He needed to be sat on, and he sure didn't like that." Jobs later said, "I never yelled at anyone more than I yelled at Scotty."

An early showdown came over employee badge numbers. Scott assigned #1 to Wozniak and #2 to Jobs. Not surprisingly, Jobs demanded to be #1. "I wouldn't let him have it, because that would stoke his ego even more," said Scott. Jobs threw a tantrum, even cried. Finally, he proposed a solution. He would have badge #0. Scott relented, at least for the purpose of the badge, but the Bank of America required a positive integer for its payroll system and Jobs's remained #2.

There was a more fundamental disagreement that went beyond personal petulance. Jay Elliot, who was hired by Jobs after a chance meeting in a restaurant, noted Jobs's salient trait: "His obsession is a passion for the product, a passion for product perfection." Mike Scott, on the other hand, never let a passion for the perfect take precedence over pragmatism. The design of the Apple II case was one of many examples. The Pantone company, which Apple used to specify colors for its plastic, had more than two thousand shades of beige. "None of them were good enough for Steve," Scott marveled. "He wanted to create a different shade, and I had to stop him." When the time came to tweak the design of the case, Jobs spent days agonizing over just how rounded the corners should be. "I didn't care how rounded they were," said Scott, "I just wanted it decided." Another dispute was over engineering benches. Scott wanted a standard gray; Jobs insisted on special-order benches that were pure white. All of this finally led to a showdown in front of Markkula about whether Jobs or Scott had the power to sign purchase orders; Markkula sided with Scott. Jobs also insisted that Apple be different in how it treated customers. He wanted a one-year warranty to come with the Apple II. This flabbergasted Scott; the usual warranty was ninety days. Again Jobs dissolved into tears during one of their arguments over the issue. They walked around the parking lot to calm down, and Scott decided to relent on this one.

Wozniak began to rankle at Jobs's style. "Steve was too tough on people. I wanted our company to feel like a family where we all had fun and shared whatever we made." Jobs, for his part, felt that Wozniak simply would not grow up. "He was very childlike. He did a great version of BASIC, but then never could buckle down and write the floating-point BASIC we needed, so we ended up later having to make a deal with Microsoft. He was just too unfocused."

But for the time being the personality clashes were manageable, mainly because the company was doing so well. Ben Rosen, the analyst whose newsletters shaped the opinions of the tech world, became an enthusiastic proselytizer for the Apple II. An independent developer came up with the first spreadsheet and personal finance program for personal computers, VisiCalc, and for a while it was available only on the Apple II, turning the computer into something that businesses and families could justify buying. The company began attracting influential new investors. The pioneering venture capitalist Arthur Rock had initially been unimpressed when Markkula sent Jobs to see him. "He looked as if he had just come back from seeing that guru he had in India," Rock recalled, "and he kind of smelled that way too." But after Rock scoped out the Apple II, he made an investment and joined the board.

The Apple II would be marketed, in various models, for the next sixteen years, with close to six million sold. More than any other machine, it launched the personal computer industry. Wozniak deserves the historic credit for the design of its awe-inspiring circuit board and related operating software, which was one of the era's great feats of solo invention. But Jobs was the one who integrated Wozniak's boards into a friendly package, from the power supply to the sleek case. He also created the company that sprang up around Wozniak's machines. As Regis McKenna later said, "Woz designed a great machine, but it would be sitting in hobby shops today were it not for Steve Jobs." Nevertheless most people considered the Apple II to be Wozniak's creation. That would spur Jobs to pursue the next great advance, one that he could call his own.

CHAPTER SEVEN: CHRISANN AND LISA

HE WHO IS ABANDONED . . .

Ever since they had lived together in a cabin during the summer after he graduated from high school, Chrisann Brennan had woven in and out of Jobs's life. When he returned from India in 1974, they spent time together at Robert Friedland's farm. "Steve invited me up there, and we were just young and easy and free," she recalled. "There was an energy there that went to my heart."

When they moved back to Los Altos, their relationship drifted into being, for the most part, merely friendly. He lived at home and worked at Atari; she had a small apartment and spent a lot of time at Kobun Chino's Zen center. By early 1975 she had begun a relationship with a mutual friend, Greg Calhoun. "She was with Greg, but went back to Steve occasionally," according to Elizabeth Holmes. "That was pretty much the way it was with all of us. We were sort of shifting back and forth; it was the seventies, after all."

Calhoun had been at Reed with Jobs, Friedland, Kottke, and Holmes. Like the others, he became deeply involved with Eastern spirituality, dropped out of Reed, and found his way to Friedland's farm. There he moved into an eight- by twenty-foot chicken coop that he converted into a little house by raising it onto cinderblocks and building a sleeping loft inside. In the spring of 1975 Brennan moved in with him, and the next year they decided to make their own pilgrimage to India. Jobs advised Calhoun not to take Brennan with him, saying that she would interfere with his spiritual quest, but they went together anyway. "I was just so impressed by what happened to Steve on his trip to India that I wanted to go there," she said.

Theirs was a serious trip, beginning in March 1976 and lasting almost a year. At one point they ran out of money, so Calhoun hitchhiked to Iran to teach English in Tehran. Brennan stayed in India, and when Calhoun's teaching stint was over they hitchhiked to meet each other in the middle, in Afghanistan. The world was a very different place back then.

After a while their relationship frayed, and they returned from India separately. By the summer of 1977 Brennan had moved back to Los Altos, where she lived for a while in a tent on the grounds of Kobun Chino's Zen center. By this time Jobs had moved out of his parents' house and was renting a $600 per month suburban ranch house in Cupertino with Daniel Kottke. It was an odd scene of free-spirited hippie types living in a tract house they dubbed Rancho Suburbia. "It was a four-bedroom house, and we occasionally rented one of the bedrooms out to all sorts of crazy people, including a stripper for a while," recalled Jobs. Kottke couldn't quite figure out why Jobs had not just gotten his own house, which he could have afforded by then. "I think he just wanted to have a roommate," Kottke speculated.

Even though her relationship with Jobs was sporadic, Brennan soon moved in as well. This made for a set of living arrangements worthy of a French farce. The house had two big bedrooms and two tiny ones. Jobs, not surprisingly, commandeered the largest of them, and Brennan (who was not really living with him) moved into the other big bedroom. "The two middle rooms were like for babies, and I didn't want either of them, so I moved into the living room and slept on a foam pad," said Kottke. They turned one of the small rooms into space for meditating and dropping acid, like the attic space they had used at Reed. It was filled with foam packing material from Apple boxes. "Neighborhood kids used to come over and we would toss them in it and it was great fun," said Kottke, "but then Chrisann brought home some cats who peed in the foam, and then we had to get rid of it."

Living in the house at times rekindled the physical relationship between Brennan and Jobs, and within a few months she was pregnant. "Steve and I were in and out of a relationship for five years before I got pregnant," she said. "We didn't know how to be together and we didn't know how to be apart." When Greg Calhoun hitchhiked from Colorado to visit them on Thanksgiving 1977, Brennan told him the news: "Steve and I got back together, and now I'm pregnant, but now we are on again and off again, and I don't know what to do."

Calhoun noticed that Jobs was disconnected from the whole situation. He even tried to convince Calhoun to stay with them and come to work at Apple. "Steve was just not dealing with Chris-

145

ann or the pregnancy," he recalled. "He could be very engaged with you in one moment, but then very disengaged. There was a side to him that was frighteningly cold."

When Jobs did not want to deal with a distraction, he sometimes just ignored it, as if he could will it out of existence. At times he was able to distort reality not just for others but even for himself. In the case of Brennan's pregnancy, he simply shut it out of his mind. When confronted, he would deny that he knew he was the father, even though he admitted that he had been sleeping with her. "I wasn't sure it was my kid, because I was pretty sure I wasn't the only one she was sleeping with," he told me later. "She and I were not really even going out when she got pregnant. She just had a room in our house." Brennan had no doubt that Jobs was the father. She had not been involved with Greg or any other men at the time.

Was he lying to himself, or did he not know that he was the father? "I just think he couldn't access that part of his brain or the idea of being responsible," Kottke said. Elizabeth Holmes agreed: "He considered the option of parenthood and considered the option of not being a parent, and he decided to believe the latter. He had other plans for his life."

There was no discussion of marriage. "I knew that she was not the person I wanted to marry, and we would never be happy, and it wouldn't last long," Jobs later said. "I was all in favor of her getting an abortion, but she didn't know what to do. She thought about it repeatedly and decided not to, or I don't know that she ever really decided — I

146

think time just decided for her." Brennan told me that it was her choice to have the baby: "He said he was fine with an abortion but never pushed for it." Interestingly, given his own background, he was adamantly against one option. "He strongly discouraged me putting the child up for adoption," she said.

There was a disturbing irony. Jobs and Brennan were both twenty-three, the same age that Joanne Schieble and Abdulfattah Jandali had been when they had Jobs. He had not yet tracked down his biological parents, but his adoptive parents had told him some of their tale. "I didn't know then about this coincidence of our ages, so it didn't affect my discussions with Chrisann," he later said. He dismissed the notion that he was somehow following his biological father's pattern of getting his girlfriend pregnant when he was twenty-three, but he did admit that the ironic resonance gave him pause. "When I did find out that he was twenty-three when he got Joanne pregnant with me, I thought, whoa!"

The relationship between Jobs and Brennan quickly deteriorated. "Chrisann would get into this kind of victim mode, when she would say that Steve and I were ganging up on her," Kottke recalled. "Steve would just laugh and not take her seriously." Brennan was not, as even she later admitted, very emotionally stable. She began breaking plates, throwing things, trashing the house, and writing obscene words in charcoal on the wall. She said that Jobs kept provoking her with his callousness: "He was an enlightened being who was cruel." Kottke was caught in the middle. "Daniel

147

didn't have that DNA of ruthlessness, so he was a bit flipped by Steve's behavior," according to Brennan. "He would go from 'Steve's not treating you right' to laughing at me with Steve."

Robert Friedland came to her rescue. "He heard that I was pregnant, and he said to come on up to the farm to have the baby," she recalled. "So I did." Elizabeth Holmes and other friends were still living there, and they found an Oregon midwife to help with the delivery. On May 17, 1978, Brennan gave birth to a baby girl. Three days later Jobs flew up to be with them and help name the new baby. The practice on the commune was to give children Eastern spiritual names, but Jobs insisted that she had been born in America and ought to have a name that fit. Brennan agreed. They named her Lisa Nicole Brennan, not giving her the last name Jobs. And then he left to go back to work at Apple. "He didn't want to have anything to do with her or with me," said Brennan.

She and Lisa moved to a tiny, dilapidated house in back of a home in Menlo Park. They lived on welfare because Brennan did not feel up to suing for child support. Finally, the County of San Mateo sued Jobs to try to prove paternity and get him to take financial responsibility. At first Jobs was determined to fight the case. His lawyers wanted Kottke to testify that he had never seen them in bed together, and they tried to line up evidence that Brennan had been sleeping with other men. "At one point I yelled at Steve on the phone, 'You know that is not true,' " Brennan recalled. "He was going to drag me through court with a little baby and try to prove I was a whore and that anyone could have been the father of that baby."

A year after Lisa was born, Jobs agreed to take a paternity test. Brennan's family was surprised, but Jobs knew that Apple would soon be going public and he decided it was best to get the issue resolved. DNA tests were new, and the one that Jobs took was done at UCLA. "I had read about DNA testing, and I was happy to do it to get things settled," he said. The results were pretty dispositive. "Probability of paternity . . . is 94.41%," the report read. The California courts ordered Jobs to start paying $385 a month in child support, sign an agreement admitting paternity, and reimburse the county $5,856 in back welfare payments. He was given visitation rights but for a long time didn't exercise them.

Even then Jobs continued at times to warp the reality around him. "He finally told us on the board," Arthur Rock recalled, "but he kept insisting that there was a large probability that he wasn't the father. He was delusional." He told a reporter for *Time*, Michael Moritz, that when you analyzed the statistics, it was clear that "28% of the male population in the United States could be the father." It was not only a false claim but an odd one. Worse yet, when Chrisann Brennan later heard what he said, she mistakenly thought that Jobs was hyperbolically claiming that she might have slept with 28% of the men in the United States. "He was trying to paint me as a slut or a whore," she recalled. "He spun the whore image onto me in order to not take responsibility."

Years later Jobs was remorseful for the way he

behaved, one of the few times in his life he admitted as much:

I wish I had handled it differently. I could not see myself as a father then, so I didn't face up to it. But when the test results showed she was my daughter, it's not true that I doubted it. I agreed to support her until she was eighteen and give some money to Chrisann as well. I found a house in Palo Alto and fixed it up and let them live there rent-free. Her mother found her great schools which I paid for. I tried to do the right thing. But if I could do it over, I would do a better job.

Once the case was resolved, Jobs began to move on with his life — maturing in some respects, though not all. He put aside drugs, eased away from being a strict vegan, and cut back the time he spent on Zen retreats. He began getting stylish haircuts and buying suits and shirts from the upscale San Francisco haberdashery Wilkes Bashford. And he settled into a serious relationship with one of Regis McKenna's employees, a beautiful Polynesian-Polish woman named Barbara Jasinski.

There was still, to be sure, a childlike rebellious streak in him. He, Jasinski, and Kottke liked to go skinny-dipping in Felt Lake on the edge of Interstate 280 near Stanford, and he bought a 1966 BMW R60/2 motorcycle that he adorned with orange tassels on the handlebars. He could also still be bratty. He belittled waitresses and frequently returned food with the proclamation that it was "garbage." At the company's first Halloween party, in 1979, he dressed in robes as Jesus Christ, an act

150

of semi-ironic self-awareness that he considered funny but that caused a lot of eye rolling. Even his initial stirrings of domesticity had some quirks. He bought a proper house in the Los Gatos hills, which he adorned with a Maxfield Parrish painting, a Braun coffeemaker, and Henckels knives. But because he was so obsessive when it came to selecting furnishings, it remained mostly barren, lacking beds or chairs or couches. Instead his bedroom had a mattress in the center, framed pictures of Einstein and Maharaj-ji on the walls, and an Apple II on the floor.

CHAPTER EIGHT: XEROX
AND LISA
GRAPHICAL USER INTERFACES

A NEW BABY

The Apple II took the company from Jobs's garage to the pinnacle of a new industry. Its sales rose dramatically, from 2,500 units in 1977 to 210,000 in 1981. But Jobs was restless. The Apple II could not remain successful forever, and he knew that, no matter how much he had done to package it, from power cord to case, it would always be seen as Wozniak's masterpiece. He needed his own machine. More than that, he wanted a product that would, in his words, make a dent in the universe.

At first he hoped that the Apple III would play that role. It would have more memory, the screen would display eighty characters across rather than forty, and it would handle uppercase and lowercase letters. Indulging his passion for industrial design, Jobs decreed the size and shape of the external case, and he refused to let anyone alter it, even as committees of engineers added more components to the circuit boards. The result was piggybacked boards with poor connectors that frequently failed. When the Apple III began shipping in May 1980, it flopped. Randy Wigginton, one of the engineers, summed it up: "The Apple III was kind of like a baby conceived during a group orgy, and later everybody had this bad headache, and there's this bastard child, and everyone says, 'It's not mine.' "

By then Jobs had distanced himself from the Apple III and was thrashing about for ways to produce something more radically different. At first he flirted with the idea of touchscreens, but he found himself frustrated. At one demonstration of the technology, he arrived late, fidgeted awhile, then abruptly cut off the engineers in the middle of their presentation with a brusque "Thank you." They were confused. "Would you like us to leave?" one asked. Jobs said yes, then berated his colleagues for wasting his time.

Then he and Apple hired two engineers from Hewlett-Packard to conceive a totally new computer. The name Jobs chose for it would have caused even the most jaded psychiatrist to do a double take: the Lisa. Other computers had been named after daughters of their designers, but Lisa was a daughter Jobs had abandoned and had not yet fully admitted was his. "Maybe he was doing it out of guilt," said Andrea Cunningham, who worked at Regis McKenna on public relations for the project. "We had to come up with an acronym so that we could claim it was not named after Lisa the child." The one they reverse-engineered was "local integrated systems architecture," and despite being meaningless it became the official explanation for the name. Among the engineers it was referred to as "Lisa: invented stupid acronym." Years later, when I asked about the name, Jobs admitted simply, "Obviously it was named for my daughter."

The Lisa was conceived as a $2,000 machine based on a sixteen-bit microprocessor, rather than the eight-bit one used in the Apple II. Without the wizardry of Wozniak, who was still working quietly on the Apple II, the engineers began producing a straightforward computer with a conventional text display, unable to push the powerful microprocessor to do much exciting stuff. Jobs began to grow impatient with how boring it was turning out to be.

There was, however, one programmer who was infusing the project with some life: Bill Atkinson. He was a doctoral student in neuroscience who had experimented with his fair share of acid. When he was asked to come work for Apple, he declined. But then Apple sent him a nonrefundable plane ticket, and he decided to use it and let Jobs try to persuade him. "We are inventing the future," Jobs told him at the end of a three-hour pitch. "Think about surfing on the front edge of a wave. It's really exhilarating. Now think about dog-paddling at the tail end of that wave. It wouldn't be anywhere near as much fun. Come down here and make a dent in the universe." Atkinson did.

With his shaggy hair and droopy moustache that did not hide the animation in his face, Atkinson had some of Woz's ingenuity along with Jobs's passion for awesome products. His first job was to develop a program to track a stock portfolio by auto-dialing the Dow Jones service, getting quotes, then hanging up. "I had to create it fast because there was a magazine ad for the Apple II showing a hubby at the kitchen table looking at an Apple screen filled with graphs of stock prices, and his wife is beaming at him — but there wasn't such a program, so I had to create one." Next he created for the Apple II a version of Pascal, a high-level programming language. Jobs had resisted, thinking that BASIC was all the Apple II needed, but he told Atkinson, "Since you're so passionate about it, I'll give you six days to prove me wrong." He did, and Jobs respected him ever after.

By the fall of 1979 Apple was breeding three ponies to be potential successors to the Apple II workhorse. There was the ill-fated Apple III. There was the Lisa project, which was beginning to disappoint Jobs. And somewhere off Jobs's radar screen, at least for the moment, there was a small skunkworks project for a low-cost machine that was being developed by a colorful employee named Jef Raskin, a former professor who had taught Bill Atkinson. Raskin's goal was to make an inexpensive "computer for the masses" that would be like an appliance — a self-contained unit with computer, keyboard, monitor, and software all together — and have a graphical interface. He tried to turn his colleagues at Apple on to a cutting-edge research center, right in Palo Alto, that was pioneering such ideas.

Xerox PARC

The Xerox Corporation's Palo Alto Research Center, known as Xerox PARC, had been established in 1970 to create a spawning ground for digital ideas. It was safely located, for better and for worse, three thousand miles from the commercial pressures of Xerox corporate headquarters in Connecticut. Among its visionaries was the scientist Alan Kay,

155

who had two great maxims that Jobs embraced: "The best way to predict the future is to invent it" and "People who are serious about software should make their own hardware." Kay pushed the vision of a small personal computer, dubbed the "Dynabook," that would be easy enough for children to use. So Xerox PARC's engineers began to develop user-friendly graphics that could replace all of the command lines and DOS prompts that made computer screens intimidating. The metaphor they came up with was that of a desktop. The screen could have many documents and folders on it, and you could use a mouse to point and click on the one you wanted to use.

This graphical user interface — or GUI, pronounced "gooey" — was facilitated by another concept pioneered at Xerox PARC: bitmapping. Until then, most computers were character-based. You would type a character on a keyboard, and the computer would generate that character on the screen, usually in glowing greenish phosphor against a dark background. Since there were a limited number of letters, numerals, and symbols, it didn't take a whole lot of computer code or processing power to accomplish this. In a bitmap system, on the other hand, each and every pixel on the screen is controlled by bits in the computer's memory. To render something on the screen, such as a letter, the computer has to tell each pixel to be light or dark or, in the case of color displays, what color to be. This uses a lot of computing power, but it permits gorgeous graphics, fonts, and gee-whiz screen displays.

Bitmapping and graphical interfaces became

156

features of Xerox PARC's prototype computers, such as the Alto, and its object-oriented programming language, Smalltalk. Jef Raskin decided that these features were the future of computing. So he began urging Jobs and other Apple colleagues to go check out Xerox PARC.

Raskin had one problem: Jobs regarded him as an insufferable theorist or, to use Jobs's own more precise terminology, "a shithead who sucks." So Raskin enlisted his friend Atkinson, who fell on the other side of Jobs's shithead/genius division of the world, to convince Jobs to take an interest in what was happening at Xerox PARC. What Raskin didn't know was that Jobs was working on a more complex deal. Xerox's venture capital division wanted to be part of the second round of Apple financing during the summer of 1979. Jobs made an offer: "I will let you invest a million dollars in Apple if you will open the kimono at PARC." Xerox accepted. It agreed to show Apple its new technology and in return got to buy 100,000 shares at about $10 each.

By the time Apple went public a year later, Xerox's $1 million worth of shares were worth $17.6 million. But Apple got the better end of the bargain. Jobs and his colleagues went to see Xerox PARC's technology in December 1979 and, when Jobs realized he hadn't been shown enough, got an even fuller demonstration a few days later. Larry Tesler was one of the Xerox scientists called upon to do the briefings, and he was thrilled to show off the work that his bosses back east had never seemed to appreciate. But the other briefer, Adele Goldberg, was appalled that her company seemed willing to give away its crown jewels. "It was incredibly stupid, completely nuts, and I fought to prevent giving Jobs much of anything," she recalled.

Goldberg got her way at the first briefing. Jobs, Raskin, and the Lisa team leader John Couch were ushered into the main lobby, where a Xerox Alto had been set up. "It was a very controlled show of a few applications, primarily a word-processing one," Goldberg said. Jobs wasn't satisfied, and he called Xerox headquarters demanding more.

So he was invited back a few days later, and this time he brought a larger team that included Bill Atkinson and Bruce Horn, an Apple programmer who had worked at Xerox PARC. They both knew what to look for. "When I arrived at work, there was a lot of commotion, and I was told that Jobs and a bunch of his programmers were in the conference room," said Goldberg. One of her engineers was trying to keep them entertained with more displays of the word-processing program. But Jobs was growing impatient. "Let's stop this bullshit!" he kept shouting. So the Xerox folks huddled privately and decided to open the kimono a bit more, but only slowly. They agreed that Tesler could show off Smalltalk, the programming language, but he would demonstrate only what was known as the "unclassified" version. "It will dazzle [Jobs] and he'll never know he didn't get the confidential disclosure," the head of the team told Goldberg.

They were wrong. Atkinson and others had read some of the papers published by Xerox PARC, so they knew they were not getting a full description.

158

Jobs phoned the head of the Xerox venture capital division to complain; a call immediately came back from corporate headquarters in Connecticut decreeing that Jobs and his group should be shown everything. Goldberg stormed out in a rage.

When Tesler finally showed them what was truly under the hood, the Apple folks were astonished. Atkinson stared at the screen, examining each pixel so closely that Tesler could feel the breath on his neck. Jobs bounced around and waved his arms excitedly. "He was hopping around so much I don't know how he actually saw most of the demo, but he did, because he kept asking questions," Tesler recalled. "He was the exclamation point for every step I showed." Jobs kept saying that he couldn't believe that Xerox had not commercialized the technology. "You're sitting on a gold mine," he shouted. "I can't believe Xerox is not taking advantage of this."

The Smalltalk demonstration showed three amazing features. One was how computers could be networked; the second was how object-oriented programming worked. But Jobs and his team paid little attention to these attributes because they were so amazed by the third feature, the graphical interface that was made possible by a bitmapped screen. "It was like a veil being lifted from my eyes," Jobs recalled. "I could see what the future of computing was destined to be."

When the Xerox PARC meeting ended after more than two hours, Jobs drove Bill Atkinson back to the Apple office in Cupertino. He was speeding, and so were his mind and mouth. "This is it!" he shouted, emphasizing each word. "We've got to do it!" It was the breakthrough he had been looking for: bringing computers to the people, with the cheerful but affordable design of an Eichler home and the ease of use of a sleek kitchen appliance.

"How long would this take to implement?" he asked.

"I'm not sure," Atkinson replied. "Maybe six months." It was a wildly optimistic assessment, but also a motivating one.

"GREAT ARTISTS STEAL"

The Apple raid on Xerox PARC is sometimes described as one of the biggest heists in the chronicles of industry. Jobs occasionally endorsed this view, with pride. As he once said, "Picasso had a saying — 'good artists copy, great artists steal' — and we have always been shameless about stealing great ideas."

Another assessment, also sometimes endorsed by Jobs, is that what transpired was less a heist by Apple than a fumble by Xerox. "They were copierheads who had no clue about what a computer could do," he said of Xerox's management. "They just grabbed defeat from the greatest victory in the computer industry. Xerox could have owned the entire computer industry."

Both assessments contain a lot of truth, but there is more to it than that. There falls a shadow, as T. S. Eliot noted, between the conception and the creation. In the annals of innovation, new ideas are only part of the equation. Execution is just as important.

Jobs and his engineers significantly improved the graphical interface ideas they saw at Xerox PARC, and then were able to implement them in ways that Xerox never could accomplish. For example, the Xerox mouse had three buttons, was complicated, cost $300 apiece, and didn't roll around smoothly; a few days after his second Xerox PARC visit, Jobs went to a local industrial design firm, IDEO, and told one of its founders, Dean Hovey, that he wanted a simple single-button model that cost $15, "and I want to be able to use it on Formica and my blue jeans." Hovey complied.

The improvements were in not just the details but the entire concept. The mouse at Xerox PARC could not be used to drag a window around the screen. Apple's engineers devised an interface so you could not only drag windows and files around, you could even drop them into folders. The Xerox system required you to select a command in order to do anything, ranging from resizing a window to changing the extension that located a file. The Apple system transformed the desktop metaphor into virtual reality by allowing you to directly touch, manipulate, drag, and relocate things. And Apple's engineers worked in tandem with its designers — with Jobs spurring them on daily — to improve the desktop concept by adding delightful icons and menus that pulled down from a bar atop each window and the capability to open files and folders with a double click.

It's not as if Xerox executives ignored what their scientists had created at PARC. In fact they did try to capitalize on it, and in the process they showed why good execution is as important as good ideas. In 1981, well before the Apple Lisa or Macintosh, they introduced the Xerox Star, a machine that featured their graphical user interface, mouse, bit-mapped display, windows, and desktop metaphor. But it was clunky (it could take minutes to save a large file), costly ($16,595 at retail stores), and aimed mainly at the networked office market. It flopped; only thirty thousand were ever sold.

Jobs and his team went to a Xerox dealer to look at the Star as soon as it was released. But he deemed it so worthless that he told his colleagues they couldn't spend the money to buy one. "We were very relieved," he recalled. "We knew they hadn't done it right, and that we could — at a fraction of the price." A few weeks later he called Bob Belleville, one of the hardware designers on the Xerox Star team. "Everything you've ever done in your life is shit," Jobs said, "so why don't you come work for me?" Belleville did, and so did Larry Tesler.

In his excitement, Jobs began to take over the daily management of the Lisa project, which was being run by John Couch, the former HP engineer. Ignoring Couch, he dealt directly with Atkinson and Tesler to insert his own ideas, especially on Lisa's graphical interface design. "He would call me at all hours, 2 a.m. or 5 a.m.," said Tesler. "I loved it. But it upset my bosses at the Lisa division." Jobs was told to stop making out-of-channel calls. He held himself back for a while, but not for long.

One important showdown occurred when Atkinson decided that the screen should have a white background rather than a dark one. This would allow an attribute that both Atkinson and Jobs wanted: WYSIWYG, pronounced "wiz-ee-wig,"

162

an acronym for "What you see is what you get." What you saw on the screen was what you'd get when you printed it out. "The hardware team screamed bloody murder," Atkinson recalled. "They said it would force us to use a phosphor that was a lot less persistent and would flicker more." So Atkinson enlisted Jobs, who came down on his side. The hardware folks grumbled, but then went off and figured it out. "Steve wasn't much of an engineer himself, but he was very good at assessing people's answers. He could tell whether the engineers were defensive or unsure of themselves."

One of Atkinson's amazing feats (which we are so accustomed to nowadays that we rarely marvel at it) was to allow the windows on a screen to overlap so that the "top" one clipped into the ones "below" it. Atkinson made it possible to move these windows around, just like shuffling papers on a desk, with those below becoming visible or hidden as you moved the top ones. Of course, on a computer screen there are no layers of pixels underneath the pixels that you see, so there are no windows actually lurking underneath the ones that appear to be on top. To create the illusion of overlapping windows requires complex coding that involves what are called "regions." Atkinson pushed himself to make this trick work because he thought he had seen this capability during his visit to Xerox PARC. In fact the folks at PARC had never accomplished it, and they later told him they were amazed that he had done so. "I got a feeling for the empowering aspect of naïveté," Atkinson said. "Because I didn't know it couldn't be done, I was enabled to do it." He was working so hard that one morning, in a daze, he drove his Corvette into a parked truck and nearly killed himself. Jobs immediately drove to the hospital to see him. "We were pretty worried about you," he said when Atkinson regained consciousness. Atkinson gave him a pained smile and replied, "Don't worry, I still remember regions."

Jobs also had a passion for smooth scrolling. Documents should not lurch line by line as you scroll through them, but instead should flow. "He was adamant that everything on the interface had a good feeling to the user," Atkinson said. They also wanted a mouse that could easily move the cursor in any direction, not just up-down/left-right. This required using a ball rather than the usual two wheels. One of the engineers told Atkinson that there was no way to build such a mouse commercially. After Atkinson complained to Jobs over dinner, he arrived at the office the next day to discover that Jobs had fired the engineer. When his replacement met Atkinson, his first words were, "I can build the mouse."

Atkinson and Jobs became best friends for a while, eating together at the Good Earth most nights. But John Couch and the other professional engineers on his Lisa team, many of them buttoned-down HP types, resented Jobs's meddling and were infuriated by his frequent insults. There was also a clash of visions. Jobs wanted to build a VolksLisa, a simple and inexpensive product for the masses. "There was a tug-of-war between people like me, who wanted a lean machine, and those from HP, like Couch, who were aiming for the corporate market," Jobs recalled.

Both Mike Scott and Mike Markkula were intent on bringing some order to Apple and became increasingly concerned about Jobs's disruptive behavior. So in September 1980, they secretly plotted a reorganization. Couch was made the undisputed manager of the Lisa division. Jobs lost control of the computer he had named after his daughter. He was also stripped of his role as vice president for research and development. He was made non-executive chairman of the board. This position allowed him to remain Apple's public face, but it meant that he had no operating control. That hurt. "I was upset and felt abandoned by Markkula," he said. "He and Scotty felt I wasn't up to running the Lisa division. I brooded about it a lot."

CHAPTER NINE: GOING PUBLIC
A MAN OF WEALTH AND FAME

OPTIONS

When Mike Markkula joined Jobs and Wozniak to turn their fledgling partnership into the Apple Computer Co. in January 1977, they valued it at $5,309. Less than four years later they decided it was time to take it public. It would become the most oversubscribed initial public offering since that of Ford Motors in 1956. By the end of December 1980, Apple would be valued at $1.79 billion. Yes, *billion*. In the process it would make three hundred people millionaires.

Daniel Kottke was not one of them. He had been Jobs's soul mate in college, in India, at the All One Farm, and in the rental house they shared during the Chrisann Brennan crisis. He joined Apple when it was headquartered in Jobs's garage, and he still worked there as an hourly employee. But he was not at a high enough level to be cut in on the stock options that were awarded before the IPO. "I totally trusted Steve, and I assumed he would take care of me like I'd taken care of him, so I didn't push," said Kottke. The official reason he wasn't given stock options was that he was an hourly technician, not a salaried engineer, which was the cutoff level for options. Even so, he could have justifiably been given "founder's stock," but Jobs decided not to. "Steve is the opposite of loyal," according to Andy Hertzfeld, an early Apple engineer who has nevertheless remained friends with him. "He's anti-loyal. He has to abandon the people he is close to."

Kottke decided to press his case with Jobs by hovering outside his office and catching him to make a plea. But at each encounter, Jobs brushed him off. "What was really so difficult for me is that Steve never told me I wasn't eligible," recalled Kottke. "He owed me that as a friend. When I would ask him about stock, he would tell me I had to talk to my manager." Finally, almost six months after the IPO, Kottke worked up the courage to march into Jobs's office and try to hash out the issue. But when he got in to see him, Jobs was so cold that Kottke froze. "I just got choked up and began to cry and just couldn't talk to him," Kottke recalled. "Our friendship was all gone. It was so sad."

Rod Holt, the engineer who had built the power supply, was getting a lot of options, and he tried to turn Jobs around. "We have to do something for your buddy Daniel," he said, and he suggested they each give him some of their own options. "Whatever you give him, I will match it," said Holt. Replied Jobs, "Okay. I will give him zero."

Wozniak, not surprisingly, had the opposite attitude. Before the shares went public, he decided to sell, at a very low price, two thousand of his options to forty different midlevel employees. Most of his beneficiaries made enough to buy a home.

Wozniak bought a dream home for himself and his new wife, but she soon divorced him and kept the house. He also later gave shares outright to employees he felt had been shortchanged, including Kottke, Fernandez, Wigginton, and Espinosa. Everyone loved Wozniak, all the more so after his generosity, but many also agreed with Jobs that he was "awfully naïve and childlike." A few months later a United Way poster showing a destitute man went up on a company bulletin board. Someone scrawled on it "Woz in 1990."

Jobs was not naïve. He had made sure his deal with Chrisann Brennan was signed before the IPO occurred.

Jobs was the public face of the IPO, and he helped choose the two investment banks handling it: the traditional Wall Street firm Morgan Stanley and the untraditional boutique firm Hambrecht & Quist in San Francisco. "Steve was very irreverent toward the guys from Morgan Stanley, which was a pretty uptight firm in those days," recalled Bill Hambrecht. Morgan Stanley planned to price the offering at $18, even though it was obvious the shares would quickly shoot up. "Tell me what happens to this stock that we priced at eighteen?" Jobs asked the bankers. "Don't you sell it to your good customers? If so, how can you charge me a 7% commission?" Hambrecht recognized that there was a basic unfairness in the system, and he later went on to formulate the idea of a reverse auction to price shares before an IPO.

Apple went public the morning of December 12, 1980. By then the bankers had priced the stock at $22 a share. It went to $29 the first day. Jobs had

168

come into the Hambrecht & Quist office just in time to watch the opening trades. At age twenty-five, he was now worth $256 million.

BABY YOU'RE A RICH MAN

Before and after he was rich, and indeed throughout a life that included being both broke and a billionaire, Steve Jobs's attitude toward wealth was complex. He was an antimaterialistic hippie who capitalized on the inventions of a friend who wanted to give them away for free, and he was a Zen devotee who made a pilgrimage to India and then decided that his calling was to create a business. And yet somehow these attitudes seemed to weave together rather than conflict.

He had a great love for some material objects, especially those that were finely designed and crafted, such as Porsche and Mercedes cars, Henckels knives and Braun appliances, BMW motorcycles and Ansel Adams prints, Bösendorfer pianos and Bang & Olufsen audio equipment. Yet the houses he lived in, no matter how rich he became, tended not to be ostentatious and were furnished so simply they would have put a Shaker to shame. Neither then nor later would he travel with an entourage, keep a personal staff, or even have security protection. He bought a nice car, but always drove himself. When Markkula asked Jobs to join him in buying a Lear jet, he declined (though he eventually would demand of Apple a Gulfstream to use). Like his father, he could be flinty when bargaining with suppliers, but he didn't allow a craving for profits to take precedence over his passion for building great products.

169

Thirty years after Apple went public, he reflected on what it was like to come into money suddenly:

I never worried about money. I grew up in a middle-class family, so I never thought I would starve. And I learned at Atari that I could be an okay engineer, so I always knew I could get by. I was voluntarily poor when I was in college and India, and I lived a pretty simple life even when I was working. So I went from fairly poor, which was wonderful, because I didn't have to worry about money, to being incredibly rich, when I also didn't have to worry about money.

I watched people at Apple who made a lot of money and felt they had to live differently. Some of them bought a Rolls-Royce and various houses, each with a house manager and then someone to manage the house managers. Their wives got plastic surgery and turned into these bizarre people. This was not how I wanted to live. It's crazy. I made a promise to myself that I'm not going to let this money ruin my life.

He was not particularly philanthropic. He briefly set up a foundation, but he discovered that it was annoying to have to deal with the person he had hired to run it, who kept talking about "venture" philanthropy and how to "leverage" giving. Jobs became contemptuous of people who made a display of philanthropy or thinking they could reinvent it. Earlier he had quietly sent in a $5,000 check to help launch Larry Brilliant's Seva Foundation to fight diseases of poverty, and he even agreed to join the board. But when Brilliant brought some board members, including Wavy Gravy and Jerry Garcia, to Apple right after its IPO to solicit a donation, Jobs was not forthcoming. He instead worked on finding ways that a donated Apple II and a VisiCalc program could make it easier for the foundation to do a survey it was planning on blindness in Nepal.

His biggest personal gift was to his parents, Paul and Clara Jobs, to whom he gave about $750,000 worth of stock. They sold some to pay off the mortgage on their Los Altos home, and their son came over for the little celebration. "It was the first time in their lives they didn't have a mortgage," Jobs recalled. "They had a handful of their friends over for the party, and it was really nice." Still, they didn't consider buying a nicer house. "They weren't interested in that," Jobs said. "They had a life they were happy with." Their only splurge was to take a Princess cruise each year. The one through the Panama Canal "was the big one for my dad," according to Jobs, because it reminded him of when his Coast Guard ship went through on its way to San Francisco to be decommissioned.

With Apple's success came fame for its poster boy. *Inc.* became the first magazine to put him on its cover, in October 1981. "This man has changed business forever," it proclaimed. It showed Jobs with a neatly trimmed beard and well-styled long hair, wearing blue jeans and a dress shirt with a blazer that was a little too satiny. He was leaning on an Apple II and looking directly into the camera with the mesmerizing stare he had picked up from Robert Friedland. "When Steve Jobs speaks, it is with the gee-whiz enthusiasm of someone who

171

sees the future and is making sure it works," the magazine reported.

Time followed in February 1982 with a package on young entrepreneurs. The cover was a painting of Jobs, again with his hypnotic stare. Jobs, said the main story, "practically singlehanded created the personal computer industry." The accompanying profile, written by Michael Moritz, noted, "At 26, Jobs heads a company that six years ago was located in a bedroom and garage of his parents' house, but this year it is expected to have sales of $600 million. . . . As an executive, Jobs has sometimes been petulant and harsh on subordinates. Admits he: 'I've got to learn to keep my feelings private.' "

Despite his new fame and fortune, he still fancied himself a child of the counterculture. On a visit to a Stanford class, he took off his Wilkes Bashford blazer and his shoes, perched on top of a table, and crossed his legs into a lotus position. The students asked questions, such as when Apple's stock price would rise, which Jobs brushed off. Instead he spoke of his passion for future products, such as someday making a computer as small as a book. When the business questions tapered off, Jobs turned the tables on the well-groomed students. "How many of you are virgins?" he asked. There were nervous giggles. "How many of you have taken LSD?" More nervous laughter, and only one or two hands went up. Later Jobs would complain about the new generation of kids, who seemed to him more materialistic and careerist than his own. "When I went to school, it was right after the sixties and before this general wave of practical pur-

172

posefulness had set in," he said. "Now students aren't even thinking in idealistic terms, or at least nowhere near as much." His generation, he said, was different. "The idealistic wind of the sixties is still at our backs, though, and most of the people I know who are my age have that ingrained in them forever."

CHAPTER TEN: THE MAC IS BORN

YOU SAY YOU WANT A REVOLUTION

JEF RASKIN'S BABY

Jef Raskin was the type of character who could enthrall Steve Jobs — or annoy him. As it turned out, he did both. A philosophical guy who could be both playful and ponderous, Raskin had studied computer science, taught music and visual arts, conducted a chamber opera company, and organized guerrilla theater. His 1967 doctoral thesis at U.C. San Diego argued that computers should have graphical rather than text-based interfaces. When he got fed up with teaching, he rented a hot air balloon, flew over the chancellor's house, and shouted down his decision to quit.

When Jobs was looking for someone to write a manual for the Apple II in 1976, he called Raskin, who had his own little consulting firm. Raskin went to the garage, saw Wozniak beavering away at a workbench, and was convinced by Jobs to write the manual for $50. Eventually he became the manager of Apple's publications department. One of Raskin's dreams was to build an inexpensive computer for the masses, and in 1979 he convinced Mike Markkula to put him in charge of a small development project code-named "Annie" to do just that. Since Raskin thought it was sexist to name computers after women, he redubbed the project in honor of his favorite type of apple, the McIntosh. But he changed the spelling in order not to conflict with the name of the audio equipment maker McIntosh Laboratory. The proposed computer became known as the Macintosh.

Raskin envisioned a machine that would sell for $1,000 and be a simple appliance, with screen and keyboard and computer all in one unit. To keep the cost down, he proposed a tiny five-inch screen and a very cheap (and underpowered) microprocessor, the Motorola 6809. Raskin fancied himself a philosopher, and he wrote his thoughts in an ever-expanding notebook that he called "The Book of Macintosh." He also issued occasional manifestos. One of these was called "Computers by the Millions," and it began with an aspiration: "If personal computers are to be truly personal, it will have to be as likely as not that a family, picked at random, will own one."

Throughout 1979 and early 1980 the Macintosh project led a tenuous existence. Every few months it would almost get killed off, but each time Raskin managed to cajole Markkula into granting clemency. It had a research team of only four engineers located in the original Apple office space next to the Good Earth restaurant, a few blocks from the company's new main building. The work space was filled with enough toys and radio-controlled model airplanes (Raskin's passion) to make it look like a day care center for geeks. Every now and then work would cease for a loosely organized game of Nerf ball tag. Andy Hertzfeld recalled,

"This inspired everyone to surround their work area with barricades made out of cardboard, to provide cover during the game, making part of the office look like a cardboard maze."

The star of the team was a blond, cherubic, and psychologically intense self-taught young engineer named Burrell Smith, who worshipped the code work of Wozniak and tried to pull off similar dazzling feats. Atkinson discovered Smith working in Apple's service department and, amazed at his ability to improvise fixes, recommended him to Raskin. Smith would later succumb to schizophrenia, but in the early 1980s he was able to channel his manic intensity into weeklong binges of engineering brilliance.

Jobs was enthralled by Raskin's vision, but not by his willingness to make compromises to keep down the cost. At one point in the fall of 1979 Jobs told him instead to focus on building what he repeatedly called an "insanely great" product. "Don't worry about price, just specify the computer's abilities," Jobs told him. Raskin responded with a sarcastic memo. It spelled out everything you would want in the proposed computer: a high-resolution color display, a printer that worked without a ribbon and could produce graphics in color at a page per second, unlimited access to the ARPA net, and the capability to recognize speech and synthesize music, "even simulate Caruso singing with the Mormon tabernacle choir, with variable reverberation." The memo concluded, "Starting with the abilities desired is nonsense. We must start both with a price goal, and a set of abilities, and keep an eye on today's and the immediate future's technology." In other words, Raskin had little patience for Jobs's belief that you could distort reality if you had enough passion for your product.

Thus they were destined to clash, especially after Jobs was ejected from the Lisa project in September 1980 and began casting around for someplace else to make his mark. It was inevitable that his gaze would fall on the Macintosh project. Raskin's manifestos about an inexpensive machine for the masses, with a simple graphic interface and clean design, stirred his soul. And it was also inevitable that once Jobs set his sights on the Macintosh project, Raskin's days were numbered. "Steve started acting on what he thought we should do, Jef started brooding, and it instantly was clear what the outcome would be," recalled Joanna Hoffman, a member of the Mac team.

The first conflict was over Raskin's devotion to the underpowered Motorola 6809 microprocessor. Once again it was a clash between Raskin's desire to keep the Mac's price under $1,000 and Jobs's determination to build an insanely great machine. So Jobs began pushing for the Mac to switch to the more powerful Motorola 68000, which is what the Lisa was using. Just before Christmas 1980, he challenged Burrell Smith, without telling Raskin, to make a redesigned prototype that used the more powerful chip. As his hero Wozniak would have done, Smith threw himself into the task around the clock, working nonstop for three weeks and employing all sorts of breathtaking programming leaps. When he succeeded, Jobs was able to force the switch to the Motorola 68000, and Raskin had to brood and recalculate the cost of the Mac.

There was something larger at stake. The cheaper microprocessor that Raskin wanted would not have been able to accommodate all of the gee-whiz graphics — windows, menus, mouse, and so on — that the team had seen on the Xerox PARC visits. Raskin had convinced everyone to go to Xerox PARC, and he liked the idea of a bitmapped display and windows, but he was not as charmed by all the cute graphics and icons, and he absolutely detested the idea of using a point-and-click mouse rather than the keyboard. "Some of the people on the project became enamored of the quest to do everything with the mouse," he later groused. "Another example is the absurd application of icons. An icon is a symbol equally incomprehensible in all human languages. There's a reason why humans invented phonetic languages."

Raskin's former student Bill Atkinson sided with Jobs. They both wanted a powerful processor that could support whizzier graphics and the use of a mouse. "Steve had to take the project away from Jef," Atkinson said. "Jef was pretty firm and stubborn, and Steve was right to take it over. The world got a better result."

The disagreements were more than just philosophical; they became clashes of personality. "I think that he likes people to jump when he says jump," Raskin once said. "I felt that he was untrustworthy, and that he does not take kindly to being found wanting. He doesn't seem to like people who see him without a halo." Jobs was equally dismissive of Raskin. "Jef was really pompous," he said. "He didn't know much about interfaces. So I decided to nab some of his people who were really good, like Atkinson, bring in some of my own, take the thing over and build a less expensive Lisa, not some piece of junk."

Some on the team found Jobs impossible to work with. "Jobs seems to introduce tension, politics, and hassles rather than enjoying a buffer from those distractions," one engineer wrote in a memo to Raskin in December 1980. "I thoroughly enjoy talking with him, and I admire his ideas, practical perspective, and energy. But I just don't feel that he provides the trusting, supportive, relaxed environment that I need."

But many others realized that despite his temperamental failings, Jobs had the charisma and corporate clout that would lead them to "make a dent in the universe." Jobs told the staff that Raskin was just a dreamer, whereas he was a doer and would get the Mac done in a year. It was clear he wanted vindication for having been ousted from the Lisa group, and he was energized by competition. He publicly bet John Couch $5,000 that the Mac would ship before the Lisa. "We can make a computer that's cheaper and better than the Lisa, and get it out first," he told the team.

Jobs asserted his control of the group by canceling a brown-bag lunch seminar that Raskin was scheduled to give to the whole company in February 1981. Raskin happened to go by the room anyway and discovered that there were a hundred people there waiting to hear him; Jobs had not bothered to notify anyone else about his cancellation order. So Raskin went ahead and gave a talk.

That incident led Raskin to write a blistering memo to Mike Scott, who once again found himself in the difficult position of being a president trying to manage a company's temperamental cofounder and major stockholder. It was titled "Working for/with Steve Jobs," and in it Raskin asserted:

> He is a dreadful manager. . . . I have always liked Steve, but I have found it impossible to work for him. . . . Jobs regularly misses appointments. This is so well-known as to be almost a running joke. . . . He acts without thinking and with bad judgment. . . . He does not give credit where due. . . . Very often, when told of a new idea, he will immediately attack it and say that it is worthless or even stupid, and tell you that it was a waste of time to work on it. This alone is bad management, but if the idea is a good one he will soon be telling people about it as though it was his own.

That afternoon Scott called in Jobs and Raskin for a showdown in front of Markkula. Jobs started crying. He and Raskin agreed on only one thing: Neither could work for the other one. On the Lisa project, Scott had sided with Couch. This time he decided it was best to let Jobs win. After all, the Mac was a minor development project housed in a distant building that could keep Jobs occupied away from the main campus. Raskin was told to take a leave of absence. "They wanted to humor me and give me something to do, which was fine," Jobs recalled. "It was like going back to the garage for me. I had my own ragtag team and I was in control."

Raskin's ouster may not have seemed fair, but it

ended up being good for the Macintosh. Raskin wanted an appliance with little memory, an anemic processor, a cassette tape, no mouse, and minimal graphics. Unlike Jobs, he might have been able to keep the price down to close to $1,000, and that may have helped Apple win market share. But he could not have pulled off what Jobs did, which was to create and market a machine that would transform personal computing. In fact we can see where the road not taken led. Raskin was hired by Canon to build the machine he wanted. "It was the Canon Cat, and it was a total flop," Atkinson said. "Nobody wanted it. When Steve turned the Mac into a compact version of the Lisa, it made it into a computing platform instead of a consumer electronic device."*

TEXACO TOWERS

A few days after Raskin left, Jobs appeared at the cubicle of Andy Hertzfeld, a young engineer on the Apple II team, who had a cherubic face and impish demeanor similar to his pal Burrell Smith's. Hertzfeld recalled that most of his colleagues were afraid of Jobs "because of his spontaneous temper tantrums and his proclivity to tell everyone exactly what he thought, which often wasn't very favorable." But Hertzfeld was excited by him. "Are you any good?" Jobs asked the moment he walked in. "We only want really good people working on the Mac, and I'm not sure you're good enough." Hertzfeld knew how to answer. "I told him that

* Raskin died of pancreatic cancer in 2005, not long after Jobs was diagnosed with the disease.

181

yes, I thought that I was pretty good."

Jobs left, and Hertzfeld went back to his work. Later that afternoon he looked up to see Jobs peering over the wall of his cubicle. "I've got good news for you," he said. "You're working on the Mac team now. Come with me."

Hertzfeld replied that he needed a couple more days to finish the Apple II product he was in the middle of. "What's more important than working on the Macintosh?" Jobs demanded. Hertzfeld explained that he needed to get his Apple II DOS program in good enough shape to hand it over to someone. "You're just wasting your time with that!" Jobs replied. "Who cares about the Apple II? The Apple II will be dead in a few years. The Macintosh is the future of Apple, and you're going to start on it now!" With that, Jobs yanked out the power cord to Hertzfeld's Apple II, causing the code he was working on to vanish. "Come with me," Jobs said. "I'm going to take you to your new desk." Jobs drove Hertzfeld, computer and all, in his silver Mercedes to the Macintosh offices. "Here's your new desk," he said, plopping him in a space next to Burrell Smith. "Welcome to the Mac team!" The desk had been Raskin's. In fact Raskin had left so hastily that some of the drawers were still filled with his flotsam and jetsam, including model airplanes.

Jobs's primary test for recruiting people in the spring of 1981 to be part of his merry band of pirates was making sure they had a passion for the product. He would sometimes bring candidates into a room where a prototype of the Mac was covered by a cloth, dramatically unveil it, and watch. "If their eyes lit up, if they went right for the mouse and started pointing and clicking, Steve would smile and hire them," recalled Andrea Cunningham. "He wanted them to say 'Wow!' "

Bruce Horn was one of the programmers at Xerox PARC. When some of his friends, such as Larry Tesler, decided to join the Macintosh group, Horn considered going there as well. But he got a good offer, and a $15,000 signing bonus, to join another company. Jobs called him on a Friday night. "You have to come into Apple tomorrow morning," he said. "I have a lot of stuff to show you." Horn did, and Jobs hooked him. "Steve was so passionate about building this amazing device that would change the world," Horn recalled. "By sheer force of his personality, he changed my mind." Jobs showed Horn exactly how the plastic would be molded and would fit together at perfect angles, and how good the board was going to look inside. "He wanted me to see that this whole thing was going to happen and it was thought out from end to end. Wow, I said, I don't see that kind of passion every day. So I signed up."

Jobs even tried to reengage Wozniak. "I resented the fact that he had not been doing much, but then I thought, hell, I wouldn't be here without his brilliance," Jobs later told me. But as soon as Jobs was starting to get him interested in the Mac, Wozniak crashed his new single-engine Beechcraft while attempting a takeoff near Santa Cruz. He barely survived and ended up with partial amnesia. Jobs spent time at the hospital, but when Wozniak recovered he decided it was time to take a break from Apple. Ten years after dropping out of Berkeley, he decided to return there to finally get his degree, enrolling under the name of Rocky Raccoon Clark.

In order to make the project his own, Jobs decided it should no longer be code-named after Raskin's favorite apple. In various interviews, Jobs had been referring to computers as a bicycle for the mind; the ability of humans to create a bicycle allowed them to move more efficiently than even a condor, and likewise the ability to create computers would multiply the efficiency of their minds. So one day Jobs decreed that henceforth the Macintosh should be known instead as the Bicycle. This did not go over well. "Burrell and I thought this was the silliest thing we ever heard, and we simply refused to use the new name," recalled Hertzfeld. Within a month the idea was dropped.

By early 1981 the Mac team had grown to about twenty, and Jobs decided that they should have bigger quarters. So he moved everyone to the second floor of a brown-shingled, two-story building about three blocks from Apple's main offices. It was next to a Texaco station and thus became known as Texaco Towers. In order to make the office more lively, he told the team to buy a stereo system. "Burrell and I ran out and bought a silver, cassette-based boom box right away, before he could change his mind," recalled Hertzfeld.

Jobs's triumph was soon complete. A few weeks after winning his power struggle with Raskin to run the Mac division, he helped push out Mike Scott as Apple's president. Scotty had become more and more erratic, alternately bullying and nurturing. He finally lost most of his support among the employees when he surprised them by imposing a round of layoffs that he handled with atypical ruthlessness. In addition, he had begun to suffer a variety of afflictions, ranging from eye infections to narcolepsy. When Scott was on vacation in Hawaii, Markkula called together the top managers to ask if he should be replaced. Most of them, including Jobs and John Couch, said yes. So Markkula took over as an interim and rather passive president, and Jobs found that he now had full rein to do what he wanted with the Mac division.

Chapter Eleven: The Reality Distortion Field

PLAYING BY HIS OWN SET OF RULES

When Andy Hertzfeld joined the Macintosh team, he got a briefing from Bud Tribble, the other software designer, about the huge amount of work that still needed to be done. Jobs wanted it finished by January 1982, less than a year away. "That's crazy," Hertzfeld said. "There's no way." Tribble said that Jobs would not accept any contrary facts. "The best way to describe the situation is a term from *Star Trek,*" Tribble explained. "Steve has a reality distortion field." When Hertzfeld looked puzzled, Tribble elaborated. "In his presence, reality is malleable. He can convince anyone of practically anything. It wears off when he's not around, but it makes it hard to have realistic schedules."

Tribble recalled that he adopted the phrase from the "Menagerie" episodes of *Star Trek,* "in which the aliens create their own new world through sheer mental force." He meant the phrase to be a compliment as well as a caution: "It was dangerous to get caught in Steve's distortion field, but it was what led him to actually be able to change reality."

At first Hertzfeld thought that Tribble was exaggerating, but after two weeks of working with Jobs, he became a keen observer of the phenomenon. "The reality distortion field was a confounding mélange of a charismatic rhetorical style, indomitable will, and eagerness to bend any fact to fit the purpose at hand," he said.

There was little that could shield you from the force, Hertzfeld discovered. "Amazingly, the reality distortion field seemed to be effective even if you were acutely aware of it. We would often discuss potential techniques for grounding it, but after a while most of us gave up, accepting it as a force of nature." After Jobs decreed that the sodas in the office refrigerator be replaced by Odwalla organic orange and carrot juices, someone on the team had T-shirts made. "Reality Distortion Field," they said on the front, and on the back, "It's in the juice!"

To some people, calling it a reality distortion field was just a clever way to say that Jobs tended to lie. But it was in fact a more complex form of dissembling. He would assert something — be it a fact about world history or a recounting of who suggested an idea at a meeting — without even considering the truth. It came from willfully defying reality, not only to others but to himself. "He can deceive himself," said Bill Atkinson. "It allowed him to con people into believing his vision, because he has personally embraced and internalized it."

A lot of people distort reality, of course. When Jobs did so, it was often a tactic for accomplishing something. Wozniak, who was as congenitally honest as Jobs was tactical, marveled at how effective it could be. "His reality distortion is when he has an illogical vision of the future, such as telling me that I could design the Breakout game in just a few days. You realize that it can't be true, but he somehow makes it true."

When members of the Mac team got ensnared in his reality distortion field, they were almost hypnotized. "He reminded me of Rasputin," said Debi Coleman. "He laser-beamed in on you and didn't blink. It didn't matter if he was serving purple Kool-Aid. You drank it." But like Wozniak, she believed that the reality distortion field was empowering: It enabled Jobs to inspire his team to change the course of computer history with a fraction of the resources of Xerox or IBM. "It was a self-fulfilling distortion," she claimed. "You did the impossible, because you didn't realize it was impossible."

At the root of the reality distortion was Jobs's belief that the rules didn't apply to him. He had some evidence for this; in his childhood, he had often been able to bend reality to his desires. Rebelliousness and willfulness were ingrained in his character. He had the sense that he was special, a chosen one, an enlightened one. "He thinks there are a few people who are special — people like Einstein and Gandhi and the gurus he met in India — and he's one of them," said Hertzfeld. "He told Chrisann this. Once he even hinted to me that he was enlightened. It's almost like Nietzsche." Jobs never studied Nietzsche, but the philosopher's concept of the will to power and the special nature of the *Über*man came naturally to him. As Nietzsche wrote in *Thus Spoke Zarathustra,* "The spirit now wills his own will, and he who had been lost to the world now conquers the world." If reality did not comport with his will, he would ignore it, as he had done with the birth of his daughter and would do years later, when first diagnosed with cancer. Even in small everyday rebellions, such as not putting a license plate on his car and parking it in handicapped spaces, he acted as if he were not subject to the strictures around him.

Another key aspect of Jobs's worldview was his binary way of categorizing things. People were either "enlightened" or "an asshole." Their work was either "the best" or "totally shitty." Bill Atkinson, the Mac designer who fell on the good side of these dichotomies, described what it was like:

It was difficult working under Steve, because there was a great polarity between gods and shitheads. If you were a god, you were up on a pedestal and could do no wrong. Those of us who were considered to be gods, as I was, knew that we were actually mortal and made bad engineering decisions and farted like any person, so we were always afraid that we would get knocked off our pedestal. The ones who were shitheads, who were brilliant engineers working very hard, felt there was no way they could get appreciated and rise above their status.

But these categories were not immutable, for Jobs could rapidly reverse himself. When briefing Hertzfeld about the reality distortion field, Tribble specifically warned him about Jobs's tendency to resemble high-voltage alternating current. "Just because he tells you that something is awful or great, it doesn't necessarily mean he'll feel that way tomorrow," Tribble explained. "If you tell him a new idea, he'll usually tell you that he thinks it's stupid. But then, if he actually likes it, exactly one week later, he'll come back to you and propose your idea to you, as if he thought of it."

The audacity of this pirouette technique would have dazzled Diaghilev. "If one line of argument failed to persuade, he would deftly switch to another," Hertzfeld said. "Sometimes, he would throw you off balance by suddenly adopting your position as his own, without acknowledging that he ever thought differently." That happened repeatedly to Bruce Horn, the programmer who, with Tesler, had been lured from Xerox PARC. "One week I'd tell him about an idea that I had, and he would say it was crazy," recalled Horn. "The next week, he'd come and say, 'Hey I have this great idea' — and it would be my idea! You'd call him on it and say, 'Steve, I told you that a week ago,' and he'd say, 'Yeah, yeah, yeah' and just move right along."

It was as if Jobs's brain circuits were missing a device that would modulate the extreme spikes of impulsive opinions that popped into his mind. So in dealing with him, the Mac team adopted an audio concept called a "low pass filter." In processing his input, they learned to reduce the amplitude of his high-frequency signals. That served to smooth out the data set and provide a less jittery moving average of his evolving attitudes. "After a few cycles of him taking alternating extreme positions," said Hertzfeld, "we would learn to low pass filter his signals and not react to the extremes."

Was Jobs's unfiltered behavior caused by a lack of emotional sensitivity? No. Almost the opposite. He was very emotionally attuned, able to read people and know their psychological strengths and vulnerabilities. He could stun an unsuspecting victim with an emotional towel-snap, perfectly aimed. He intuitively knew when someone was faking it or truly knew something. This made him masterful at cajoling, stroking, persuading, flattering, and intimidating people. "He had the uncanny capacity to know exactly what your weak point is, know what will make you feel small, to make you cringe," Joanna Hoffman said. "It's a common trait in people who are charismatic and know how to manipulate people. Knowing that he can crush you makes you feel weakened and eager for his approval, so then he can elevate you and put you on a pedestal and own you."

Ann Bowers became an expert at dealing with Jobs's perfectionism, petulance, and prickliness. She had been the human resources director at Intel, but had stepped aside after she married its cofounder Bob Noyce. She joined Apple in 1980 and served as a calming mother figure who would step in after one of Jobs's tantrums. She would go to his office, shut the door, and gently lecture him. "I know, I know," he would say. "Well, then, please stop doing it," she would insist. Bowers recalled, "He would be good for a while, and then a week or so later I would get a call again." She realized that he could barely contain himself. "He had these huge expectations, and if people didn't deliver, he couldn't stand it. He couldn't control himself. I could understand why Steve would get upset, and he was usually right, but it had a hurtful effect. It created a fear factor. He was self-aware, but that didn't always modify his behavior."

Jobs became close to Bowers and her husband, and he would drop in at their Los Gatos Hills home unannounced. She would hear his motorcycle in the distance and say, "I guess we have Steve for dinner again." For a while she and Noyce were like a surrogate family. "He was so bright and also so needy. He needed a grown-up, a father figure, which Bob became, and I became like a mother figure."

There were some upsides to Jobs's demanding and wounding behavior. People who were not crushed ended up being stronger. They did better work, out of both fear and an eagerness to please. "His behavior can be emotionally draining, but if you survive, it works," Hoffman said. You could also push back — sometimes — and not only survive but thrive. That didn't always work; Raskin tried it, succeeded for a while, and then was destroyed. But if you were calmly confident, if Jobs sized you up and decided that you knew what you were doing, he would respect you. In both his personal and his professional life over the years, his inner circle tended to include many more strong people than toadies.

The Mac team knew that. Every year, beginning in 1981, it gave out an award to the person who did the best job of standing up to him. The award was partly a joke, but also partly real, and Jobs knew about it and liked it. Joanna Hoffman won the first year. From an Eastern European refugee family, she had a strong temper and will. One day, for example, she discovered that Jobs had changed her marketing projections in a way she found totally

192

reality-distorting. Furious, she marched to his office. "As I'm climbing the stairs, I told his assistant I am going to take a knife and stab it into his heart," she recounted. Al Eisenstat, the corporate counsel, came running out to restrain her. "But Steve heard me out and backed down."

Hoffman won the award again in 1982. "I remember being envious of Joanna, because she would stand up to Steve and I didn't have the nerve yet," said Debi Coleman, who joined the Mac team that year. "Then, in 1983, I got the award. I had learned you had to stand up for what you believe, which Steve respected. I started getting promoted by him after that." Eventually she rose to become head of manufacturing.

One day Jobs barged into the cubicle of one of Atkinson's engineers and uttered his usual "This is shit." As Atkinson recalled, "The guy said, 'No it's not, it's actually the best way,' and he explained to Steve the engineering trade-offs he'd made." Jobs backed down. Atkinson taught his team to put Jobs's words through a translator. "We learned to interpret 'This is shit' to actually be a question that means, 'Tell me why this is the best way to do it.' " But the story had a coda, which Atkinson also found instructive. Eventually the engineer found an even better way to perform the function that Jobs had criticized. "He did it better because Steve had challenged him," said Atkinson, "which shows you can push back on him but should also listen, for he's usually right."

Jobs's prickly behavior was partly driven by his perfectionism and his impatience with those who made compromises in order to get a product out on time and on budget. "He could not make trade-offs well," said Atkinson. "If someone didn't care to make their product perfect, they were a bozo." At the West Coast Computer Faire in April 1981, for example, Adam Osborne released the first truly portable personal computer. It was not great — it had a five-inch screen and not much memory — but it worked well enough. As Osborne famously declared, "Adequacy is sufficient. All else is super-fluous." Jobs found that approach to be morally appalling, and he spent days making fun of Osborne. "This guy just doesn't get it," Jobs repeat-edly railed as he wandered the Apple corridors. "He's not making art, he's making shit."

One day Jobs came into the cubicle of Larry Kenyon, an engineer who was working on the Macintosh operating system, and complained that it was taking too long to boot up. Kenyon started to explain, but Jobs cut him off. "If it could save a person's life, would you find a way to shave ten seconds off the boot time?" he asked. Kenyon allowed that he probably could. Jobs went to a whiteboard and showed that if there were five mil-lion people using the Mac, and it took ten seconds extra to turn it on every day, that added up to three hundred million or so hours per year that people would save, which was the equivalent of at least one hundred lifetimes saved per year. "Larry was suitably impressed, and a few weeks later he came back and it booted up twenty-eight seconds faster," Atkinson recalled. "Steve had a way of motivating by looking at the bigger picture."

The result was that the Macintosh team came to share Jobs's passion for making a great product, not just a profitable one. "Jobs thought of himself as an artist, and he encouraged the design team to think of ourselves that way too," said Hertzfeld. "The goal was never to beat the competition, or to make a lot of money. It was to do the greatest thing possible, or even a little greater." He once took the team to see an exhibit of Tiffany glass at the Metropolitan Museum in Manhattan because he believed they could learn from Louis Tiffany's example of creating great art that could be mass-produced. Recalled Bud Tribble, "We said to ourselves, 'Hey, if we're going to make things in our lives, we might as well make them beautiful.'"

Was all of his stormy and abusive behavior necessary? Probably not, nor was it justified. There were other ways to have motivated his team. Even though the Macintosh would turn out to be great, it was way behind schedule and way over budget because of Jobs's impetuous interventions. There was also a cost in brutalized human feelings, which caused much of the team to burn out. "Steve's contributions could have been made without so many stories about him terrorizing folks," Wozniak said. "I like being more patient and not having so many conflicts. I think a company can be a good family. If the Macintosh project had been run my way, things probably would have been a mess. But I think if it had been a mix of both our styles, it would have been better than just the way Steve did it."

But even though Jobs's style could be demoralizing, it could also be oddly inspiring. It infused Apple employees with an abiding passion to create groundbreaking products and a belief that they

195

could accomplish what seemed impossible. They had T-shirts made that read "90 hours a week and loving it!" Out of a fear of Jobs mixed with an incredibly strong urge to impress him, they exceeded their own expectations. "I've learned over the years that when you have really good people you don't have to baby them," Jobs later explained. "By expecting them to do great things, you can get them to do great things. The original Mac team taught me that A-plus players like to work together, and they don't like it if you tolerate B work. Ask any member of that Mac team. They will tell you it was worth the pain."

Most of them agree. "He would shout at a meeting, 'You asshole, you never do anything right,' " Debi Coleman recalled. "It was like an hourly occurrence. Yet I consider myself the absolute luckiest person in the world to have worked with him."

Chapter Twelve: The Design
REAL ARTISTS SIMPLIFY

A Bauhaus Aesthetic

Unlike most kids who grew up in Eichler homes, Jobs knew what they were and why they were so wonderful. He liked the notion of simple and clean modernism produced for the masses. He also loved listening to his father describe the styling intricacies of various cars. So from the beginning at Apple, he believed that great industrial design — a colorfully simple logo, a sleek case for the Apple II — would set the company apart and make its products distinctive.

The company's first office, after it moved out of his family garage, was in a small building it shared with a Sony sales office. Sony was famous for its signature style and memorable product designs, so Jobs would drop by to study the marketing material. "He would come in looking scruffy and fondle the product brochures and point out design features," said Dan'l Lewin, who worked there. "Every now and then, he would ask, 'Can I take this brochure?' " By 1980, he had hired Lewin.

His fondness for the dark, industrial look of Sony receded around June 1981, when he began attending the annual International Design Conference in

Aspen. The meeting that year focused on Italian style, and it featured the architect-designer Mario Bellini, the filmmaker Bernardo Bertolucci, the car maker Sergio Pininfarina, and the Fiat heiress and politician Susanna Agnelli. "I had come to revere the Italian designers, just like the kid in *Breaking Away* reveres the Italian bikers," recalled Jobs, "so it was an amazing inspiration."

In Aspen he was exposed to the spare and functional design philosophy of the Bauhaus movement, which was enshrined by Herbert Bayer in the buildings, living suites, sans serif font typography, and furniture on the Aspen Institute campus. Like his mentors Walter Gropius and Ludwig Mies van der Rohe, Bayer believed that there should be no distinction between fine art and applied industrial design. The modernist International Style championed by the Bauhaus taught that design should be simple, yet have an expressive spirit. It emphasized rationality and functionality by employing clean lines and forms. Among the maxims preached by Mies and Gropius were "God is in the details" and "Less is more." As with Eichler homes, the artistic sensibility was combined with the capability for mass production.

Jobs publicly discussed his embrace of the Bauhaus style in a talk he gave at the 1983 design conference, the theme of which was "The Future Isn't What It Used to Be." He predicted the passing of the Sony style in favor of Bauhaus simplicity. "The current wave of industrial design is Sony's high-tech look, which is gunmetal gray, maybe paint it black, do weird stuff to it," he said. "It's easy to do that. But it's not great." He proposed an alternative, born of the Bauhaus, that was more true to the function and nature of the products. "What we're going to do is make the products high-tech, and we're going to package them cleanly so that you know they're high-tech. We will fit them in a small package, and then we can make them beautiful and white, just like Braun does with its electronics."

He repeatedly emphasized that Apple's products would be clean and simple. "We will make them bright and pure and honest about being high-tech, rather than a heavy industrial look of black, black, black, black, like Sony," he preached. "So that's our approach. Very simple, and we're really shooting for Museum of Modern Art quality. The way we're running the company, the product design, the advertising, it all comes down to this: Let's make it simple. Really simple." Apple's design mantra would remain the one featured on its first brochure: "Simplicity is the ultimate sophistication."

Jobs felt that design simplicity should be linked to making products easy to use. Those goals do not always go together. Sometimes a design can be so sleek and simple that a user finds it intimidating or unfriendly to navigate. "The main thing in our design is that we have to make things intuitively obvious," Jobs told the crowd of design mavens. For example, he extolled the desktop metaphor he was creating for the Macintosh. "People know how to deal with a desktop intuitively. If you walk into an office, there are papers on the desk. The one on the top is the most important. People know how to switch priority. Part of the reason we model our

199

computers on metaphors like the desktop is that we can leverage this experience people already have."

Speaking at the same time as Jobs that Wednesday afternoon, but in a smaller seminar room, was Maya Lin, twenty-three, who had been catapulted into fame the previous November when her Vietnam Veterans Memorial was dedicated in Washington, D.C. They struck up a close friendship, and Jobs invited her to visit Apple. "I came to work with Steve for a week," Lin recalled. "I asked him, 'Why do computers look like clunky TV sets? Why don't you make something thin? Why not a flat laptop?' " Jobs replied that this was indeed his goal, as soon as the technology was ready.

At that time there was not much exciting happening in the realm of industrial design, Jobs felt. He had a Richard Sapper lamp, which he admired, and he also liked the furniture of Charles and Ray Eames and the Braun products of Dieter Rams. But there were no towering figures energizing the world of industrial design the way that Raymond Loewy and Herbert Bayer had done. "There really wasn't much going on in industrial design, particularly in Silicon Valley, and Steve was very eager to change that," said Lin. "His design sensibility is sleek but not slick, and it's playful. He embraced minimalism, which came from his Zen devotion to simplicity, but he avoided allowing that to make his products cold. They stayed fun. He's passionate and super-serious about design, but at the same time there's a sense of play."

As Jobs's design sensibilities evolved, he became particularly attracted to the Japanese style and began hanging out with its stars, such as Issey

200

Miyake and I. M. Pei. His Buddhist training was a big influence. "I have always found Buddhism, Japanese Zen Buddhism in particular, to be aesthetically sublime," he said. "The most sublime thing I've ever seen are the gardens around Kyoto. I'm deeply moved by what that culture has produced, and it's directly from Zen Buddhism."

LIKE A PORSCHE

Jef Raskin's vision for the Macintosh was that it would be like a boxy carry-on suitcase, which would be closed by flipping up the keyboard over the front screen. When Jobs took over the project, he decided to sacrifice portability for a distinctive design that wouldn't take up much space on a desk. He plopped down a phone book and declared, to the horror of the engineers, that it shouldn't have a footprint larger than that. So his design team of Jerry Manock and Terry Oyama began working on ideas that had the screen above the computer box, with a keyboard that was detachable.

One day in March 1981, Andy Hertzfeld came back to the office from dinner to find Jobs hovering over their one Mac prototype in intense discussion with the creative services director, James Ferris. "We need it to have a classic look that won't go out of style, like the Volkswagen Beetle," Jobs said. From his father he had developed an appreciation for the contours of classic cars.

"No, that's not right," Ferris replied. "The lines should be voluptuous, like a Ferrari."

"Not a Ferrari, that's not right either," Jobs countered. "It should be more like a Porsche!" Jobs owned a Porsche 928 at the time. When Bill

Atkinson was over one weekend, Jobs brought him outside to admire the car. "Great art stretches the taste, it doesn't follow tastes," he told Atkinson. He also admired the design of the Mercedes. "Over the years, they've made the lines softer but the details starker," he said one day as he walked around the parking lot. "That's what we have to do with the Macintosh."

Oyama drafted a preliminary design and had a plaster model made. The Mac team gathered around for the unveiling and expressed their thoughts. Hertzfeld called it "cute." Others also seemed satisfied. Then Jobs let loose a blistering burst of criticism. "It's way too boxy, it's got to be more curvaceous. The radius of the first chamfer needs to be bigger, and I don't like the size of the bevel." With his new fluency in industrial design lingo, Jobs was referring to the angular or curved edge connecting the sides of the computer. But then he gave a resounding compliment. "It's a start," he said.

Every month or so, Manock and Oyama would present a new iteration based on Jobs's previous criticisms. The latest plaster model would be dramatically unveiled, and all the previous attempts would be lined up next to it. That not only helped them gauge the design's evolution, but it prevented Jobs from insisting that one of his suggestions had been ignored. "By the fourth model, I could barely distinguish it from the third one," said Hertzfeld, "but Steve was always critical and decisive, saying he loved or hated a detail that I could barely perceive."

One weekend Jobs went to Macy's in Palo Alto

and again spent time studying appliances, especially the Cuisinart. He came bounding into the Mac office that Monday, asked the design team to go buy one, and made a raft of new suggestions based on its lines, curves, and bevels.

Jobs kept insisting that the machine should look friendly. As a result, it evolved to resemble a human face. With the disk drive built in below the screen, the unit was taller and narrower than most computers, suggesting a head. The recess near the base evoked a gentle chin, and Jobs narrowed the strip of plastic at the top so that it avoided the Neanderthal forehead that made the Lisa subtly unattractive. The patent for the design of the Apple case was issued in the name of Steve Jobs as well as Manock and Oyama. "Even though Steve didn't draw any of the lines, his ideas and inspiration made the design what it is," Oyama later said. "To be honest, we didn't know what it meant for a computer to be 'friendly' until Steve told us."

Jobs obsessed with equal intensity about the look of what would appear on the screen. One day Bill Atkinson burst into Texaco Towers all excited. He had just come up with a brilliant algorithm that could draw circles and ovals onscreen quickly. The math for making circles usually required calculating square roots, which the 68000 microprocessor didn't support. But Atkinson did a workaround based on the fact that the sum of a sequence of odd numbers produces a sequence of perfect squares (for example, $1 + 3 = 4$, $1 + 3 + 5 = 9$, etc.). Hertzfeld recalled that when Atkinson fired up his demo, everyone was impressed except Jobs. "Well, circles and ovals are good," he said, "but how about drawing rectangles with rounded corners?"

"I don't think we really need it," said Atkinson, who explained that it would be almost impossible to do. "I wanted to keep the graphics routines lean and limit them to the primitives that truly needed to be done," he recalled.

"Rectangles with rounded corners are everywhere!" Jobs said, jumping up and getting more intense. "Just look around this room!" He pointed out the whiteboard and the tabletop and other objects that were rectangular with rounded corners. "And look outside, there's even more, practically everywhere you look!" He dragged Atkinson out for a walk, pointing out car windows and billboards and street signs. "Within three blocks, we found seventeen examples," said Jobs. "I started pointing them out everywhere until he was completely convinced."

"When he finally got to a No Parking sign, I said, 'Okay, you're right, I give up. We need to have a rounded-corner rectangle as a primitive!'" Hertzfeld recalled, "Bill returned to Texaco Towers the following afternoon, with a big smile on his face. His demo was now drawing rectangles with beautifully rounded corners blisteringly fast." The dialogue boxes and windows on the Lisa and the Mac, and almost every other subsequent computer, ended up being rendered with rounded corners.

At the calligraphy class he had audited at Reed, Jobs learned to love typefaces, with all of their serif and sans serif variations, proportional spacing, and leading. "When we were designing the first Macintosh computer, it all came back to me," he later said of that class. Because the Mac was bitmapped, it was possible to devise an endless array of fonts, ranging from the elegant to the wacky, and render them pixel by pixel on the screen.

To design these fonts, Hertzfeld recruited a high school friend from suburban Philadelphia, Susan Kare. They named the fonts after the stops on Philadelphia's Main Line commuter train: Overbrook, Merion, Ardmore, and Rosemont. Jobs found the process fascinating. Late one afternoon he stopped by and started brooding about the font names. They were "little cities that nobody's ever heard of," he complained. "They ought to be *world-class* cities!" The fonts were renamed Chicago, New York, Geneva, London, San Francisco, Toronto, and Venice.

Markkula and some others could never quite appreciate Jobs's obsession with typography. "His knowledge of fonts was remarkable, and he kept insisting on having great ones," Markkula recalled. "I kept saying, 'Fonts?!? Don't we have more important things to do?'" In fact the delightful assortment of Macintosh fonts, when combined with laser-writer printing and great graphics capabilities, would help launch the desktop publishing industry and be a boon for Apple's bottom line. It also introduced all sorts of regular folks, ranging from high school journalists to moms who edited PTA newsletters, to the quirky joy of knowing about fonts, which was once reserved for printers, grizzled editors, and other ink-stained wretches.

Kare also developed the icons, such as the trash can for discarding files, that helped define graphical interfaces. She and Jobs hit it off because they shared an instinct for simplicity along with a desire to make the Mac whimsical. "He usually came in at the end of every day," she said. "He'd always want to know what was new, and he's always had good taste and a good sense for visual details." Sometimes he came in on Sunday morning, so Kare made it a point to be there working. Every now and then, she would run into a problem. He rejected one of her renderings of a rabbit, an icon for speeding up the mouse-click rate, saying that the furry creature looked "too gay."

Jobs lavished similar attention on the title bars atop windows and documents. He had Atkinson and Kare do them over and over again as he agonized over their look. He did not like the ones on the Lisa because they were too black and harsh. He wanted the ones on the Mac to be smoother, to have pinstripes. "We must have gone through twenty different title bar designs before he was happy," Atkinson recalled. At one point Kare and Atkinson complained that he was making them spend too much time on tiny little tweaks to the title bar when they had bigger things to do. Jobs erupted. "Can you imagine looking at that every day?" he shouted. "It's not just a little thing, it's something we have to do right."

Chris Espinosa found one way to satisfy Jobs's design demands and control-freak tendencies. One of Wozniak's youthful acolytes from the days in the garage, Espinosa had been convinced to drop out of Berkeley by Jobs, who argued that he would always have a chance to study, but only one chance to work on the Mac. On his own, he decided to design a calculator for the computer. "We all gathered around as Chris showed the calculator to

Steve and then held his breath, waiting for Steve's reaction," Hertzfeld recalled.

"Well, it's a start," Jobs said, "but basically, it stinks. The background color is too dark, some lines are the wrong thickness, and the buttons are too big." Espinosa kept refining it in response to Jobs's critiques, day after day, but with each iteration came new criticisms. So finally one afternoon, when Jobs came by, Espinosa unveiled his inspired solution: "The Steve Jobs Roll Your Own Calculator Construction Set." It allowed the user to tweak and personalize the look of the calculator by changing the thickness of the lines, the size of the buttons, the shading, the background, and other attributes. Instead of just laughing, Jobs plunged in and started to play around with the look to suit his tastes. After about ten minutes he got it the way he liked. His design, not surprisingly, was the one that shipped on the Mac and remained the standard for fifteen years.

Although his focus was on the Macintosh, Jobs wanted to create a consistent design language for all Apple products. So he set up a contest to choose a world-class designer who would be for Apple what Dieter Rams was for Braun. The project was code-named Snow White, not because of his preference for the color but because the products to be designed were code-named after the seven dwarfs. The winner was Hartmut Esslinger, a German designer who was responsible for the look of Sony's Trinitron televisions. Jobs flew to the Black Forest region of Bavaria to meet him and was impressed not only with Esslinger's passion but also his spirited way of driving his Mercedes at more than one hundred miles per hour.

Even though he was German, Esslinger proposed that there should be a "born-in-America gene for Apple's DNA" that would produce a "California global" look, inspired by "Hollywood and music, a bit of rebellion, and natural sex appeal." His guiding principle was "Form follows emotion," a play on the familiar maxim that form follows function. He produced forty models of products to demonstrate the concept, and when Jobs saw them he proclaimed, "Yes, this is it!" The Snow White look, which was adopted immediately for the Apple IIc, featured white cases, tight rounded curves, and lines of thin grooves for both ventilation and decoration. Jobs offered Esslinger a contract on the condition that he move to California. They shook hands and, in Esslinger's not-so-modest words, "that handshake launched one of the most decisive collaborations in the history of industrial design." Esslinger's firm, frogdesign,* opened in Palo Alto in mid-1983 with a $1.2 million annual contract to work for Apple, and from then on every Apple product has included the proud declaration "Designed in California."

---

* The firm changed its name from frogdesign to frog design in 2000 and moved to San Francisco. Esslinger picked the original name not merely because frogs have the ability to metamorphose, but as a salute to its roots in the (f)ederal (r)epublic (o)f (g)ermany. He said that "the lowercase letters offered a nod to the Bauhaus notion of a non-hierarchical language, reinforcing the company's ethos of democratic partnership."

From his father Jobs had learned that a hallmark of passionate craftsmanship is making sure that even the aspects that will remain hidden are done beautifully. One of the most extreme — and telling — implementations of that philosophy came when he scrutinized the printed circuit board that would hold the chips and other components deep inside the Macintosh. No consumer would ever see it, but Jobs began critiquing it on aesthetic grounds. "That part's really pretty," he said. "But look at the memory chips. That's ugly. The lines are too close together."

One of the new engineers interrupted and asked why it mattered. "The only thing that's important is how well it works. Nobody is going to see the PC board."

Jobs reacted typically. "I want it to be as beautiful as possible, even if it's inside the box. A great carpenter isn't going to use lousy wood for the back of a cabinet, even though nobody's going to see it." In an interview a few years later, after the Macintosh came out, Jobs again reiterated that lesson from his father: "When you're a carpenter making a beautiful chest of drawers, you're not going to use a piece of plywood on the back, even though it faces the wall and nobody will ever see it. You'll know it's there, so you're going to use a beautiful piece of wood on the back. For you to sleep well at night, the aesthetic, the quality, has to be carried all the way through."

From Mike Markkula he had learned the importance of packaging and presentation. People do judge a book by its cover, so for the box of the Macintosh, Jobs chose a full-color design and kept trying to make it look better. "He got the guys to redo it fifty times," recalled Alain Rossmann, a member of the Mac team who married Joanna Hoffman. "It was going to be thrown in the trash as soon as the consumer opened it, but he was obsessed by how it looked." To Rossmann, this showed a lack of balance; money was being spent on expensive packaging while they were trying to save money on the memory chips. But for Jobs, each detail was essential to making the Macintosh amazing.

When the design was finally locked in, Jobs called the Macintosh team together for a ceremony. "Real artists sign their work," he said. So he got out a sheet of drafting paper and a Sharpie pen and had all of them sign their names. The signatures were engraved inside each Macintosh. No one would ever see them, but the members of the team knew that their signatures were inside, just as they knew that the circuit board was laid out as elegantly as possible. Jobs called them each up by name, one at a time. Burrell Smith went first. Jobs waited until last, after all forty-five of the others. He found a place right in the center of the sheet and signed his name in lowercase letters with a grand flair. Then he toasted them with champagne. "With moments like this, he got us seeing our work as art," said Atkinson.

CHAPTER THIRTEEN: BUILDING THE MAC

THE JOURNEY IS THE REWARD

COMPETITION

When IBM introduced its personal computer in August 1981, Jobs had his team buy one and dissect it. Their consensus was that it sucked. Chris Espinosa called it "a half-assed, hackneyed attempt," and there was some truth to that. It used old-fashioned command-line prompts and didn't support bitmapped graphical displays. Apple became cocky, not realizing that corporate technology managers might feel more comfortable buying from an established company like IBM rather than one named after a piece of fruit. Bill Gates happened to be visiting Apple headquarters for a meeting on the day the IBM PC was announced. "They didn't seem to care," he said. "It took them a year to realize what had happened."

Reflecting its cheeky confidence, Apple took out a full-page ad in the *Wall Street Journal* with the headline "Welcome, IBM. Seriously." It cleverly positioned the upcoming computer battle as a two-way contest between the spunky and rebellious Apple and the establishment Goliath IBM, conveniently relegating to irrelevance companies such as Commodore, Tandy, and Osborne that were doing

211

just as well as Apple.

Throughout his career, Jobs liked to see himself as an enlightened rebel pitted against evil empires, a Jedi warrior or Buddhist samurai fighting the forces of darkness. IBM was his perfect foil. He cleverly cast the upcoming battle not as a mere business competition, but as a spiritual struggle. "If, for some reason, we make some giant mistakes and IBM wins, my personal feeling is that we are going to enter sort of a computer Dark Ages for about twenty years," he told an interviewer. "Once IBM gains control of a market sector, they almost always stop innovation." Even thirty years later, reflecting back on the competition, Jobs cast it as a holy crusade: "IBM was essentially Microsoft at its worst. They were not a force for innovation; they were a force for evil. They were like ATT or Microsoft or Google is."

Unfortunately for Apple, Jobs also took aim at another perceived competitor to his Macintosh: the company's own Lisa. Partly it was psychological. He had been ousted from that group, and now he wanted to beat it. He also saw healthy rivalry as a way to motivate his troops. That's why he bet John Couch $5,000 that the Mac would ship before the Lisa. The problem was that the rivalry became unhealthy. Jobs repeatedly portrayed his band of engineers as the cool kids on the block, in contrast to the plodding HP engineer types working on the Lisa.

More substantively, when he moved away from Jef Raskin's plan for an inexpensive and under-powered portable appliance and reconceived the

Mac as a desktop machine with a graphical user interface, it became a scaled-down version of the Lisa that would likely undercut it in the marketplace.

Larry Tesler, who managed application software for the Lisa, realized that it would be important to design both machines to use many of the same software programs. So to broker peace, he arranged for Smith and Hertzfeld to come to the Lisa work space and demonstrate the Mac prototype. Twenty-five engineers showed up and were listening politely when, halfway into the presentation, the door burst open. It was Rich Page, a volatile engineer who was responsible for much of the Lisa's design. "The Macintosh is going to destroy the Lisa!" he shouted. "The Macintosh is going to ruin Apple!" Neither Smith nor Hertzfeld responded, so Page continued his rant. "Jobs wants to destroy Lisa because we wouldn't let him control it," he said, looking as if he were about to cry. "Nobody's going to buy a Lisa because they know the Mac is coming! But you don't care!" He stormed out of the room and slammed the door, but a moment later he barged back in briefly. "I know it's not your fault," he said to Smith and Hertzfeld. "Steve Jobs is the problem. Tell Steve that he's destroying Apple!"

Jobs did indeed make the Macintosh into a low-cost competitor to the Lisa, one with incompatible software. Making matters worse was that neither machine was compatible with the Apple II. With no one in overall charge at Apple, there was no chance of keeping Jobs in harness.

Jobs's reluctance to make the Mac compatible with the architecture of the Lisa was motivated by more than rivalry or revenge. There was a philosophical component, one that was related to his penchant for control. He believed that for a computer to be truly great, its hardware and its software had to be tightly linked. When a computer was open to running software that also worked on other computers, it would end up sacrificing some functionality. The best products, he believed, were "whole widgets" that were designed end-to-end, with the software closely tailored to the hardware and vice versa. This is what would distinguish the Macintosh, which had an operating system that worked only on its own hardware, from the environment that Microsoft was creating, in which its operating system could be used on hardware made by many different companies.

"Jobs is a strong-willed, elitist artist who doesn't want his creations mutated inauspiciously by unworthy programmers," explained ZDNet's editor Dan Farber. "It would be as if someone off the street added some brush strokes to a Picasso painting or changed the lyrics to a Dylan song." In later years Jobs's whole-widget approach would distinguish the iPhone, iPod, and iPad from their competitors. It resulted in awesome products. But it was not always the best strategy for dominating a market. "From the first Mac to the latest iPhone, Jobs's systems have always been sealed shut to prevent consumers from meddling and modifying them," noted Leander Kahney, author of *Cult of the Mac*.

Jobs's desire to control the user experience had

been at the heart of his debate with Wozniak over whether the Apple II would have slots that allow a user to plug expansion cards into a computer's motherboard and thus add some new functionality. Wozniak won that argument: The Apple II had eight slots. But this time around it would be Jobs's machine, not Wozniak's, and the Macintosh would have limited slots. You wouldn't even be able to open the case and get to the motherboard. For a hobbyist or hacker, that was uncool. But for Jobs, the Macintosh was for the masses. He wanted to give them a controlled experience.

"It reflects his personality, which is to want control," said Berry Cash, who was hired by Jobs in 1982 to be a market strategist at Texaco Towers. "Steve would talk about the Apple II and complain, 'We don't have control, and look at all these crazy things people are trying to do to it. That's a mistake I'll never make again.' " He went so far as to design special tools so that the Macintosh case could not be opened with a regular screwdriver. "We're going to design this thing so nobody but Apple employees can get inside this box," he told Cash.

Jobs also decided to eliminate the cursor arrow keys on the Macintosh keyboard. The only way to move the cursor was to use the mouse. It was a way of forcing old-fashioned users to adapt to point-and-click navigation, even if they didn't want to. Unlike other product developers, Jobs did not believe the customer was always right; if they wanted to resist using a mouse, they were wrong.

There was one other advantage, he believed, to eliminating the cursor keys: It forced outside software developers to write programs specially for the Mac operating system, rather than merely writing generic software that could be ported to a variety of computers. That made for the type of tight vertical integration between application software, operating systems, and hardware devices that Jobs liked.

Jobs's desire for end-to-end control also made him allergic to proposals that Apple license the Macintosh operating system to other office equipment manufacturers and allow them to make Macintosh clones. The new and energetic Macintosh marketing director Mike Murray proposed a licensing program in a confidential memo to Jobs in May 1982. "We would like the Macintosh user environment to become an industry standard," he wrote. "The hitch, of course, is that now one must buy Mac hardware in order to get this user environment. Rarely (if ever) has one company been able to create and maintain an industry-wide standard that cannot be shared with other manufacturers." His proposal was to license the Macintosh operating system to Tandy. Because Tandy's Radio Shack stores went after a different type of customer, Murray argued, it would not severely cannibalize Apple sales. But Jobs was congenitally averse to such a plan. His approach meant that the Macintosh remained a controlled environment that met his standards, but it also meant that, as Murray feared, it would have trouble securing its place as an industry standard in a world of IBM clones.

MACHINES OF THE YEAR

As 1982 drew to a close, Jobs came to believe that he was going to be *Time*'s Man of the Year. He arrived at Texaco Towers one day with the magazine's San Francisco bureau chief, Michael Moritz, and encouraged colleagues to give Moritz interviews. But Jobs did not end up on the cover. Instead the magazine chose "the Computer" as the topic for the year-end issue and called it "the Machine of the Year."

Accompanying the main story was a profile of Jobs, which was based on the reporting done by Moritz and written by Jay Cocks, an editor who usually handled rock music for the magazine. "With his smooth sales pitch and a blind faith that would have been the envy of the early Christian martyrs, it is Steven Jobs, more than anyone, who kicked open the door and let the personal computer move in," the story proclaimed. It was a richly reported piece, but also harsh at times — so harsh that Moritz (after he wrote a book about Apple and went on to be a partner in the venture firm Sequoia Capital with Don Valentine) repudiated it by complaining that his reporting had been "siphoned, filtered, and poisoned with gossipy benzene by an editor in New York whose regular task was to chronicle the wayward world of rock-and-roll music." The article quoted Bud Tribble on Jobs's "reality distortion field" and noted that he "would occasionally burst into tears at meetings." Perhaps the best quote came from Jef Raskin. Jobs, he declared, "would have made an excellent King of France."

To Jobs's dismay, the magazine made public the existence of the daughter he had forsaken, Lisa Brennan. He knew that Kottke had been the one to tell the magazine about Lisa, and he berated him in the Mac group work space in front of a half dozen people. "When the *Time* reporter asked me if Steve had a daughter named Lisa, I said 'Of course,' " Kottke recalled. "Friends don't let friends deny that they're the father of a child. I'm not going to let my friend be a jerk and deny paternity. He was really angry and felt violated and told me in front of everyone that I had betrayed him."

But what truly devastated Jobs was that he was not, after all, chosen as the Man of the Year. As he later told me:

> *Time* decided they were going to make me Man of the Year, and I was twenty-seven, so I actually cared about stuff like that. I thought it was pretty cool. They sent out Mike Moritz to write a story. We're the same age, and I had been very successful, and I could tell he was jealous and there was an edge to him. He wrote this terrible hatchet job. So the editors in New York get this story and say, "We can't make this guy Man of the Year." That really hurt. But it was a good lesson. It taught me to never get too excited about things like that, since the media is a circus anyway. They FedExed me the magazine, and I remember opening the package, thoroughly expecting to see my mug on the cover, and it was this computer sculpture thing. I thought, "Huh?" And then I read the article, and it was so awful that I actually cried.

In fact there's no reason to believe that Moritz was jealous or that he intended his reporting to be unfair. Nor was Jobs ever slated to be Man of the

Year, despite what he thought. That year the top editors (I was then a junior editor there) decided early on to go with the computer rather than a person, and they commissioned, months in advance, a piece of art from the famous sculptor George Segal to be a gatefold cover image. Ray Cave was then the magazine's editor. "We never considered Jobs," he said. "You couldn't personify the computer, so that was the first time we decided to go with an inanimate object. We never searched around for a face to be put on the cover."

Apple launched the Lisa in January 1983 — a full year before the Mac was ready — and Jobs paid his $5,000 wager to Couch. Even though he was not part of the Lisa team, Jobs went to New York to do publicity for it in his role as Apple's chairman and poster boy.

He had learned from his public relations consultant Regis McKenna how to dole out exclusive interviews in a dramatic manner. Reporters from anointed publications were ushered in sequentially for their hour with him in his Carlyle Hotel suite, where a Lisa computer was set on a table and surrounded by cut flowers. The publicity plan called for Jobs to focus on the Lisa and not mention the Macintosh, because speculation about it could undermine the Lisa. But Jobs couldn't help himself. In most of the stories based on his interviews that day — in *Time, Business Week,* the *Wall Street Journal,* and *Fortune* — the Macintosh was mentioned. "Later this year Apple will introduce a less powerful, less expensive version of Lisa, the Macintosh," *Fortune* reported. "Jobs himself has directed that project." *Business Week* quoted him as saying, "When it comes out, Mac is going to be the most incredible computer in the world." He also admitted that the Mac and the Lisa would not be compatible. It was like launching the Lisa with the kiss of death.

The Lisa did indeed die a slow death. Within two years it would be discontinued. "It was too expensive, and we were trying to sell it to big companies when our expertise was selling to consumers," Jobs later said. But there was a silver lining for Jobs: Within months of Lisa's launch, it became clear that Apple had to pin its hopes on the Macintosh instead.

LET'S BE PIRATES!

As the Macintosh team grew, it moved from Texaco Towers to the main Apple buildings on Bandley Drive, finally settling in mid-1983 into Bandley 3. It had a modern atrium lobby with video games, which Burrell Smith and Andy Hertzfeld chose, and a Toshiba compact disc stereo system with MartinLogan speakers and a hundred CDs. The software team was visible from the lobby in a fishbowl-like glass enclosure, and the kitchen was stocked daily with Odwalla juices. Over time the atrium attracted even more toys, most notably a Bösendorfer piano and a BMW motorcycle that Jobs felt would inspire an obsession with lapidary craftsmanship.

Jobs kept a tight rein on the hiring process. The goal was to get people who were creative, wickedly smart, and slightly rebellious. The software team would make applicants play Defender, Smith's favorite video game. Jobs would ask his usual offbeat questions to see how well the applicant could think in unexpected situations. One day he, Hertzfeld, and Smith interviewed a candidate for software manager who, it became clear as soon as he walked in the room, was too uptight and conventional to manage the wizards in the fishbowl. Jobs began to toy with him mercilessly. "How old were you when you lost your virginity?" he asked.

The candidate looked baffled. "What did you say?"

"Are you a virgin?" Jobs asked. The candidate sat there flustered, so Jobs changed the subject. "How many times have you taken LSD?" Hertzfeld recalled, "The poor guy was turning varying shades of red, so I tried to change the subject and asked a straightforward technical question." But when the candidate droned on in his response, Jobs broke in. "Gobble, gobble, gobble, gobble," he said, cracking up Smith and Hertzfeld.

"I guess I'm not the right guy," the poor man said as he got up to leave.

For all of his obnoxious behavior, Jobs also had the ability to instill in his team an esprit de corps. After tearing people down, he would find ways to lift them up and make them feel that being part of the Macintosh project was an amazing mission. Every six months he would take most of his team on a two-day retreat at a nearby resort.

The retreat in September 1982 was at the Pajaro Dunes near Monterey. Fifty or so members of the Mac division sat in the lodge facing a fireplace. Jobs sat on top of a table in front of them. He spoke quietly for a while, then walked to an easel and began posting his thoughts.

The first was "Don't compromise." It was an injunction that would, over time, be both helpful and harmful. Most technology teams made trade-offs. The Mac, on the other hand, would end up being as "insanely great" as Jobs and his acolytes could possibly make it — but it would not ship for another sixteen months, way behind schedule. After mentioning a scheduled completion date, he told them, "It would be better to miss than to turn out the wrong thing." A different type of project manager, willing to make some trade-offs, might try to lock in dates after which no changes could be made. Not Jobs. He displayed another maxim: "It's not done until it ships."

Another chart contained a kōan-like phrase that he later told me was his favorite maxim: "The journey is the reward." The Mac team, he liked to emphasize, was a special corps with an exalted mission. Someday they would all look back on their journey together and, forgetting or laughing off the painful moments, would regard it as a magical high point in their lives.

At the end of the presentation someone asked whether he thought they should do some market research to see what customers wanted. "No," he replied, "because customers don't know what they want until we've shown them." Then he pulled out a device that was about the size of a desk diary. "Do you want to see something neat?" When he flipped it open, it turned out to be a mock-up of a computer that could fit on your lap, with a keyboard and screen hinged together like a notebook.

"This is my dream of what we will be making in the mid- to late eighties," he said. They were building a company that would invent the future.

For the next two days there were presentations by various team leaders and the influential computer industry analyst Ben Rosen, with a lot of time in the evenings for pool parties and dancing. At the end, Jobs stood in front of the assemblage and gave a soliloquy. "As every day passes, the work fifty people are doing here is going to send a giant ripple through the universe," he said. "I know I might be a little hard to get along with, but this is the most fun thing I've done in my life." Years later most of those in the audience would be able to laugh about the "little hard to get along with" episodes and agree with him that creating that giant ripple was the most fun they had in their lives.

The next retreat was at the end of January 1983, the same month the Lisa launched, and there was a shift in tone. Four months earlier Jobs had written on his flip chart: "Don't compromise." This time one of the maxims was "Real artists ship." Nerves were frayed. Atkinson had been left out of the publicity interviews for the Lisa launch, and he marched into Jobs's hotel room and threatened to quit. Jobs tried to minimize the slight, but Atkinson refused to be mollified. Jobs got annoyed. "I don't have time to deal with this now," he said. "I have sixty other people out there who are pouring their hearts into the Macintosh, and they're waiting for me to start the meeting." With that he brushed past Atkinson to go address the faithful.

Jobs proceeded to give a rousing speech in which he claimed that he had resolved the dispute with

McIntosh audio labs to use the Macintosh name. (In fact the issue was still being negotiated, but the moment called for a bit of the old reality distortion field.) He pulled out a bottle of mineral water and symbolically christened the prototype onstage. Down the hall, Atkinson heard the loud cheer, and with a sigh joined the group. The ensuing party featured skinny-dipping in the pool, a bonfire on the beach, and loud music that lasted all night, which caused the hotel, La Playa in Carmel, to ask them never to come back.

Another of Jobs's maxims at the retreat was "It's better to be a pirate than to join the navy." He wanted to instill a rebel spirit in his team, to have them behave like swashbucklers who were proud of their work but willing to commandeer from others. As Susan Kare put it, "He meant, 'Let's have a renegade feeling to our group. We can move fast. We can get things done.' " To celebrate Jobs's birthday a few weeks later, the team paid for a billboard on the road to Apple headquarters. It read: "Happy 28th Steve. The Journey is the Reward. — The Pirates."

One of the Mac team's programmers, Steve Capps, decided this new spirit warranted hoisting a Jolly Roger. He cut a patch of black cloth and had Kare paint a skull and crossbones on it. The eye patch she put on the skull was an Apple logo. Late one Sunday night Capps climbed to the roof of their newly built Bandley 3 building and hoisted the flag on a scaffolding pole that the construction workers had left behind. It waved proudly for a few weeks, until members of the Lisa team, in a late-night foray, stole the flag and sent their Mac rivals a ransom note. Capps led a raid to recover it and was able to wrestle it from a secretary who was guarding it for the Lisa team. Some of the grown-ups overseeing Apple worried that Jobs's buccaneer spirit was getting out of hand. "Flying that flag was really stupid," said Arthur Rock. "It was telling the rest of the company they were no good." But Jobs loved it, and he made sure it waved proudly all the way through to the completion of the Mac project. "We were the renegades, and we wanted people to know it," he recalled.

Veterans of the Mac team had learned that they could stand up to Jobs. If they knew what they were talking about, he would tolerate the pushback, even admire it. By 1983 those most familiar with his reality distortion field had discovered something further: They could, if necessary, just quietly disregard what he decreed. If they turned out to be right, he would appreciate their renegade attitude and willingness to ignore authority. After all, that's what he did.

By far the most important example of this involved the choice of a disk drive for the Macintosh. Apple had a corporate division that built mass-storage devices, and it had developed a disk-drive system, code-named Twiggy, that could read and write onto those thin, delicate 5 1/4-inch floppy disks that older readers (who also remember Twiggy the model) will recall. But by the time the Lisa was ready to ship in the spring of 1983, it was clear that the Twiggy was buggy. Because the Lisa also came with a hard-disk drive, this was not a complete disaster. But the Mac had no hard disk, so it faced a crisis. "The Mac

225

team was beginning to panic," said Hertzfeld. "We were using a single Twiggy drive, and we didn't have a hard disk to fall back on."

The team discussed the problem at the January 1983 retreat, and Debi Coleman gave Jobs data about the Twiggy failure rate. A few days later he drove to Apple's factory in San Jose to see the Twiggy being made. More than half were rejected. Jobs erupted. With his face flushed, he began shouting and sputtering about firing everyone who worked there. Bob Belleville, the head of the Mac engineering team, gently guided him to the parking lot, where they could take a walk and talk about alternatives.

One possibility that Belleville had been exploring was to use a new 3 1/2-inch disk drive that Sony had developed. The disk was cased in sturdier plastic and could fit into a shirt pocket. Another option was to have a clone of Sony's 3 1/2-inch disk drive manufactured by a smaller Japanese supplier, the Alps Electronics Co., which had been supplying disk drives for the Apple II. Alps had already licensed the technology from Sony, and if they could build their own version in time it would be much cheaper.

Jobs and Belleville, along with Apple veteran Rod Holt (the guy Jobs enlisted to design the first power supply for the Apple II), flew to Japan to figure out what to do. They took the bullet train from Tokyo to visit the Alps facility. The engineers there didn't even have a working prototype, just a crude model. Jobs thought it was great, but Belleville was appalled. There was no way, he thought, that Alps could have it ready for the Mac within a year.

As they proceeded to visit other Japanese companies, Jobs was on his worst behavior. He wore jeans and sneakers to meetings with Japanese managers in dark suits. When they formally handed him little gifts, as was the custom, he often left them behind, and he never reciprocated with gifts of his own. He would sneer when rows of engineers lined up to greet him, bow, and politely offer their products for inspection. Jobs hated both the devices and the obsequiousness. "What are you showing me *this* for?" he snapped at one stop. "This is a piece of crap! *Anybody* could build a better drive than this." Although most of his hosts were appalled, some seemed amused. They had heard tales of his obnoxious style and brash behavior, and now they were getting to see it in full display.

The final stop was the Sony factory, located in a drab suburb of Tokyo. To Jobs, it looked messy and inelegant. A lot of the work was done by hand. He hated it. Back at the hotel, Belleville argued for going with the Sony disk drive. It was ready to use. Jobs disagreed. He decided that they would work with Alps to produce their own drive, and he ordered Belleville to cease all work with Sony.

Belleville decided it was best to partially ignore Jobs, and he asked a Sony executive to get its disk drive ready for use in the Macintosh. If and when it became clear that Alps could not deliver on time, Apple would switch to Sony. So Sony sent over the engineer who had developed the drive, Hidetoshi Komoto, a Purdue graduate who fortunately possessed a good sense of humor about his clandestine task.

Whenever Jobs would come from his corporate

office to visit the Mac team's engineers — which was almost every afternoon — they would hurriedly find somewhere for Komoto to hide. At one point Jobs ran into him at a newsstand in Cupertino and recognized him from the meeting in Japan, but he didn't suspect anything. The closest call was when Jobs came bustling onto the Mac work space unexpectedly one day while Komoto was sitting in one of the cubicles. A Mac engineer grabbed him and pointed him to a janitorial closet. "Quick, hide in this closet. Please! Now!" Komoto looked confused, Hertzfeld recalled, but he jumped up and did as told. He had to stay in the closet for five minutes, until Jobs left. The Mac engineers apologized. "No problem," he replied. "But American business practices, they are very strange. Very strange."

Belleville's prediction came true. In May 1983 the folks at Alps admitted it would take them at least eighteen more months to get their clone of the Sony drive into production. At a retreat in Pajaro Dunes, Markkula grilled Jobs on what he was going to do. Finally, Belleville interrupted and said that he might have an alternative to the Alps drive ready soon. Jobs looked baffled for just a moment, and then it became clear to him why he'd glimpsed Sony's top disk designer in Cupertino. "You son of a bitch!" Jobs said. But it was not in anger. There was a big grin on his face. As soon as he realized what Belleville and the other engineers had done behind his back, said Hertzfeld, "Steve swallowed his pride and thanked them for disobeying him and doing the right thing." It was, after all, what he would have done in their situation.

CHAPTER FOURTEEN:
ENTER SCULLEY
THE PEPSI CHALLENGE

THE COURTSHIP

Mike Markkula had never wanted to be Apple's president. He liked designing his new houses, flying his private plane, and living high off his stock options; he did not relish adjudicating conflict or curating high-maintenance egos. He had stepped into the role reluctantly, after he felt compelled to ease out Mike Scott, and he promised his wife the gig would be temporary. By the end of 1982, after almost two years, she gave him an order: Find a replacement right away.

Jobs knew that he was not ready to run the company himself, even though there was a part of him that wanted to try. Despite his arrogance, he could be self-aware. Markkula agreed; he told Jobs that he was still a bit too rough-edged and immature to be Apple's president. So they launched a search for someone from the outside.

The person they most wanted was Don Estridge, who had built IBM's personal computer division from scratch and launched a PC that, even though Jobs and his team disparaged it, was now outselling Apple's. Estridge had sheltered his division in Boca Raton, Florida, safely removed from the corporate mentality of Armonk, New York. Like Jobs, he was driven and inspiring, but unlike Jobs, he had the ability to allow others to think that his brilliant ideas were their own. Jobs flew to Boca Raton with the offer of a $1 million salary and a $1 million signing bonus, but Estridge turned him down. He was not the type who would jump ship to join the enemy. He also enjoyed being part of the establishment, a member of the Navy rather than a pirate. He was discomforted by Jobs's tales of ripping off the phone company. When asked where he worked, he loved to be able to answer "IBM."

So Jobs and Markkula enlisted Gerry Roche, a gregarious corporate headhunter, to find someone else. They decided not to focus on technology executives; what they needed was a consumer marketer who knew advertising and had the corporate polish that would play well on Wall Street. Roche set his sights on the hottest consumer marketing wizard of the moment, John Sculley, president of the Pepsi-Cola division of PepsiCo, whose Pepsi Challenge campaign had been an advertising and publicity triumph. When Jobs gave a talk to Stanford business students, he heard good things about Sculley, who had spoken to the class earlier. So he told Roche he would be happy to meet him.

Sculley's background was very different from Jobs's. His mother was an Upper East Side Manhattan matron who wore white gloves when she went out, and his father was a proper Wall Street lawyer. Sculley was sent off to St. Mark's School, then got his undergraduate degree from Brown and a business degree from Wharton. He had risen through the ranks at PepsiCo as an innovative marketer and advertiser, with little passion for product development or information technology.

Sculley flew to Los Angeles to spend Christmas with his two teenage children from a previous marriage. He took them to visit a computer store, where he was struck by how poorly the products were marketed. When his kids asked why he was so interested, he said he was planning to go up to Cupertino to meet Steve Jobs. They were totally blown away. They had grown up among movie stars, but to them Jobs was a true celebrity. It made Sculley take more seriously the prospect of being hired as his boss.

When he arrived at Apple headquarters, Sculley was startled by the unassuming offices and casual atmosphere. "Most people were less formally dressed than PepsiCo's maintenance staff," he noted. Over lunch Jobs picked quietly at his salad, but when Sculley declared that most executives found computers more trouble than they were worth, Jobs clicked into evangelical mode. "We want to change the way people use computers," he said.

On the flight home Sculley outlined his thoughts. The result was an eight-page memo on marketing computers to consumers and business executives. It was a bit sophomoric in parts, filled with underlined phrases, diagrams, and boxes, but it revealed his newfound enthusiasm for figuring out ways to sell something more interesting than soda. Among his recommendations: "Invest in in-store merchandizing that romances the consumer with Apple's potential to enrich their life!" He was still

231

reluctant to leave Pepsi, but Jobs intrigued him. "I was taken by this young, impetuous genius and thought it would be fun to get to know him a little better," he recalled.

So Sculley agreed to meet again when Jobs next came to New York, which happened to be for the January 1983 Lisa introduction at the Carlyle Hotel. After the full day of press sessions, the Apple team was surprised to see an unscheduled visitor come into the suite. Jobs loosened his tie and introduced Sculley as the president of Pepsi and a potential big corporate customer. As John Couch demonstrated the Lisa, Jobs chimed in with bursts of commentary, sprinkled with his favorite words, "revolutionary" and "incredible," claiming it would change the nature of human interaction with computers.

They then headed off to the Four Seasons restaurant, a shimmering haven of elegance and power. As Jobs ate a special vegan meal, Sculley described Pepsi's marketing successes. The Pepsi Generation campaign, he said, sold not a product but a lifestyle and an optimistic outlook. "I think Apple's got a chance to create an Apple Generation." Jobs enthusiastically agreed. The Pepsi Challenge campaign, in contrast, focused on the product; it combined ads, events, and public relations to stir up buzz. The ability to turn the introduction of a new product into a moment of national excitement was, Jobs noted, what he and Regis McKenna wanted to do at Apple.

When they finished talking, it was close to midnight. "This has been one of the most exciting evenings in my whole life," Jobs said as Sculley walked him back to the Carlyle. "I can't tell you how much fun I've had." When he finally got home to Greenwich, Connecticut, that night, Sculley had trouble sleeping. Engaging with Jobs was a lot more fun than negotiating with bottlers. "It stimulated me, roused my long-held desire to be an architect of ideas," he later noted. The next morning Roche called Sculley. "I don't know what you guys did last night, but let me tell you, Steve Jobs is ecstatic," he said.

And so the courtship continued, with Sculley playing hard but not impossible to get. Jobs flew east for a visit one Saturday in February and took a limo up to Greenwich. He found Sculley's newly built mansion ostentatious, with its floor-to-ceiling windows, but he admired the three hundred-pound custom-made oak doors that were so carefully hung and balanced that they swung open with the touch of a finger. "Steve was fascinated by that because he is, as I am, a perfectionist," Sculley recalled. Thus began the somewhat unhealthy process of a star-struck Sculley perceiving in Jobs qualities that he fancied in himself.

Sculley usually drove a Cadillac, but, sensing his guest's taste, he borrowed his wife's Mercedes 450SL convertible to take Jobs to see Pepsi's 144-acre corporate headquarters, which was as lavish as Apple's was austere. To Jobs, it epitomized the difference between the feisty new digital economy and the Fortune 500 corporate establishment. A winding drive led through manicured fields and a sculpture garden (including pieces by Rodin, Moore, Calder, and Giacometti) to a concrete-and-glass building designed by Edward

Durell Stone. Sculley's huge office had a Persian rug, nine windows, a small private garden, a hideaway study, and its own bathroom. When Jobs saw the corporate fitness center, he was astonished that executives had an area, with its own whirlpool, separate from that of the regular employees. "That's weird," he said. Sculley hastened to agree. "As a matter of fact, I was against it, and I go over and work out sometimes in the employees' area," he said.

Their next meeting was a few weeks later in Cupertino, when Sculley stopped on his way back from a Pepsi bottlers' convention in Hawaii. Mike Murray, the Macintosh marketing manager, took charge of preparing the team for the visit, but he was not clued in on the real agenda. "PepsiCo could end up purchasing literally thousands of Macs over the next few years," he exulted in a memo to the Macintosh staff. "During the past year, Mr. Sculley and a certain Mr. Jobs have become friends. Mr. Sculley is considered to be one of the best marketing heads in the big leagues; as such, let's give him a good time here."

Jobs wanted Sculley to share his excitement about the Macintosh. "This product means more to me than anything I've done," he said. "I want you to be the first person outside of Apple to see it." He dramatically pulled the prototype out of a vinyl bag and gave a demonstration. Sculley found Jobs as memorable as his machine. "He seemed more a showman than a businessman. Every move seemed calculated, as if it was rehearsed, to create an occasion of the moment."

Jobs had asked Hertzfeld and the gang to prepare

a special screen display for Sculley's amusement. "He's really smart," Jobs said. "You wouldn't believe how smart he is." The explanation that Sculley might buy a lot of Macintoshes for Pepsi "sounded a little bit fishy to me," Hertzfeld recalled, but he and Susan Kare created a screen of Pepsi caps and cans that danced around with the Apple logo. Hertzfeld was so excited he began waving his arms around during the demo, but Sculley seemed underwhelmed. "He asked a few questions, but he didn't seem all that interested," Hertzfeld recalled. He never ended up warming to Sculley. "He was incredibly phony, a complete poseur," he later said. "He pretended to be interested in technology, but he wasn't. He was a marketing guy, and that is what marketing guys are: paid poseurs."

Matters came to a head when Jobs visited New York in March 1983 and was able to convert the courtship into a blind and blinding romance. "I really think you're the guy," Jobs said as they walked through Central Park. "I want you to come and work with me. I can learn so much from you." Jobs, who had cultivated father figures in the past, knew just how to play to Sculley's ego and insecurities. It worked. "I was smitten by him," Sculley later admitted. "Steve was one of the brightest people I'd ever met. I shared with him a passion for ideas."

Sculley, who was interested in art history, steered them toward the Metropolitan Museum for a little test of whether Jobs was really willing to learn from others. "I wanted to see how well he could take coaching in a subject where he had no background," he recalled. As they strolled through the

Greek and Roman antiquities, Sculley expounded on the difference between the Archaic sculpture of the sixth century B.C. and the Periclean sculptures a century later. Jobs, who loved to pick up historical nuggets he never learned in college, seemed to soak it in. "I gained a sense that I could be a teacher to a brilliant student," Sculley recalled. Once again he indulged the conceit that they were alike: "I saw in him a mirror image of my younger self. I, too, was impatient, stubborn, arrogant, impetuous. My mind exploded with ideas, often to the exclusion of everything else. I, too, was intolerant of those who couldn't live up to my demands."

As they continued their long walk, Sculley confided that on vacations he went to the Left Bank in Paris to draw in his sketchbook; if he hadn't become a businessman, he would be an artist. Jobs replied that if he weren't working with computers, he could see himself as a poet in Paris. They continued down Broadway to Colony Records on Forty-ninth Street, where Jobs showed Sculley the music he liked, including Bob Dylan, Joan Baez, Ella Fitzgerald, and the Windham Hill jazz artists. Then they walked all the way back up to the San Remo on Central Park West and Seventy-fourth, where Jobs was planning to buy a two-story tower penthouse apartment.

The consummation occurred outside the penthouse on one of the terraces, with Sculley sticking close to the wall because he was afraid of heights. First they discussed money. "I told him I needed $1 million in salary, $1 million for a sign-up bonus," said Sculley. Jobs claimed that would be doable. "Even if I have to pay for it out of my own pocket," he said. "We'll have to solve those problems, because you're the best person I've ever met. I know you're perfect for Apple, and Apple deserves the best." He added that never before had he worked for someone he really respected, but he knew that Sculley was the person who could teach him the most. Jobs gave him his unblinking stare.

Sculley uttered one last demurral, a token suggestion that maybe they should just be friends and he could offer Jobs advice from the sidelines. "Any time you're in New York, I'd love to spend time with you." He later recounted the climactic moment: "Steve's head dropped as he stared at his feet. After a weighty, uncomfortable pause, he issued a challenge that would haunt me for days. 'Do you want to spend the rest of your life selling sugared water, or do you want a chance to change the world?'"

Sculley felt as if he had been punched in the stomach. There was no response possible other than to acquiesce. "He had an uncanny ability to always get what he wanted, to size up a person and know exactly what to say to reach a person," Sculley recalled. "I realized for the first time in four months that I couldn't say no." The winter sun was beginning to set. They left the apartment and walked back across the park to the Carlyle.

THE HONEYMOON

Sculley arrived in California just in time for the May 1983 Apple management retreat at Pajaro Dunes. Even though he had left all but one of his dark suits back in Greenwich, he was still having trouble adjusting to the casual atmosphere. In the

237

front of the meeting room, Jobs sat on the floor in the lotus position absentmindedly playing with the toes of his bare feet. Sculley tried to impose an agenda; he wanted to discuss how to differentiate their products — the Apple II, Apple III, Lisa, and Mac — and whether it made sense to organize the company around product lines or markets or functions. But the discussion descended into a free-for-all of random ideas, complaints, and debates.

At one point Jobs attacked the Lisa team for producing an unsuccessful product. "Well," someone shot back, "you haven't delivered the Macintosh! Why don't you wait until you get a product out before you start being critical?" Sculley was astonished. At Pepsi no one would have challenged the chairman like that. "Yet here, everyone began pig-piling on Steve." It reminded him of an old joke he had heard from one of the Apple ad salesmen: "What's the difference between Apple and the Boy Scouts? The Boy Scouts have adult supervision."

In the midst of the bickering, a small earthquake began to rumble the room. "Head for the beach," someone shouted. Everyone ran through the door to the water. Then someone else shouted that the previous earthquake had produced a tidal wave, so they all turned and ran the other way. "The indecision, the contradictory advice, the specter of natural disaster, only foreshadowed what was to come," Sculley later wrote.

One Saturday morning Jobs invited Sculley and his wife, Leezy, over for breakfast. He was then living in a nice but unexceptional Tudor-style home in Los Gatos with his girlfriend, Barbara Jasinski, a smart and reserved beauty who worked for Regis McKenna. Leezy had brought a pan and made vegetarian omelets. (Jobs had edged away from his strict vegan diet for the time being.) "I'm sorry I don't have much furniture," Jobs apologized. "I just haven't gotten around to it." It was one of his enduring quirks: His exacting standards of craftsmanship combined with a Spartan streak made him reluctant to buy any furnishings that he wasn't passionate about. He had a Tiffany lamp, an antique dining table, and a laser disc video attached to a Sony Trinitron, but foam cushions on the floor rather than sofas and chairs. Sculley smiled and mistakenly thought that it was similar to his own "frantic and Spartan life in a cluttered New York City apartment" early in his own career.

Jobs confided in Sculley that he believed he would die young, and therefore he needed to accomplish things quickly so that he would make his mark on Silicon Valley history. "We all have a short period of time on this earth," he told the Sculleys as they sat around the table that morning. "We probably only have the opportunity to do a few things really great and do them well. None of us has any idea how long we're going to be here, nor do I, but my feeling is I've got to accomplish a lot of these things while I'm young."

Jobs and Sculley would talk dozens of times a day in the early months of their relationship. "Steve and I became soul mates, near constant companions," Sculley said. "We tended to speak in half sentences and phrases." Jobs flattered Sculley. When he dropped by to hash something out, he would say something like "You're the only one who will understand." They would tell each other repeatedly, indeed so often that it should have been worrying, how happy they were to be with each other and working in tandem. And at every opportunity Sculley would find similarities with Jobs and point them out:

> We could complete each other's sentences because we were on the same wavelength. Steve would rouse me from sleep at 2 a.m. with a phone call to chat about an idea that suddenly crossed his mind. "Hi! It's me," he'd harmlessly say to the dazed listener, totally unaware of the time. I curiously had done the same in my Pepsi days. Steve would rip apart a presentation he had to give the next morning, throwing out slides and text. So had I as I struggled to turn public speaking into an important management tool during my early days at Pepsi. As a young executive, I was always impatient to get things done and often felt I could do them better myself. So did Steve. Sometimes I felt as if I was watching Steve playing me in a movie. The similarities were uncanny, and they were behind the amazing symbiosis we developed.

This was self-delusion, and it was a recipe for disaster. Jobs began to sense it early on. "We had different ways of looking at the world, different views on people, different values," Jobs recalled. "I began to realize this a few months after he arrived. He didn't learn things very quickly, and the people he wanted to promote were usually bozos."

Yet Jobs knew that he could manipulate Sculley by encouraging his belief that they were so alike.

And the more he manipulated Sculley, the more contemptuous of him he became. Canny observers in the Mac group, such as Joanna Hoffman, soon realized what was happening and knew that it would make the inevitable breakup more explosive. "Steve made Sculley feel like he was exceptional," she said. "Sculley had never felt that. Sculley became infatuated, because Steve projected on him a whole bunch of attributes that he didn't really have. When it became clear that Sculley didn't match all of these projections, Steve's distortion of reality had created an explosive situation."

The ardor eventually began to cool on Sculley's side as well. Part of his weakness in trying to manage a dysfunctional company was his desire to please other people, one of many traits that he did not share with Jobs. He was a polite person; this caused him to recoil at Jobs's rudeness to their fellow workers. "We would go to the Mac building at eleven at night," he recalled, "and they would bring him code to show. In some cases he wouldn't even look at it. He would just take it and throw it back at them. I'd say, 'How can you turn it down?' And he would say, 'I know they can do better.'" Sculley tried to coach him. "You've got to learn to hold things back," he told him at one point. Jobs would agree, but it was not in his nature to filter his feelings through a gauze.

Sculley began to believe that Jobs's mercurial personality and erratic treatment of people were rooted deep in his psychological makeup, perhaps the reflection of a mild bipolarity. There were big mood swings; sometimes he would be ecstatic, at other times he was depressed. At times he would launch into brutal tirades without warning, and Sculley would have to calm him down. "Twenty minutes later, I would get another call and be told to come over because Steve is losing it again," he said.

Their first substantive disagreement was over how to price the Macintosh. It had been conceived as a $1,000 machine, but Jobs's design changes had pushed up the cost so that the plan was to sell it at $1,995. However, when Jobs and Sculley began making plans for a huge launch and marketing push, Sculley decided that they needed to charge $500 more. To him, the marketing costs were like any other production cost and needed to be factored into the price. Jobs resisted, furiously. "It will destroy everything we stand for," he said. "I want to make this a revolution, not an effort to squeeze out profits." Sculley said it was a simple choice: He could have the $1,995 price or he could have the marketing budget for a big launch, but not both.

"You're not going to like this," Jobs told Hertzfeld and the other engineers, "but Sculley is insisting that we charge $2,495 for the Mac instead of $1,995." Indeed the engineers were horrified. Hertzfeld pointed out that they were designing the Mac for people like themselves, and overpricing it would be a "betrayal" of what they stood for. So Jobs promised them, "Don't worry, I'm not going to let him get away with it!" But in the end, Sculley prevailed. Even twenty-five years later Jobs seethed when recalling the decision: "It's the main reason the Macintosh sales slowed and Microsoft got to dominate the market." The decision made him feel that he was losing control of his product and company, and this was as dangerous as making a tiger feel cornered.

CHAPTER FIFTEEN: THE LAUNCH
A DENT IN THE UNIVERSE

REAL ARTISTS SHIP

The high point of the October 1983 Apple sales conference in Hawaii was a skit based on a TV show called *The Dating Game.* Jobs played emcee, and his three contestants, whom he had convinced to fly to Hawaii, were Bill Gates and two other software executives, Mitch Kapor and Fred Gibbons. As the show's jingly theme song played, the three took their stools. Gates, looking like a high school sophomore, got wild applause from the 750 Apple salesmen when he said, "During 1984, Microsoft expects to get half of its revenues from software for the Macintosh." Jobs, clean-shaven and bouncy, gave a toothy smile and asked if he thought that the Macintosh's new operating system would become one of the industry's new standards. Gates answered, "To create a new standard takes not just making something that's a little bit different, it takes something that's really new and captures people's imagination. And the Macintosh, of all the machines I've ever seen, is the only one that meets that standard."

But even as Gates was speaking, Microsoft was edging away from being primarily a collaborator with Apple to being more of a competitor. It would continue to make application software, like Microsoft Word, for Apple, but a rapidly increasing share of its revenue would come from the operating system it had written for the IBM personal computer. The year before, 279,000 Apple IIs were sold, compared to 240,000 IBM PCs and its clones. But the figures for 1983 were coming in starkly different: 420,000 Apple IIs versus 1.3 million IBMs and its clones. And both the Apple III and the Lisa were dead in the water.

Just when the Apple sales force was arriving in Hawaii, this shift was hammered home on the cover of *Business Week*. Its headline: "Personal Computers: And the Winner Is . . . IBM." The story inside detailed the rise of the IBM PC. "The battle for market supremacy is already over," the magazine declared. "In a stunning blitz, IBM has taken more than 26% of the market in two years, and is expected to account for half the world market by 1985. An additional 25% of the market will be turning out IBM-compatible machines."

That put all the more pressure on the Macintosh, due out in January 1984, three months away, to save the day against IBM. At the sales conference Jobs decided to play the showdown to the hilt. He took the stage and chronicled all the missteps made by IBM since 1958, and then in ominous tones described how it was now trying to take over the market for personal computers: "Will Big Blue dominate the entire computer industry? The entire information age? Was George Orwell right about 1984?" At that moment a screen came down from the ceiling and showed a preview of an upcoming sixty-second television ad for the Macintosh. In a few months it was destined to make advertising history, but in the meantime it served its purpose of rallying Apple's demoralized sales force. Jobs had always been able to draw energy by imagining himself as a rebel pitted against the forces of darkness. Now he was able to energize his troops with the same vision.

There was one more hurdle: Hertzfeld and the other wizards had to finish writing the code for the Macintosh. It was due to start shipping on Monday, January 16. One week before that, the engineers concluded they could not make that deadline.

Jobs was at the Grand Hyatt in Manhattan, preparing for the press previews, so a Sunday morning conference call was scheduled. The software manager calmly explained the situation to Jobs, while Hertzfeld and the others huddled around the speakerphone holding their breath. All they needed was an extra two weeks. The initial shipments to the dealers could have a version of the software labeled "demo," and these could be replaced as soon as the new code was finished at the end of the month. There was a pause. Jobs did not get angry; instead he spoke in cold, somber tones. He told them they were really great. So great, in fact, that he knew they could get this done. "There's no way we're slipping!" he declared. There was a collective gasp in the Bandley building work space. "You guys have been working on this stuff for months now, another couple weeks isn't going to make that much of a difference. You may as well get it over with. I'm going to ship the code a week from Monday, with your names on it."

"Well, we've got to finish it," Steve Capps said. And so they did. Once again, Jobs's reality distortion field pushed them to do what they had thought impossible. On Friday Randy Wigginton brought in a huge bag of chocolate-covered espresso beans for the final three all-nighters. When Jobs arrived at work at 8:30 a.m. that Monday, he found Hertzfeld sprawled nearly comatose on the couch. They talked for a few minutes about a remaining tiny glitch, and Jobs decreed that it wasn't a problem. Hertzfeld dragged himself to his blue Volkswagen Rabbit (license plate: MACWIZ) and drove home to bed. A short while later Apple's Fremont factory began to roll out boxes emblazoned with the colorful line drawings of the Macintosh. Real artists ship, Jobs had declared, and now the Macintosh team had.

THE "1984" AD

In the spring of 1983, when Jobs had begun to plan for the Macintosh launch, he asked for a commercial that was as revolutionary and astonishing as the product they had created. "I want something that will stop people in their tracks," he said. "I want a thunderclap." The task fell to the Chiat/Day advertising agency, which had acquired the Apple account when it bought the advertising side of Regis McKenna's business. The person put in charge was a lanky beach bum with a bushy beard, wild hair, goofy grin, and twinkling eyes named Lee Clow, who was the creative director of the agency's office in the Venice Beach section of Los Angeles. Clow was savvy and fun, in a laid-back yet focused way, and he forged a bond with Jobs that would last three decades.

Clow and two of his team, the copywriter Steve Hayden and the art director Brent Thomas, had been toying with a tagline that played off the George Orwell novel: "Why 1984 won't be like *1984.*" Jobs loved it, and asked them to develop it for the Macintosh launch. So they put together a storyboard for a sixty-second ad that would look like a scene from a sci-fi movie. It featured a rebellious young woman outrunning the Orwellian thought police and throwing a sledgehammer into a screen showing a mind-controlling speech by Big Brother.

The concept captured the zeitgeist of the personal computer revolution. Many young people, especially those in the counterculture, had viewed computers as instruments that could be used by Orwellian governments and giant corporations to sap individuality. But by the end of the 1970s, they were also being seen as potential tools for personal empowerment. The ad cast Macintosh as a warrior for the latter cause — a cool, rebellious, and heroic company that was the only thing standing in the way of the big evil corporation's plan for world domination and total mind control.

Jobs liked that. Indeed the concept for the ad had a special resonance for him. He fancied himself a rebel, and he liked to associate himself with the values of the ragtag band of hackers and pirates he recruited to the Macintosh group. Even though he had left the apple commune in Oregon to start the Apple corporation, he still wanted to be viewed as a denizen of the counterculture rather than the

248

corporate culture.

But he also realized, deep inside, that he had increasingly abandoned the hacker spirit. Some might even accuse him of selling out. When Wozniak held true to the Homebrew ethic by sharing his design for the Apple I for free, it was Jobs who insisted that they sell the boards instead. He was also the one who, despite Wozniak's reluctance, wanted to turn Apple into a corporation and not freely distribute stock options to the friends who had been in the garage with them. Now he was about to launch the Macintosh, a machine that violated many of the principles of the hacker's code: It was overpriced; it would have no slots, which meant that hobbyists could not plug in their own expansion cards or jack into the motherboard to add their own new functions; and it took special tools just to open the plastic case. It was a closed and controlled system, like something designed by Big Brother rather than by a hacker.

So the "1984" ad was a way of reaffirming, to himself and to the world, his desired self-image. The heroine, with a drawing of a Macintosh emblazoned on her pure white tank top, was a renegade out to foil the establishment. By hiring Ridley Scott, fresh off the success of *Blade Runner,* as the director, Jobs could attach himself and Apple to the cyberpunk ethos of the time. With the ad, Apple could identify itself with the rebels and hackers who thought differently, and Jobs could reclaim his right to identify with them as well.

Sculley was initially skeptical when he saw the storyboards, but Jobs insisted that they needed something revolutionary. He was able to get an unprecedented budget of $750,000 just to film the ad, which they planned to premiere during the Super Bowl. Ridley Scott made it in London using dozens of real skinheads among the enthralled masses listening to Big Brother on the screen. A female discus thrower was chosen to play the heroine. Using a cold industrial setting dominated by metallic gray hues, Scott evoked the dystopian aura of *Blade Runner.* Just at the moment when Big Brother announces "We shall prevail!" the heroine's hammer smashes the screen and it vaporizes in a flash of light and smoke.

When Jobs previewed the ad for the Apple sales force at the meeting in Hawaii, they were thrilled. So he screened it for the board at its December 1983 meeting. When the lights came back on in the boardroom, everyone was mute. Philip Schlein, the CEO of Macy's California, had his head on the table. Mike Markkula stared silently; at first it seemed he was overwhelmed by the power of the ad. Then he spoke: "Who wants to move to find a new agency?" Sculley recalled, "Most of them thought it was the worst commercial they had ever seen." Sculley himself got cold feet. He asked Chiat/Day to sell off the two commercial spots — one sixty seconds, the other thirty — that they had purchased.

Jobs was beside himself. One evening Wozniak, who had been floating into and out of Apple for the previous two years, wandered into the Macintosh building. Jobs grabbed him and said, "Come over here and look at this." He pulled out a VCR and played the ad. "I was astounded," Woz recalled. "I thought it was the most incredible thing." When

Jobs said the board had decided not to run it during the Super Bowl, Wozniak asked what the cost of the time slot was. Jobs told him $800,000. With his usual impulsive goodness, Wozniak immediately offered, "Well, I'll pay half if you will."

He ended up not needing to. The agency was able to sell off the thirty-second time slot, but in an act of passive defiance it didn't sell the longer one. "We told them that we couldn't sell the sixty-second slot, though in truth we didn't try," recalled Lee Clow. Sculley, perhaps to avoid a showdown with either the board or Jobs, decided to let Bill Campbell, the head of marketing, figure out what to do. Campbell, a former football coach, decided to throw the long bomb. "I think we ought to go for it," he told his team.

Early in the third quarter of Super Bowl XVIII, the dominant Raiders scored a touchdown against the Redskins and, instead of an instant replay, television screens across the nation went black for an ominous two full seconds. Then an eerie black-and-white image of drones marching to spooky music began to fill the screen. More than ninety-six million people watched an ad that was unlike any they'd seen before. At its end, as the drones watched in horror the vaporizing of Big Brother, an announcer calmly intoned, "On January 24th, Apple Computer will introduce Macintosh. And you'll see why 1984 won't be like '1984.'"

It was a sensation. That evening all three networks and fifty local stations aired news stories about the ad, giving it a viral life unprecedented in the pre–YouTube era. It would eventually be selected by both *TV Guide* and *Advertising Age* as the greatest commercial of all time.

PUBLICITY BLAST

Over the years Steve Jobs would become the grand master of product launches. In the case of the Macintosh, the astonishing Ridley Scott ad was just one of the ingredients. Another part of the recipe was media coverage. Jobs found ways to ignite blasts of publicity that were so powerful the frenzy would feed on itself, like a chain reaction. It was a phenomenon that he would be able to replicate whenever there was a big product launch, from the Macintosh in 1984 to the iPad in 2010. Like a conjurer, he could pull the trick off over and over again, even after journalists had seen it happen a dozen times and knew how it was done. Some of the moves he had learned from Regis McKenna, who was a pro at cultivating and stroking prideful reporters. But Jobs had his own intuitive sense of how to stoke the excitement, manipulate the competitive instincts of journalists, and trade exclusive access for lavish treatment.

In December 1983 he took his elfin engineering wizards, Andy Hertzfeld and Burrell Smith, to New York to visit *Newsweek* to pitch a story on "the kids who created the Mac." After giving a demo of the Macintosh, they were taken upstairs to meet Katharine Graham, the legendary proprietor, who had an insatiable interest in whatever was new. Afterward the magazine sent its technology columnist and a photographer to spend time in Palo Alto with Hertzfeld and Smith. The result was a flattering and smart four-page profile of the two of them, with pictures that made them look like cherubim of a new age. The article quoted Smith saying what he wanted to do next: "I want to build the computer of the 90's. Only I want to do it tomorrow." The article also described the mix of volatility and charisma displayed by his boss: "Jobs sometimes defends his ideas with highly vocal displays of temper that aren't always bluster; rumor has it that he has threatened to fire employees for insisting that his computers should have cursor keys, a feature that Jobs considers obsolete. But when he is on his best behavior, Jobs is a curious blend of charm and impatience, oscillating between shrewd reserve and his favorite expression of enthusiasm: 'Insanely great.' "

The technology writer Steven Levy, who was then working for *Rolling Stone,* came to interview Jobs, who urged him to convince the magazine's publisher to put the Macintosh team on the cover of the magazine. "The chances of Jann Wenner agreeing to displace Sting in favor of a bunch of computer nerds were approximately one in a googolplex," Levy thought, correctly. Jobs took Levy to a pizza joint and pressed the case: *Rolling Stone* was "on the ropes, running crummy articles, looking desperately for new topics and new audiences. The Mac could be its salvation!" Levy pushed back. *Rolling Stone* was actually good, he said, and he asked Jobs if he had read it recently. Jobs said that he had, an article about MTV that was "a piece of shit." Levy replied that he had written that article. Jobs, to his credit, didn't back away from the assessment. Instead he turned philosophical as he talked about the Macintosh. We are constantly benefiting from advances that went before us and taking things that people before us developed, he said. "It's a wonderful, ecstatic feeling to create something that puts it back in the pool of human experience and knowledge."

Levy's story didn't make it to the cover. But in the future, every major product launch that Jobs was involved in — at NeXT, at Pixar, and years later when he returned to Apple — would end up on the cover of either *Time, Newsweek,* or *Business Week.*

JANUARY 24, 1984

On the morning that he and his teammates completed the software for the Macintosh, Andy Hertzfeld had gone home exhausted and expected to stay in bed for at least a day. But that afternoon, after only six hours of sleep, he drove back to the office. He wanted to check in to see if there had been any problems, and most of his colleagues had done the same. They were lounging around, dazed but excited, when Jobs walked in. "Hey, pick yourselves up off the floor, you're not done yet!" he announced. "We need a demo for the intro!" His plan was to dramatically unveil the Macintosh in front of a large audience and have it show off some of its features to the inspirational theme from *Chariots of Fire.* "It needs to be done by the weekend, to be ready for the rehearsals," he added. They all groaned, Hertzfeld recalled, "but as we talked we realized that it would be fun to cook up something impressive."

The launch event was scheduled for the Apple annual stockholders' meeting on January 24 —

eight days away — at the Flint Auditorium of De Anza Community College. The television ad and the frenzy of press preview stories were the first two components in what would become the Steve Jobs playbook for making the introduction of a new product seem like an epochal moment in world history. The third component was the public unveiling of the product itself, amid fanfare and flourishes, in front of an audience of adoring faithful mixed with journalists who were primed to be swept up in the excitement.

Hertzfeld pulled off the remarkable feat of writing a music player in two days so that the computer could play the *Chariots of Fire* theme. But when Jobs heard it, he judged it lousy, so they decided to use a recording instead. At the same time, Jobs was thrilled with a speech generator that turned text into spoken words with a charming electronic accent, and he decided to make it part of the demo. "I want the Macintosh to be the first computer to introduce itself!" he insisted.

At the rehearsal the night before the launch, nothing was working well. Jobs hated the way the animation scrolled across the Macintosh screen, and he kept ordering tweaks. He also was dissatisfied with the stage lighting, and he directed Sculley to move from seat to seat to give his opinion as various adjustments were made. Sculley had never thought much about variations of stage lighting and gave the type of tentative answers a patient might give an eye doctor when asked which lens made the letters clearer. The rehearsals and changes went on for five hours, well into the night. "He was driving people insane, getting mad at the stagehands for every glitch in the presentation," Sculley recalled. "I thought there was no way we were going to get it done for the show the next morning."

Most of all, Jobs fretted about his presentation. Sculley fancied himself a good writer, so he suggested changes in Jobs's script. Jobs recalled being slightly annoyed, but their relationship was still in the phase when he was lathering on flattery and stroking Sculley's ego. "I think of you just like Woz and Markkula," he told Sculley. "You're like one of the founders of the company. They founded the company, but you and I are founding the future." Sculley lapped it up.

The next morning the 2,600-seat auditorium was mobbed. Jobs arrived in a double-breasted blue blazer, a starched white shirt, and a pale green bow tie. "This is the most important moment in my entire life," he told Sculley as they waited backstage for the program to begin. "I'm really nervous. You're probably the only person who knows how I feel about this." Sculley grasped his hand, held it for a moment, and whispered "Good luck."

As chairman of the company, Jobs went onstage first to start the shareholders' meeting. He did so with his own form of an invocation. "I'd like to open the meeting," he said, "with a twenty-year-old poem by Dylan — that's Bob Dylan." He broke into a little smile, then looked down to read from the second verse of "The Times They Are a-Changin'." His voice was high-pitched as he raced through the ten lines, ending with "For the loser now / Will be later to win / For the times they are a-changin'." That song was the anthem that kept the multimillionaire board chairman in touch with his counterculture self-image. He had a boot-leg copy of his favorite version, which was from the live concert Dylan performed, with Joan Baez, on Halloween 1964 at Lincoln Center's Philharmonic Hall.

Sculley came onstage to report on the company's earnings, and the audience started to become restless as he droned on. Finally, he ended with a personal note. "The most important thing that has happened to me in the last nine months at Apple has been a chance to develop a friendship with Steve Jobs," he said. "For me, the rapport we have developed means an awful lot."

The lights dimmed as Jobs reappeared onstage and launched into a dramatic version of the battle cry he had delivered at the Hawaii sales conference. "It is 1958," he began. "IBM passes up a chance to buy a young fledgling company that has invented a new technology called xerography. Two years later, Xerox was born, and IBM has been kicking them-selves ever since." The crowd laughed. Hertzfeld had heard versions of the speech both in Hawaii and elsewhere, but he was struck by how this time it was pulsing with more passion. After recounting other IBM missteps, Jobs picked up the pace and the emotion as he built toward the present:

It is now 1984. It appears that IBM wants it all. Apple is perceived to be the only hope to offer IBM a run for its money. Dealers, after initially welcoming IBM with open arms, now fear an IBM-dominated and -controlled future and are turning back to Apple as the only force who can ensure their future freedom. IBM wants it all, and is aiming its guns at its last obstacle to industry control, Apple. Will Big Blue dominate the entire computer industry? The entire information age? Was George Orwell right?

As he built to the climax, the audience went from murmuring to applauding to a frenzy of cheering and chanting. But before they could answer the Orwell question, the auditorium went black and the "1984" commercial appeared on the screen. When it was over, the entire audience was on its feet cheering.

With a flair for the dramatic, Jobs walked across the dark stage to a small table with a cloth bag on it. "Now I'd like to show you Macintosh in person," he said. He took out the computer, keyboard, and mouse, hooked them together deftly, then pulled one of the new 3 1/2-inch floppies from his shirt pocket. The theme from *Chariots of Fire* began to play. Jobs held his breath for a moment, because the demo had not worked well the night before. But this time it ran flawlessly. The word "MACINTOSH" scrolled horizontally onscreen, then underneath it the words "Insanely great" appeared in script, as if being slowly written by hand. Not used to such beautiful graphic displays, the audience quieted for a moment. A few gasps could be heard. And then, in rapid succession, came a series of screen shots: Bill Atkinson's QuickDraw graphics package followed by displays of different fonts, documents, charts, drawings, a chess game, a spreadsheet, and a rendering of Steve Jobs with a thought bubble containing a Macintosh.

When it was over, Jobs smiled and offered a treat.

"We've done a lot of talking about Macintosh recently," he said. "But today, for the first time ever, I'd like to let Macintosh speak for itself." With that, he strolled back over to the computer, pressed the button on the mouse, and in a vibrato but endearing electronic deep voice, Macintosh became the first computer to introduce itself. "Hello. I'm Macintosh. It sure is great to get out of that bag," it began. The only thing it didn't seem to know how to do was to wait for the wild cheering and shrieks that erupted. Instead of basking for a moment, it barreled ahead. "Unaccustomed as I am to public speaking, I'd like to share with you a maxim I thought of the first time I met an IBM mainframe: Never trust a computer you can't lift." Once again the roar almost drowned out its final lines. "Obviously, I can talk. But right now I'd like to sit back and listen. So it is with considerable pride that I introduce a man who's been like a father to me, Steve Jobs."

Pandemonium erupted, with people in the crowd jumping up and down and pumping their fists in a frenzy. Jobs nodded slowly, a tight-lipped but broad smile on his face, then looked down and started to choke up. The ovation continued for five minutes.

After the Macintosh team returned to Bandley 3 that afternoon, a truck pulled into the parking lot and Jobs had them all gather next to it. Inside were a hundred new Macintosh computers, each personalized with a plaque. "Steve presented them one at a time to each team member, with a handshake and a smile, as the rest of us stood around cheering," Hertzfeld recalled. It had been a grueling ride, and many egos had been bruised by Jobs's obnoxious and rough management style. But neither Raskin nor Wozniak nor Sculley nor anyone else at the company could have pulled off the creation of the Macintosh. Nor would it likely have emerged from focus groups and committees. On the day he unveiled the Macintosh, a reporter from *Popular Science* asked Jobs what type of market research he had done. Jobs responded by scoffing, "Did Alexander Graham Bell do any market research before he invented the telephone?"

Chapter Sixteen: Gates and Jobs
When Orbits Intersect

The Macintosh Partnership

In astronomy, a binary system occurs when the orbits of two stars are linked because of their gravitational interaction. There have been analogous situations in history, when an era is shaped by the relationship and rivalry of two orbiting superstars: Albert Einstein and Niels Bohr in twentieth-century physics, for example, or Thomas Jefferson and Alexander Hamilton in early American governance. For the first thirty years of the personal computer age, beginning in the late 1970s, the defining binary star system was composed of two high-energy college dropouts both born in 1955.

Bill Gates and Steve Jobs, despite their similar ambitions at the confluence of technology and business, had very different personalities and backgrounds. Gates's father was a prominent Seattle lawyer, his mother a civic leader on a variety of prestigious boards. He became a tech geek at the area's finest private school, Lakeside High, but he was never a rebel, hippie, spiritual seeker, or member of the counterculture. Instead of a Blue Box to rip off the phone company, Gates created for his school a program for scheduling classes,

261

which helped him get into ones with the right girls, and a car-counting program for local traffic engineers. He went to Harvard, and when he decided to drop out it was not to find enlightenment with an Indian guru but to start a computer software company.

Gates was good at computer coding, unlike Jobs, and his mind was more practical, disciplined, and abundant in analytic processing power. Jobs was more intuitive and romantic and had a greater instinct for making technology usable, design delightful, and interfaces friendly. He had a passion for perfection, which made him fiercely demanding, and he managed by charisma and scattershot intensity. Gates was more methodical; he held tightly scheduled product review meetings where he would cut to the heart of issues with lapidary skill. Both could be rude, but with Gates — who early in his career seemed to have a typical geek's flirtation with the fringes of the Asperger's scale — the cutting behavior tended to be less personal, based more on intellectual incisiveness than emotional callousness. Jobs would stare at people with a burning, wounding intensity; Gates sometimes had trouble making eye contact, but he was fundamentally humane.

"Each one thought he was smarter than the other one, but Steve generally treated Bill as someone who was slightly inferior, especially in matters of taste and style," said Andy Hertzfeld. "Bill looked down on Steve because he couldn't actually program." From the beginning of their relationship, Gates was fascinated by Jobs and slightly envious of his mesmerizing effect on people. But he also found him "fundamentally odd" and "weirdly flawed as a human being," and he was put off by Jobs's rudeness and his tendency to be "either in the mode of saying you were shit or trying to seduce you." For his part, Jobs found Gates unnervingly narrow. "He'd be a broader guy if he had dropped acid once or gone off to an ashram when he was younger," Jobs once declared.

Their differences in personality and character would lead them to opposite sides of what would become the fundamental divide in the digital age. Jobs was a perfectionist who craved control and indulged in the uncompromising temperament of an artist; he and Apple became the exemplars of a digital strategy that tightly integrated hardware, software, and content into a seamless package. Gates was a smart, calculating, and pragmatic analyst of business and technology; he was open to licensing Microsoft's operating system and software to a variety of manufacturers.

After thirty years Gates would develop a grudging respect for Jobs. "He really never knew much about technology, but he had an amazing instinct for what works," he said. But Jobs never reciprocated by fully appreciating Gates's real strengths. "Bill is basically unimaginative and has never invented anything, which is why I think he's more comfortable now in philanthropy than technology," Jobs said, unfairly. "He just shamelessly ripped off other people's ideas."

When the Macintosh was first being developed, Jobs went up to visit Gates at his office near Seattle. Microsoft had written some applications for the Apple II, including a spreadsheet program called Multiplan, and Jobs wanted to excite Gates and Co. about doing even more for the forthcoming Macintosh. Sitting in Gates's conference room, Jobs spun an enticing vision of a computer for the masses, with a friendly interface, which would be churned out by the millions in an automated California factory. His description of the dream factory sucking in the California silicon components and turning out finished Macintoshes caused the Microsoft team to code-name the project "Sand." They even reverse-engineered it into an acronym, for "Steve's amazing new device."

Gates had launched Microsoft by writing a version of BASIC, a programming language, for the Altair. Jobs wanted Microsoft to write a version of BASIC for the Macintosh, because Wozniak — despite much prodding by Jobs — had never enhanced his version of the Apple II's BASIC to handle floating-point numbers. In addition, Jobs wanted Microsoft to write application software — such as word processing and spreadsheet programs — for the Macintosh. At the time, Jobs was a king and Gates still a courtier: In 1982 Apple's annual sales were $1 billion, while Microsoft's were a mere $32 million. Gates signed on to do graphical versions of a new spreadsheet called Excel, a word-processing program called Word, and BASIC.

Gates frequently went to Cupertino for demonstrations of the Macintosh operating system, and he was not very impressed. "I remember the first time we went down, Steve had this app where it was just things bouncing around on the screen," he said. "That was the only app that ran." Gates was also put off by Jobs's attitude. "It was kind of a weird seduction visit, where Steve was saying, 'We don't really need you and we're doing this great thing, and it's under the cover.' He's in his Steve Jobs sales mode, but kind of the sales mode that also says, 'I don't need you, but I might let you be involved.' "

The Macintosh pirates found Gates hard to take. "You could tell that Bill Gates was not a very good listener. He couldn't bear to have anyone explain how something worked to him — he had to leap ahead instead and guess about how he thought it would work," Hertzfeld recalled. They showed him how the Macintosh's cursor moved smoothly across the screen without flickering. "What kind of hardware do you use to draw the cursor?" Gates asked. Hertzfeld, who took great pride that they could achieve their functionality solely using software, replied, "We don't have any special hardware for it!" Gates insisted that it was necessary to have special hardware to move the cursor that way. "So what do you say to somebody like that?" Bruce Horn, one of the Macintosh engineers, later said. "It made it clear to me that Gates was not the kind of person that would understand or appreciate the elegance of a Macintosh."

Despite their mutual wariness, both teams were excited by the prospect that Microsoft would create graphical software for the Macintosh that would take personal computing into a new realm, and they went to dinner at a fancy restaurant to celebrate. Microsoft soon dedicated a large team to the task. "We had more people working on the Mac than he did," Gates said. "He had about fourteen or fifteen people. We had like twenty people. We really bet our life on it." And even though Jobs thought that they didn't exhibit much taste, the Microsoft programmers were persistent. "They came out with applications that were terrible," Jobs recalled, "but they kept at it and they made them better." Eventually Jobs became so enamored of Excel that he made a secret bargain with Gates: If Microsoft would make Excel exclusively for the Macintosh for two years, and not make a version for IBM PCs, then Jobs would shut down his team working on a version of BASIC for the Macintosh and instead indefinitely license Microsoft's BASIC. Gates smartly took the deal, which infuriated the Apple team whose project got canceled and gave Microsoft a lever in future negotiations.

For the time being, Gates and Jobs forged a bond. That summer they went to a conference hosted by the industry analyst Ben Rosen at a Playboy Club retreat in Lake Geneva, Wisconsin, where nobody knew about the graphical interfaces that Apple was developing. "Everybody was acting like the IBM PC was everything, which was nice, but Steve and I were kind of smiling that, hey, we've got something," Gates recalled. "And he's kind of leaking, but nobody actually caught on." Gates became a regular at Apple retreats. "I went to every luau," said Gates. "I was part of the crew."

Gates enjoyed his frequent visits to Cupertino, where he got to watch Jobs interact erratically with his employees and display his obsessions. "Steve was in his ultimate pied piper mode, proclaiming how the Mac will change the world and overworking people like mad, with incredible tensions and complex personal relationships." Sometimes Jobs would begin on a high, then lapse into sharing his fears with Gates. "We'd go down Friday night, have dinner, and Steve would just be promoting that everything is great. Then the second day, without fail, he'd be kind of, 'Oh shit, is this thing going to sell, oh God, I have to raise the price, I'm sorry I did that to you, and my team is a bunch of idiots.' "

Gates saw Jobs's reality distortion field at play when the Xerox Star was launched. At a joint team dinner one Friday night, Jobs asked Gates how many Stars had been sold thus far. Gates said six hundred. The next day, in front of Gates and the whole team, Jobs said that three hundred Stars had been sold, forgetting that Gates had just told everyone it was actually six hundred. "So his whole team starts looking at me like, 'Are you going to tell him that he's full of shit?' " Gates recalled. "And in that case I didn't take the bait." On another occasion Jobs and his team were visiting Microsoft and having dinner at the Seattle Tennis Club. Jobs launched into a sermon about how the Macintosh and its software would be so easy to use that there would be no manuals. "It was like anybody who ever thought that there would be a manual for any Mac application was the greatest idiot," said Gates. "And we were like, 'Does he really mean it? Should we not tell him that we have people who are actually working on manuals?' "

After a while the relationship became bumpier. The original plan was to have some of the Microsoft applications — such as Excel, Chart, and File — carry the Apple logo and come bundled with the purchase of a Macintosh. "We were going to get $10 per app, per machine," said Gates. But this arrangement upset competing software makers. In addition, it seemed that some of Microsoft's programs might be late. So Jobs invoked a provision in his deal with Microsoft and decided not to bundle its software; Microsoft would have to scramble to distribute its software as products sold directly to consumers.

Gates went along without much complaint. He was already getting used to the fact that, as he put it, Jobs could "play fast and loose," and he suspected that the unbundling would actually help Microsoft. "We could make more money selling our software separately," Gates said. "It works better that way if you're willing to think you're going to have reasonable market share." Microsoft ended up making its software for various other platforms, and it began to give priority to the IBM PC version of Microsoft Word rather than the Macintosh version. In the end, Jobs's decision to back out of the bundling deal hurt Apple more than it did Microsoft.

When Excel for the Macintosh was released, Jobs and Gates unveiled it together at a press dinner at New York's Tavern on the Green. Asked if Microsoft would make a version of it for IBM PCs, Gates did not reveal the bargain he had made with Jobs but merely answered that "in time" that might happen. Jobs took the microphone. "I'm sure 'in time' we'll all be dead," he joked.

THE BATTLE OF THE GUI

At that time, Microsoft was producing an operating system, known as DOS, which it licensed to IBM and compatible computers. It was based on an old-fashioned command line interface that confronted users with surly little prompts such as C:\>. As Jobs and his team began to work closely with Microsoft, they grew worried that it would copy Macintosh's graphical user interface. Andy Hertzfeld noticed that his contact at Microsoft was asking detailed questions about how the Macintosh operating system worked. "I told Steve that I suspected that Microsoft was going to clone the Mac," he recalled.

They were right to worry. Gates believed that graphical interfaces were the future, and that Microsoft had just as much right as Apple did to copy what had been developed at Xerox PARC. As he freely admitted later, "We sort of say, 'Hey, we believe in graphics interfaces, we saw the Xerox Alto too.' "

In their original deal, Jobs had convinced Gates to agree that Microsoft would not create graphical software for anyone other than Apple until a year after the Macintosh shipped in January 1983. Unfortunately for Apple, it did not provide for the possibility that the Macintosh launch would be delayed for a year. So Gates was within his rights when, in November 1983, he revealed that Microsoft planned to develop a new operating system for IBM PCs featuring a graphical interface with windows, icons, and a mouse for point-and-click navigation. It would be called Windows. Gates hosted a Jobs-like product announcement, the most lavish thus far in Microsoft's history, at the Helmsley Palace Hotel in New York.

Jobs was furious. He knew there was little he

269

could do about it — Microsoft's deal with Apple not to do competing graphical software was running out — but he lashed out nonetheless. "Get Gates down here immediately," he ordered Mike Boich, who was Apple's evangelist to other software companies. Gates arrived, alone and willing to discuss things with Jobs. "He called me down to get pissed off at me," Gates recalled. "I went down to Cupertino, like a command performance. I told him, 'We're doing Windows.' I said to him, 'We're betting our company on graphical interfaces.' "

They met in Jobs's conference room, where Gates found himself surrounded by ten Apple employees who were eager to watch their boss assail him. Jobs didn't disappoint his troops. "You're ripping us off!" he shouted. "I trusted you, and now you're stealing from us!" Hertzfeld recalled that Gates just sat there coolly, looking Steve in the eye, before hurling back, in his squeaky voice, what became a classic zinger. "Well, Steve, I think there's more than one way of looking at it. I think it's more like we both had this rich neighbor named Xerox and I broke into his house to steal the TV set and found out that you had already stolen it."

Gates's two-day visit provoked the full range of Jobs's emotional responses and manipulation techniques. It also made clear that the Apple-Microsoft symbiosis had become a scorpion dance, with both sides circling warily, knowing that a sting by either could cause problems for both. After the confrontation in the conference room, Gates quietly gave Jobs a private demo of what was being planned for Windows. "Steve didn't know what to say," Gates recalled. "He could either say, 'Oh, this is a violation of something,' but he didn't. He chose to say, 'Oh, it's actually really a piece of shit.' " Gates was thrilled, because it gave him a chance to calm Jobs down for a moment. "I said, 'Yes, it's a nice little piece of shit.' " So Jobs went through a gamut of other emotions. "During the course of this meeting, he's just ruder than shit," Gates said. "And then there's a part where he's almost crying, like, 'Oh, just give me a chance to get this thing off.' " Gates responded by becoming very calm. "I'm good at when people are emotional, I'm kind of less emotional."

As he often did when he wanted to have a serious conversation, Jobs suggested they go on a long walk. They trekked the streets of Cupertino, back and forth to De Anza college, stopping at a diner and then walking some more. "We had to take a walk, which is not one of my management techniques," Gates said. "That was when he began saying things like, 'Okay, okay, but don't make it too much like what we're doing.' "

As it turned out, Microsoft wasn't able to get Windows 1.0 ready for shipping until the fall of 1985. Even then, it was a shoddy product. It lacked the elegance of the Macintosh interface, and it had tiled windows rather than the magical clipping of overlapping windows that Bill Atkinson had devised. Reviewers ridiculed it and consumers spurned it. Nevertheless, as is often the case with Microsoft products, persistence eventually made Windows better and then dominant.

Jobs never got over his anger. "They just ripped us off completely, because Gates has no shame," Jobs told me almost thirty years later. Upon hearing this, Gates responded, "If he believes that, he really has entered into one of his own reality distortion fields." In a legal sense, Gates was right, as courts over the years have subsequently ruled. And on a practical level, he had a strong case as well. Even though Apple made a deal for the right to use what it saw at Xerox PARC, it was inevitable that other companies would develop similar graphical interfaces. As Apple found out, the "look and feel" of a computer interface design is a hard thing to protect.

And yet Jobs's dismay was understandable. Apple had been more innovative, imaginative, elegant in execution, and brilliant in design. But even though Microsoft created a crudely copied series of products, it would end up winning the war of operating systems. This exposed an aesthetic flaw in how the universe worked: The best and most innovative products don't always win. A decade later, this truism caused Jobs to let loose a rant that was somewhat arrogant and over-the-top, but also had a whiff of truth to it. "The only problem with Microsoft is they just have no taste, they have absolutely no taste," he said. "I don't mean that in a small way. I mean that in a big way, in the sense that they don't think of original ideas and they don't bring much culture into their product."

CHAPTER SEVENTEEN: ICARUS
WHAT GOES UP . . .

FLYING HIGH

The launch of the Macintosh in January 1984 propelled Jobs into an even higher orbit of celebrity, as was evident during a trip to Manhattan he took at the time. He went to a party that Yoko Ono threw for her son, Sean Lennon, and gave the nine-year-old a Macintosh. The boy loved it. The artists Andy Warhol and Keith Haring were there, and they were so enthralled by what they could create with the machine that the contemporary art world almost took an ominous turn. "I drew a circle," Warhol exclaimed proudly after using QuickDraw. Warhol insisted that Jobs take a computer to Mick Jagger. When Jobs arrived at the rock star's townhouse, Jagger seemed baffled. He didn't quite know who Jobs was. Later Jobs told his team, "I think he was on drugs. Either that or he's brain-damaged." Jagger's daughter Jade, however, took to the computer immediately and started drawing with MacPaint, so Jobs gave it to her instead.

He bought the top-floor duplex apartment that he'd shown Sculley in the San Remo on Manhattan's Central Park West and hired James Freed of I. M. Pei's firm to renovate it, but he never moved in. (He would later sell it to Bono for $15 million.) He also bought an old Spanish colonial–style fourteen-bedroom mansion in Woodside, in the hills above Palo Alto, that had been built by a copper baron, which he moved into but never got around to furnishing.

At Apple his status revived. Instead of seeking ways to curtail Jobs's authority, Sculley gave him more: The Lisa and Macintosh divisions were folded together, with Jobs in charge. He was flying high, but this did not serve to make him more mellow. Indeed there was a memorable display of his brutal honesty when he stood in front of the combined Lisa and Macintosh teams to describe how they would be merged. His Macintosh group leaders would get all of the top positions, he said, and a quarter of the Lisa staff would be laid off. "You guys failed," he said, looking directly at those who had worked on the Lisa. "You're a B team. B players. Too many people here are B or C players, so today we are releasing some of you to have the opportunity to work at our sister companies here in the valley."

Bill Atkinson, who had worked on both teams, thought it was not only callous, but unfair. "These people had worked really hard and were brilliant engineers," he said. But Jobs had latched onto what he believed was a key management lesson from his Macintosh experience: You have to be ruthless if you want to build a team of A players. "It's too easy, as a team grows, to put up with a few B players, and they then attract a few more B players, and soon you will even have some C players," he recalled. "The Macintosh experience taught me

274

that A players like to work only with other A players, which means you can't indulge B players."

For the time being, Jobs and Sculley were able to convince themselves that their friendship was still strong. They professed their fondness so effusively and often that they sounded like high school sweethearts at a Hallmark card display. The first anniversary of Sculley's arrival came in May 1984, and to celebrate Jobs lured him to a dinner party at Le Mouton Noir, an elegant restaurant in the hills southwest of Cupertino. To Sculley's surprise, Jobs had gathered the Apple board, its top managers, and even some East Coast investors. As they all congratulated him during cocktails, Sculley recalled, "a beaming Steve stood in the background, nodding his head up and down and wearing a Cheshire Cat smile on his face." Jobs began the dinner with a fulsome toast. "The happiest two days for me were when Macintosh shipped and when John Sculley agreed to join Apple," he said. "This has been the greatest year I've ever had in my whole life, because I've learned so much from John." He then presented Sculley with a montage of memorabilia from the year.

In response, Sculley effused about the joys of being Jobs's partner for the past year, and he concluded with a line that, for different reasons, everyone at the table found memorable. "Apple has one leader," he said, "Steve and me." He looked across the room, caught Jobs's eye, and watched him smile. "It was as if we were communicating with each other," Sculley recalled. But he also noticed that Arthur Rock and some of the others were looking quizzical, perhaps even skeptical. They were worried that Jobs was completely rolling him. They had hired Sculley to control Jobs, and now it was clear that Jobs was the one in control. "Sculley was so eager for Steve's approval that he was unable to stand up to him," Rock recalled.

Keeping Jobs happy and deferring to his expertise may have seemed like a smart strategy to Sculley. But he failed to realize that it was not in Jobs's nature to share control. Deference did not come naturally to him. He began to become more vocal about how he thought the company should be run. At the 1984 business strategy meeting, for example, he pushed to make the company's centralized sales and marketing staffs bid on the right to provide their services to the various product divisions. (This would have meant, for example, that the Macintosh group could decide not to use Apple's marketing team and instead create one of its own.) No one else was in favor, but Jobs kept trying to ram it through. "People were looking to me to take control, to get him to sit down and shut up, but I didn't," Sculley recalled. As the meeting broke up, he heard someone whisper, "Why doesn't Sculley shut him up?"

When Jobs decided to build a state-of-the-art factory in Fremont to manufacture the Macintosh, his aesthetic passions and controlling nature kicked into high gear. He wanted the machinery to be painted in bright hues, like the Apple logo, but he spent so much time going over paint chips that Apple's manufacturing director, Matt Carter, finally just installed them in their usual beige and gray. When Jobs took a tour, he ordered that the machines be repainted in the bright colors he wanted. Carter objected; this was precision equipment, and repainting the machines could cause problems. He turned out to be right. One of the most expensive machines, which got painted bright blue, ended up not working properly and was dubbed "Steve's folly." Finally Carter quit. "It took so much energy to fight him, and it was usually over something so pointless that finally I had enough," he recalled.

Jobs tapped as a replacement Debi Coleman, the spunky but good-natured Macintosh financial officer who had once won the team's annual award for the person who best stood up to Jobs. But she knew how to cater to his whims when necessary. When Apple's art director, Clement Mok, informed her that Jobs wanted the walls to be pure white, she protested, "You can't paint a factory pure white. There's going to be dust and stuff all over." Mok replied, "There's no white that's too white for Steve." She ended up going along. With its pure white walls and its bright blue, yellow, and red machines, the factory floor "looked like an Alexander Calder showcase," said Coleman.

When asked about his obsessive concern over the look of the factory, Jobs said it was a way to ensure a passion for perfection:

I'd go out to the factory, and I'd put on a white glove to check for dust. I'd find it everywhere — on machines, on the tops of the racks, on the floor. And I'd ask Debi to get it cleaned. I told her I thought we should be able to eat off the floor of the factory. Well, this drove Debi up the wall. She didn't understand why. And I couldn't articulate it back then. See, I'd been very influenced by what I'd seen in Japan. Part of what I greatly admired there — and part of what we were lacking in our factory — was a sense of teamwork and discipline. If we didn't have the discipline to keep that place spotless, then we weren't going to have the discipline to keep all these machines running.

One Sunday morning Jobs brought his father to see the factory. Paul Jobs had always been fastidious about making sure that his craftsmanship was exacting and his tools in order, and his son was proud to show that he could do the same. Coleman came along to give the tour. "Steve was, like, beaming," she recalled. "He was so proud to show his father this creation." Jobs explained how everything worked, and his father seemed truly admiring. "He kept looking at his father, who touched everything and loved how clean and perfect everything looked."

Things were not quite as sweet when Danielle Mitterrand toured the factory. The Cuba-admiring wife of France's socialist president François Mitterrand asked a lot of questions, through her translator, about the working conditions, while Jobs, who had grabbed Alain Rossmann to serve as his translator, kept trying to explain the advanced robotics and technology. After Jobs talked about the just-in-time production schedules, she asked about overtime pay. He was annoyed, so he described how automation helped him keep down labor costs, a subject he knew would not delight her. "Is it hard work?" she asked. "How much vacation time do they get?" Jobs couldn't contain himself. "If she's so interested in their welfare," he said to her translator, "tell her she can come work here any time." The translator turned pale and said nothing. After a moment Rossmann stepped in to say, in French, "M. Jobs says he thanks you for your visit and your interest in the factory." Neither Jobs nor Madame Mitterrand knew what happened, Rossmann recalled, but her translator looked very relieved.

Afterward, as he sped his Mercedes down the freeway toward Cupertino, Jobs fumed to Rossmann about Madame Mitterrand's attitude. At one point he was going just over 100 miles per hour when a policeman stopped him and began writing a ticket. After a few minutes, as the officer scribbled away, Jobs honked. "Excuse me?" the policeman said. Jobs replied, "I'm in a hurry." Amazingly, the officer didn't get mad. He simply finished writing the ticket and warned that if Jobs was caught going over 55 again he would be sent to jail. As soon as the policeman left, Jobs got back on the road and accelerated to 100. "He absolutely believed that the normal rules didn't apply to him," Rossmann marveled.

His wife, Joanna Hoffman, saw the same thing when she accompanied Jobs to Europe a few months after the Macintosh was launched. "He was just completely obnoxious and thinking he could get away with anything," she recalled. In Paris she had arranged a formal dinner with French software developers, but Jobs suddenly decided he didn't want to go. Instead he shut the car door on Hoffman and told her he was going to see the poster artist Folon instead. "The developers were so pissed off they wouldn't shake our hands," she said.

In Italy, he took an instant dislike to Apple's general manager, a soft rotund guy who had come from a conventional business. Jobs told him bluntly that he was not impressed with his team or his sales strategy. "You don't deserve to be able to sell the Mac," Jobs said coldly. But that was mild compared to his reaction to the restaurant the hapless manager had chosen. Jobs demanded a vegan meal, but the waiter very elaborately proceeded to dish out a sauce filled with sour cream. Jobs got so nasty that Hoffman had to threaten him. She whispered that if he didn't calm down, she was going to pour her hot coffee on his lap.

The most substantive disagreements Jobs had on the European trip concerned sales forecasts. Using his reality distortion field, Jobs was always pushing his team to come up with higher projections. He kept threatening the European managers that he wouldn't give them any allocations unless they projected bigger forecasts. They insisted on being realistic, and Hoffmann had to referee. "By the end of the trip, my whole body was shaking uncontrollably," Hoffman recalled.

It was on this trip that Jobs first got to know Jean-Louis Gassée, Apple's manager in France. Gassée was among the few to stand up successfully to Jobs on the trip. "He has his own way with the truth," Gassée later remarked. "The only way to deal with him was to out-bully him." When Jobs made his usual threat about cutting down on France's allocations if Gassée didn't jack up sales projections, Gassée got angry. "I remember grabbing his lapel and telling him to stop, and then he backed down. I used to be an angry man myself. I am a recovering assaholic. So I could recognize that in Steve."

Gassée was impressed, however, at how Jobs could turn on the charm when he wanted to. François Mitterrand had been preaching the gospel of *informatique pour tous* — computing for all — and various academic experts in technology, such as Marvin Minsky and Nicholas Negroponte, came over to sing in the choir. Jobs gave a talk to the group at the Hotel Bristol and painted a picture of how France could move ahead if it put computers in all of its schools. Paris also brought out the romantic in him. Both Gassée and Negroponte tell tales of him pining over women while there.

FALLING

After the burst of excitement that accompanied the release of Macintosh, its sales began to taper off in the second half of 1984. The problem was a fundamental one: It was a dazzling but woefully slow and underpowered computer, and no amount of hoopla could mask that. Its beauty was that its user interface looked like a sunny playroom rather than a somber dark screen with sickly green pulsating letters and surly command lines. But that led to its greatest weakness: A character on a text-based display took less than a byte of code, whereas when the Mac drew a letter, pixel by pixel in any elegant font you wanted, it required twenty or thirty times more memory. The Lisa handled this by shipping with more than 1,000K RAM, whereas the Macintosh made do with 128K.

Another problem was the lack of an internal hard disk drive. Jobs had called Joanna Hoffman a "Xerox bigot" when she fought for such a storage device. He insisted that the Macintosh have just one floppy disk drive. If you wanted to copy data, you could end up with a new form of tennis elbow from having to swap floppy disks in and out of the single drive. In addition, the Macintosh lacked a fan, another example of Jobs's dogmatic stubbornness. Fans, he felt, detracted from the calm of a computer. This caused many component failures and earned the Macintosh the nickname "the beige toaster," which did not enhance its popularity. It was so seductive that it had sold well enough for the first few months, but when people became more aware of its limitations, sales fell. As Hoffman later lamented, "The reality distortion field can serve as a spur, but then reality itself hits."

At the end of 1984, with Lisa sales virtually nonexistent and Macintosh sales falling below ten thousand a month, Jobs made a shoddy, and atypical, decision out of desperation. He decided to take the inventory of unsold Lisas, graft on a Macintosh-emulation program, and sell them as a new product, the "Macintosh XL." Since the Lisa had been discontinued and would not be restarted, it was an unusual instance of Jobs producing something that he did not believe in. "I was furious because the Mac XL wasn't real," said Hoffman. "It was just to blow the excess Lisas out the door. It sold well, and then we had to discontinue the horrible hoax, so I resigned."

The dark mood was evident in the ad that was developed in January 1985, which was supposed to reprise the anti-IBM sentiment of the resonant "1984" ad. Unfortunately there was a fundamental difference: The first ad had ended on a heroic, optimistic note, but the storyboards presented by Lee Clow and Jay Chiat for the new ad, titled "Lemmings," showed dark-suited, blindfolded corporate managers marching off a cliff to their death. From the beginning both Jobs and Sculley were uneasy. It didn't seem as if it would convey a positive or glorious image of Apple, but instead would merely insult every manager who had bought an IBM.

Jobs and Sculley asked for other ideas, but the agency folks pushed back. "You guys didn't want to run '1984' last year," one of them said. According to Sculley, Lee Clow added, "I will put my whole reputation, everything, on this commercial." When the filmed version, done by Ridley Scott's brother Tony, came in, the concept looked even worse. The mindless managers marching off the cliff were singing a funeral-paced version of the *Snow White* song "Heigh-ho, Heigh-ho," and the dreary filmmaking made it even more depressing than the storyboards portended. "I can't believe you're going to insult businesspeople across America by running that," Debi Coleman yelled at Jobs when she saw the ad. At the marketing meetings, she stood up to make her point about how much she hated it. "I literally put a resignation letter on his desk. I wrote it on my Mac. I thought it was an affront to corporate managers. We were just beginning to get a toehold with desktop publishing."

Nevertheless Jobs and Sculley bent to the agency's entreaties and ran the commercial during the Super Bowl. They went to the game together at Stanford Stadium with Sculley's wife, Leezy (who couldn't stand Jobs), and Jobs's new girlfriend, Tina Redse. When the commercial was shown near the end of the fourth quarter of a dreary game, the fans watched on the overhead screen and had little reaction. Across the country, most of the response was negative. "It insulted the very people Apple was trying to reach," the president of a market research firm told *Fortune*. Apple's marketing manager suggested afterward that the company might want to buy an ad in the *Wall Street Journal* apologizing. Jay Chiat threatened that if Apple did that his agency would buy the facing page and apologize for the apology.

Jobs's discomfort, with both the ad and the situation at Apple in general, was on display when he traveled to New York in January to do another round of one-on-one press interviews. Andy Cunningham, from Regis McKenna's firm, was in charge of hand-holding and logistics at the Carlyle. When Jobs arrived, he told her that his suite needed to be completely redone, even though it was 10 p.m. and the meetings were to begin the next day. The piano was not in the right place; the strawberries were the wrong type. But his biggest objection was that he didn't like the flowers. He wanted calla lilies. "We got into a big fight on what a calla lily is," Cunningham recalled. "I know what they are, because I had them at my wedding, but he insisted on having a different type of lily and said I was 'stupid' because I didn't know what a real calla lily was." So Cunningham went out and, this being New York, was able to find a place open at midnight where she could get the lilies he wanted. By the time they got the room rearranged, Jobs started objecting to what she was wearing. "That suit's disgusting," he told her. Cunningham knew that at times he just simmered with undirected anger, so she tried to calm him down. "Look, I know you're angry, and I know how you feel," she said.

"You have no fucking idea how I feel," he shot back, "no fucking idea what it's like to be me."

THIRTY YEARS OLD

Turning thirty is a milestone for most people, especially those of the generation that proclaimed it would never trust anyone over that age. To celebrate his own thirtieth, in February 1985, Jobs threw a lavishly formal but also playful — black tie and tennis shoes — party for one thousand in the ballroom of the St. Francis Hotel in San Francisco. The invitation read, "There's an old Hindu saying that goes, 'In the first 30 years of your life, you make your habits. For the last 30 years of your life, your habits make you.' Come help me celebrate mine."

One table featured software moguls, including Bill Gates and Mitch Kapor. Another had old friends such as Elizabeth Holmes, who brought as her date a woman dressed in a tuxedo. Andy Hertzfeld and Burrell Smith had rented tuxes and wore floppy tennis shoes, which made it all the more memorable when they danced to the Strauss waltzes played by the San Francisco Symphony Orchestra.

Ella Fitzgerald provided the entertainment, as Bob Dylan had declined. She sang mainly from her standard repertoire, though occasionally tailoring a song like "The Girl from Ipanema" to be about the boy from Cupertino. When she asked for some requests, Jobs called out a few. She concluded with a slow rendition of "Happy Birthday."

Sculley came to the stage to propose a toast to "technology's foremost visionary." Wozniak also came up and presented Jobs with a framed copy of the Zaltair hoax from the 1977 West Coast Computer Faire, where the Apple II had been introduced. The venture capitalist Don Valentine marveled at the change in the decade since that time. "He went from being a Ho Chi Minh look-alike, who said never trust anyone over thirty, to a person who gives himself a fabulous thirtieth birthday with Ella Fitzgerald," he said.

Many people had picked out special gifts for a person who was not easy to shop for. Debi Coleman, for example, found a first edition of F. Scott Fitzgerald's *The Last Tycoon*. But Jobs, in an act that was odd yet not out of character, left all of the gifts in a hotel room. Wozniak and some of the Apple veterans, who did not take to the goat cheese and salmon mousse that was served, met after the party and went out to eat at a Denny's.

"It's rare that you see an artist in his 30s or 40s able to really contribute something amazing," Jobs said wistfully to the writer David Sheff, who published a long and intimate interview in *Playboy* the month he turned thirty. "Of course, there are some people who are innately curious, forever little kids in their awe of life, but they're rare." The interview touched on many subjects, but Jobs's most poignant ruminations were about growing old and facing the future:

> Your thoughts construct patterns like scaffolding in your mind. You are really etching chemical patterns. In most cases, people get stuck in those patterns, just like grooves in a record, and they never get out of them.
>
> I'll always stay connected with Apple. I hope that throughout my life I'll sort of have the thread of my life and the thread of Apple weave in and out of each other, like a tapestry. There may be a few years when I'm not there, but I'll always come back. . . .
>
> If you want to live your life in a creative way, as an artist, you have to not look back too much. You have to be willing to take whatever you've done and whoever you were and throw them away.
>
> The more the outside world tries to reinforce an image of you, the harder it is to continue to be an artist, which is why a lot of times, artists have to say, "Bye. I have to go. I'm going crazy and I'm getting out of here." And they go and hibernate somewhere. Maybe later they re-emerge a little differently.

With each of those statements, Jobs seemed to have a premonition that his life would soon be changing. Perhaps the thread of his life would indeed weave in and out of the thread of Apple's. Perhaps it was time to throw away some of what he had been. Perhaps it was time to say "Bye, I have to go," and then reemerge later, thinking differently.

Andy Hertzfeld had taken a leave of absence after the Macintosh came out in 1984. He needed to recharge his batteries and get away from his supervisor, Bob Belleville, whom he didn't like. One day he learned that Jobs had given out bonuses of up to $50,000 to engineers on the Macintosh team. So he went to Jobs to ask for one. Jobs responded that Belleville had decided not to give the bonuses to people who were on leave. Hertzfeld later heard that the decision had actually been made by Jobs, so he confronted him. At first Jobs equivocated, then he said, "Well, let's assume what you are saying is true. How does that change things?" Hertzfeld said that if Jobs was withholding the bonus as a reason for him to come back, then he wouldn't come back as a matter of principle. Jobs relented, but it left Hertzfeld with a bad taste.

When his leave was coming to an end, Hertzfeld made an appointment to have dinner with Jobs, and they walked from his office to an Italian restaurant a few blocks away. "I really want to return," he told Jobs. "But things seem really messed up right now." Jobs was vaguely annoyed and distracted, but Hertzfeld plunged ahead. "The software team is completely demoralized and has hardly done a thing for months, and Burrell is so frustrated that he won't last to the end of the year."

At that point Jobs cut him off. "You don't know what you're talking about!" he said. "The Macintosh team is doing great, and I'm having the best time of my life right now. You're just completely out of touch." His stare was withering, but he also tried to look amused at Hertzfeld's assessment.

"If you really believe that, I don't think there's any way that I can come back," Hertzfeld replied glumly. "The Mac team that I want to come back to doesn't even exist anymore."

"The Mac team had to grow up, and so do you," Jobs replied. "I want you to come back, but if you don't want to, that's up to you. You don't matter as much as you think you do, anyway."

Hertzfeld didn't come back.

By early 1985 Burrell Smith was also ready to leave. He had worried that it would be hard to quit if Jobs tried to talk him out of it; the reality distortion field was usually too strong for him to resist. So he plotted with Hertzfeld how he could break free of it. "I've got it!" he told Hertzfeld one day. "I know the perfect way to quit that will nullify the reality distortion field. I'll just walk into Steve's office, pull down my pants, and urinate on his desk. What could he say to that? It's guaranteed to work." The betting on the Mac team was that even brave Burrell Smith would not have the gumption to do that. When he finally decided he had to make his break, around the time of Jobs's birthday bash, he made an appointment to see Jobs. He was surprised to find Jobs smiling broadly when he walked in. "Are you gonna do it? Are you really gonna do it?" Jobs asked. He had heard about the plan.

Smith looked at him. "Do I have to? I'll do it if I have to." Jobs gave him a look, and Smith decided it wasn't necessary. So he resigned less dramatically and walked out on good terms.

He was quickly followed by another of the great Macintosh engineers, Bruce Horn. When Horn went in to say good-bye, Jobs told him, "Everything that's wrong with the Mac is your fault."

Horn responded, "Well, actually, Steve, a lot of things that are right with the Mac are my fault, and I had to fight like crazy to get those things in."

"You're right," admitted Jobs. "I'll give you 15,000 shares to stay." When Horn declined the offer, Jobs showed his warmer side. "Well, give me a hug," he said. And so they hugged.

But the biggest news that month was the departure from Apple, yet again, of its cofounder, Steve Wozniak. Wozniak was then quietly working as a midlevel engineer in the Apple II division, serving as a humble mascot of the roots of the company and staying as far away from management and corporate politics as he could. He felt, with justification, that Jobs was not appreciative of the Apple II, which remained the cash cow of the company and accounted for 70% of its sales at Christmas 1984. "People in the Apple II group were being treated as very unimportant by the rest of the company," he later said. "This was despite the fact that the Apple II was by far the largest-selling product in our company for ages, and would be for years to come." He even roused himself to do something out of character; he picked up the phone one day and called Sculley, berating him for lavishing so much attention on Jobs and the Macintosh division.

Frustrated, Wozniak decided to leave quietly to start a new company that would make a universal remote control device he had invented. It would control your television, stereo, and other electronic devices with a simple set of buttons that you could easily program. He informed the head of engineering at the Apple II division, but he didn't feel he was important enough to go out of channels and tell Jobs or Markkula. So Jobs first heard about it when the news leaked in the *Wall Street Journal*. In his earnest way, Wozniak had openly answered the reporter's questions when he called. Yes, he said, he felt that Apple had been giving short shrift to the Apple II division. "Apple's direction has been horrendously wrong for five years," he said.

Less than two weeks later Wozniak and Jobs traveled together to the White House, where Ronald Reagan presented them with the first National Medal of Technology. The president quoted what President Rutherford Hayes had said when first shown a telephone — "An amazing invention, but who would ever want to use one?" — and then quipped, "I thought at the time that he might be mistaken." Because of the awkward situation surrounding Wozniak's departure, Apple did not throw a celebratory dinner. So Jobs and Wozniak went for a walk afterward and ate at a sandwich shop. They chatted amiably, Wozniak recalled, and avoided any discussion of their disagreements.

Wozniak wanted to make the parting amicable. It was his style. So he agreed to stay on as a part-time Apple employee at a $20,000 salary and represent the company at events and trade shows. That could have been a graceful way to drift apart. But Jobs could not leave well enough alone. One Saturday, a few weeks after they had visited Washington together, Jobs went to the new Palo Alto studios of Hartmut Esslinger, whose company frogdesign had moved there to handle its design work for Apple. There he happened to see sketches that the firm had made for Wozniak's new remote control device, and he flew into a rage. Apple had a clause in its contract that gave it the right to bar frog-design from working on other computer-related projects, and Jobs invoked it. "I informed them," he recalled, "that working with Woz wouldn't be acceptable to us."

When the *Wall Street Journal* heard what happened, it got in touch with Wozniak, who, as usual, was open and honest. He said that Jobs was punishing him. "Steve Jobs has a hate for me, probably because of the things I said about Apple," he told the reporter. Jobs's action was remarkably petty, but it was also partly caused by the fact that he understood, in ways that others did not, that the look and style of a product served to brand it. A device that had Wozniak's name on it and used the same design language as Apple's products might be mistaken for something that Apple had produced. "It's not personal," Jobs told the newspaper, explaining that he wanted to make sure that Wozniak's remote wouldn't look like something made by Apple. "We don't want to see our design language used on other products. Woz has to find his own resources. He can't leverage off Apple's resources; we can't treat him specially."

Jobs volunteered to pay for the work that frog-design had already done for Wozniak, but even so the executives at the firm were taken aback. When Jobs demanded that they send him the drawings done for Wozniak or destroy them, they refused. Jobs had to send them a letter invoking Apple's contractual right. Herbert Pfeifer, the design director of the firm, risked Jobs's wrath by publicly dismissing his claim that the dispute with Wozniak was not personal. "It's a power play," Pfeifer told the *Journal*. "They have personal problems between them."

Hertzfeld was outraged when he heard what Jobs had done. He lived about twelve blocks from Jobs, who sometimes would drop by on his walks. "I got so furious about the Wozniak remote episode that when Steve next came over, I wouldn't let him in the house," Hertzfeld recalled. "He knew he was wrong, but he tried to rationalize, and maybe in his distorted reality he was able to." Wozniak, always a teddy bear even when annoyed, hired another design firm and even agreed to stay on Apple's retainer as a spokesman.

SHOWDOWN, SPRING 1985

There were many reasons for the rift between Jobs and Sculley in the spring of 1985. Some were merely business disagreements, such as Sculley's attempt to maximize profits by keeping the Macintosh price high when Jobs wanted to make it more affordable. Others were weirdly psychological and stemmed from the torrid and unlikely infatuation they initially had with each other. Sculley had painfully craved Jobs's affection, Jobs had eagerly sought a father figure and mentor, and when the ardor began to cool there was an emotional backwash. But at its core, the growing breach had two fundamental causes, one on each side.

For Jobs, the problem was that Sculley never became a product person. He didn't make the effort, or show the capacity, to understand the fine points of what they were making. On the contrary, he found Jobs's passion for tiny technical tweaks and design details to be obsessive and counterproductive. He had spent his career selling sodas and snacks whose recipes were largely irrelevant to him. He wasn't naturally passionate about products, which was among the most damning sins that Jobs could imagine. "I tried to educate him about the details of engineering," Jobs recalled, "but he had no idea how products are created, and after a while it just turned into arguments. But I learned that my perspective was right. Products are everything." He came to see Sculley as clueless, and his contempt was exacerbated by Sculley's hunger for his affection and delusions that they were very similar.

For Sculley, the problem was that Jobs, when he was no longer in courtship or manipulative mode, was frequently obnoxious, rude, selfish, and nasty to other people. He found Jobs's boorish behavior as despicable as Jobs found Sculley's lack of passion for product details. Sculley was kind, caring, and polite to a fault. At one point they were planning to meet with Xerox's vice chair Bill Glavin, and Sculley begged Jobs to behave. But as soon as they sat down, Jobs told Glavin, "You guys don't have any clue what you're doing," and the meeting broke up. "I'm sorry, but I couldn't help myself," Jobs told Sculley. It was one of many such cases. As Atari's Al Alcorn later observed, "Sculley believed in keeping people happy and worrying about relationships. Steve didn't give a shit about that. But he did care about the product in a way that Sculley never could, and he was able to avoid having too many bozos working at Apple by insulting anyone who wasn't an A player."

The board became increasingly alarmed at the turmoil, and in early 1985 Arthur Rock and some other disgruntled directors delivered a stern lecture to both. They told Sculley that he was supposed to be running the company, and he should start doing so with more authority and less eagerness to be pals with Jobs. They told Jobs that he was supposed to be fixing the mess at the Macintosh division and not telling other divisions how to do their job. Afterward Jobs retreated to his office and typed on his Macintosh, "I will not criticize the rest of the organization, I will not criticize the rest of the organization . . ."

As the Macintosh continued to disappoint — sales in March 1985 were only 10% of the budget forecast — Jobs holed up in his office fuming or wandered the halls berating everyone else for the problems. His mood swings became worse, and so did his abuse of those around him. Middle-level managers began to rise up against him. The marketing chief Mike Murray sought a private meeting with Sculley at an industry conference. As they were going up to Sculley's hotel room, Jobs spotted them and asked to come along. Murray asked him not to. He told Sculley that Jobs was wreaking havoc and had to be removed from managing the Macintosh division. Sculley replied that he was not yet resigned to having a showdown with Jobs. Murray later sent a memo directly to Jobs criticizing the way he treated colleagues and denouncing "management by character assassination."

For a few weeks it seemed as if there might be a solution to the turmoil. Jobs became fascinated by a flat-screen technology developed by a firm near Palo Alto called Woodside Design, run by an eccentric engineer named Steve Kitchen. He also was impressed by another startup that made a touch-screen display that could be controlled by your finger, so you didn't need a mouse. Together these might help fulfill Jobs's vision of creating a "Mac in a book." On a walk with Kitchen, Jobs spotted a building in nearby Menlo Park and declared that they should open a skunkworks facility to work on these ideas. It could be called AppleLabs and Jobs could run it, going back to the joy of having a small team and developing a great new product.

Sculley was thrilled by the possibility. It would solve most of his management issues, moving Jobs back to what he did best and getting rid of his disruptive presence in Cupertino. Sculley also had a candidate to replace Jobs as manager of the Macintosh division: Jean-Louis Gassée, Apple's chief in France, who had suffered through Jobs's visit there. Gassée flew to Cupertino and said he would take the job if he got a guarantee that he would run the division rather than work under Jobs. One of the board members, Phil Schlein of Macy's, tried to convince Jobs that he would be better off thinking up new products and inspiring a passionate little team.

But after some reflection, Jobs decided that was not the path he wanted. He declined to cede control to Gassée, who wisely went back to Paris to avoid the power clash that was becoming inevitable. For the rest of the spring, Jobs vacillated. There were times when he wanted to assert himself as a corporate manager, even writing a memo urging cost savings by eliminating free beverages and first-class air travel, and other times when he agreed with those who were encouraging him to go off and run a new AppleLabs R&D group.

In March Murray let loose with another memo that he marked "Do not circulate" but gave to multiple colleagues. "In my three years at Apple, I've never observed so much confusion, fear, and dysfunction as in the past 90 days," he began. "We are perceived by the rank and file as a boat without a rudder, drifting away into foggy oblivion." Murray had been on both sides of the fence; at times he conspired with Jobs to undermine Sculley, but in this memo he laid the blame on Jobs. "Whether the *cause of* or *because of* the dysfunction, Steve Jobs now controls a seemingly impenetrable power base."

At the end of that month, Sculley finally worked up the nerve to tell Jobs that he should give up running the Macintosh division. He walked over to Jobs's office one evening and brought the human resources manager, Jay Elliot, to make the confrontation more formal. "There is no one who admires your brilliance and vision more than I do," Sculley began. He had uttered such flatteries before, but this time it was clear that there would be a brutal "but" punctuating the thought. And there was. "But this is really not going to work," he declared. The flatteries punctured by "buts" continued. "We have developed a great friendship with each other," he said, "but I have lost confidence in your ability to run the Macintosh division." He also berated Jobs for badmouthing him as a bozo behind his back.

Jobs looked stunned and countered with an odd challenge, that Sculley should help and coach him more: "You've got to spend more time with me." Then he lashed back. He told Sculley he knew nothing about computers, was doing a terrible job running the company, and had disappointed Jobs ever since coming to Apple. Then he began to cry. Sculley sat there biting his fingernails.

"I'm going to bring this up with the board," Sculley declared. "I'm going to recommend that you step down from your operating position of running the Macintosh division. I want you to know that." He urged Jobs not to resist and to agree instead to work on developing new technologies and products.

Jobs jumped from his seat and turned his intense stare on Sculley. "I don't believe you're going to do that," he said. "If you do that, you're going to destroy the company."

Over the next few weeks Jobs's behavior fluctuated wildly. At one moment he would be talking about going off to run AppleLabs, but in the next moment he would be enlisting support to have Sculley ousted. He would reach out to Sculley, then lash out at him behind his back, sometimes on the same night. One night at 9 he called Apple's general counsel Al Eisenstat to say he was losing confidence in Sculley and needed his help convincing the board to fire him; at 11 the same night, he phoned Sculley to say, "You're terrific, and I just want you to know I love working with you."

At the board meeting on April 11, Sculley officially reported that he wanted to ask Jobs to step down as the head of the Macintosh division and focus instead on new product development. Arthur Rock, the most crusty and independent of the board members, then spoke. He was fed up with both of them: with Sculley for not having the guts to take command over the past year, and with Jobs for "acting like a petulant brat." The board needed to get this dispute behind them, and to do so it should meet privately with each of them.

Sculley left the room so that Jobs could present first. Jobs insisted that Sculley was the problem because he had no understanding of computers. Rock responded by berating Jobs. In his growling voice, he said that Jobs had been behaving foolishly for a year and had no right to be managing a division. Even Jobs's strongest supporter, Phil Schlein, tried to talk him into stepping aside gracefully to run a research lab for the company.

When it was Sculley's turn to meet privately with the board, he gave an ultimatum: "You can back me, and then I take responsibility for running the company, or we can do nothing, and you're going to have to find yourselves a new CEO." If given the authority, he said, he would not move abruptly, but would ease Jobs into the new role over the next few months. The board unanimously sided with Sculley. He was given the authority to remove Jobs whenever he felt the timing was right. As Jobs waited outside the boardroom, knowing full well that he was losing, he saw Del Yocam, a longtime colleague, and hugged him.

After the board made its decision, Sculley tried to be conciliatory. Jobs asked that the transition occur slowly, over the next few months, and Sculley agreed. Later that evening Sculley's executive assistant, Nanette Buckhout, called Jobs to see how he was doing. He was still in his office, shell-shocked. Sculley had already left, and Jobs came over to talk to her. Once again he began oscillating wildly in his attitude toward Sculley. "Why did John do this to me?" he said. "He betrayed me." Then he swung the other way. Perhaps he should take some time away to work on restoring his relationship with Sculley, he said. "John's friendship is more important than anything else, and I think maybe that's what I should do, concentrate on our friendship."

PLOTTING A COUP

Jobs was not good at taking no for an answer. He went to Sculley's office in early May 1985 and asked for more time to show that he could manage the Macintosh division. He would prove himself as an operations guy, he promised. Sculley didn't back down. Jobs next tried a direct challenge: He asked Sculley to resign. "I think you really lost your stride," Jobs told him. "You were really great the first year, and everything went wonderful. But something happened." Sculley, who generally was even-tempered, lashed back, pointing out that Jobs had been unable to get Macintosh software developed, come up with new models, or win customers. The meeting degenerated into a shouting match about who was the worse manager. After Jobs stalked out, Sculley turned away from the glass wall of his office, where others had been looking in on the meeting, and wept.

Matters began to come to a head on Tuesday, May 14, when the Macintosh team made its quarterly review presentation to Sculley and other Apple corporate leaders. Jobs still had not relinquished control of the division, and he was defiant when he arrived in the corporate boardroom with his team. He and Sculley began by clashing over what the division's mission was. Jobs said it was to sell more Macintosh machines. Sculley said it was to serve the interests of the Apple company as a whole. As usual there was little cooperation among the divisions; for one thing, the Macintosh team was planning new disk drives that were different from those being developed by the Apple II division. The debate, according to the minutes, took a full hour.

Jobs then described the projects under way: a more powerful Mac, which would take the place of the discontinued Lisa; and software called File-Server, which would allow Macintosh users to share files on a network. Sculley learned for the first time that these projects were going to be late. He gave a cold critique of Murray's marketing record, Belleville's missed engineering deadlines, and Jobs's overall management. Despite all this, Jobs ended the meeting with a plea to Sculley, in front of all the others there, to be given one more chance to prove he could run a division. Sculley refused.

That night Jobs took his Macintosh team out to dinner at Nina's Café in Woodside. Jean-Louis Gassée was in town because Sculley wanted him to prepare to take over the Macintosh division, and Jobs invited him to join them. Belleville proposed a toast "to those of us who really understand what the world according to Steve Jobs is all about."

That phrase — "the world according to Steve" — had been used dismissively by others at Apple who belittled the reality warp he created. After the others left, Belleville sat with Jobs in his Mercedes and urged him to organize a battle to the death with Sculley.

Months earlier, Apple had gotten the right to export computers to China, and Jobs had been invited to sign a deal in the Great Hall of the People over the 1985 Memorial Day weekend. He had told Sculley, who decided he wanted to go himself, which was just fine with Jobs. Jobs decided to use Sculley's absence to execute his coup. Throughout the week leading up to Memorial Day, he took a lot of people on walks to share his plans. "I'm going to launch a coup while John is in China," he told Mike Murray.

SEVEN DAYS IN MAY

Thursday, May 23: At his regular Thursday meeting with his top lieutenants in the Macintosh division, Jobs told his inner circle about his plan to oust Sculley. He also confided in the corporate human resources director, Jay Elliot, who told him bluntly that the proposed rebellion wouldn't work. Elliot had talked to some board members and urged them to stand up for Jobs, but he discovered that most of the board was with Sculley, as were most members of Apple's senior staff. Yet Jobs barreled ahead. He even revealed his plans to Gassée on a walk around the parking lot, despite the fact that Gassée had come from Paris to take his job. "I made the mistake of telling Gassée," Jobs wryly conceded years later.

That evening Apple's general counsel Al Eisenstat had a small barbecue at his home for Sculley, Gassée, and their wives. When Gassée told Eisenstat what Jobs was plotting, he recommended that Gassée inform Sculley. "Steve was trying to raise a cabal and have a coup to get rid of John," Gassée recalled. "In the den of Al Eisenstat's house, I put my index finger lightly on John's breastbone and said, 'If you leave tomorrow for China, you could be ousted. Steve's plotting to get rid of you.' "

Friday, May 24: Sculley canceled his trip and decided to confront Jobs at the executive staff meeting on Friday morning. Jobs arrived late, and he saw that his usual seat next to Sculley, who sat at the head of the table, was taken. He sat instead at the far end. He was dressed in a well-tailored suit and looked energized. Sculley looked pale. He announced that he was dispensing with the agenda to confront the issue on everyone's mind. "It's come to my attention that you'd like to throw me out of the company," he said, looking directly at Jobs. "I'd like to ask you if that's true."

Jobs was not expecting this. But he was never shy about indulging in brutal honesty. His eyes narrowed, and he fixed Sculley with his unblinking stare. "I think you're bad for Apple, and I think you're the wrong person to run the company," he replied, coldly and slowly. "You really should leave this company. You don't know how to operate and never have." He accused Sculley of not understanding the product development process, and then he added a self-centered swipe: "I wanted you here to help me grow, and you've been ineffective

303

in helping me."

As the rest of the room sat frozen, Sculley finally lost his temper. A childhood stutter that had not afflicted him for twenty years started to return. "I don't trust you, and I won't tolerate a lack of trust," he stammered. When Jobs claimed that he would be better than Sculley at running the company, Sculley took a gamble. He decided to poll the room on that question. "He pulled off this clever maneuver," Jobs recalled, still smarting twenty-five years later. "It was at the executive committee meeting, and he said, 'It's me or Steve, who do you vote for?' He set the whole thing up so that you'd kind of have to be an idiot to vote for me."

Suddenly the frozen onlookers began to squirm. Del Yocam had to go first. He said he loved Jobs, wanted him to continue to play some role in the company, but he worked up the nerve to conclude, with Jobs staring at him, that he "respected" Sculley and would support him to run the company. Eisenstat faced Jobs directly and said much the same thing: He liked Jobs but was supporting Sculley. Regis McKenna, who sat in on senior staff meetings as an outside consultant, was more direct. He looked at Jobs and told him he was not yet ready to run the company, something he had told him before. Others sided with Sculley as well. For Bill Campbell, it was particularly tough. He was fond of Jobs and didn't particularly like Sculley. His voice quavered a bit as he told Jobs he had decided to support Sculley, and he urged the two of them to work it out and find some role for Jobs to play in the company. "You can't let Steve leave this company," he told Sculley.

Jobs looked shattered. "I guess I know where things stand," he said, and bolted out of the room. No one followed.

He went back to his office, gathered his long-time loyalists on the Macintosh staff, and started to cry. He would have to leave Apple, he said. As he started to walk out the door, Debi Coleman restrained him. She and the others urged him to settle down and not do anything hasty. He should take the weekend to regroup. Perhaps there was a way to prevent the company from being torn apart.

Sculley was devastated by his victory. Like a wounded warrior, he retreated to Eisenstat's office and asked the corporate counsel to go for a ride. When they got into Eisenstat's Porsche, Sculley lamented, "I don't know whether I can go through with this." When Eisenstat asked what he meant, Sculley responded, "I think I'm going to resign."

"You can't," Eisenstat protested. "Apple will fall apart."

"I'm going to resign," Sculley declared. "I don't think I'm right for the company."

"I think you're copping out," Eisenstat replied. "You've got to stand up to him." Then he drove Sculley home.

Sculley's wife was surprised to see him back in the middle of the day. "I've failed," he said to her forlornly. She was a volatile woman who had never liked Jobs or appreciated her husband's infatuation with him. So when she heard what had happened, she jumped into her car and sped over to Jobs's office. Informed that he had gone to the Good Earth restaurant, she marched over there and confronted him in the parking lot as he was coming out with loyalists on his Macintosh team.

"Steve, can I talk to you?" she said. His jaw dropped. "Do you have any idea what a privilege it has been even to know someone as fine as John Sculley?" she demanded. He averted his gaze. "Can't you look me in the eyes when I'm talking to you?" she asked. But when Jobs did so — giving her his practiced, unblinking stare — she recoiled. "Never mind, don't look at me," she said. "When I look into most people's eyes, I see a soul. When I look into your eyes, I see a bottomless pit, an empty hole, a dead zone." Then she walked away.

Saturday, May 25: Mike Murray drove to Jobs's house in Woodside to offer some advice: He should consider accepting the role of being a new product visionary, starting AppleLabs, and getting away from headquarters. Jobs seemed willing to consider it. But first he would have to restore peace with Sculley. So he picked up the telephone and surprised Sculley with an olive branch. Could they meet the following afternoon, Jobs asked, and take a walk together in the hills above Stanford University. They had walked there in the past, in happier times, and maybe on such a walk they could work things out.

Jobs did not know that Sculley had told Eisenstat he wanted to quit, but by then it didn't matter. Overnight, he had changed his mind and decided to stay. Despite the blowup the day before, he was still eager for Jobs to like him. So he agreed to meet the next afternoon.

If Jobs was prepping for conciliation, it didn't show in the choice of movie he wanted to see with

Murray that night. He picked *Patton,* the epic of the never-surrender general. But he had lent his copy of the tape to his father, who had once ferried troops for the general, so he drove to his childhood home with Murray to retrieve it. His parents weren't there, and he didn't have a key. They walked around the back, checked for unlocked doors or windows, and finally gave up. The video store didn't have a copy of *Patton* in stock, so in the end he had to settle for watching the 1983 film adaptation of Harold Pinter's *Betrayal.*

Sunday, May 26: As planned, Jobs and Sculley met in back of the Stanford campus on Sunday afternoon and walked for several hours amid the rolling hills and horse pastures. Jobs reiterated his plea that he should have an operational role at Apple. This time Sculley stood firm. It won't work, he kept saying. Sculley urged him to take the role of being a product visionary with a lab of his own, but Jobs rejected this as making him into a mere "figurehead." Defying all connection to reality, he countered with the proposal that Sculley give up control of the entire company to him. "Why don't you become chairman and I'll become president and chief executive officer?" he suggested. Sculley was struck by how earnest he seemed.

"Steve, that doesn't make any sense," Sculley replied. Jobs then proposed that they split the duties of running the company, with him handling the product side and Sculley handling marketing and business. But the board had not only emboldened Sculley, it had ordered him to bring Jobs to heel. "One person has got to run the company," he re-

307

plied. "I've got the support and you don't."

On his way home, Jobs stopped at Mike Markkula's house. He wasn't there, so Jobs left a message asking him to come to dinner the following evening. He would also invite the core of loyalists from his Macintosh team. He hoped that they could persuade Markkula of the folly of siding with Sculley.

Monday, May 27: Memorial Day was sunny and warm. The Macintosh team loyalists — Debi Coleman, Mike Murray, Susan Barnes, and Bob Belleville — got to Jobs's Woodside home an hour before the scheduled dinner so they could plot strategy. Sitting on the patio as the sun set, Coleman told Jobs that he should accept Sculley's offer to be a product visionary and help start up AppleLabs. Of all the inner circle, Coleman was the most willing to be realistic. In the new organization plan, Sculley had tapped her to run the manufacturing division because he knew that her loyalty was to Apple and not just to Jobs. Some of the others were more hawkish. They wanted to urge Markkula to support a reorganization plan that put Jobs in charge.

When Markkula showed up, he agreed to listen with one proviso: Jobs had to keep quiet. "I seriously wanted to hear the thoughts of the Macintosh team, not watch Jobs enlist them in a rebellion," he recalled. As it turned cooler, they went inside the sparsely furnished mansion and sat by a fireplace. Instead of letting it turn into a gripe session, Markkula made them focus on very specific management issues, such as what had caused the problem in producing the FileServer software and why the Macintosh distribution system had not responded well to the change in demand. When they were finished, Markkula bluntly declined to back Jobs. "I said I wouldn't support his plan, and that was the end of that," Markkula recalled. "Sculley was the boss. They were mad and emotional and putting together a revolt, but that's not how you do things."

Tuesday, May 28: His ire stoked by hearing from Markkula that Jobs had spent the previous evening trying to subvert him, Sculley walked over to Jobs's office on Tuesday morning. He had talked to the board, he said, and he had its support. He wanted Jobs out. Then he drove to Markkula's house, where he gave a presentation of his reorganization plans. Markkula asked detailed questions, and at the end he gave Sculley his blessing. When he got back to his office, Sculley called the other members of the board, just to make sure he still had their backing. He did.

At that point he called Jobs to make sure he understood. The board had given final approval of his reorganization plan, which would proceed that week. Gassée would take over control of Jobs's beloved Macintosh as well as other products, and there was no other division for Jobs to run. Sculley was still somewhat conciliatory. He told Jobs that he could stay on with the title of board chairman and be a product visionary with no operational duties. But by this point, even the idea of starting a skunkworks such as AppleLabs was no longer on the table.

It finally sank in. Jobs realized there was no appeal, no way to warp the reality. He broke down in tears and started making phone calls — to Bill Campbell, Jay Elliot, Mike Murray, and others. Murray's wife, Joyce, was on an overseas call when Jobs phoned, and the operator broke in saying it was an emergency. It better be important, she told the operator. "It is," she heard Jobs say. When her husband got on the phone, Jobs was crying. "It's over," he said. Then he hung up.

Murray was worried that Jobs was so despondent he might do something rash, so he called back. There was no answer, so he drove to Woodside. No one came to the door when he knocked, so he went around back and climbed up some exterior steps and looked in the bedroom. Jobs was lying there on a mattress in his unfurnished room. He let Murray in and they talked until almost dawn.

Wednesday, May 29: Jobs finally got hold of a tape of *Patton,* which he watched Wednesday evening, but Murray prevented him from getting stoked up for another battle. Instead he urged Jobs to come in on Friday for Sculley's announcement of the reorganization plan. There was no option left other than to play the good soldier rather than the renegade commander.

LIKE A ROLLING STONE

Jobs slipped quietly into the back row of the auditorium to listen to Sculley explain to the troops the new order of battle. There were a lot of sideways glances, but few people acknowledged him and none came over to provide public displays of

310

affection. He stared without blinking at Sculley, who would remember "Steve's look of contempt" years later. "It's unyielding," Sculley recalled, "like an X-ray boring inside your bones, down to where you're soft and destructibly mortal." For a moment, standing onstage while pretending not to notice Jobs, Sculley thought back to a friendly trip they had taken a year earlier to Cambridge, Massachusetts, to visit Jobs's hero, Edwin Land. He had been dethroned from the company he created, Polaroid, and Jobs had said to Sculley in disgust, "All he did was blow a lousy few million and they took his company away from him." Now, Sculley reflected, he was taking Jobs's company away from him.

As Sculley went over the organizational chart, he introduced Gassée as the new head of a combined Macintosh and Apple II product group. On the chart was a small box labeled "chairman" with no lines connecting to it, not to Sculley or to anyone else. Sculley briefly noted that in that role, Jobs would play the part of "global visionary." But he didn't acknowledge Jobs's presence. There was a smattering of awkward applause.

Jobs stayed home for the next few days, blinds drawn, his answering machine on, seeing only his girlfriend, Tina Redse. For hours on end he sat there playing his Bob Dylan tapes, especially "The Times They Are a-Changin.'" He had recited the second verse the day he unveiled the Macintosh to the Apple shareholders sixteen months earlier. That verse ended nicely: "For the loser now / Will be later to win. . . ."

A rescue squad from his former Macintosh posse

arrived to dispel the gloom on Sunday night, led by Andy Hertzfeld and Bill Atkinson. Jobs took a while to answer their knock, and then he led them to a room next to the kitchen that was one of the few places with any furniture. With Redse's help, he served some vegetarian food he had ordered. "So what really happened?" Hertzfeld asked. "Is it really as bad as it looks?"

"No, it's worse." Jobs grimaced. "It's much worse than you can imagine." He blamed Sculley for betraying him, and said that Apple would not be able to manage without him. His role as chairman, he complained, was completely ceremonial. He was being ejected from his Bandley 3 office to a small and almost empty building he nicknamed "Siberia." Hertzfeld turned the topic to happier days, and they began to reminisce about the past.

Earlier that week, Dylan had released a new album, *Empire Burlesque,* and Hertzfeld brought a copy that they played on Jobs's high-tech turntable. The most notable track, "When the Night Comes Falling from the Sky," with its apocalyptic message, seemed appropriate for the evening, but Jobs didn't like it. It sounded almost disco, and he gloomily argued that Dylan had been going downhill since *Blood on the Tracks*. So Hertzfeld moved the needle to the last song on the album, "Dark Eyes," which was a simple acoustic number featuring Dylan alone on guitar and harmonica. It was slow and mournful and, Hertzfeld hoped, would remind Jobs of the earlier Dylan tracks he so loved. But Jobs didn't like that song either and had no desire to hear the rest of the album.

Jobs's overwrought reaction was understandable.

312

Sculley had once been a father figure to him. So had Mike Markkula. So had Arthur Rock. That week all three had abandoned him. "It gets back to the deep feeling of being rejected at an early age," his friend and lawyer George Riley later said. "It's a deep part of his own mythology, and it defines to himself who he is." Jobs recalled years later, "I felt like I'd been punched, the air knocked out of me and I couldn't breathe."

Losing the support of Arthur Rock was especially painful. "Arthur had been like a father to me," Jobs said. "He took me under his wing." Rock had taught him about opera, and he and his wife, Toni, had been his hosts in San Francisco and Aspen. "I remember driving into San Francisco one time, and I said to him, 'God, that Bank of America building is ugly,' and he said, 'No, it's the best,' and he proceeded to lecture me, and he was right of course." Years later Jobs's eyes welled with tears as he recounted the story: "He chose Sculley over me. That really threw me for a loop. I never thought he would abandon me."

Making matters worse was that his beloved company was now in the hands of a man he considered a bozo. "The board felt that I couldn't run a company, and that was their decision to make," he said. "But they made one mistake. They should have separated the decision of what to do with me and what to do with Sculley. They should have fired Sculley, even if they didn't think I was ready to run Apple." Even as his personal gloom slowly lifted, his anger at Sculley, his feeling of betrayal, deepened.

The situation worsened when Sculley told a

group of analysts that he considered Jobs irrelevant to the company, despite his title as chairman. "From an operations standpoint, there is no role either today or in the future for Steve Jobs," he said. "I don't know what he'll do." The blunt comment shocked the group, and a gasp went through the auditorium.

Perhaps getting away to Europe would help, Jobs thought. So in June he went to Paris, where he spoke at an Apple event and went to a dinner honoring Vice President George H. W. Bush. From there he went to Italy, where he drove the hills of Tuscany with Redse and bought a bike so he could spend time riding by himself. In Florence he soaked in the architecture of the city and the texture of the building materials. Particularly memorable were the paving stones, which came from Il Casone quarry near the Tuscan town of Firenzuola. They were a calming bluish gray. Twenty years later he would decide that the floors of most major Apple stores would be made of this sandstone.

The Apple II was just going on sale in Russia, so Jobs headed off to Moscow, where he met up with Al Eisenstat. Because there was a problem getting Washington's approval for some of the required export licenses, they visited the commercial attaché at the American embassy in Moscow, Mike Merwin. He warned them that there were strict laws against sharing technology with the Soviets. Jobs was annoyed. At the Paris trade show, Vice President Bush had encouraged him to get computers into Russia in order to "foment revolution from below." Over dinner at a Georgian restaurant that specialized in shish kebab, Jobs continued his rant.

"How could you suggest this violates American law when it so obviously benefits our interests?" he asked Merwin. "By putting Macs in the hands of Russians, they could print all their newspapers."

Jobs also showed his feisty side in Moscow by insisting on talking about Trotsky, the charismatic revolutionary who fell out of favor and was ordered assassinated by Stalin. At one point the KGB agent assigned to him suggested he tone down his fervor. "You don't want to talk about Trotsky," he said. "Our historians have studied the situation, and we don't believe he's a great man anymore." That didn't help. When they got to the state university in Moscow to speak to computer students, Jobs began his speech by praising Trotsky. He was a revolutionary Jobs could identify with.

Jobs and Eisenstat attended the July Fourth party at the American embassy, and in his thank-you letter to Ambassador Arthur Hartman, Eisenstat noted that Jobs planned to pursue Apple's ventures in Russia more vigorously in the coming year. "We are tentatively planning on returning to Moscow in September." For a moment it looked as if Sculley's hope that Jobs would turn into a "global visionary" for the company might come to pass. But it was not to be. Something much different was in store for September.

Chapter Eighteen: NeXT
PROMETHEUS UNBOUND

The Pirates Abandon Ship

Upon his return from Europe in August 1985, while he was casting about for what to do next, Jobs called the Stanford biochemist Paul Berg to discuss the advances that were being made in gene splicing and recombinant DNA. Berg described how difficult it was to do experiments in a biology lab, where it could take weeks to nurture an experiment and get a result. "Why don't you simulate them on a computer?" Jobs asked. Berg replied that computers with such capacities were too expensive for university labs. "Suddenly, he was excited about the possibilities," Berg recalled. "He had it in his mind to start a new company. He was young and rich, and had to find something to do with the rest of his life."

Jobs had already been canvassing academics to ask what their workstation needs were. It was something he had been interested in since 1983, when he had visited the computer science department at Brown to show off the Macintosh, only to be told that it would take a far more powerful machine to do anything useful in a university lab. The dream of academic researchers was to have a workstation that was both powerful and personal. As head of the Macintosh division, Jobs had launched a project to build such a machine, which was dubbed the Big Mac. It would have a UNIX operating system but with the friendly Macintosh interface. But after Jobs was ousted from the Macintosh division, his replacement, Jean-Louis Gassée, canceled the Big Mac.

When that happened, Jobs got a distressed call from Rich Page, who had been engineering the Big Mac's chip set. It was the latest in a series of conversations that Jobs was having with disgruntled Apple employees urging him to start a new company and rescue them. Plans to do so began to jell over Labor Day weekend, when Jobs spoke to Bud Tribble, the original Macintosh software chief, and floated the idea of starting a company to build a powerful but personal workstation. He also enlisted two other Macintosh division employees who had been talking about leaving, the engineer George Crow and the controller Susan Barnes.

That left one key vacancy on the team: a person who could market the new product to universities. The obvious candidate was Dan'l Lewin, who at Apple had organized a consortium of universities to buy Macintosh computers in bulk. Besides missing two letters in his first name, Lewin had the chiseled good looks of Clark Kent and a Princetonian's polish. He and Jobs shared a bond: Lewin had written a Princeton thesis on Bob Dylan and charismatic leadership, and Jobs knew something about both of those topics.

Lewin's university consortium had been a godsend to the Macintosh group, but he had become frustrated after Jobs left and Bill Campbell had reorganized marketing in a way that reduced the role of direct sales to universities. He had been meaning to call Jobs when, that Labor Day weekend, Jobs called first. He drove to Jobs's unfurnished mansion, and they walked the grounds while discussing the possibility of creating a new company. Lewin was excited, but not ready to commit. He was going to Austin with Campbell the following week, and he wanted to wait until then to decide. Upon his return, he gave his answer: He was in. The news came just in time for the September 13 Apple board meeting.

Although Jobs was still nominally the board's chairman, he had not been to any meetings since he lost power. He called Sculley, said he was going to attend, and asked that an item be added to the end of the agenda for a "chairman's report." He didn't say what it was about, and Sculley assumed it would be a criticism of the latest reorganization. Instead, when his turn came to speak, Jobs described to the board his plans to start a new company. "I've been thinking a lot, and it's time for me to get on with my life," he began. "It's obvious that I've got to do something. I'm thirty years old." Then he referred to some prepared notes to describe his plan to create a computer for the higher education market. The new company would not be competitive with Apple, he promised, and he would take with him only a handful of non-key personnel. He offered to resign as chairman of Apple, but he expressed hope that they could work together. Perhaps Apple would want to buy the distribution rights to his product, he suggested, or license Macintosh software to it.

Mike Markkula rankled at the possibility that Jobs would hire anyone from Apple. "Why would you take anyone at all?" he asked.

"Don't get upset," Jobs assured him and the rest of the board. "These are very low-level people that you won't miss, and they will be leaving anyway."

The board initially seemed disposed to wish Jobs well in his venture. After a private discussion, the directors even proposed that Apple take a 10% stake in the new company and that Jobs remain on the board.

That night Jobs and his five renegades met again at his house for dinner. He was in favor of taking the Apple investment, but the others convinced him it was unwise. They also agreed that it would be best if they resigned all at once, right away. Then they could make a clean break.

So Jobs wrote a formal letter telling Sculley the names of the five who would be leaving, signed it in his spidery lowercase signature, and drove to Apple the next morning to hand it to him before his 7:30 staff meeting.

"Steve, these are not low-level people," Sculley said.

"Well, these people were going to resign any-way," Jobs replied. "They are going to be handing in their resignations by nine this morning."

From Jobs's perspective, he had been honest. The five were not division managers or members of Sculley's top team. They had all felt diminished, in fact, by the company's new organization. But from Sculley's perspective, these were important players; Page was an Apple Fellow, and Lewin was a key to the higher education market. In addition, they knew about the plans for Big Mac; even though it had been shelved, this was still proprietary information. Nevertheless Sculley was sanguine. Instead of pushing the point, he asked Jobs to remain on the board. Jobs replied that he would think about it.

But when Sculley walked into his 7:30 staff meeting and told his top lieutenants who was leaving, there was an uproar. Most of them felt that Jobs had breached his duties as chairman and displayed stunning disloyalty to the company. "We should expose him for the fraud that he is so that people here stop regarding him as a messiah," Campbell shouted, according to Sculley.

Campbell admitted that, although he later became a great Jobs defender and supportive board member, he was ballistic that morning. "I was fucking furious, especially about him taking Dan'l Lewin," he recalled. "Dan'l had built the relationships with the universities. He was always muttering about how hard it was to work with Steve, and then he left." Campbell was so angry that he walked out of the meeting to call Lewin at home. When his wife said he was in the shower, Campbell said, "I'll wait." A few minutes later, when she said he was still in the shower, Campbell again said, "I'll wait." When Lewin finally came on the phone, Campbell asked him if it was true. Lewin acknowledged it was. Campbell hung up without saying another word.

After hearing the fury of his senior staff, Sculley surveyed the members of the board. They likewise felt that Jobs had misled them with his pledge that he would not raid important employees. Arthur Rock was especially angry. Even though he had sided with Sculley during the Memorial Day showdown, he had been able to repair his paternal relationship with Jobs. Just the week before, he had invited Jobs to bring his girlfriend up to San Francisco so that he and his wife could meet her, and the four had a nice dinner in Rock's Pacific Heights home. Jobs had not mentioned the new company he was forming, so Rock felt betrayed when he heard about it from Sculley. "He came to the board and lied to us," Rock growled later. "He told us he was thinking of forming a company when in fact he had already formed it. He said he was going to take a few middle-level people. It turned out to be five senior people." Markkula, in his subdued way, was also offended. "He took some top executives he had secretly lined up before he left. That's not the way you do things. It was ungentlemanly."

Over the weekend both the board and the executive staff convinced Sculley that Apple would have to declare war on its cofounder. Markkula issued a formal statement accusing Jobs of acting "in direct contradiction to his statements that he wouldn't recruit any key Apple personnel for his company." He added ominously, "We are evaluating what possible actions should be taken." Campbell was quoted in the *Wall Street Journal* as saying he "was stunned and shocked" by Jobs's behavior.

Jobs had left his meeting with Sculley thinking that things might proceed smoothly, so he had kept quiet. But after reading the newspapers, he felt that he had to respond. He phoned a few favored reporters and invited them to his home for private briefings the next day. Then he called Andy Cunningham, who had handled his publicity at Regis McKenna. "I went over to his unfurnished mansiony place in Woodside," she recalled, "and I found him huddled in the kitchen with his five colleagues and a few reporters hanging outside on the lawn." Jobs told her that he was going to do a full-fledged press conference and started spewing some of the derogatory things he was going to say. Cunningham was appalled. "This is going to reflect badly on you," she told him. Finally he backed down. He decided that he would give the reporters a copy of the resignation letter and limit any on-the-record comments to a few bland statements.

Jobs had considered just mailing in his letter of resignation, but Susan Barnes convinced him that this would be too contemptuous. Instead he drove it to Markkula's house, where he also found Al Eisenstat. There was a tense conversation for about fifteen minutes; then Barnes, who had been waiting outside, came to the door to retrieve him before he said anything he would regret. He left behind the letter, which he had composed on a Macintosh and printed on the new LaserWriter:

September 17, 1985

Dear Mike:

This morning's papers carried suggestions that Apple is considering removing me as Chairman. I don't know the source of these reports but they are both misleading to the public and unfair to me.

322

You will recall that at last Thursday's Board meeting I stated I had decided to start a new venture and I tendered my resignation as Chairman.

The Board declined to accept my resignation and asked me to defer it for a week. I agreed to do so in light of the encouragement the Board offered with regard to the proposed new venture and the indications that Apple would invest in it. On Friday, after I told John Sculley who would be joining me, he confirmed Apple's willingness to discuss areas of possible collaboration between Apple and my new venture.

Subsequently the Company appears to be adopting a hostile posture toward me and the new venture. Accordingly, I must insist upon the immediate acceptance of my resignation. . . .

As you know, the company's recent reorganization left me with no work to do and no access even to regular management reports. I am but 30 and want still to contribute and achieve.

After what we have accomplished together, I would wish our parting to be both amicable and dignified.

Yours sincerely, steven p. jobs

When a guy from the facilities team went to Jobs's office to pack up his belongings, he saw a picture frame on the floor. It contained a photograph of Jobs and Sculley in warm conversation, with an inscription from seven months earlier: "Here's to Great Ideas, Great Experiences, and a Great Friendship! John." The glass frame was shattered. Jobs had hurled it across the room before leaving. From that day, he never spoke to Sculley again.

Apple's stock went up a full point, or almost 7%, when Jobs's resignation was announced. "East Coast stockholders always worried about California flakes running the company," explained the editor of a tech stock newsletter. "Now with both Wozniak and Jobs out, those shareholders are relieved." But Nolan Bushnell, the Atari founder who had been an amused mentor ten years earlier, told *Time* that Jobs would be badly missed. "Where is Apple's inspiration going to come from? Is Apple going to have all the romance of a new brand of Pepsi?"

After a few days of failed efforts to reach a settlement with Jobs, Sculley and the Apple board decided to sue him "for breaches of fiduciary obligations." The suit spelled out his alleged transgressions:

Notwithstanding his fiduciary obligations to Apple, Jobs, while serving as the Chairman of Apple's Board of Directors and an officer of Apple and pretending loyalty to the interests of Apple . . .

(a) secretly planned the formation of an enterprise to compete with Apple;

(b) secretly schemed that his competing enterprise would wrongfully take advantage of and utilize Apple's plan to design, develop and market the Next Generation Product . . .

(c) secretly lured away key employees of Apple.

At the time, Jobs owned 6.5 million shares of Apple stock, 11% of the company, worth more than $100 million. He began to sell his shares, and within five months had dumped them all, retaining only one share so he could attend shareholder meetings if he wanted. He was furious, and that was reflected in his passion to start what was, no matter how he spun it, a rival company. "He was angry at Apple," said Joanna Hoffman, who briefly went to work for the new company. "Aiming at the educational market, where Apple was strong, was simply Steve being vengeful. He was doing it for revenge."

Jobs, of course, didn't see it that way. "I haven't got any sort of odd chip on my shoulder," he told *Newsweek*. Once again he invited his favorite reporters over to his Woodside home, and this time he did not have Andy Cunningham there urging him to be circumspect. He dismissed the allegation that he had improperly lured the five colleagues from Apple. "These people all called me," he told the gaggle of journalists who were milling around in his unfurnished living room. "They were thinking of leaving the company. Apple has a way of neglecting people."

He decided to cooperate with a *Newsweek* cover in order to get his version of the story out, and the interview he gave was revealing. "What I'm best at doing is finding a group of talented people and making things with them," he told the magazine. He said that he would always harbor affection for Apple. "I'll always remember Apple like any man remembers the first woman he's fallen in love with." But he was also willing to fight with its management if need be. "When someone calls you a thief in public, you have to respond." Apple's threat to sue him was outrageous. It was also sad. It showed that Apple was no longer a confident, rebellious company. "It's hard to think that a $2 billion company with 4,300 employees couldn't compete with six people in blue jeans."

To try to counter Jobs's spin, Sculley called Wozniak and urged him to speak out. "Steve can be an insulting and hurtful guy," he told *Time* that week. He revealed that Jobs had asked him to join his new firm — it would have been a sly way to land another blow against Apple's current management — but he wanted no part of such games and had not returned Jobs's phone call. To the *San Francisco Chronicle,* he recounted how Jobs had blocked frogdesign from working on his remote control under the pretense that it might compete with Apple products. "I look forward to a great product and I wish him success, but his integrity I cannot trust," Wozniak said.

TO BE ON YOUR OWN

"The best thing ever to happen to Steve is when we fired him, told him to get lost," Arthur Rock later said. The theory, shared by many, is that the tough love made him wiser and more mature. But it's not that simple. At the company he founded after being ousted from Apple, Jobs was able to indulge all of his instincts, both good and bad. He was unbound. The result was a series of spectacular products that were dazzling market flops. *This* was the true learning experience. What prepared him for the great success he would have in Act III was not his ouster from his Act I at Apple but his brilliant failures in Act II.

The first instinct that he indulged was his passion for design. The name he chose for his new company was rather straightforward: Next. In order to make it more distinctive, he decided he needed a world-class logo. So he courted the dean of corporate logos, Paul Rand. At seventy-one, the Brooklyn-born graphic designer had already created some of the best-known logos in business, including those of *Esquire,* IBM, Westinghouse, ABC, and UPS. He was under contract to IBM, and his supervisors there said that it would obviously be a conflict for him to create a logo for another computer company. So Jobs picked up the phone and called IBM's CEO, John Akers. Akers was out of town, but Jobs was so persistent that he was finally put through to Vice Chairman Paul Rizzo. After two days, Rizzo concluded that it was futile to resist Jobs, and he gave permission for Rand to do the work.

Rand flew out to Palo Alto and spent time walking with Jobs and listening to his vision. The computer would be a cube, Jobs pronounced. He loved that shape. It was perfect and simple. So Rand decided that the logo should be a cube as well, one that was tilted at a 28° angle. When Jobs asked for a number of options to consider, Rand declared that he did not create different *options* for clients. "I will solve your problem, and you will pay me," he told Jobs. "You can use what I produce, or not, but I will not do options, and either way you will pay me."

Jobs admired that kind of thinking, so he made what was quite a gamble. The company would pay an astonishing $100,000 flat fee to get *one* design.

"There was a clarity in our relationship," Jobs said. "He had a purity as an artist, but he was astute at solving business problems. He had a tough exterior, and had perfected the image of a curmudgeon, but he was a teddy bear inside." It was one of Jobs's highest praises: purity as an artist.

It took Rand just two weeks. He flew back to deliver the result to Jobs at his Woodside house. First they had dinner, then Rand handed him an elegant and vibrant booklet that described his thought process. On the final spread, Rand presented the logo he had chosen. "In its design, color arrangement, and orientation, the logo is a study in contrasts," his booklet proclaimed. "Tipped at a jaunty angle, it brims with the informality, friendliness, and spontaneity of a Christmas seal and the authority of a rubber stamp." The word "next" was split into two lines to fill the square face of the cube, with only the "e" in lowercase. That letter stood out, Rand's booklet explained, to connote "education, excellence . . . e = mc^2."

It was often hard to predict how Jobs would react to a presentation. He could label it shitty or brilliant; one never knew which way he might go. But with a legendary designer such as Rand, the chances were that Jobs would embrace the proposal. He stared at the final spread, looked up at Rand, and then hugged him. They had one minor disagreement: Rand had used a dark yellow for the "e" in the logo, and Jobs wanted him to change it to a brighter and more traditional yellow. Rand banged his fist on the table and declared, "I've been doing this for fifty years, and I know what I'm doing." Jobs relented.

328

The company had not only a new logo, but a new name. No longer was it Next. It was NeXT. Others might not have understood the need to obsess over a logo, much less pay $100,000 for one. But for Jobs it meant that NeXT was starting life with a world-class feel and identity, even if it hadn't yet designed its first product. As Markkula had taught him, a great company must be able to impute its values from the first impression it makes.

As a bonus, Rand agreed to design a personal calling card for Jobs. He came up with a colorful type treatment, which Jobs liked, but they ended up having a lengthy and heated disagreement about the placement of the period after the "P" in Steven P. Jobs. Rand had placed the period to the right of the "P.", as it would appear if set in lead type. Steve preferred the period to be nudged to the left, under the curve of the "P.", as is possible with digital typography. "It was a fairly large argument about something relatively small," Susan Kare recalled. On this one Jobs prevailed.

In order to translate the NeXT logo into the look of real products, Jobs needed an industrial designer he trusted. He talked to a few possibilities, but none of them impressed him as much as the wild Bavarian he had imported to Apple: Hartmut Esslinger, whose frogdesign had set up shop in Silicon Valley and who, thanks to Jobs, had a lucrative contract with Apple. Getting IBM to permit Paul Rand to do work for NeXT was a small miracle willed into existence by Jobs's belief that reality can be distorted. But that was a snap compared to the likelihood that he could convince Apple to permit Esslinger to work for NeXT.

This did not keep Jobs from trying. At the beginning of November 1985, just five weeks after Apple filed suit against him, Jobs wrote to Eisenstat and asked for a dispensation. "I spoke with Hartmut Esslinger this weekend and he suggested I write you a note expressing why I wish to work with him and frogdesign on the new products for NeXT," he said. Astonishingly, Jobs's argument was that he did not know what Apple had in the works, but Esslinger did. "NeXT has no knowledge as to the current or future directions of Apple's product designs, nor do other design firms we might deal with, so it is possible to inadvertently design similar looking products. It is in both Apple's and NeXT's best interest to rely on Hartmut's professionalism to make sure this does not occur." Eisenstat recalled being flabbergasted by Jobs's audacity, and he replied curtly. "I have previously expressed my concern on behalf of Apple that you are engaged in a business course which involves your utilization of Apple's confidential business information," he wrote. "Your letter does not alleviate my concern in any way. In fact it heightens my concern because it states that you have 'no knowledge as to the current or future directions of Apple's product designs,' a statement which is not true." What made the request all the more astonishing to Eisenstat was that it was Jobs who, just a year earlier, had forced frogdesign to abandon its work on Wozniak's remote control device.

Jobs realized that in order to work with Esslinger (and for a variety of other reasons), it would be necessary to resolve the lawsuit that Apple had filed. Fortunately Sculley was willing. In January 1986 they reached an out-of-court agreement involving no financial damages. In return for Apple's dropping its suit, NeXT agreed to a variety of restrictions: Its product would be marketed as a high-end workstation, it would be sold directly to colleges and universities, and it would not ship before March 1987. Apple also insisted that the NeXT machine "not use an operating system compatible with the Macintosh," though it could be argued that Apple would have been better served by insisting on just the opposite.

After the settlement Jobs continued to court Esslinger until the designer decided to wind down his contract with Apple. That allowed frogdesign to work with NeXT at the end of 1986. Esslinger insisted on having free rein, just as Paul Rand had. "Sometimes you have to use a big stick with Steve," he said. Like Rand, Esslinger was an artist, so Jobs was willing to grant him indulgences he denied other mortals.

Jobs decreed that the computer should be an absolutely perfect cube, with each side exactly a foot long and every angle precisely 90 degrees. He liked cubes. They had gravitas but also the slight whiff of a toy. But the NeXT cube was a Jobsian example of design desires trumping engineering considerations. The circuit boards, which fitted nicely into the traditional pizza-box shape, had to be reconfigured and stacked in order to nestle into a cube.

Even worse, the perfection of the cube made it hard to manufacture. Most parts that are cast in molds have angles that are slightly greater than pure 90 degrees, so that it's easier to get them out of the mold (just as it is easier to get a cake out of a pan that has angles slightly greater than 90 degrees). But Esslinger dictated, and Jobs enthusiastically agreed, that there would be no such "draft angles" that would ruin the purity and perfection of the cube. So the sides had to be produced separately, using molds that cost $650,000, at a specialty machine shop in Chicago. Jobs's passion for perfection was out of control. When he noticed a tiny line in the chassis caused by the molds, something that any other computer maker would accept as unavoidable, he flew to Chicago and convinced the die caster to start over and do it perfectly. "Not a lot of die casters expect a celebrity to fly in," noted one of the engineers. Jobs also had the company buy a $150,000 sanding machine to remove all lines where the mold faces met and insisted that the magnesium case be a matte black, which made it more susceptible to showing blemishes.

Jobs had always indulged his obsession that the unseen parts of a product should be crafted as beautifully as its façade, just as his father had taught him when they were building a fence. This too he took to extremes when he found himself unfettered at NeXT. He made sure that the screws inside the machine had expensive plating. He even insisted that the matte black finish be coated onto the inside of the cube's case, even though only repairmen would see it.

Joe Nocera, then writing for *Esquire,* captured Jobs's intensity at a NeXT staff meeting:

It's not quite right to say that he is sitting through this staff meeting, because Jobs doesn't sit through much of anything; one of the ways he dominates is through sheer movement. One moment he's kneeling in his chair; the next minute he's slouching in it; the next he has leaped out of his chair entirely and is scribbling on the blackboard directly behind him. He is full of mannerisms. He bites his nails. He stares with unnerving earnestness at whoever is speaking. His hands, which are slightly and inexplicably yellow, are in constant motion.

What particularly struck Nocera was Jobs's "almost willful lack of tact." It was more than just an inability to hide his opinions when others said something he thought dumb; it was a conscious readiness, even a perverse eagerness, to put people down, humiliate them, show he was smarter. When Dan'l Lewin handed out an organization chart, for example, Jobs rolled his eyes. "These charts are bullshit," he interjected. Yet his moods still swung wildly, as at Apple. A finance person came into the meeting and Jobs lavished praise on him for a "really, really great job on this"; the previous day Jobs had told him, "This deal is crap."

One of NeXT's first ten employees was an interior designer for the company's first headquarters, in Palo Alto. Even though Jobs had leased a building that was new and nicely designed, he had it completely gutted and rebuilt. Walls were replaced by glass, the carpets were replaced by light hardwood flooring. The process was repeated when NeXT moved to a bigger space in Redwood City in 1989. Even though the building was brand-new, Jobs insisted that the elevators be moved so that the entrance lobby would be more dramatic. As a centerpiece, Jobs commissioned I. M. Pei to design a grand staircase that seemed to float in the air. The contractor said it couldn't be built. Jobs said it could, and it was. Years later Jobs would make such staircases a feature at Apple's signature stores.

THE COMPUTER

During the early months of NeXT, Jobs and Dan'l Lewin went on the road, often accompanied by a few colleagues, to visit campuses and solicit opinions. At Harvard they met with Mitch Kapor, the chairman of Lotus software, over dinner at Harvest restaurant. When Kapor began slathering butter on his bread, Jobs asked him, "Have you ever heard of serum cholesterol?" Kapor responded, "I'll make you a deal. You stay away from commenting on my dietary habits, and I will stay away from the subject of your personality." It was meant humorously, but as Kapor later commented, "Human relationships were not his strong suit." Lotus agreed to write a spreadsheet program for the NeXT operating system.

Jobs wanted to bundle useful content with the machine, so Michael Hawley, one of the engineers, developed a digital dictionary. He learned that a friend of his at Oxford University Press had been involved in the typesetting of a new edition of Shakespeare's works. That meant that there was probably a computer tape he could get his hands on and, if so, incorporate it into the NeXT's memory. "So I called up Steve, and he said that would be awesome, and we flew over to Oxford together." On a beautiful spring day in 1986, they met in the publishing house's grand building in the heart of Oxford, where Jobs made an offer of $2,000 plus 74 cents for every computer sold in order to have the rights to Oxford's edition of Shakespeare. "It will be all gravy to you," he argued. "You will be ahead of the parade. It's never been done before." They agreed in principle and then went out to play skittles over beer at a nearby pub where Lord Byron used to drink. By the time it launched, the NeXT would also include a dictionary, a thesaurus, and the *Oxford Dictionary of Quotations,* making it one of the pioneers of the concept of searchable electronic books.

Instead of using off-the-shelf chips for the NeXT, Jobs had his engineers design custom ones that integrated a variety of functions on one chip. That would have been hard enough, but Jobs made it almost impossible by continually revising the functions he wanted it to do. After a year it became clear that this would be a major source of delay.

He also insisted on building his own fully automated and futuristic factory, just as he had for the Macintosh; he had not been chastened by that experience. This time too he made the same mistakes, only more excessively. Machines and robots were painted and repainted as he compulsively revised his color scheme. The walls were museum white, as they had been at the Macintosh factory, and there were $20,000 black leather chairs and a custom-made staircase, just as in the corporate headquarters. He insisted that the machinery on the 165-foot assembly line be configured to move the circuit boards from right to left as they got built, so that the process would look better to visitors who watched from the viewing gallery. Empty circuit boards were fed in at one end and twenty minutes later, untouched by humans, came out the other end as completed boards. The process followed the Japanese principle known as *kanban,* in which each machine performs its task only when the next machine is ready to receive another part.

Jobs had not tempered his way of dealing with employees. "He applied charm or public humiliation in a way that in most cases proved to be pretty effective," Tribble recalled. But sometimes it wasn't. One engineer, David Paulsen, put in ninety-hour weeks for the first ten months at NeXT. He quit when "Steve walked in one Friday afternoon and told us how unimpressed he was with what we were doing." When *Business Week* asked him why he treated employees so harshly, Jobs said it made the company better. "Part of my responsibility is to be a yardstick of quality. Some people aren't used to an environment where excellence is expected." But he still had his spirit and charisma. There were plenty of field trips, visits by akido masters, and off-site retreats. And he still exuded the pirate flag spunkiness. When Apple fired Chiat/Day, the ad firm that had done the "1984" ad and taken out the newspaper ad saying "Welcome IBM — seriously," Jobs took out a full-page ad in the *Wall Street Journal* proclaiming, "Congratulations Chiat/Day — Seriously . . . Because I can guarantee you: there is life after Apple."

Perhaps the greatest similarity to his days at Apple was that Jobs brought with him his reality distortion field. It was on display at the company's first retreat at Pebble Beach in late 1985. There

Jobs pronounced that the first NeXT computer would be shipped in just eighteen months. It was already clear that this date was impossible, but he blew off a suggestion from one engineer that they be realistic and plan on shipping in 1988. "If we do that, the world isn't standing still, the technology window passes us by, and all the work we've done we have to throw down the toilet," he argued.

Joanna Hoffman, the veteran of the Macintosh team who was among those willing to challenge Jobs, did so. "Reality distortion has motivational value, and I think that's fine," she said as Jobs stood at a whiteboard. "However, when it comes to setting a date in a way that affects the design of the product, then we get into real deep shit." Jobs didn't agree: "I think we have to drive a stake in the ground somewhere, and I think if we miss this window, then our credibility starts to erode." What he did not say, even though it was suspected by all, was that if their targets slipped they might run out of money. Jobs had pledged $7 million of his own funds, but at their current burn rate that would run out in eighteen months if they didn't start getting some revenue from shipped products.

Three months later, when they returned to Pebble Beach for their next retreat, Jobs began his list of maxims with "The honeymoon is over." By the time of the third retreat, in Sonoma in September 1986, the timetable was gone, and it looked as though the company would hit a financial wall.

Perot to the Rescue

In late 1986 Jobs sent out a proposal to venture capital firms offering a 10% stake in NeXT for $3 million. That put a valuation on the entire company of $30 million, a number that Jobs had pulled out of thin air. Less than $7 million had gone into the company thus far, and there was little to show for it other than a neat logo and some snazzy offices. It had no revenue or products, nor any on the horizon. Not surprisingly, the venture capitalists all passed on the offer to invest.

There was, however, one cowboy who was dazzled. Ross Perot, the bantam Texan who had founded Electronic Data Systems, then sold it to General Motors for $2.4 billion, happened to watch a PBS documentary, *The Entrepreneurs,* which had a segment on Jobs and NeXT in November 1986. He instantly identified with Jobs and his gang, so much so that, as he watched them on television, he said, "I was finishing their sentences for them." It was a line eerily similar to one Sculley had often used. Perot called Jobs the next day and offered, "If you ever need an investor, call me."

Jobs did indeed need one, badly. But he was careful not to show it. He waited a week before calling back. Perot sent some of his analysts to size up NeXT, but Jobs took care to deal directly with Perot. One of his great regrets in life, Perot later said, was that he had not bought Microsoft, or a large stake in it, when a very young Bill Gates had come to visit him in Dallas in 1979. By the time Perot called Jobs, Microsoft had just gone public with a $1 billion valuation. Perot had missed out on the opportunity to make a lot of money and have a fun adventure. He was eager not to make that mistake again.

Jobs made an offer to Perot that was three times

more costly than had quietly been offered to venture capitalists a few months earlier. For $20 million, Perot would get 16% of the equity in the company, after Jobs put in another $5 million. That meant the company would be valued at about $126 million. But money was not a major consideration for Perot. After a meeting with Jobs, he declared that he was in. "I pick the jockeys, and the jockeys pick the horses and ride them," he told Jobs. "You guys are the ones I'm betting on, so you figure it out."

Perot brought to NeXT something that was almost as valuable as his $20 million lifeline: He was a quotable, spirited cheerleader for the company, who could lend it an air of credibility among grown-ups. "In terms of a startup company, it's one that carries the least risk of any I've seen in 25 years in the computer industry," he told the *New York Times.* "We've had some sophisticated people see the hardware — it blew them away. Steve and his whole NeXT team are the darnedest bunch of perfectionists I've ever seen."

Perot also traveled in rarefied social and business circles that complemented Jobs's own. He took Jobs to a black-tie dinner dance in San Francisco that Gordon and Ann Getty gave for King Juan Carlos I of Spain. When the king asked Perot whom he should meet, Perot immediately produced Jobs. They were soon engaged in what Perot later described as "electric conversation," with Jobs animatedly describing the next wave in computing. At the end the king scribbled a note and handed it to Jobs. "What happened?" Perot asked. Jobs answered, "I sold him a computer."

These and other stories were incorporated into the mythologized story of Jobs that Perot told wherever he went. At a briefing at the National Press Club in Washington, he spun Jobs's life story into a Texas-size yarn about a young man

so poor he couldn't afford to go to college, working in his garage at night, playing with computer chips, which was his hobby, and his dad — who looks like a character out of a Norman Rockwell painting — comes in one day and said, "Steve, either make something you can sell or go get a job." Sixty days later, in a wooden box that his dad made for him, the first Apple computer was created. And this high school graduate literally changed the world.

The one phrase that was true was the one about Paul Jobs's looking like someone in a Rockwell painting. And perhaps the last phrase, the one about Jobs changing the world. Certainly Perot believed that. Like Sculley, he saw himself in Jobs. "Steve's like me," Perot told the *Washington Post*'s David Remnick. "We're weird in the same way. We're soul mates."

GATES AND NEXT

Bill Gates was not a soul mate. Jobs had convinced him to produce software applications for the Macintosh, which had turned out to be hugely profitable for Microsoft. But Gates was one person who was resistant to Jobs's reality distortion field, and as a result he decided not to create software tailored for the NeXT platform. Gates went to California to get periodic demonstrations, but each time he came away unimpressed. "The Macintosh was truly unique, but I personally don't understand what is so unique about Steve's new computer," he told *Fortune.*

Part of the problem was that the rival titans were congenitally unable to be deferential to each other. When Gates made his first visit to NeXT's Palo Alto headquarters, in the summer of 1987, Jobs kept him waiting for a half hour in the lobby, even though Gates could see through the glass walls that Jobs was walking around having casual conversations. "I'd gone down to NeXT and I had the Odwalla, the most expensive carrot juice, and I'd never seen tech offices so lavish," Gates recalled, shaking his head with just a hint of a smile. "And Steve comes a half hour late to the meeting."

Jobs's sales pitch, according to Gates, was simple. "We did the Mac together," Jobs said. "How did that work for you? Very well. Now, we're going to do this together and this is going to be great."

But Gates was brutal to Jobs, just as Jobs could be to others. "This machine is crap," he said. "The optical disk has too low latency, the fucking case is too expensive. This thing is ridiculous." He decided then, and reaffirmed on each subsequent visit, that it made no sense for Microsoft to divert resources from other projects to develop applications for NeXT. Worse yet, he repeatedly said so publicly, which made others less likely to spend time developing for NeXT. "Develop for it? I'll piss on it," he told *InfoWorld.*

When they happened to meet in the hallway at a conference, Jobs started berating Gates for his refusal to do software for NeXT. "When you get a market, I will consider it," Gates replied. Jobs got angry. "It was a screaming battle, right in front of everybody," recalled Adele Goldberg, the Xerox PARC engineer. Jobs insisted that NeXT was the next wave of computing. Gates, as he often did, got more expressionless as Jobs got more heated. He finally just shook his head and walked away.

Beneath their personal rivalry — and occasional grudging respect — was their basic philosophical difference. Jobs believed in an end-to-end integration of hardware and software, which led him to build a machine that was not compatible with others. Gates believed in, and profited from, a world in which different companies made machines that were compatible with one another; their hardware ran a standard operating system (Microsoft's Windows) and could all use the same software apps (such as Microsoft's Word and Excel). "His product comes with an interesting feature called incompatibility," Gates told the *Washington Post.* "It doesn't run any of the existing software. It's a super-nice computer. I don't think if I went out to design an incompatible computer I would have done as well as he did."

At a forum in Cambridge, Massachusetts, in 1989, Jobs and Gates appeared sequentially, laying out their competing worldviews. Jobs spoke about how new waves come along in the computer industry every few years. Macintosh had launched a revolutionary new approach with the graphical interface; now NeXT was doing it with object-oriented programming tied to a powerful new machine based on an optical disk. Every major software vendor realized they had to be part of this new wave, he said, "except Microsoft." When Gates came up, he reiterated his belief that Jobs's end-to-end control of the software and the hardware was destined for failure, just as Apple had failed in competing against the Microsoft Windows standard. "The hardware market and the software market are separate," he said. When asked about the great design that could come from Jobs's approach, Gates gestured to the NeXT prototype that was still sitting onstage and sneered, "If you want black, I'll get you a can of paint."

IBM

Jobs came up with a brilliant jujitsu maneuver against Gates, one that could have changed the balance of power in the computer industry forever. It required Jobs to do two things that were against his nature: licensing out his software to another hardware maker and getting into bed with IBM. He had a pragmatic streak, albeit a tiny one, so he was able to overcome his reluctance. But his heart was never fully in it, which is why the alliance would turn out to be short-lived.

It began at a party, a truly memorable one, for the seventieth birthday of the *Washington Post* publisher Katharine Graham in June 1987 in Washington. Six hundred guests attended, including President Ronald Reagan. Jobs flew in from California and IBM's chairman John Akers from New York. It was the first time they had met. Jobs took the opportunity to bad-mouth Microsoft and attempt to wean IBM from using its Windows operating system. "I couldn't resist telling him I

343

thought IBM was taking a giant gamble betting its entire software strategy on Microsoft, because I didn't think its software was very good," Jobs recalled.

To Jobs's delight, Akers replied, "How would you like to help us?" Within a few weeks Jobs showed up at IBM's Armonk, New York, headquarters with his software engineer Bud Tribble. They put on a demo of NeXT, which impressed the IBM engineers. Of particular significance was NeXT-STEP, the machine's object-oriented operating system. "NeXTSTEP took care of a lot of trivial programming chores that slow down the software development process," said Andrew Heller, the general manager of IBM's workstation unit, who was so impressed by Jobs that he named his new-born son Steve.

The negotiations lasted into 1988, with Jobs becoming prickly over tiny details. He would stalk out of meetings over disagreements about colors or design, only to be calmed down by Tribble or Lewin. He didn't seem to know which frightened him more, IBM or Microsoft. In April Perot decided to play host for a mediating session at his Dallas headquarters, and a deal was struck: IBM would license the current version of the NeXT-STEP software, and if the managers liked it, they would use it on some of their workstations. IBM sent to Palo Alto a 125-page contract. Jobs tossed it down without reading it. "You don't get it," he said as he walked out of the room. He demanded a simpler contract of only a few pages, which he got within a week.

Jobs wanted to keep the arrangement secret from

Bill Gates until the big unveiling of the NeXT computer, scheduled for October. But IBM insisted on being forthcoming. Gates was furious. He realized this could wean IBM off its dependence on Microsoft operating systems. "NeXT-STEP isn't compatible with anything," he raged to IBM executives.

At first Jobs seemed to have pulled off Gates's worst nightmare. Other computer makers that were beholden to Microsoft's operating systems, most notably Compaq and Dell, came to ask Jobs for the right to clone NeXT and license NeXT-STEP. There were even offers to pay a lot more if NeXT would get out of the hardware business altogether.

That was too much for Jobs, at least for the time being. He cut off the clone discussions. And he began to cool toward IBM. The chill became reciprocal. When the person who made the deal at IBM moved on, Jobs went to Armonk to meet his replacement, Jim Cannavino. They cleared the room and talked one-on-one. Jobs demanded more money to keep the relationship going and to license newer versions of NeXTSTEP to IBM. Cannavino made no commitments, and he subsequently stopped returning Jobs's phone calls. The deal lapsed. NeXT got a bit of money for a licensing fee, but it never got the chance to change the world.

THE LAUNCH, OCTOBER 1988

Jobs had perfected the art of turning product launches into theatrical productions, and for the world premiere of the NeXT computer — on October 12, 1988, in San Francisco's Symphony Hall — he wanted to outdo himself. He needed to blow away the doubters. In the weeks leading up to the event, he drove up to San Francisco almost every day to hole up in the Victorian house of Susan Kare, NeXT's graphic designer, who had done the original fonts and icons for the Macintosh. She helped prepare each of the slides as Jobs fretted over everything from the wording to the right hue of green to serve as the background color. "I like that green," he said proudly as they were doing a trial run in front of some staffers. "Great green, great green," they all murmured in assent.

No detail was too small. Jobs went over the invitation list and even the lunch menu (mineral water, croissants, cream cheese, bean sprouts). He picked out a video projection company and paid it $60,000 for help. And he hired the postmodernist theater producer George Coates to stage the show. Coates and Jobs decided, not surprisingly, on an austere and radically simple stage look. The unveiling of the black perfect cube would occur on a starkly minimalist stage setting with a black background, a table covered by a black cloth, a black veil draped over the computer, and a simple vase of flowers. Because neither the hardware nor the operating system was actually ready, Jobs was urged to do a simulation. But he refused. Knowing it would be like walking a tightrope without a net, he decided to do the demonstration live.

More than three thousand people showed up at the event, lining up two hours before curtain time. They were not disappointed, at least by the show. Jobs was onstage for three hours, and he again proved to be, in the words of Andrew Pollack of the *New York Times,* "the Andrew Lloyd Webber of product introductions, a master of stage flair and special effects." Wes Smith of the *Chicago Tribune* said the launch was "to product demonstrations what Vatican II was to church meetings."

Jobs had the audience cheering from his opening line: "It's great to be back." He began by recounting the history of personal computer architecture, and he promised that they would now witness an event "that occurs only once or twice in a decade — a time when a new architecture is rolled out that is going to change the face of computing." The NeXT software and hardware were designed, he said, after three years of consulting with universities across the country. "What we realized was that higher ed wants a personal mainframe."

As usual there were superlatives. The product was "incredible," he said, "the best thing we could have imagined." He praised the beauty of even the parts unseen. Balancing on his fingertips the foot-square circuit board that would be nestled in the foot-cube box, he enthused, "I hope you get a chance to look at this a little later. It's the most beautiful printed circuit board I've ever seen in my life." He then showed how the computer could play speeches — he featured King's "I Have a Dream" and Kennedy's "Ask Not" — and send email with audio attachments. He leaned into the microphone on the computer to record one of his own. "Hi, this is Steve, sending a message on a pretty historic day." Then he asked those in the audience to add "a round of applause" to the message, and they did.

One of Jobs's management philosophies was that it is crucial, every now and then, to roll the dice and "bet the company" on some new idea or technology. At the NeXT launch, he boasted of an example that, as it turned out, would not be a wise gamble: having a high-capacity (but slow) optical read/write disk and no floppy disk as a backup. "Two years ago we made a decision," he said. "We saw some new technology and we made a decision to risk our company."

Then he turned to a feature that would prove more prescient. "What we've done is made the first real digital books," he said, noting the inclusion of the Oxford edition of Shakespeare and other tomes. "There has not been an advancement in the state of the art of printed book technology since Gutenberg."

At times he could be amusingly aware of his own foibles, and he used the electronic book demonstration to poke fun at himself. "A word that's sometimes used to describe me is 'mercurial,' " he said, then paused. The audience laughed knowingly, especially those in the front rows, which were filled with NeXT employees and former members of the Macintosh team. Then he pulled up the word in the computer's dictionary and read the first definition: "Of or relating to, or born under the planet Mercury." Scrolling down, he said, "I think the third one is the one they mean: 'Characterized by unpredictable changeableness of mood.' " There was a bit more laughter. "If we scroll down the thesaurus, though, we see that the antonym is 'saturnine.' Well what's that? By simply double-clicking on it, we immediately look that up in the dictionary, and here it is: 'Cold and steady in moods. Slow to act or change. Of a gloomy or surly disposition.' " A little smile came across his face as he waited for the ripple of laughter. "Well," he concluded, "I don't think 'mercurial' is so bad after all." After the applause, he used the quotations book to make a more subtle point, about his reality distortion field. The quote he chose was from Lewis Carroll's *Through the Looking Glass.* After Alice laments that no matter how hard she tries she can't believe impossible things, the White Queen retorts, "Why, sometimes I've believed as many as six impossible things before breakfast." Especially from the front rows, there was a roar of knowing laughter.

All of the good cheer served to sugarcoat, or distract attention from, the bad news. When it came time to announce the price of the new machine, Jobs did what he would often do in product demonstrations: reel off the features, describe them as being "worth thousands and thousands of dollars," and get the audience to imagine how expensive it really should be. Then he announced what he hoped would seem like a low price: "We're going to be charging higher education a single price of $6,500." From the faithful, there was scattered applause. But his panel of academic advisors had long pushed to keep the price to between $2,000 and $3,000, and they thought that Jobs had promised to do so. Some of them were appalled. This was especially true once they discovered that the optional printer would cost another $2,000, and the slowness of the optical disk would make the purchase of a $2,500 external hard disk advisable.

There was another disappointment that he tried to downplay: "Early next year, we will have our 0.9 release, which is for software developers and aggressive end users." There was a bit of nervous laughter. What he was saying was that the real release of the machine and its software, known as the 1.0 release, would not actually be happening in early 1989. In fact he didn't set a hard date. He merely suggested it would be sometime in the second quarter of that year. At the first NeXT retreat back in late 1985, he had refused to budge, despite Joanna Hoffman's pushback, from his commitment to have the machine finished in early 1987. Now it was clear it would be more than two years later.

The event ended on a more upbeat note, literally. Jobs brought onstage a violinist from the San Francisco Symphony who played Bach's A Minor Violin Concerto in a duet with the NeXT computer onstage. People erupted in jubilant applause. The price and the delayed release were forgotten in the frenzy. When one reporter asked him immediately afterward why the machine was going to be so late, Jobs replied, "It's not late. It's five years ahead of its time."

As would become his standard practice, Jobs offered to provide "exclusive" interviews to anointed publications in return for their promising to put the story on the cover. This time he went one "exclusive" too far, though it didn't really hurt. He agreed to a request from *Business Week*'s Katie Hafner for exclusive access to him before the launch, but he also made a similar deal with *Newsweek* and then with *Fortune.* What he didn't consider was that one of *Fortune*'s top editors, Susan Fraker, was married to *Newsweek*'s editor Maynard Parker. At the *Fortune* story conference, when they were talking excitedly about their exclusive, Fraker mentioned that she happened to know that *Newsweek* had also been promised an exclusive, and it would be coming out a few days before *Fortune.* So Jobs ended up that week on only two magazine covers. *Newsweek* used the cover line "Mr. Chips" and showed him leaning on a beautiful NeXT, which it proclaimed to be "the most exciting machine in years." *Business Week* showed him looking angelic in a dark suit, fingertips pressed together like a preacher or professor. But Hafner pointedly reported on the manipulation that surrounded her exclusive. "NeXT carefully parceled out interviews with its staff and suppliers, monitoring them with a censor's eye," she wrote. "That strategy worked, but at a price: Such maneuvering — self-serving and relentless — displayed the side of Steve Jobs that so hurt him at Apple. The trait that most stands out is Jobs's need to control events."

When the hype died down, the reaction to the NeXT computer was muted, especially since it was not yet commercially available. Bill Joy, the brilliant and wry chief scientist at rival Sun Microsystems, called it "the first Yuppie workstation," which was not an unalloyed compliment. Bill Gates, as might be expected, continued to be publicly dismissive. "Frankly, I'm disappointed," he told the *Wall Street Journal.* "Back in 1981, we were truly excited by the Macintosh when Steve showed it to us, because when you put it side-by-side with another computer, it was unlike anything anybody had ever seen before." The NeXT machine was not like that. "In the grand scope of things, most of these features are truly trivial." He said that Microsoft would continue its plans not to write software for the NeXT. Right after the announcement event, Gates wrote a parody email to his staff. "All reality has been completely suspended," it began. Looking back at it, Gates laughs that it may have been "the best email I ever wrote."

When the NeXT computer finally went on sale in mid-1989, the factory was primed to churn out ten thousand units a month. As it turned out, sales were about four hundred a month. The beautiful factory robots, so nicely painted, remained mostly idle, and NeXT continued to hemorrhage cash.

CHAPTER NINETEEN: PIXAR
TECHNOLOGY MEETS ART

LUCASFILM'S COMPUTER DIVISION

When Jobs was losing his footing at Apple in the summer of 1985, he went for a walk with Alan Kay, who had been at Xerox PARC and was then an Apple Fellow. Kay knew that Jobs was interested in the intersection of creativity and technology, so he suggested they go see a friend of his, Ed Catmull, who was running the computer division of George Lucas's film studio. They rented a limo and rode up to Marin County to the edge of Lucas's Sky-walker Ranch, where Catmull and his little computer division were based. "I was blown away, and I came back and tried to convince Sculley to buy it for Apple," Jobs recalled. "But the folks running Apple weren't interested, and they were busy kicking me out anyway."

The Lucasfilm computer division made hardware and software for rendering digital images, and it also had a group of computer animators making shorts, which was led by a talented cartoon-loving executive named John Lasseter. Lucas, who had completed his first *Star Wars* trilogy, was embroiled in a contentious divorce, and he needed to sell off the division. He told Catmull to find a

353

buyer as soon as possible.

After a few potential purchasers balked in the fall of 1985, Catmull and his colleague Alvy Ray Smith decided to seek investors so that they could buy the division themselves. So they called Jobs, arranged another meeting, and drove down to his Woodside house. After railing for a while about the perfidies and idiocies of Sculley, Jobs proposed that he buy their Lucasfilm division outright. Catmull and Smith demurred: They wanted an investor, not a new owner. But it soon became clear that there was a middle ground: Jobs could buy a majority of the division and serve as chairman but allow Catmull and Smith to run it.

"I wanted to buy it because I was really into computer graphics," Jobs recalled. "I realized they were way ahead of others in combining art and technology, which is what I've always been interested in." He offered to pay Lucas $5 million plus invest another $5 million to capitalize the division as a stand-alone company. That was far less than Lucas had been asking, but the timing was right. They decided to negotiate a deal.

The chief financial officer at Lucasfilm found Jobs arrogant and prickly, so when it came time to hold a meeting of all the players, he told Catmull, "We have to establish the right pecking order." The plan was to gather everyone in a room with Jobs, and then the CFO would come in a few minutes late to establish that he was the person running the meeting. "But a funny thing happened," Catmull recalled. "Steve started the meeting on time without the CFO, and by the time the CFO walked in Steve was already in control of the meeting."

Jobs met only once with George Lucas, who warned him that the people in the division cared more about making animated movies than they did about making computers. "You know, these guys are hell-bent on animation," Lucas told him. Lucas later recalled, "I did warn him that was basically Ed and John's agenda. I think in his heart he bought the company because that was his agenda too."

The final agreement was reached in January 1986. It provided that, for his $10 million investment, Jobs would own 70% of the company, with the rest of the stock distributed to Ed Catmull, Alvy Ray Smith, and the thirty-eight other founding employees, down to the receptionist. The division's most important piece of hardware was called the Pixar Image Computer, and from it the new company took its name.

For a while Jobs let Catmull and Smith run Pixar without much interference. Every month or so they would gather for a board meeting, usually at NeXT headquarters, where Jobs would focus on the finances and strategy. Nevertheless, by dint of his personality and controlling instincts, Jobs was soon playing a stronger role. He spewed out a stream of ideas — some reasonable, others wacky — about what Pixar's hardware and software could become. And on his occasional visits to the Pixar offices, he was an inspiring presence. "I grew up a Southern Baptist, and we had revival meetings with mesmerizing but corrupt preachers," recounted Alvy Ray Smith. "Steve's got it: the power of the tongue and the web of words that catches people up. We were aware of this when we had board meetings, so we developed signals — nose scratching or ear tugs — for when someone had been caught up in Steve's distortion field and he needed to be tugged back to reality."

Jobs had always appreciated the virtue of integrating hardware and software, which is what Pixar did with its Image Computer and rendering software. It also produced creative content, such as animated films and graphics. All three elements benefited from Jobs's combination of artistic creativity and technological geekiness. "Silicon Valley folks don't really respect Hollywood creative types, and the Hollywood folks think that tech folks are people you hire and never have to meet," Jobs later said. "Pixar was one place where both cultures were respected."

Initially the revenue was supposed to come from the hardware side. The Pixar Image Computer sold for $125,000. The primary customers were animators and graphic designers, but the machine also soon found specialized markets in the medical industry (CAT scan data could be rendered in three-dimensional graphics) and intelligence fields (for rendering information from reconnaissance flights and satellites). Because of the sales to the National Security Agency, Jobs had to get a security clearance, which must have been fun for the FBI agent assigned to vet him. At one point, a Pixar executive recalled, Jobs was called by the investigator to go over the drug use questions, which he answered unabashedly. "The last time I used that . . . ," he would say, or on occasion he would answer that no, he had actually never tried that particular drug.

Jobs pushed Pixar to build a lower-cost version of the computer that would sell for around $30,000. He insisted that Hartmut Esslinger design it, despite protests by Catmull and Smith about his fees. It ended up looking like the original Pixar Image Computer, which was a cube with a round dimple in the middle, but it had Esslinger's signature thin grooves.

Jobs wanted to sell Pixar's computers to a mass market, so he had the Pixar folks open up sales offices — for which he approved the design — in major cities, on the theory that creative people would soon come up with all sorts of ways to use the machine. "My view is that people are creative animals and will figure out clever new ways to use tools that the inventor never imagined," he later said. "I thought that would happen with the Pixar computer, just as it did with the Mac." But the machine never took hold with regular consumers. It cost too much, and there were not many software programs for it.

On the software side, Pixar had a rendering program, known as Reyes (Renders everything you ever saw), for making 3-D graphics and images. After Jobs became chairman, the company created a new language and interface, named RenderMan, that it hoped would become a standard for 3-D graphics rendering, just as Adobe's PostScript was for laser printing.

As he had with the hardware, Jobs decided that they should try to find a mass market, rather than just a specialized one, for the software they made. He was never content to aim only at the corporate or high-end specialized markets. "He would have

357

these great visions of how RenderMan could be for everyman," recalled Pam Kerwin, Pixar's marketing director. "He kept coming up with ideas about how ordinary people would use it to make amazing 3-D graphics and photorealistic images." The Pixar team would try to dissuade him by saying that RenderMan was not as easy to use as, say, Excel or Adobe Illustrator. Then Jobs would go to a whiteboard and show them how to make it simpler and more user-friendly. "We would be nodding our heads and getting excited and say, 'Yes, yes, this will be great!' " Kerwin recalled. "And then he would leave and we would consider it for a moment and then say, 'What the heck was he thinking!' He was so weirdly charismatic that you almost had to get deprogrammed after you talked to him." As it turned out, average consumers were not craving expensive software that would let them render realistic images. RenderMan didn't take off.

There was, however, one company that was eager to automate the rendering of animators' drawings into color images for film. When Roy Disney led a board revolution at the company that his uncle Walt had founded, the new CEO, Michael Eisner, asked what role he wanted. Disney said that he would like to revive the company's venerable but fading animation department. One of his first initiatives was to look at ways to computerize the process, and Pixar won the contract. It created a package of customized hardware and software known as CAPS, Computer Animation Production System. It was first used in 1988 for the final scene of *The Little Mermaid,* in which King Triton waves good-bye to Ariel. Disney bought dozens of

Pixar Image Computers as CAPS became an integral part of its production.

ANIMATION

The digital animation business at Pixar — the group that made little animated films — was originally just a sideline, its main purpose being to show off the hardware and software of the company. It was run by John Lasseter, a man whose childlike face and demeanor masked an artistic perfectionism that rivaled that of Jobs. Born in Hollywood, Lasseter grew up loving Saturday morning cartoon shows. In ninth grade, he wrote a report on the history of Disney Studios, and he decided then how he wished to spend his life.

When he graduated from high school, Lasseter enrolled in the animation program at the California Institute of the Arts, founded by Walt Disney. In his summers and spare time, he researched the Disney archives and worked as a guide on the Jungle Cruise ride at Disneyland. The latter experience taught him the value of timing and pacing in telling a story, an important but difficult concept to master when creating, frame by frame, animated footage. He won the Student Academy Award for the short he made in his junior year, *Lady and the Lamp,* which showed his debt to Disney films and foreshadowed his signature talent for infusing inanimate objects such as lamps with human personalities. After graduation he took the job for which he was destined: as an animator at Disney Studios.

Except it didn't work out. "Some of us younger guys wanted to bring *Star Wars*–level quality to the art of animation, but we were held in check," Lasseter recalled. "I got disillusioned, then I got caught in a feud between two bosses, and the head animation guy fired me." So in 1984 Ed Catmull and Alvy Ray Smith were able to recruit him to work where *Star Wars*–level quality was being defined, Lucasfilm. It was not certain that George Lucas, already worried about the cost of his computer division, would really approve of hiring a full-time animator, so Lasseter was given the title "interface designer."

After Jobs came onto the scene, he and Lasseter began to share their passion for graphic design. "I was the only guy at Pixar who was an artist, so I bonded with Steve over his design sense," Lasseter said. He was a gregarious, playful, and huggable man who wore flowery Hawaiian shirts, kept his office cluttered with vintage toys, and loved cheeseburgers. Jobs was a prickly, whip-thin vegetarian who favored austere and uncluttered surroundings. But they were actually well-suited for each other. Lasseter was an artist, so Jobs treated him deferentially, and Lasseter viewed Jobs, correctly, as a patron who could appreciate artistry and knew how it could be interwoven with technology and commerce.

Jobs and Catmull decided that, in order to show off their hardware and software, Lasseter should produce another short animated film in 1986 for SIGGRAPH, the annual computer graphics conference. At the time, Lasseter was using the Luxo lamp on his desk as a model for graphic rendering, and he decided to turn Luxo into a lifelike character. A friend's young child inspired him to add Luxo Jr., and he showed a few test frames to another animator, who urged him to make sure he told a story. Lasseter said he was making only a short, but the animator reminded him that a story can be told even in a few seconds. Lasseter took the lesson to heart. *Luxo Jr.* ended up being just over two minutes; it told the tale of a parent lamp and a child lamp pushing a ball back and forth until the ball bursts, to the child's dismay.

Jobs was so excited that he took time off from the pressures at NeXT to fly down with Lasseter to SIGGRAPH, which was being held in Dallas that August. "It was so hot and muggy that when we'd walk outside the air hit us like a tennis racket," Lasseter recalled. There were ten thousand people at the trade show, and Jobs loved it. Artistic creativity energized him, especially when it was connected to technology.

There was a long line to get into the auditorium where the films were being screened, so Jobs, not one to wait his turn, fast-talked their way in first. *Luxo Jr.* got a prolonged standing ovation and was named the best film. "Oh, wow!" Jobs exclaimed at the end. "I really get this, I get what it's all about." As he later explained, "Our film was the only one that had art to it, not just good technology. Pixar was about making that combination, just as the Macintosh had been."

Luxo Jr. was nominated for an Academy Award, and Jobs flew down to Los Angeles to be there for the ceremony. It didn't win, but Jobs became committed to making new animated shorts each year, even though there was not much of a business rationale for doing so. As times got tough at

Pixar, he would sit through brutal budget-cutting meetings showing no mercy. Then Lasseter would ask that the money they had just saved be used for his next film, and Jobs would agree.

TIN TOY

Not all of Jobs's relationships at Pixar were as good. His worst clash came with Catmull's cofounder, Alvy Ray Smith. From a Baptist background in rural north Texas, Smith became a free-spirited hippie computer imaging engineer with a big build, big laugh, and big personality — and occasionally an ego to match. "Alvy just glows, with a high color, friendly laugh, and a whole bunch of groupies at conferences," said Pam Kerwin. "A personality like Alvy's was likely to ruffle Steve. They are both visionaries and high energy and high ego. Alvy is not as willing to make peace and overlook things as Ed was."

Smith saw Jobs as someone whose charisma and ego led him to abuse power. "He was like a televangelist," Smith said. "He wanted to control people, but I would not be a slave to him, which is why we clashed. Ed was much more able to go with the flow." Jobs would sometimes assert his dominance at a meeting by saying something outrageous or untrue. Smith took great joy in calling him on it, and he would do so with a large laugh and a smirk. This did not endear him to Jobs.

One day at a board meeting, Jobs started berating Smith and other top Pixar executives for the delay in getting the circuit boards completed for the new version of the Pixar Image Computer. At the time, NeXT was also very late in complet-

362

ing its own computer boards, and Smith pointed that out: "Hey, you're even later with your NeXT boards, so quit jumping on us." Jobs went ballistic, or in Smith's phrase, "totally nonlinear." When Smith was feeling attacked or confrontational, he tended to lapse into his southwestern accent. Jobs started parodying it in his sarcastic style. "It was a bully tactic, and I exploded with everything I had," Smith recalled. "Before I knew it, we were in each other's faces — about three inches apart — screaming at each other."

Jobs was very possessive about control of the whiteboard during a meeting, so the burly Smith pushed past him and started writing on it. "You can't do that!" Jobs shouted.

"What?" responded Smith, "I can't write on your whiteboard? Bullshit." At that point Jobs stormed out.

Smith eventually resigned to form a new company to make software for digital drawing and image editing. Jobs refused him permission to use some code he had created while at Pixar, which further inflamed their enmity. "Alvy eventually got what he needed," said Catmull, "but he was very stressed for a year and developed a lung infection." In the end it worked out well enough; Microsoft eventually bought Smith's company, giving him the distinction of being a founder of one company that was sold to Jobs and another that was sold to Gates.

Ornery in the best of times, Jobs became particularly so when it became clear that all three Pixar endeavors — hardware, software, and animated content — were losing money. "I'd get these plans, and in the end I kept having to put in more money," he recalled. He would rail, but then write the check. Having been ousted at Apple and flailing at NeXT, he couldn't afford a third strike.

To stem the losses, he ordered a round of deep layoffs, which he executed with his typical empathy deficiency. As Pam Kerwin put it, he had "neither the emotional nor financial runway to be decent to people he was letting go." Jobs insisted that the firings be done immediately, with no severance pay. Kerwin took Jobs on a walk around the parking lot and begged that the employees be given at least two weeks notice. "Okay," he shot back, "but the notice is retroactive from two weeks ago." Catmull was in Moscow, and Kerwin put in frantic calls to him. When he returned, he was able to institute a meager severance plan and calm things down just a bit.

At one point the members of the Pixar animation team were trying to convince Intel to let them make some of its commercials, and Jobs became impatient. During a meeting, in the midst of berating an Intel marketing director, he picked up the phone and called CEO Andy Grove directly. Grove, still playing mentor, tried to teach Jobs a lesson: He supported his Intel manager. "I stuck by my employee," he recalled. "Steve doesn't like to be treated like a supplier."

Grove also played mentor when Jobs proposed that Pixar give Intel suggestions on how to improve the capacity of its processors to render 3-D graphics. When the engineers at Intel accepted the offer, Jobs sent an email back saying Pixar would need to be paid for its advice. Intel's chief engineer replied, "We have not entered into any financial arrangement in exchange for good ideas for our microprocessors in the past and have no intention for the future." Jobs forwarded the answer to Grove, saying that he found the engineer's response to be "extremely arrogant, given Intel's dismal showing in understanding computer graphics." Grove sent Jobs a blistering reply, saying that sharing ideas is "what friendly companies and friends do for each other." Grove added that he had often freely shared ideas with Jobs in the past and that Jobs should not be so mercenary. Jobs relented. "I have many faults, but one of them is not ingratitude," he responded. "Therefore, I have changed my position 180 degrees — we will freely help. Thanks for the clearer perspective."

Pixar was able to create some powerful software products aimed at average consumers, or at least those average consumers who shared Jobs's passion for designing things. Jobs still hoped that the ability to make super-realistic 3-D images at home would become part of the desktop publishing craze. Pixar's Showplace, for example, allowed users to change the shadings on the 3-D objects they created so that they could display them from various angles with appropriate shadows. Jobs thought it was incredibly compelling, but most consumers were content to live without it. It was a case where his passions misled him: The software had so many amazing features that it lacked the simplicity Jobs usually demanded. Pixar couldn't compete with Adobe, which was making software that was less sophisticated but far less complicated and expensive.

Even as Pixar's hardware and software product lines foundered, Jobs kept protecting the animation group. It had become for him a little island of magical artistry that gave him deep emotional pleasure, and he was willing to nurture it and bet on it. In the spring of 1988 cash was running so short that he convened a meeting to decree deep spending cuts across the board. When it was over, Lasseter and his animation group were almost too afraid to ask Jobs about authorizing some extra money for another short. Finally, they broached the topic and Jobs sat silent, looking skeptical. It would require close to $300,000 more out of his pocket. After a few minutes, he asked if there were any storyboards. Catmull took him down to the animation offices, and once Lasseter started his show — displaying his boards, doing the voices, showing his passion for his product — Jobs started to warm up.

The story was about Lasseter's love, classic toys. It was told from the perspective of a toy one-man band named Tinny, who meets a baby that charms and terrorizes him. Escaping under the couch, Tinny finds other frightened toys, but when the baby hits his head and cries, Tinny goes back out to cheer him up.

Jobs said he would provide the money. "I believed in what John was doing," he later said. "It was art. He cared, and I cared. I always said yes." His only comment at the end of Lasseter's presentation was, "All I ask of you, John, is to make it great."

Tin Toy went on to win the 1988 Academy Award for animated short films, the first computer-generated film to do so. To celebrate, Jobs took Lasseter and his team to Greens, a vegetarian restaurant in San Francisco. Lasseter grabbed the Oscar, which was in the center of the table, held it aloft, and toasted Jobs by saying, "All you asked is that we make a great movie."

The new team at Disney — Michael Eisner the CEO and Jeffrey Katzenberg in the film division — began a quest to get Lasseter to come back. They liked *Tin Toy,* and they thought that something more could be done with animated stories of toys that come alive and have human emotions. But Lasseter, grateful for Jobs's faith in him, felt that Pixar was the only place where he could create a new world of computer-generated animation. He told Catmull, "I can go to Disney and be a director, or I can stay here and make history." So Disney began talking about making a production deal with Pixar. "Lasseter's shorts were really breathtaking both in storytelling and in the use of technology," recalled Katzenberg. "I tried so hard to get him to Disney, but he was loyal to Steve and Pixar. So if you can't beat them, join them. We decided to look for ways we could join up with Pixar and have them make a film about toys for us."

By this point Jobs had poured close to $50 million of his own money into Pixar — more than half of what he had pocketed when he cashed out of Apple — and he was still losing money at NeXT. He was hard-nosed about it; he forced all Pixar employees to give up their options as part of his agreement to add another round of personal funding in 1991. But he was also a romantic in his love for what artistry and technology could do together. His belief that ordinary consumers would love to do 3-D modeling on Pixar software turned out to be wrong, but that was soon replaced by an instinct that turned out to be right: that combining great art and digital technology would transform animated films more than anything had since 1937, when Walt Disney had given life to Snow White.

Looking back, Jobs said that, had he known more, he would have focused on animation sooner and not worried about pushing the company's hardware or software applications. On the other hand, had he known the hardware and software would never be profitable, he would not have taken over Pixar. "Life kind of snookered me into doing that, and perhaps it was for the better."

CHAPTER TWENTY:
A REGULAR GUY
LOVE IS JUST A FOUR-LETTER WORD

JOAN BAEZ

In 1982, when he was still working on the Macintosh, Jobs met the famed folksinger Joan Baez through her sister Mimi Fariña, who headed a charity that was trying to get donations of computers for prisons. A few weeks later he and Baez had lunch in Cupertino. "I wasn't expecting a lot, but she was really smart and funny," he recalled. At the time, he was nearing the end of his relationship with Barbara Jasinski. They had vacationed in Hawaii, shared a house in the Santa Cruz mountains, and even gone to one of Baez's concerts together. As his relationship with Jasinski flamed out, Jobs began getting more serious with Baez. He was twenty-seven and Baez was forty-one, but for a few years they had a romance. "It turned into a serious relationship between two accidental friends who became lovers," Jobs recalled in a somewhat wistful tone.

Elizabeth Holmes, Jobs's friend from Reed College, believed that one of the reasons he went out with Baez — other than the fact that she was beautiful and funny and talented — was that she had once been the lover of Bob Dylan. "Steve loved that connection to Dylan," she later said. Baez and Dylan had been lovers in the early 1960s, and they toured as friends after that, including with the Rolling Thunder Revue in 1975. (Jobs had the bootlegs of those concerts.)

When she met Jobs, Baez had a fourteen-year-old son, Gabriel, from her marriage to the antiwar activist David Harris. At lunch she told Jobs she was trying to teach Gabe how to type. "You mean on a typewriter?" Jobs asked. When she said yes, he replied, "But a typewriter is antiquated."

"If a typewriter is antiquated, what does that make me?" she asked. There was an awkward pause. As Baez later told me, "As soon as I said it, I realized the answer was so obvious. The question just hung in the air. I was just horrified."

Much to the astonishment of the Macintosh team, Jobs burst into the office one day with Baez and showed her the prototype of the Macintosh. They were dumbfounded that he would reveal the computer to an outsider, given his obsession with secrecy, but they were even more blown away to be in the presence of Joan Baez. He gave Gabe an Apple II, and he later gave Baez a Macintosh. On visits Jobs would show off the features he liked. "He was sweet and patient, but he was so advanced in his knowledge that he had trouble teaching me," she recalled.

He was a sudden multimillionaire; she was a world-famous celebrity, but sweetly down-to-earth and not all that wealthy. She didn't know what to make of him then, and still found him puzzling when she talked about him almost thirty years later. At one dinner early in their relationship, Jobs started talking about Ralph Lauren and his Polo Shop, which she admitted she had never visited. "There's a beautiful red dress there that would be perfect for you," he said, and then drove her to the store in the Stanford Mall. Baez recalled, "I said to myself, far out, terrific, I'm with one of the world's richest men and he wants me to have this beautiful dress." When they got to the store, Jobs bought a handful of shirts for himself and showed her the red dress. "You ought to buy it," he said. She was a little surprised, and told him she couldn't really afford it. He said nothing, and they left. "Wouldn't you think if someone had talked like that the whole evening, that they were going to get it for you?" she asked me, seeming genuinely puzzled about the incident. "The mystery of the red dress is in your hands. I felt a bit strange about it." He would give her computers, but not a dress, and when he brought her flowers he made sure to say they were left over from an event in the office. "He was both romantic and afraid to be romantic," she said.

When he was working on the NeXT computer, he went to Baez's house in Woodside to show her how well it could produce music. "He had it play a Brahms quartet, and he told me eventually computers would sound better than humans playing it, even get the innuendo and the cadences better," Baez recalled. She was revolted by the idea. "He was working himself up into a fervor of delight while I was shrinking into a rage and thinking, How could you defile music like that?"

Jobs would confide in Debi Coleman and Jo-anna Hoffman about his relationship with Baez and worry about whether he could marry someone who had a teenage son and was probably past the point of wanting to have more children. "At times he would belittle her as being an 'issues' singer and not a true 'political' singer like Dylan," said Hoffman. "She was a strong woman, and he wanted to show he was in control. Plus, he always said he wanted to have a family, and with her he knew that he wouldn't."

And so, after about three years, they ended their romance and drifted into becoming just friends. "I thought I was in love with her, but I really just liked her a lot," he later said. "We weren't destined to be together. I wanted kids, and she didn't want any more." In her 1989 memoir, Baez wrote about her breakup with her husband and why she never remarried: "I belonged alone, which is how I have been since then, with occasional interruptions that are mostly picnics." She did add a nice acknowledgment at the end of the book to "Steve Jobs for forcing me to use a word processor by putting one in my kitchen."

FINDING JOANNE AND MONA

When Jobs was thirty-one, a year after his ouster from Apple, his mother Clara, who was a smoker, was stricken with lung cancer. He spent time by her deathbed, talking to her in ways he had rarely done in the past and asking some questions he had refrained from raising before. "When you and Dad got married, were you a virgin?" he asked. It was hard for her to talk, but she forced a smile. That's when she told him that she had been married before, to a man who never made it back from the war. She also filled in some of the details of how she and Paul Jobs had come to adopt him.

Soon after that, Jobs succeeded in tracking down the woman who had put him up for adoption. His quiet quest to find her had begun in the early 1980s, when he hired a detective who had failed to come up with anything. Then Jobs noticed the name of a San Francisco doctor on his birth certificate. "He was in the phone book, so I gave him a call," Jobs recalled. The doctor was no help. He claimed that his records had been destroyed in a fire. That was not true. In fact, right after Jobs called, the doctor wrote a letter, sealed it in an envelope, and wrote on it, "To be delivered to Steve Jobs on my death." When he died a short time later, his widow sent the letter to Jobs. In it, the doctor explained that his mother had been an unmarried graduate student from Wisconsin named Joanne Schieble.

It took another few weeks and the work of another detective to track her down. After giving him up, Joanne had married his biological father, Abdulfattah "John" Jandali, and they had another child, Mona. Jandali abandoned them five years later, and Joanne married a colorful ice-skating instructor, George Simpson. That marriage didn't last long either, and in 1970 she began a meandering journey that took her and Mona (both of them now using the last name Simpson) to Los Angeles.

Jobs had been reluctant to let Paul and Clara, whom he considered his real parents, know about his search for his birth mother. With a sensitivity that was unusual for him, and which showed the deep affection he felt for his parents, he worried that they might be offended. So he never contacted Joanne Simpson until after Clara Jobs died in early

1986. "I never wanted them to feel like I didn't consider them my parents, because they were totally my parents," he recalled. "I loved them so much that I never wanted them to know of my search, and I even had reporters keep it quiet when any of them found out." When Clara died, he decided to tell Paul Jobs, who was perfectly comfortable and said he didn't mind at all if Steve made contact with his biological mother.

So one day Jobs called Joanne Simpson, said who he was, and arranged to come down to Los Angeles to meet her. He later claimed it was mainly out of curiosity. "I believe in environment more than heredity in determining your traits, but still you have to wonder a little about your biological roots," he said. He also wanted to reassure Joanne that what she had done was all right. "I wanted to meet my biological mother mostly to see if she was okay and to thank her, because I'm glad I didn't end up as an abortion. She was twenty-three and she went through a lot to have me."

Joanne was overcome with emotion when Jobs arrived at her Los Angeles house. She knew he was famous and rich, but she wasn't exactly sure why. She immediately began to pour out her emotions. She had been pressured to sign the papers putting him up for adoption, she said, and did so only when told that he was happy in the house of his new parents. She had always missed him and suffered about what she had done. She apologized over and over, even as Jobs kept reassuring her that he understood, and that things had turned out just fine.

Once she calmed down, she told Jobs that he

had a full sister, Mona Simpson, who was then an aspiring novelist in Manhattan. She had never told Mona that she had a brother, and that day she broke the news, or at least part of it, by telephone. "You have a brother, and he's wonderful, and he's famous, and I'm going to bring him to New York so you can meet him," she said. Mona was in the throes of finishing a novel about her mother and their peregrination from Wisconsin to Los Angeles, *Anywhere but Here.* Those who've read it will not be surprised that Joanne was somewhat quirky in the way she imparted to Mona the news about her brother. She refused to say who he was — only that he had been poor, had gotten rich, was good-looking and famous, had long dark hair, and lived in California. Mona then worked at the *Paris Review,* George Plimpton's literary journal housed on the ground floor of his townhouse near Manhattan's East River. She and her coworkers began a guessing game on who her brother might be. John Travolta? That was one of the favorite guesses. Other actors were also hot prospects. At one point someone did toss out a guess that "maybe it's one of those guys who started Apple computer," but no one could recall their names.

The meeting occurred in the lobby of the St. Regis Hotel. "He was totally straightforward and lovely, just a normal and sweet guy," Mona recalled. They all sat and talked for a few minutes, then he took his sister for a long walk, just the two of them. Jobs was thrilled to find that he had a sibling who was so similar to him. They were both intense in their artistry, observant of their surroundings, and sensitive yet strong-willed. When they went to dinner together, they noticed the same architectural details and talked about them excitedly afterward. "My sister's a writer!" he exulted to colleagues at Apple when he found out.

When Plimpton threw a party for *Anywhere but Here* in late 1986, Jobs flew to New York to accompany Mona to it. They grew increasingly close, though their friendship had the complexities that might be expected, considering who they were and how they had come together. "Mona was not completely thrilled at first to have me in her life and have her mother so emotionally affectionate toward me," he later said. "As we got to know each other, we became really good friends, and she is my family. I don't know what I'd do without her. I can't imagine a better sister. My adopted sister, Patty, and I were never close." Mona likewise developed a deep affection for him, and at times could be very protective, although she would later write an edgy novel about him, *A Regular Guy,* that described his quirks with discomforting accuracy.

One of the few things they would argue about was her clothes. She dressed like a struggling novelist, and he would berate her for not wearing clothes that were "fetching enough." At one point his comments so annoyed her that she wrote him a letter: "I am a young writer, and this is my life, and I'm not trying to be a model anyway." He didn't answer. But shortly after, a box arrived from the store of Issey Miyake, the Japanese fashion designer whose stark and technology-influenced style made him one of Jobs's favorites. "He'd gone shopping for me," she later said, "and he'd picked out great things, exactly my size, in flattering colors." There was one pantsuit that he had particularly liked, and the shipment included three of them, all identical. "I still remember those first suits I sent Mona," he said. "They were linen pants and tops in a pale grayish green that looked beautiful with her reddish hair."

THE LOST FATHER

In the meantime, Mona Simpson had been trying to track down their father, who had wandered off when she was five. Through Ken Auletta and Nick Pileggi, prominent Manhattan writers, she was introduced to a retired New York cop who had formed his own detective agency. "I paid him what little money I had," Simpson recalled, but the search was unsuccessful. Then she met another private eye in California, who was able to find an address for Abdulfattah Jandali in Sacramento through a Department of Motor Vehicles search. Simpson told her brother and flew out from New York to see the man who was apparently their father.

Jobs had no interest in meeting him. "He didn't treat me well," he later explained. "I don't hold anything against him — I'm happy to be alive. But what bothers me most is that he didn't treat Mona well. He abandoned her." Jobs himself had abandoned his own illegitimate daughter, Lisa, and now was trying to restore their relationship, but that complexity did not soften his feelings toward Jandali. Simpson went to Sacramento alone.

"It was very intense," Simpson recalled. She found her father working in a small restaurant. He

377

seemed happy to see her, yet oddly passive about the entire situation. They talked for a few hours, and he recounted that, after he left Wisconsin, he had drifted away from teaching and gotten into the restaurant business.

Jobs had asked Simpson not to mention him, so she didn't. But at one point her father casually remarked that he and her mother had had another baby, a boy, before she had been born. "What happened to him?" she asked. He replied, "We'll never see that baby again. That baby's gone." Simpson recoiled but said nothing.

An even more astonishing revelation occurred when Jandali was describing the previous restaurants that he had run. There had been some nice ones, he insisted, fancier than the Sacramento joint they were then sitting in. He told her, somewhat emotionally, that he wished she could have seen him when he was managing a Mediterranean restaurant north of San Jose. "That was a wonderful place," he said. "All of the successful technology people used to come there. *Even Steve Jobs.*" Simpson was stunned. "Oh, yeah, he used to come in, and he was a sweet guy, and a big tipper," her father added. Mona was able to refrain from blurting out, *Steve Jobs is your son!*

When the visit was over, she called Jobs surreptitiously from the pay phone at the restaurant and arranged to meet him at the Espresso Roma café in Berkeley. Adding to the personal and family drama, he brought along Lisa, now in grade school, who lived with her mother, Chrisann. When they all arrived at the café, it was close to 10 p.m., and Simpson poured forth the tale. Jobs

378

was understandably astonished when she mentioned the restaurant near San Jose. He could recall being there and even meeting the man who was his biological father. "It was amazing," he later said of the revelation. "I had been to that restaurant a few times, and I remember meeting the owner. He was Syrian. Balding. We shook hands."

Nevertheless Jobs still had no desire to see him. "I was a wealthy man by then, and I didn't trust him not to try to blackmail me or go to the press about it," he recalled. "I asked Mona not to tell him about me."

She never did, but years later Jandali saw his relationship to Jobs mentioned online. (A blogger noticed that Simpson had listed Jandali as her father in a reference book and figured out he must be Jobs's father as well.) By then Jandali was married for a fourth time and working as a food and beverage manager at the Boomtown Resort and Casino just west of Reno, Nevada. When he brought his new wife, Roscille, to visit Simpson in 2006, he raised the topic. "What is this thing about Steve Jobs?" he asked. She confirmed the story, but added that she thought Jobs had no interest in meeting him. Jandali seemed to accept that. "My father is thoughtful and a beautiful storyteller, but he is very, very passive," Simpson said. "He never contacted Steve."

Simpson turned her search for Jandali into a basis for her second novel, *The Lost Father,* published in 1992. (Jobs convinced Paul Rand, the designer who did the NeXT logo, to design the cover, but according to Simpson, "It was God-awful and we never used it.") She also tracked down various members of the Jandali family, in Homs and in America, and in 2011 was writing a novel about her Syrian roots. The Syrian ambassador in Washington threw a dinner for her that included a cousin and his wife who then lived in Florida and had flown up for the occasion.

Simpson assumed that Jobs would eventually meet Jandali, but as time went on he showed even less interest. In 2010, when Jobs and his son, Reed, went to a birthday dinner for Simpson at her Los Angeles house, Reed spent some time looking at pictures of his biological grandfather, but Jobs ignored them. Nor did he seem to care about his Syrian heritage. When the Middle East would come up in conversation, the topic did not engage him or evoke his typical strong opinions, even after Syria was swept up in the 2011 Arab Spring uprisings. "I don't think anybody really knows what we should be doing over there," he said when I asked whether the Obama administration should be intervening more in Egypt, Libya, and Syria. "You're fucked if you do and you're fucked if you don't."

Jobs did retain a friendly relationship with his biological mother, Joanne Simpson. Over the years she and Mona would often spend Christmas at Jobs's house. The visits could be sweet, but also emotionally draining. Joanne would sometimes break into tears, say how much she had loved him, and apologize for giving him up. It turned out all right, Jobs would reassure her. As he told her one Christmas, "Don't worry. I had a great childhood. I turned out okay."

LISA

Lisa Brennan, however, did not have a great childhood. When she was young, her father almost never came to see her. "I didn't want to be a father, so I wasn't," Jobs later said, with only a touch of remorse in his voice. Yet occasionally he felt the tug. One day, when Lisa was three, Jobs was driving near the house he had bought for her and Chrisann, and he decided to stop. Lisa didn't know who he was. He sat on the doorstep, not venturing inside, and talked to Chrisann. The scene was repeated once or twice a year. Jobs would come by unannounced, talk a little bit about Lisa's school options or other issues, then drive off in his Mercedes.

But by the time Lisa turned eight, in 1986, the visits were occurring more frequently. Jobs was no longer immersed in the grueling push to create the Macintosh or in the subsequent power struggles with Sculley. He was at NeXT, which was calmer, friendlier, and headquartered in Palo Alto, near where Chrisann and Lisa lived. In addition, by the time she was in third grade, it was clear that Lisa was a smart and artistic kid, who had already been singled out by her teachers for her writing ability. She was spunky and high-spirited and had a little of her father's defiant attitude. She also looked a bit like him, with arched eyebrows and a faintly Middle Eastern angularity. One day, to the surprise of his colleagues, he brought her by the office. As she turned cartwheels in the corridor, she squealed, "Look at me!"

Avie Tevanian, a lanky and gregarious engineer at NeXT who had become Jobs's friend, remembers that every now and then, when they were going out to dinner, they would stop by Chrisann's house to pick up Lisa. "He was very sweet to her," Tevanian recalled. "He was a vegetarian, and so was Chrisann, but she wasn't. He was fine with that. He suggested she order chicken, and she did."

Eating chicken became her little indulgence as she shuttled between two parents who were vegetarians with a spiritual regard for natural foods. "We bought our groceries — our puntarella, quinoa, celeriac, carob-covered nuts — in yeasty-smelling stores where the women didn't dye their hair," she later wrote about her time with her mother. "But we sometimes tasted foreign treats. A few times we bought a hot, seasoned chicken from a gourmet shop with rows and rows of chickens turning on spits, and ate it in the car from the foil-lined paper bag with our fingers." Her father, whose dietary fixations came in fanatic waves, was more fastidious about what he ate. She watched him spit out a mouthful of soup one day after learning that it contained butter. After loosening up a bit while at Apple, he was back to being a strict vegan. Even at a young age Lisa began to realize his diet obsessions reflected a life philosophy, one in which asceticism and minimalism could heighten subsequent sensations. "He believed that great harvests came from arid sources, pleasure from restraint," she noted. "He knew the equations that most people didn't know: Things led to their opposites."

In a similar way, the absence and coldness of her father made his occasional moments of warmth so much more intensely gratifying. "I didn't live with him, but he would stop by our house some days, a deity among us for a few tingling moments or hours," she recalled. Lisa soon became interesting enough that he would take walks with her. He would also go rollerblading with her on the quiet streets of old Palo Alto, often stopping at the houses of Joanna Hoffman and Andy Hertzfeld. The first time he brought her around to see Hoffman, he just knocked on the door and announced, "This is Lisa." Hoffman knew right away. "It was obvious she was his daughter," she told me. "Nobody has that jaw. It's a signature jaw." Hoffman, who suffered from not knowing her own divorced father until she was ten, encouraged Jobs to be a better father. He followed her advice, and later thanked her for it.

Once he took Lisa on a business trip to Tokyo, and they stayed at the sleek and businesslike Okura Hotel. At the elegant downstairs sushi bar, Jobs ordered large trays of unagi sushi, a dish he loved so much that he allowed the warm cooked eel to pass muster as vegetarian. The pieces were coated with fine salt or a thin sweet sauce, and Lisa remembered later how they dissolved in her mouth. So, too, did the distance between them. As she later wrote, "It was the first time I'd felt, with him, so relaxed and content, over those trays of meat; the excess, the permission and warmth after the cold salads, meant a once inaccessible space had opened. He was less rigid with himself, even human under the great ceilings with the little chairs, with the meat, and me."

But it was not always sweetness and light. Jobs was as mercurial with Lisa as he was with almost everyone, cycling between embrace and abandonment. On one visit he would be playful; on the next he would be cold; often he was not there at all. "She was always unsure of their relationship," according to Hertzfeld. "I went to a birthday party of hers, and Steve was supposed to come, and he was very, very, late. She got extremely anxious and disappointed. But when he finally did come, she totally lit up."

Lisa learned to be temperamental in return. Over the years their relationship would be a roller coaster, with each of the low points elongated by their shared stubbornness. After a falling-out, they could go for months not speaking to each other. Neither one was good at reaching out, apologizing, or making the effort to heal, even when he was wrestling with repeated health problems. One day in the fall of 2010 he was wistfully going through a box of old snapshots with me, and paused over one that showed him visiting Lisa when she was young. "I probably didn't go over there enough," he said. Since he had not spoken to her all that year, I asked if he might want to reach out to her with a call or email. He looked at me blankly for a moment, then went back to riffling through other old photographs.

THE ROMANTIC

When it came to women, Jobs could be deeply romantic. He tended to fall in love dramatically, share with friends every up and down of a relationship, and pine in public whenever he was away from his current girlfriend. In the summer of 1983 he went to a small dinner party in Silicon Valley with Joan Baez and sat next to an undergraduate at the University of Pennsylvania named Jennifer Egan, who was not quite sure who he was. By then he and Baez had realized that they weren't destined to be forever young together, and Jobs found himself fascinated by Egan, who was working on a San Francisco weekly during her summer vacation. He tracked her down, gave her a call, and took her to Café Jacqueline, a little bistro near Telegraph Hill that specialized in vegetarian soufflés.

They dated for a year, and Jobs often flew east to visit her. At a Boston Macworld event, he told a large gathering how much in love he was and thus needed to rush out to catch a plane for Philadelphia to see his girlfriend. The audience was enchanted. When he was visiting New York, she would take the train up to stay with him at the Carlyle or at Jay Chiat's Upper East Side apartment, and they would eat at Café Luxembourg, visit (repeatedly) the apartment in the San Remo he was planning to remodel, and go to movies or (once at least) the opera.

He and Egan also spoke for hours on the phone many nights. One topic they wrestled with was his belief, which came from his Buddhist studies, that it was important to avoid attachment to material objects. Our consumer desires are unhealthy, he told her, and to attain enlightenment you need to develop a life of nonattachment and nonmaterialism. He even sent her a tape of Kobun Chino, his Zen teacher, lecturing about the problems caused by craving and obtaining things. Egan pushed back. Wasn't he defying that philosophy, she asked, by making computers and other products that people coveted? "He was irritated by the dichotomy, and we had exuberant debates about it," Egan recalled.

In the end Jobs's pride in the objects he made overcame his sensibility that people should eschew being attached to such possessions. When the Macintosh came out in January 1984, Egan was staying at her mother's apartment in San Francisco during her winter break from Penn. Her mother's dinner guests were astonished one night when Steve Jobs — suddenly very famous — appeared at the door carrying a freshly boxed Macintosh and proceeded to Egan's bedroom to set it up.

Jobs told Egan, as he had a few other friends, about his premonition that he would not live a long life. That was why he was driven and impatient, he confided. "He felt a sense of urgency about all he wanted to get done," Egan later said. Their relationship tapered off by the fall of 1984, when Egan made it clear that she was still far too young to think of getting married.

Shortly after that, just as the turmoil with Sculley was beginning to build at Apple in early 1985, Jobs was heading to a meeting when he stopped at the office of a guy who was working with the Apple Foundation, which helped get computers to nonprofit organizations. Sitting in his office was a lithe, very blond woman who combined a hippie aura of natural purity with the solid sensibilities of a computer consultant. Her name was Tina Redse. "She was the most beautiful woman I'd ever seen," Jobs recalled.

He called her the next day and asked her to dinner. She said no, that she was living with a boyfriend. A few days later he took her on a walk to a nearby park and again asked her out, and this time she told her boyfriend that she wanted to go. She was very honest and open. After dinner she started to cry because she knew her life was about to be disrupted. And it was. Within a few months she had moved into the unfurnished mansion in Woodside. "She was the first person I was truly in love with," Jobs later said. "We had a very deep connection. I don't know that anyone will ever understand me better than she did."

Redse came from a troubled family, and Jobs shared with her his own pain about being put up for adoption. "We were both wounded from our childhood," Redse recalled. "He said to me that we were misfits, which is why we belonged together." They were physically passionate and prone to public displays of affection; their make-out sessions in the NeXT lobby are well remembered by employees. So too were their fights, which occurred at movie theaters and in front of visitors to Woodside. Yet he constantly praised her purity and naturalness. As the well-grounded Joanna Hoffman pointed out when discussing Jobs's infatuation with the otherworldly Redse, "Steve had a tendency to look at vulnerabilities and neuroses and turn them into spiritual attributes."

When he was being eased out at Apple in 1985, Redse traveled with him in Europe, where he was salving his wounds. Standing on a bridge over the Seine one evening, they bandied about the idea, more romantic than serious, of just staying in France, maybe settling down, perhaps indefinitely. Redse was eager, but Jobs didn't want to. He was burned but still ambitious. "I am a reflection of what I do," he told her. She recalled their Paris moment in a poignant email she sent to him twenty-five years later, after they had gone their separate ways but retained their spiritual connection:

We were on a bridge in Paris in the summer of 1985. It was overcast. We leaned against the smooth stone rail and stared at the green water rolling on below. Your world had cleaved and then it paused, waiting to rearrange itself around whatever you chose next. I wanted to run away from what had come before. I tried to convince you to begin a new life with me in Paris, to shed our former selves and let something else course through us. I wanted us to crawl through that black chasm of your broken world and emerge, anonymous and new, in simple lives where I could cook you simple dinners and we could be together every day, like children playing a sweet game with no purpose save the game itself. I like to think you considered it before you laughed and said "What could I do? I've made myself unemployable." I like to think that in that moment's hesitation before our bold futures reclaimed us, we lived that simple life together all the way into our peaceful old ages, with a brood of grandchildren around us on a farm in the south of France, quietly going about our days, warm and complete like loaves of fresh bread, our small world filled with the aroma of patience and familiarity.

The relationship lurched up and down for five years. Redse hated living in his sparsely furnished

Woodside house. Jobs had hired a hip young couple, who had once worked at Chez Panisse, as housekeepers and vegetarian cooks, and they made her feel like an interloper. She would occasionally move out to an apartment of her own in Palo Alto, especially after one of her torrential arguments with Jobs. "Neglect is a form of abuse," she once scrawled on the wall of the hallway to their bedroom. She was entranced by him, but she was also baffled by how uncaring he could be. She would later recall how incredibly painful it was to be in love with someone so self-centered. Caring deeply about someone who seemed incapable of caring was a particular kind of hell that she wouldn't wish on anyone, she said.

They were different in so many ways. "On the spectrum of cruel to kind, they are close to the opposite poles," Hertzfeld later said. Redse's kindness was manifest in ways large and small; she always gave money to street people, she volunteered to help those who (like her father) were afflicted with mental illness, and she took care to make Lisa and even Chrisann feel comfortable with her. More than anyone, she helped persuade Jobs to spend more time with Lisa. But she lacked Jobs's ambition and drive. The ethereal quality that made her seem so spiritual to Jobs also made it hard for them to stay on the same wavelength. "Their relationship was incredibly tempestuous," said Hertzfeld. "Because of both of their characters, they would have lots and lots of fights."

They also had a basic philosophical difference about whether aesthetic tastes were fundamentally individual, as Redse believed, or universal and could be taught, as Jobs believed. She accused him of being too influenced by the Bauhaus movement. "Steve believed it was our job to teach people aesthetics, to teach people what they should like," she recalled. "I don't share that perspective. I believe when we listen deeply, both within ourselves and to each other, we are able to allow what's innate and true to emerge."

When they were together for a long stretch, things did not work out well. But when they were apart, Jobs would pine for her. Finally, in the summer of 1989, he asked her to marry him. She couldn't do it. It would drive her crazy, she told friends. She had grown up in a volatile household, and her relationship with Jobs bore too many similarities to that environment. They were opposites who attracted, she said, but the combination was too combustible. "I could not have been a good wife to 'Steve Jobs,' the icon," she later explained. "I would have sucked at it on many levels. In our personal interactions, I couldn't abide his unkindness. I didn't want to hurt him, yet I didn't want to stand by and watch him hurt other people either. It was painful and exhausting."

After they broke up, Redse helped found Open-Mind, a mental health resource network in California. She happened to read in a psychiatric manual about Narcissistic Personality Disorder and decided that Jobs perfectly met the criteria. "It fits so well and explained so much of what we had struggled with, that I realized expecting him to be nicer or less self-centered was like expecting a blind man to see," she said. "It also explained some of the choices he'd made about his daughter

Lisa at that time. I think the issue is empathy — the capacity for empathy is lacking."

Redse later married, had two children, and then divorced. Every now and then Jobs would openly pine for her, even after he was happily married. And when he began his battle with cancer, she got in touch again to give support. She became very emotional whenever she recalled their relationship. "Though our values clashed and made it impossible for us to have the relationship we once hoped for," she told me, "the care and love I felt for him decades ago has continued." Similarly, Jobs suddenly started to cry one afternoon as he sat in his living room reminiscing about her. "She was one of the purest people I've ever known," he said, tears rolling down his cheeks. "There was something spiritual about her and spiritual about the connection we had." He said he always regretted that they could not make it work, and he knew that she had such regrets as well. But it was not meant to be. On that they both agreed.

CHAPTER TWENTY-ONE: FAMILY MAN

AT HOME WITH THE JOBS CLAN

LAURENE POWELL

By this point, based on his dating history, a matchmaker could have put together a composite sketch of the woman who would be right for Jobs. Smart, yet unpretentious. Tough enough to stand up to him, yet Zen-like enough to rise above turmoil. Well-educated and independent, yet ready to make accommodations for him and a family. Down-to-earth, but with a touch of the ethereal. Savvy enough to know how to manage him, but secure enough to not always need to. And it wouldn't hurt to be a beautiful, lanky blonde with an easygoing sense of humor who liked organic vegetarian food. In October 1989, after his split with Tina Redse, just such a woman walked into his life.

More specifically, just such a woman walked into his classroom. Jobs had agreed to give one of the "View from the Top" lectures at the Stanford Business School one Thursday evening. Laurene Powell was a new graduate student at the business school, and a guy in her class talked her into going to the lecture. They arrived late and all the seats were taken, so they sat in the aisle. When an usher told them they had to move, Powell took her friend down to the front row and commandeered two of the reserved seats there. Jobs was led to the one next to her when he arrived. "I looked to my right, and there was a beautiful girl there, so we started chatting while I was waiting to be introduced," Jobs recalled. They bantered a bit, and Laurene joked that she was sitting there because she had won a raffle, and the prize was that he got to take her to dinner. "He was so adorable," she later said.

After the speech Jobs hung around on the edge of the stage chatting with students. He watched Powell leave, then come back and stand at the edge of the crowd, then leave again. He bolted out after her, brushing past the dean, who was trying to grab him for a conversation. After catching up with her in the parking lot, he said, "Excuse me, wasn't there something about a raffle you won, that I'm supposed to take you to dinner?" She laughed. "How about Saturday?" he asked. She agreed and wrote down her number. Jobs headed to his car to drive up to the Thomas Fogarty winery in the Santa Cruz mountains above Woodside, where the NeXT education sales group was holding a dinner. But he suddenly stopped and turned around. "I thought, wow, I'd rather have dinner with her than the education group, so I ran back to her car and said 'How about dinner *tonight?*' " She said yes. It was a beautiful fall evening, and they walked into Palo Alto to a funky vegetarian restaurant, St. Michael's Alley, and ended up staying there for four hours. "We've been together ever since," he said.

Avie Tevanian was sitting at the winery restaurant waiting with the rest of the NeXT education group. "Steve was sometimes unreliable, but when I talked to him I realized that something special had come up," he said. As soon as Powell got home, after midnight, she called her close friend Kathryn (Kat) Smith, who was at Berkeley, and left a message on her machine. "You will not believe what just happened to me!" it said. "You will not believe who I met!" Smith called back the next morning and heard the tale. "We had known about Steve, and he was a person of interest to us, because we were business students," she recalled.

Andy Hertzfeld and a few others later speculated that Powell had been scheming to meet Jobs. "Laurene is nice, but she can be calculating, and I think she targeted him from the beginning," Hertzfeld said. "Her college roommate told me that Laurene had magazine covers of Steve and vowed she was going to meet him. If it's true that Steve was manipulated, there is a fair amount of irony there." But Powell later insisted that this wasn't the case. She went only because her friend wanted to go, and she was slightly confused as to who they were going to see. "I knew that Steve Jobs was the speaker, but the face I thought of was that of Bill Gates," she recalled. "I had them mixed up. This was 1989. He was working at NeXT, and he was not that big of a deal to me. I wasn't that enthused, but my friend was, so we went."

"There were only two women in my life that I was truly in love with, Tina and Laurene," Jobs later said. "I thought I was in love with Joan Baez, but I really just liked her a lot. It was just Tina and then Laurene."

■ ■ ■

Laurene Powell had been born in New Jersey in 1963 and learned to be self-sufficient at an early age. Her father was a Marine Corps pilot who died a hero in a crash in Santa Ana, California; he had been leading a crippled plane in for a landing, and when it hit his plane he kept flying to avoid a residential area rather than ejecting in time to save his life. Her mother's second marriage turned out to be a horrible situation, but she felt she couldn't leave because she had no means to support her large family. For ten years Laurene and her three brothers had to suffer in a tense household, keeping a good demeanor while compartmentalizing problems. She did well. "The lesson I learned was clear, that I always wanted to be self-sufficient," she said. "I took pride in that. My relationship with money is that it's a tool to be self-sufficient, but it's not something that is part of who I am."

After graduating from the University of Pennsylvania, she worked at Goldman Sachs as a fixed income trading strategist, dealing with enormous sums of money that she traded for the house account. Jon Corzine, her boss, tried to get her to stay at Goldman, but instead she decided the work was unedifying. "You could be really successful," she said, "but you're just contributing to capital formation." So after three years she quit and went to Florence, Italy, living there for eight months before enrolling in Stanford Business School.

After their Thursday night dinner, she invited Jobs over to her Palo Alto apartment on Saturday. Kat Smith drove down from Berkeley and pretended to be her roommate so she could meet him as well. Their relationship became very passionate. "They would kiss and make out," Smith said. "He was enraptured with her. He would call me on the phone and ask, 'What do you think, does she like me?' Here I am in this bizarre position of having this iconic person call me."

That New Year's Eve of 1989 the three went to Chez Panisse, the famed Alice Waters restaurant in Berkeley, along with Lisa, then eleven. Something happened at the dinner that caused Jobs and Powell to start arguing. They left separately, and Powell ended up spending the night at Kat Smith's apartment. At nine the next morning there was a knock at the door, and Smith opened it to find Jobs, standing in the drizzle holding some wildflowers he had picked. "May I come in and see Laurene?" he said. She was still asleep, and he walked into the bedroom. A couple of hours went by, while Smith waited in the living room, unable to go in and get her clothes. Finally, she put a coat on over her nightgown and went to Peet's Coffee to pick up some food. Jobs did not emerge until after noon. "Kat, can you come here for a minute?" he asked. They all gathered in the bedroom. "As you know, Laurene's father passed away, and Laurene's mother isn't here, and since you're her best friend, I'm going to ask you the question," he said. "I'd like to marry Laurene. Will you give your blessing?"

Smith clambered onto the bed and thought about it. "Is this okay with you?" she asked Powell. When she nodded yes, Smith announced, "Well, there's your answer."

It was not, however, a definitive answer. Jobs had a way of focusing on something with insane intensity for a while and then, abruptly, turning away his gaze. At work, he would focus on what he wanted to, when he wanted to, and on other matters he would be unresponsive, no matter how hard people tried to get him to engage. In his personal life, he was the same way. At times he and Powell would indulge in public displays of affection that were so intense they embarrassed everyone in their presence, including Kat Smith and Powell's mother. In the mornings at his Woodside mansion, he would wake Powell up by blasting the Fine Young Cannibals' "She Drives Me Crazy" on his tape deck. Yet at other times he would ignore her. "Steve would fluctuate between intense focus, where she was the center of the universe, to being coldly distant and focused on work," said Smith. "He had the power to focus like a laser beam, and when it came across you, you basked in the light of his attention. When it moved to another point of focus, it was very, very dark for you. It was very confusing to Laurene."

Once she had accepted his marriage proposal on the first day of 1990, he didn't mention it again for several months. Finally, Smith confronted him while they were sitting on the edge of a sandbox in Palo Alto. What was going on? Jobs replied that he needed to feel sure that Powell could handle the life he lived and the type of person he was. In September she became fed up with waiting and moved out. The following month, he gave her a diamond engagement ring, and she moved back in.

In December Jobs took Powell to his favorite vacation spot, Kona Village in Hawaii. He had started going there nine years earlier when, stressed out at Apple, he had asked his assistant to pick out a place for him to escape. At first glance, he didn't like the cluster of sparse thatched-roof bungalows nestled on a beach on the big island of Hawaii. It was a family resort, with communal eating. But within hours he had begun to view it as paradise. There was a simplicity and spare beauty that moved him, and he returned whenever he could. He especially enjoyed being there that December with Powell. Their love had matured. The night before Christmas he again declared, even more formally, that he wanted to marry her. Soon another factor would drive that decision. While in Hawaii, Powell got pregnant. "We know exactly where it happened," Jobs later said with a laugh.

THE WEDDING, MARCH 18, 1991

Powell's pregnancy did not completely settle the issue. Jobs again began balking at the idea of marriage, even though he had dramatically proposed to her both at the very beginning and the very end of 1990. Furious, she moved out of his house and back to her apartment. For a while he sulked or ignored the situation. Then he thought he might still be in love with Tina Redse; he sent her roses and tried to convince her to return to him, maybe even get married. He was not sure what he wanted, and he surprised a wide swath of friends and even acquaintances by asking them what he should do. Who was prettier, he would ask, Tina or Laurene? Who did they like better? Who should he marry? In a chapter about this in Mona Simpson's novel *A*

Regular Guy, the Jobs character "asked more than a hundred people who they thought was more beautiful." But that was fiction; in reality, it was probably fewer than a hundred.

He ended up making the right choice. As Redse told friends, she never would have survived if she had gone back to Jobs, nor would their marriage. Even though he would pine about the spiritual nature of his connection to Redse, he had a far more solid relationship with Powell. He liked her, he loved her, he respected her, and he was comfortable with her. He may not have seen her as mystical, but she was a sensible anchor for his life. "He is the luckiest guy to have landed with Laurene, who is smart and can engage him intellectually and can sustain his ups and downs and tempestuous personality," said Joanna Hoffman. "Because she's not neurotic, Steve may feel that she is not as mystical as Tina or something. But that's silly." Andy Hertzfeld agreed. "Laurene looks a lot like Tina, but she is totally different because she is tougher and armor-plated. That's why the marriage works."

Jobs understood this as well. Despite his emotional turbulence and occasional meanness, the marriage would turn out to be enduring, marked by loyalty and faithfulness, overcoming the ups and downs and jangling emotional complexities it encountered.

Avie Tevanian decided Jobs needed a bachelor's party. This was not as easy as it sounded. Jobs did not like to party and didn't have a gang of male buddies. He didn't even have a best man. So the party turned out to be just Tevanian and Richard Crandall, a computer science professor at Reed who had taken a leave to work at NeXT. Tevanian hired a limo, and when they got to Jobs's house, Powell answered the door dressed in a suit and wearing a fake moustache, saying that she wanted to come as one of the guys. It was just a joke, and soon the three bachelors, none of them drinkers, were rolling to San Francisco to see if they could pull off their own pale version of a bachelor party.

Tevanian had been unable to get reservations at Greens, the vegetarian restaurant at Fort Mason that Jobs liked, so he booked a very fancy restaurant at a hotel. "I don't want to eat here," Jobs announced as soon as the bread was placed on the table. He made them get up and walk out, to the horror of Tevanian, who was not yet used to Jobs's restaurant manners. He led them to Café Jacqueline in North Beach, the soufflé place that he loved, which was indeed a better choice. Afterward they took the limo across the Golden Gate Bridge to a bar in Sausalito, where all three ordered shots of tequila but only sipped them. "It was not great as bachelor parties go, but it was the best we could come up with for someone like Steve, and nobody else volunteered to do it," recalled Tevanian. Jobs was appreciative. He decided that he wanted Tevanian to marry his sister Mona Simpson. Though nothing came of it, the thought was a sign of affection.

Powell had fair warning of what she was getting into. As she was planning the wedding, the person who was going to do the calligraphy for the invitations came by the house to show them some op-

400

tions. There was no furniture for her to sit on, so she sat on the floor and laid out the samples. Jobs looked for a few minutes, then got up and left the room. They waited for him to come back, but he didn't. After a while Powell went to find him in his room. "Get rid of her," he said. "I can't look at her stuff. It's shit."

On March 18, 1991, Steven Paul Jobs, thirty-six, married Laurene Powell, twenty-seven, at the Ahwahnee Lodge in Yosemite National Park. Built in the 1920s, the Ahwahnee is a sprawling pile of stone, concrete, and timber designed in a style that mixed Art Deco, the Arts and Crafts movement, and the Park Service's love of huge fireplaces. Its best features are the views. It has floor-to-ceiling windows looking out on Half Dome and Yosemite Falls.

About fifty people came, including Steve's father Paul Jobs and sister Mona Simpson. She brought her fiancé, Richard Appel, a lawyer who went on to become a television comedy writer. (As a writer for *The Simpsons,* he named Homer's mother after his wife.) Jobs insisted that they all arrive by chartered bus; he wanted to control all aspects of the event.

The ceremony was in the solarium, with the snow coming down hard and Glacier Point just visible in the distance. It was conducted by Jobs's longtime Sōtō Zen teacher, Kobun Chino, who shook a stick, struck a gong, lit incense, and chanted in a mumbling manner that most guests found incomprehensible. "I thought he was drunk," said Tevanian. He wasn't. The wedding cake was in

401

the shape of Half Dome, the granite crest at the end of Yosemite Valley, but since it was strictly vegan — devoid of eggs, milk, or any refined products — more than a few of the guests found it inedible. Afterward they all went hiking, and Powell's three strapping brothers launched a snowball fight, with lots of tackling and roughhousing. "You see, Mona," Jobs said to his sister, "Laurene is descended from Joe Namath and we're descended from John Muir."

A FAMILY HOME

Powell shared her husband's interest in natural foods. While at business school, she had worked part time at Odwalla, the juice company, where she helped develop the first marketing plan. After marrying Jobs, she felt that it was important to have a career, having learned from her childhood the need to be self-sufficient. So she started her own company, Terravera, that made ready-to-eat organic meals and delivered them to stores throughout northern California.

Instead of living in the isolated and rather spooky unfurnished Woodside mansion, the couple moved into a charming and unpretentious house on a corner in a family-friendly neighborhood in old Palo Alto. It was a privileged realm — neighbors would eventually include the visionary venture capitalist John Doerr, Google's founder Larry Page, and Facebook's founder Mark Zuckerberg, along with Andy Hertzfeld and Joanna Hoffman — but the homes were not ostentatious, and there were no high hedges or long drives shielding them from view. Instead, houses were nestled on lots next to each other along flat, quiet streets flanked by wide sidewalks. "We wanted to live in a neighborhood where kids could walk to see friends," Jobs later said.

The house was not the minimalist and modernist style Jobs would have designed if he had built a home from scratch. Nor was it a large or distinctive mansion that would make people stop and take notice as they drove down his street in Palo Alto. It was built in the 1930s by a local designer named Carr Jones, who specialized in carefully crafted homes in the "storybook style" of English or French country cottages.

The two-story house was made of red brick, with exposed wood beams and a shingle roof with curved lines; it evoked a rambling Cotswold cottage, or perhaps a home where a well-to-do Hobbit might have lived. The one Californian touch was a mission-style courtyard framed by the wings of the house. The two-story vaulted-ceiling living room was informal, with a floor of tile and terra-cotta. At one end was a large triangular window leading up to the peak of the ceiling; it had stained glass when Jobs bought it, as if it were a chapel, but he replaced it with clear glass. The other renovation he and Powell made was to expand the kitchen to include a wood-burning pizza oven and room for a long wooden table that would become the family's primary gathering place. It was supposed to be a four-month renovation, but it took sixteen months because Jobs kept redoing the design. They also bought the small house behind them and razed it to make a backyard, which Powell turned into a beautiful

403

natural garden filled with a profusion of seasonal flowers along with vegetables and herbs.

Jobs became fascinated by the way Carr Jones relied on old material, including used bricks and wood from telephone poles, to provide a simple and sturdy structure. The beams in the kitchen had been used to make the molds for the concrete foundations of the Golden Gate Bridge, which was under construction when the house was built. "He was a careful craftsman who was self-taught," Jobs said as he pointed out each of the details. "He cared more about being inventive than about making money, and he never got rich. He never left California. His ideas came from reading books in the library and *Architectural Digest*."

Jobs had never furnished his Woodside house beyond a few bare essentials: a chest of drawers and a mattress in his bedroom, a card table and some folding chairs in what would have been a dining room. He wanted around him only things that he could admire, and that made it hard simply to go out and buy a lot of furniture. Now that he was living in a normal neighborhood home with a wife and soon a child, he had to make some concessions to necessity. But it was hard. They got beds, dressers, and a music system for the living room, but items like sofas took longer. "We spoke about furniture in theory for eight years," recalled Powell. "We spent a lot of time asking ourselves, 'What is the purpose of a sofa?' " Buying appliances was also a philosophical task, not just an impulse purchase. A few years later, Jobs described to *Wired* the process that went into getting a new washing machine:

It turns out that the Americans make washers and dryers all wrong. The Europeans make them much better — but they take twice as long to do clothes! It turns out that they wash them with about a quarter as much water and your clothes end up with a lot less detergent on them. Most important, they don't trash your clothes. They use a lot less soap, a lot less water, but they come out much cleaner, much softer, and they last a lot longer. We spent some time in our family talking about what's the trade-off we want to make. We ended up talking a lot about design, but also about the values of our family. Did we care most about getting our wash done in an hour versus an hour and a half? Or did we care most about our clothes feeling really soft and lasting longer? Did we care about using a quarter of the water? We spent about two weeks talking about this every night at the dinner table.

They ended up getting a Miele washer and dryer, made in Germany. "I got more thrill out of them than I have out of any piece of high tech in years," Jobs said.

The one piece of art that Jobs bought for the vaulted-ceiling living room was an Ansel Adams print of the winter sunrise in the Sierra Nevada taken from Lone Pine, California. Adams had made the huge mural print for his daughter, who later sold it. At one point Jobs's housekeeper wiped it with a wet cloth, and Jobs tracked down a person who had worked with Adams to come to the house, strip it down a layer, and restore it.

The house was so unassuming that Bill Gates was

somewhat baffled when he visited with his wife. "Do *all* of you live here?" asked Gates, who was then in the process of building a 66,000-square-foot mansion near Seattle. Even when he had his second coming at Apple and was a world-famous billionaire, Jobs had no security guards or live-in servants, and he even kept the back door unlocked during the day.

His only security problem came, sadly and strangely, from Burrell Smith, the mop-headed, cherubic Macintosh software engineer who had been Andy Hertzfeld's sidekick. After leaving Apple, Smith descended into schizophrenia. He lived in a house down the street from Hertzfeld, and as his disorder progressed he began wandering the streets naked, at other times smashing the windows of cars and churches. He was put on strong medication, but it proved difficult to calibrate. At one point when his demons returned, he began going over to the Jobs house in the evenings, throwing rocks through the windows, leaving rambling letters, and once tossing a firecracker into the house. He was arrested, but the case was dropped when he went for more treatment. "Burrell was so funny and naïve, and then one April day he suddenly snapped," Jobs recalled. "It was the weirdest, saddest thing."

Jobs was sympathetic, and often asked Hertzfeld what more he could do to help. At one point Smith was thrown in jail and refused to identify himself. When Hertzfeld found out, three days later, he called Jobs and asked for assistance in getting him released. Jobs did help, but he surprised Hertzfeld with a question: "If something similar happened to me, would you take as good care of me as you do Burrell?"

Jobs kept his mansion in Woodside, about ten miles up into the mountains from Palo Alto. He wanted to tear down the fourteen-bedroom 1925 Spanish colonial revival, and he had plans drawn up to replace it with an extremely simple, Japanese-inspired modernist home one-third the size. But for more than twenty years he engaged in a slow-moving series of court battles with preservationists who wanted the crumbling original house to be saved. (In 2011 he finally got permission to raze the house, but by then he had no desire to build a second home.)

On occasion Jobs would use the semi-abandoned Woodside home, especially its swimming pool, for family parties. When Bill Clinton was president, he and Hillary Clinton stayed in the 1950s ranch house on the property on their visits to their daughter, who was at Stanford. Since both the main house and ranch house were unfurnished, Powell would call furniture and art dealers when the Clintons were coming and pay them to furnish the houses temporarily. Once, shortly after the Monica Lewinsky flurry broke, Powell was making a final inspection of the furnishings and noticed that one of the paintings was missing. Worried, she asked the advance team and Secret Service what had happened. One of them pulled her aside and explained that it was a painting of a dress on a hanger, and given the issue of the blue dress in the Lewinsky matter they had decided to hide it. (During one of his late-night phone conversations with Jobs, Clinton asked how he should handle the Lewinsky issue. "I don't know if you did it, but if so, you've got to tell the country," Jobs told the president. There was silence on the other end of the line.)

LISA MOVES IN

In the middle of Lisa's eighth-grade year, her teachers called Jobs. There were serious problems, and it was probably best for her to move out of her mother's house. So Jobs went on a walk with Lisa, asked about the situation, and offered to let her move in with him. She was a mature girl, just turning fourteen, and she thought about it for two days. Then she said yes. She already knew which room she wanted: the one right next to her father's. When she was there once, with no one home, she had tested it out by lying down on the bare floor.

It was a tough period. Chrisann Brennan would sometimes walk over from her own house a few blocks away and yell at them from the yard. When I asked her recently about her behavior and the allegations that led to Lisa's moving out of her house, she said that she had still not been able to process in her own mind what occurred during that period. But then she wrote me a long email that she said would help explain the situation:

> Do you know how Steve was able to get the city of Woodside to allow him to tear his Woodside home down? There was a community of people who wanted to preserve his Woodside house due to its historical value, but Steve wanted to tear it down and build a home with an orchard. Steve let that house fall into so much disrepair

and decay over a number of years that there was no way to save it. The strategy he used to get what he wanted was to simply follow the line of least involvement and resistance. So by his doing nothing on the house, and maybe even leaving the windows open for years, the house fell apart. Brilliant, no? . . . In a similar way did Steve work to undermine my effectiveness AND my well being at the time when Lisa was 13 and 14 to get her to move into his house. He started with one strategy but then it moved to another easier one that was even more destructive to me and more problematic for Lisa. It may not have been of the greatest integrity, but he got what he wanted.

Lisa lived with Jobs and Powell for all four of her years at Palo Alto High School, and she began using the name Lisa Brennan-Jobs. He tried to be a good father, but there were times when he was cold and distant. When Lisa felt she had to escape, she would seek refuge with a friendly family who lived nearby. Powell tried to be supportive, and she was the one who attended most of Lisa's school events.

By the time Lisa was a senior, she seemed to be flourishing. She joined the school newspaper, *The Campanile,* and became the coeditor. Together with her classmate Ben Hewlett, grandson of the man who gave her father his first job, she exposed secret raises that the school board had given to administrators. When it came time to go to college, she knew she wanted to go east. She applied to Harvard — forging her father's signature on the application because he was out of town — and was accepted for the class entering in 1996.

At Harvard Lisa worked on the college newspaper, *The Crimson,* and then the literary magazine, *The Advocate.* After breaking up with her boyfriend, she took a year abroad at King's College, London. Her relationship with her father remained tumultuous throughout her college years. When she would come home, fights over small things — what was being served for dinner, whether she was paying enough attention to her half-siblings — would blow up, and they would not speak to each other for weeks and sometimes months. The arguments occasionally got so bad that Jobs would stop supporting her, and she would borrow money from Andy Hertzfeld or others. Hertzfeld at one point lent Lisa $20,000 when she thought that her father was not going to pay her tuition. "He was mad at me for making the loan," Hertzfeld recalled, "but he called early the next morning and had his accountant wire me the money." Jobs did not go to Lisa's Harvard graduation in 2000. He said, "She didn't even invite me."

There were, however, some nice times during those years, including one summer when Lisa came back home and performed at a benefit concert for the Electronic Frontier Foundation, an advocacy group that supports access to technology. The concert took place at the Fillmore Auditorium in San Francisco, which had been made famous by the Grateful Dead, Jefferson Airplane, and Jimi Hendrix. She sang Tracy Chapman's anthem "Talkin' bout a Revolution" ("Poor people are gonna rise up / And get their share") as her father stood in the back cradling his one-year-old daughter, Erin.

Jobs's ups and downs with Lisa continued after she moved to Manhattan as a freelance writer. Their problems were exacerbated because of Jobs's frustrations with Chrisann. He had bought a $700,000 house for Chrisann to use and put it in Lisa's name, but Chrisann convinced her to sign it over and then sold it, using the money to travel with a spiritual advisor and to live in Paris. Once the money ran out, she returned to San Francisco and became an artist creating "light paintings" and Buddhist mandalas. "I am a 'Connector' and a visionary contributor to the future of evolving humanity and the ascended Earth," she said on her website (which Hertzfeld maintained for her). "I experience the forms, color, and sound frequencies of sacred vibration as I create and live with the paintings." When Chrisann needed money for a bad sinus infection and dental problem, Jobs refused to give it to her, causing Lisa again to not speak to him for a few years. And thus the pattern would continue.

Mona Simpson used all of this, plus her imagination, as a springboard for her third novel, *A Regular Guy,* published in 1996. The book's title character is based on Jobs, and to some extent it adheres to reality: It depicts Jobs's quiet generosity to, and purchase of a special car for, a brilliant friend who had degenerative bone disease, and it accurately describes many unflattering aspects of his relationship with Lisa, including his original denial of paternity. But other parts are purely fiction; Chrisann had taught Lisa at a very early age how to drive, for example, but the book's scene of "Jane" driving a truck across the mountains alone at age five to find her father of course never happened. In addition, there are little details in the novel that, in journalist parlance, are too good to check, such as the head-snapping description of the character based on Jobs in the very first sentence: "He was a man too busy to flush toilets."

On the surface, the novel's fictional portrayal of Jobs seems harsh. Simpson describes her main character as unable "to see any need to pander to the wishes or whims of other people." His hygiene is also as dubious as that of the real Jobs. "He didn't believe in deodorant and often professed that with a proper diet and the peppermint castile soap, you would neither perspire nor smell." But the novel is lyrical and intricate on many levels, and by the end there is a fuller picture of a man who loses control of the great company he had founded and learns to appreciate the daughter he had abandoned. The final scene is of him dancing with his daughter.

Jobs later said that he never read the novel. "I heard it was about me," he told me, "and if it was about me, I would have gotten really pissed off, and I didn't want to get pissed at my sister, so I didn't read it." However, he told the *New York Times* a few months after the book appeared that he had read it and saw the reflections of himself in the main character. "About 25% of it is totally me, right down to the mannerisms," he told the reporter, Steve Lohr. "And I'm certainly not telling you which 25%." His wife said that, in fact, Jobs glanced at the book and asked her to read it for him to see what he should make of it.

Simpson sent the manuscript to Lisa before it was published, but at first she didn't read more than the opening. "In the first few pages, I was confronted with my family, my anecdotes, my things, my thoughts, myself in the character Jane," she noted. "And sandwiched between the truths was invention — lies to me, made more evident because of their dangerous proximity to the truth." Lisa was wounded, and she wrote a piece for the Harvard *Advocate* explaining why. Her first draft was very bitter, then she toned it down a bit before she published it. She felt violated by Simpson's friendship. "I didn't know, for those six years, that Mona was collecting," she wrote. "I didn't know that as I sought her consolations and took her advice, she, too, was taking." Eventually Lisa reconciled with Simpson. They went out to a coffee shop to discuss the book, and Lisa told her that she hadn't been able to finish it. Simpson told her she would like the ending. Over the years Lisa had an on-and-off relationship with Simpson, but it would be closer in some ways than the one she had with her father.

CHILDREN

When Powell gave birth in 1991, a few months after her wedding to Jobs, their child was known for two weeks as "baby boy Jobs," because settling on a name was proving only slightly less difficult than choosing a washing machine. Finally, they named him Reed Paul Jobs. His middle name was that of Jobs's father, and his first name (both Jobs and Powell insist) was chosen because it sounded good rather than because it was the name of Jobs's college.

Reed turned out to be like his father in many ways: incisive and smart, with intense eyes and a mesmerizing charm. But unlike his father, he had sweet manners and a self-effacing grace. He was creative — as a kid he liked to dress in costume and stay in character — and also a great student, interested in science. He could replicate his father's stare, but he was demonstrably affectionate and seemed not to have an ounce of cruelty in his nature.

Erin Siena Jobs was born in 1995. She was a little quieter and sometimes suffered from not getting much of her father's attention. She picked up her father's interest in design and architecture, but she also learned to keep a bit of an emotional distance, so as not to be hurt by his detachment.

The youngest child, Eve, was born in 1998, and she turned into a strong-willed, funny firecracker who, neither needy nor intimidated, knew how to handle her father, negotiate with him (and sometimes win), and even make fun of him. Her father joked that she's the one who will run Apple someday, if she doesn't become president of the United States.

Jobs developed a strong relationship with Reed, but with his daughters he was more distant. As he would with others, he would occasionally focus on them, but just as often would completely ignore them when he had other things on his mind. "He focuses on his work, and at times he has not been there for the girls," Powell said. At one point Jobs marveled to his wife at how well their kids were turning out, "especially since we're not always there for them." This amused, and slightly annoyed, Powell, because she had given up her career when Reed turned two and she decided she wanted to have more children.

In 1995 Oracle's CEO Larry Ellison threw a fortieth-birthday party for Jobs filled with tech stars and moguls. Ellison had become a close friend, and he would often take the Jobs family out on one of his many luxurious yachts. Reed started referring to him as "our rich friend," which was amusing evidence of how his father refrained from ostentatious displays of wealth. The lesson Jobs learned from his Buddhist days was that material possessions often cluttered life rather than enriched it. "Every other CEO I know has a security detail," he said. "They've even got them at their homes. It's a nutso way to live. We just decided that's not how we wanted to raise our kids."

CHAPTER TWENTY-TWO: TOY STORY

BUZZ AND WOODY TO THE RESCUE

JEFFREY KATZENBERG

"It's kind of fun to do the impossible," Walt Disney once said. That was the type of attitude that appealed to Jobs. He admired Disney's obsession with detail and design, and he felt that there was a natural fit between Pixar and the movie studio that Disney had founded.

The Walt Disney Company had licensed Pixar's Computer Animation Production System, and that made it the largest customer for Pixar's computers. One day Jeffrey Katzenberg, the head of Disney's film division, invited Jobs down to the Burbank studios to see the technology in operation. As the Disney folks were showing him around, Jobs turned to Katzenberg and asked, "Is Disney happy with Pixar?" With great exuberance, Katzenberg answered yes. Then Jobs asked, "Do you think we at Pixar are happy with Disney?" Katzenberg said he assumed so. "No, we're not," Jobs said. "We want to do a film with you. That would make us happy."

Katzenberg was willing. He admired John Lasseter's animated shorts and had tried unsuccessfully to lure him back to Disney. So Katzenberg invited the Pixar team down to discuss partnering on a film. When Catmull, Jobs, and Lasseter got settled at the conference table, Katzenberg was forthright. "John, since you won't come work for me," he said, looking at Lasseter, "I'm going to make it work this way."

Just as the Disney company shared some traits with Pixar, so Katzenberg shared some with Jobs. Both were charming when they wanted to be, and aggressive (or worse) when it suited their moods or interests. Alvy Ray Smith, on the verge of quitting Pixar, was at the meeting. "Katzenberg and Jobs impressed me as a lot alike," he recalled. "Tyrants with an amazing gift of gab." Katzenberg was delightfully aware of this. "Everybody thinks I'm a tyrant," he told the Pixar team. "I *am* a tyrant. But I'm usually right." One can imagine Jobs saying the same.

As befitted two men of equal passion, the negotiations between Katzenberg and Jobs took months. Katzenberg insisted that Disney be given the rights to Pixar's proprietary technology for making 3-D animation. Jobs refused, and he ended up winning that engagement. Jobs had his own demand: Pixar would have part ownership of the film and its characters, sharing control of both video rights and sequels. "If that's what you want," Katzenberg said, "we can just quit talking and you can leave now." Jobs stayed, conceding that point.

Lasseter was riveted as he watched the two wiry and tightly wound principals parry and thrust. "Just to see Steve and Jeffrey go at it, I was in awe," he recalled. "It was like a fencing match. They were both masters." But Katzenberg went into the match with a saber, Jobs with a mere foil. Pixar was on the verge of bankruptcy and needed a deal with Disney far more than Disney needed a deal with Pixar. Plus, Disney could afford to finance the whole enterprise, and Pixar couldn't. The result was a deal, struck in May 1991, by which Disney would own the picture and its characters outright, have creative control, and pay Pixar about 12.5% of the ticket revenues. It had the option (but not the obligation) to do Pixar's next two films and the right to make (with or without Pixar) sequels using the characters in the film. Disney could also kill the film at any time with only a small penalty.

The idea that John Lasseter pitched was called "Toy Story." It sprang from a belief, which he and Jobs shared, that products have an essence to them, a purpose for which they were made. If the object were to have feelings, these would be based on its desire to fulfill its essence. The purpose of a glass, for example, is to hold water; if it had feelings, it would be happy when full and sad when empty. The essence of a computer screen is to interface with a human. The essence of a unicycle is to be ridden in a circus. As for toys, their purpose is to be played with by kids, and thus their existential fear is of being discarded or upstaged by newer toys. So a buddy movie pairing an old favorite toy with a shiny new one would have an essential drama to it, especially when the action revolved around the toys' being separated from their kid. The original treatment began, "Everyone has had the traumatic childhood experience of losing a toy. Our story takes the toy's point of view as he loses and tries to regain the single thing most important to him: to be played with by children. This is the reason for the existence of all toys. It is the emotional foundation of their existence."

The two main characters went through many iterations before they ended up as Buzz Lightyear and Woody. Every couple of weeks, Lasseter and his team would put together their latest set of storyboards or footage to show the folks at Disney. In early screen tests, Pixar showed off its amazing technology by, for example, producing a scene of Woody rustling around on top of a dresser while the light rippling in through a Venetian blind cast shadows on his plaid shirt — an effect that would have been almost impossible to render by hand. Impressing Disney with the plot, however, was more difficult. At each presentation by Pixar, Katzenberg would tear much of it up, barking out his detailed comments and notes. And a cadre of clipboard-carrying flunkies was on hand to make sure every suggestion and whim uttered by Katzenberg received follow-up treatment.

Katzenberg's big push was to add more ediginess to the two main characters. It may be an animated movie called *Toy Story,* he said, but it should not be aimed only at children. "At first there was no drama, no real story, and no conflict," Katzenberg recalled. He suggested that Lasseter watch some classic buddy movies, such as *The Defiant Ones* and *48 Hours,* in which two characters with different attitudes are thrown together and have to bond. In addition, he kept pushing for what he called "edge," and that meant making Woody's character more jealous, mean, and belligerent toward Buzz, the new interloper in the toy box. "It's a toy-eat-toy world," Woody says at one point, after pushing Buzz out of a window.

After many rounds of notes from Katzenberg and other Disney execs, Woody had been stripped of almost all charm. In one scene he throws the other toys off the bed and orders Slinky to come help. When Slinky hesitates, Woody barks, "Who said your job was to think, spring-wiener?" Slinky then asks a question that the Pixar team members would soon be asking themselves: "Why is the cowboy so scary?" As Tom Hanks, who had signed up to be Woody's voice, exclaimed at one point, "This guy's a real jerk!"

CUT!

Lasseter and his Pixar team had the first half of the movie ready to screen by November 1993, so they brought it down to Burbank to show to Katzenberg and other Disney executives. Peter Schneider, the head of feature animation, had never been enamored of Katzenberg's idea of having outsiders make animation for Disney, and he declared it a mess and ordered that production be stopped. Katzenberg agreed. "Why is this so terrible?" he asked a colleague, Tom Schumacher. "Because it's not their movie anymore," Schumacher bluntly replied. He later explained, "They were following Katzenberg's notes, and the project had been driven completely off-track."

Lasseter realized that Schumacher was right. "I sat there and I was pretty much embarrassed with what was on the screen," he recalled. "It was a story filled with the most unhappy, mean characters that I've ever seen." He asked Disney for the chance to retreat back to Pixar and rework the script. Katzenberg was supportive.

Jobs did not insert himself much into the creative process. Given his proclivity to be in control, especially on matters of taste and design, this self-restraint was a testament to his respect for Lasseter and the other artists at Pixar — as well as for the ability of Lasseter and Catmull to keep him at bay. He did, however, help manage the relationship with Disney, and the Pixar team appreciated that. When Katzenberg and Schneider halted production on *Toy Story,* Jobs kept the work going with his own personal funding. And he took their side against Katzenberg. "He had *Toy Story* all messed up," Jobs later said. "He wanted Woody to be a bad guy, and when he shut us down we kind of kicked him out and said, 'This isn't what we want,' and did it the way we always wanted."

The Pixar team came back with a new script three months later. The character of Woody morphed from being a tyrannical boss of Andy's other toys to being their wise leader. His jealousy after the arrival of Buzz Lightyear was portrayed more sympathetically, and it was set to the strains of a Randy Newman song, "Strange Things." The scene in which Woody pushed Buzz out of the window was rewritten to make Buzz's fall the result of an accident triggered by a little trick Woody initiated involving a Luxo lamp. Katzenberg & Co. approved the new approach, and by February 1994 the film was back in production.

Katzenberg had been impressed with Jobs's focus on keeping costs under control. "Even in the early budgeting process, Steve was very eager to do it as efficiently as possible," he said. But the $17 million production budget was proving inadequate, especially given the major revision that was necessary after Katzenberg had pushed them to make Woody too edgy. So Jobs demanded more in order to complete the film right. "Listen, we made a deal," Katzenberg told him. "We gave you business control, and you agreed to do it for the amount we offered." Jobs was furious. He would call Katzenberg by phone or fly down to visit him and be, in Katzenberg's words, "as wildly relentless as only Steve can be." Jobs insisted that Disney was liable for the cost overruns because Katzenberg had so badly mangled the original concept that it required extra work to restore things. "Wait a minute!" Katzenberg shot back. "We were helping you. You got the benefit of our creative help, and now you want us to pay you for that." It was a case of two control freaks arguing about who was doing the other a favor.

Ed Catmull, more diplomatic than Jobs, was able to reach a compromise new budget. "I had a much more positive view of Jeffrey than some of the folks working on the film did," he said. But the incident did prompt Jobs to start plotting about how to have more leverage with Disney in the future. He did not like being a mere contractor; he liked being in control. That meant Pixar would have to bring its own funding to projects in the future, and it would need a new deal with Disney.

As the film progressed, Jobs became ever more excited about it. He had been talking to various companies, ranging from Hallmark to Microsoft, about selling Pixar, but watching Woody and Buzz come to life made him realize that he might be on the verge of transforming the movie industry. As scenes from the movie were finished, he watched them repeatedly and had friends come by his home to share his new passion. "I can't tell you the number of versions of *Toy Story* I saw before it came out," said Larry Ellison. "It eventually became a form of torture. I'd go over there and see the latest 10% improvement. Steve is obsessed with getting it right — both the story and the technology — and isn't satisfied with anything less than perfection."

Jobs's sense that his investments in Pixar might actually pay off was reinforced when Disney invited him to attend a gala press preview of scenes from *Pocahontas* in January 1995 in a tent in Manhattan's Central Park. At the event, Disney CEO Michael Eisner announced that *Pocahontas* would have its premiere in front of 100,000 people on eighty-foot-high screens on the Great Lawn of Central Park. Jobs was a master showman who knew how to stage great premieres, but even he was astounded by this plan. Buzz Lightyear's great exhortation — "To infinity and beyond!" — suddenly seemed worth heeding.

Jobs decided that the release of *Toy Story* that November would be the occasion to take Pixar public. Even the usually eager investment bankers were dubious and said it couldn't happen. Pixar had spent five years hemorrhaging money. But Jobs was determined. "I was nervous and argued that we should wait until after our second movie," Lasseter recalled. "Steve overruled me and said we needed the cash so we could put up half the money for our films and renegotiate the Disney deal."

To Infinity!

There were two premieres of *Toy Story* in November 1995. Disney organized one at El Capitan, a grand old theater in Los Angeles, and built a fun house next door featuring the characters. Pixar was given a handful of passes, but the evening and its celebrity guest list was very much a Disney production; Jobs did not even attend. Instead, the next night he rented the Regency, a similar theater in San Francisco, and held his own premiere. Instead of Tom Hanks and Steve Martin, the guests were Silicon Valley celebrities, such as Larry Ellison and Andy Grove. This was clearly Jobs's show; he, not Lasseter, took the stage to introduce the movie.

The dueling premieres highlighted a festering issue: Was *Toy Story* a Disney or a Pixar movie? Was Pixar merely an animation contractor helping Disney make movies? Or was Disney merely a distributor and marketer helping Pixar roll out its movies? The answer was somewhere in between. The question would be whether the egos involved, mainly those of Michael Eisner and Steve Jobs, could get to such a partnership.

The stakes were raised when *Toy Story* opened to blockbuster commercial and critical success. It recouped its cost the first weekend, with a domestic opening of $30 million, and it went on to become the top-grossing film of the year, beating *Batman Forever* and *Apollo 13,* with $192 million in receipts domestically and a total of $362 million worldwide. According to the review aggregator Rotten Tomatoes, 100% of the seventy-three critics surveyed gave it a positive review. *Time*'s Richard

Corliss called it "the year's most inventive comedy," David Ansen of *Newsweek* pronounced it a "marvel," and Janet Maslin of the *New York Times* recommended it both for children and adults as "a work of incredible cleverness in the best two-tiered Disney tradition."

The only rub for Jobs was that reviewers such as Maslin wrote of the "Disney tradition," not the emergence of Pixar. After reading her review, he decided he had to go on the offensive to raise Pixar's profile. When he and Lasseter went on the *Charlie Rose* show, Jobs emphasized that *Toy Story* was a Pixar movie, and he even tried to highlight the historic nature of a new studio being born. "Since *Snow White* was released, every major studio has tried to break into the animation business, and until now Disney was the only studio that had ever made a feature animated film that was a blockbuster," he told Rose. "Pixar has now become the second studio to do that."

Jobs made a point of casting Disney as merely the distributor of a Pixar film. "He kept saying, 'We at Pixar are the real thing and you Disney guys are shit,'" recalled Michael Eisner. "But we were the ones who made *Toy Story* work. We helped shape the movie, and we pulled together all of our divisions, from our consumer marketers to the Disney Channel, to make it a hit." Jobs came to the conclusion that the fundamental issue — Whose movie was it? — would have to be settled contractually rather than by a war of words. "After *Toy Story*'s success," he said, "I realized that we needed to cut a new deal with Disney if we were ever to build a studio and not just be a work-for-

425

hire place." But in order to sit down with Disney on an equal basis, Pixar had to bring money to the table. That required a successful IPO.

The public offering occurred exactly one week after *Toy Story*'s opening. Jobs had gambled that the movie would be successful, and the risky bet paid off, big-time. As with the Apple IPO, a celebration was planned at the San Francisco office of the lead underwriter at 7 a.m., when the shares were to go on sale. The plan had originally been for the first shares to be offered at about $14, to be sure they would sell. Jobs insisted on pricing them at $22, which would give the company more money if the offering was a success. It was, beyond even his wildest hopes. It exceeded Netscape as the biggest IPO of the year. In the first half hour, the stock shot up to $45, and trading had to be delayed because there were too many buy orders. It then went up even further, to $49, before settling back to close the day at $39.

Earlier that year Jobs had been hoping to find a buyer for Pixar that would let him merely recoup the $50 million he had put in. By the end of the day the shares he had retained — 80% of the company — were worth more than twenty times that, an astonishing $1.2 *billion*. That was about five times what he'd made when Apple went public in 1980. But Jobs told John Markoff of the *New York Times* that the money did not mean much to him. "There's no yacht in my future," he said. "I've never done this for the money."

The successful IPO meant that Pixar would no longer have to be dependent on Disney to finance

426

its movies. That was just the leverage Jobs wanted. "Because we could now fund half the cost of our movies, I could demand half the profits," he recalled. "But more important, I wanted co-branding. These were to be Pixar as well as Disney movies."

Jobs flew down to have lunch with Eisner, who was stunned at his audacity. They had a three-picture deal, and Pixar had made only one. Each side had its own nuclear weapons. After an acrimonious split with Eisner, Katzenberg had left Disney and become a cofounder, with Steven Spielberg and David Geffen, of DreamWorks SKG. If Eisner didn't agree to a new deal with Pixar, Jobs said, then Pixar would go to another studio, such as Katzenberg's, once the three-picture deal was done. In Eisner's hand was the threat that Disney could, if that happened, make its own sequels to *Toy Story,* using Woody and Buzz and all of the characters that Lasseter had created. "That would have been like molesting our children," Jobs later recalled. "John started crying when he considered that possibility."

So they hammered out a new arrangement. Eisner agreed to let Pixar put up half the money for future films and in return take half of the profits. "He didn't think we could have many hits, so he thought he was saving himself some money," said Jobs. "Ultimately that was great for us, because Pixar would have ten blockbusters in a row." They also agreed on co-branding, though that took a lot of haggling to define. "I took the position that it's a Disney movie, but eventually I relented," Eisner recalled. "We start negotiating how big the letters

427

in 'Disney' are going to be, how big is 'Pixar' going to be, just like four-year-olds." But by the beginning of 1997 they had a deal, for five films over the course of ten years, and even parted as friends, at least for the time being. "Eisner was reasonable and fair to me then," Jobs later said. "But eventually, over the course of a decade, I came to the conclusion that he was a dark man."

In a letter to Pixar shareholders, Jobs explained that winning the right to have equal branding with Disney on all the movies, as well as advertising and toys, was the most important aspect of the deal. "We want Pixar to grow into a brand that embodies the same level of trust as the Disney brand," he wrote. "But in order for Pixar to earn this trust, consumers must know that Pixar is creating the films." Jobs was known during his career for creating great products. But just as significant was his ability to create great companies with valuable brands. And he created two of the best of his era: Apple and Pixar.

A PORTFOLIO OF DIANA WALKER PHOTOS

For almost thirty years, photographer Diana Walker has had special access to her friend Steve Jobs. Here is a selection from her portfolio.

1

At his home in Cupertino, 1982: He was such a perfectionist that he had trouble buying furniture.

2 *In his kitchen: "Coming back after seven months in Indian villages, I saw the craziness of the Western world as well as its capacity for rational thought."*

3   *At Stanford, 1982: "How many of you are virgins? How many of you have taken LSD?"*

4   *With the Lisa: "Picasso had a saying—'good artists copy, great artists steal'—and we have always been shameless about stealing great ideas."*

5   *With John Sculley in Central Park, 1984: "Do you want to spend
the rest of your life selling sugared water, or do you want a
chance to change the world?"*

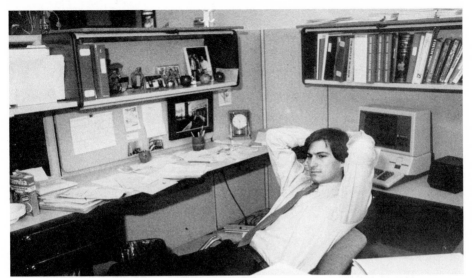

6 *In his Apple office, 1982: Asked if he wanted to do market research, he said, "No, because customers don't know what they want until we've shown them."*

7 *At NeXT, 1988: Freed from the constraints at Apple, he indulged his own best and worst instincts.*

8 *With John Lasseter, August 1997: His cherubic face and demeanor masked an artistic perfectionism that rivaled that of Jobs.*

9 *At home working on his Boston Macworld speech after regaining command of Apple, 1997: "In that craziness we see genius."*

Sealing the Microsoft deal by phone with Gates: "Bill, thank you for your support of this company. I think the world's a better place for it."

At Boston Macworld, as Gates discusses their deal: "That was my worst and stupidest staging event ever. It made me look small."

With his wife, Laurene Powell, in their backyard in Palo Alto, August 1997: She was the sensible anchor in his life.

At his home office in Palo Alto, 2004: "I like living at the intersection of the humanities and technology."

FROM THE JOBS FAMILY ALBUM

In August 2011, when Jobs was very ill, we sat in his room and went through wedding and vacation pictures for me to use in this book.

14

The wedding ceremony, 1991: Kobun Chino, Steve's Sōtō Zen teacher, shook a stick, struck a gong, lit incense, and chanted.

15

With his proud father Paul Jobs: After Steve's sister Mona tracked down their biological father, Steve refused ever to meet him.

Cutting the cake in the shape of Half Dome with Laurene and his daughter from a previous relationship, Lisa Brennan.

Laurene, Lisa, and Steve: Lisa moved into their home shortly afterward and stayed through her high school years.

Steve, Eve, Reed, Erin, and Laurene in Ravello, Italy, 2003: Even on vacation, he often withdrew into his work.

19

Dangling Eve in Foothills Park, Palo Alto: "She's a pistol and has the strongest will of any kid I've ever met. It's like payback."

20

With Laurene, Eve, Erin, and Lisa at the Corinth Canal in Greece, 2006: "For young people, this whole world is the same now."

With Erin in Kyoto, 2010: Like Reed and Lisa, she got a special trip to Japan with her father.

With Reed in Kenya, 2007: "When I was diagnosed with cancer, I made my deal with God or whatever, which was that I really wanted to see Reed graduate."

And just one more from Diana Walker: a 2004 portrait at his house in Palo Alto.

Chapter Twenty-Three:
The Second Coming

WHAT ROUGH BEAST, ITS HOUR COME ROUND AT LAST . . .

Things Fall Apart

When Jobs unveiled the NeXT computer in 1988, there was a burst of excitement. That fizzled when the computer finally went on sale the following year. Jobs's ability to dazzle, intimidate, and spin the press began to fail him, and there was a series of stories on the company's woes. "NeXT is incompatible with other computers at a time when the industry is moving toward interchangeable systems," Bart Ziegler of Associated Press reported. "Because relatively little software exists to run on NeXT, it has a hard time attracting customers."

NeXT tried to reposition itself as the leader in a new category, personal workstations, for people who wanted the power of a workstation and the friendliness of a personal computer. But those customers were by now buying them from fast-growing Sun Microsystems. Revenues for NeXT in 1990 were $28 million; Sun made $2.5 billion that year. IBM abandoned its deal to license the NeXT software, so Jobs was forced to do something against his nature: Despite his ingrained belief that hardware and software should be integrally linked, he agreed in January 1992 to license

429

the NeXTSTEP operating system to run on other computers.

One surprising defender of Jobs was Jean-Louis Gassée, who had bumped elbows with Jobs when he replaced him at Apple and subsequently been ousted himself. He wrote an article extolling the creativity of NeXT products. "NeXT might not be Apple," Gassée argued, "but Steve is still Steve." A few days later his wife answered a knock on the door and went running upstairs to tell him that Jobs was standing there. He thanked Gassée for the article and invited him to an event where Intel's Andy Grove would join Jobs in announcing that NeXTSTEP would be ported to the IBM/Intel platform. "I sat next to Steve's father, Paul Jobs, a movingly dignified individual," Gassée recalled. "He raised a difficult son, but he was proud and happy to see him onstage with Andy Grove."

A year later Jobs took the inevitable subsequent step: He gave up making the hardware altogether. This was a painful decision, just as it had been when he gave up making hardware at Pixar. He cared about all aspects of his products, but the hardware was a particular passion. He was energized by great design, obsessed over manufacturing details, and would spend hours watching his robots make his perfect machines. But now he had to lay off more than half his workforce, sell his beloved factory to Canon (which auctioned off the fancy furniture), and satisfy himself with a company that tried to license an operating system to manufacturers of uninspired machines.

■ ■ ■ ■

By the mid-1990s Jobs was finding some pleasure in his new family life and his astonishing triumph in the movie business, but he despaired about the personal computer industry. "Innovation has virtually ceased," he told Gary Wolf of *Wired* at the end of 1995. "Microsoft dominates with very little innovation. Apple lost. The desktop market has entered the dark ages."

He was also gloomy in an interview with Tony Perkins and the editors of *Red Herring*. First, he displayed the "Bad Steve" side of his personality. Soon after Perkins and his colleagues arrived, Jobs slipped out the back door "for a walk," and he didn't return for forty-five minutes. When the magazine's photographer began taking pictures, he snapped at her sarcastically and made her stop. Perkins later noted, "Manipulation, selfishness, or downright rudeness, we couldn't figure out the motivation behind his madness." When he finally settled down for the interview, he said that even the advent of the web would do little to stop Microsoft's domination. "Windows has won," he said. "It beat the Mac, unfortunately, it beat UNIX, it beat OS/2. An inferior product won."

APPLE FALLING

For a few years after Jobs was ousted, Apple was able to coast comfortably with a high profit margin based on its temporary dominance in desktop publishing. Feeling like a genius back in 1987, John Sculley had made a series of proclamations that nowadays sound embarrassing. Jobs wanted Apple "to become a wonderful consumer products company," Sculley wrote. "This was a lunatic plan. . . . Apple would never be a consumer products company. . . . We couldn't bend reality to all our dreams of changing the world. . . . High tech could not be designed and sold as a consumer product."

Jobs was appalled, and he became angry and contemptuous as Sculley presided over a steady decline in market share for Apple in the early 1990s. "Sculley destroyed Apple by bringing in corrupt people and corrupt values," Jobs later lamented. "They cared about making money — for themselves mainly, and also for Apple — rather than making great products." He felt that Sculley's drive for profits came at the expense of gaining market share. "Macintosh lost to Microsoft because Sculley insisted on milking all the profits he could get rather than improving the product and making it affordable." As a result, the profits eventually disappeared.

It had taken Microsoft a few years to replicate Macintosh's graphical user interface, but by 1990 it had come out with Windows 3.0, which began the company's march to dominance in the desktop market. Windows 95, which was released in 1995, became the most successful operating system ever, and Macintosh sales began to collapse. "Microsoft simply ripped off what other people did," Jobs later said. "Apple deserved it. After I left, it didn't invent anything new. The Mac hardly improved. It was a sitting duck for Microsoft."

His frustration with Apple was evident when he gave a talk to a Stanford Business School club at the home of a student, who asked him to sign a

Macintosh keyboard. Jobs agreed to do so if he could remove the keys that had been added to the Mac after he left. He pulled out his car keys and pried off the four arrow cursor keys, which he had once banned, as well as the top row of F1, F2, F3 . . . function keys. "I'm changing the world one keyboard at a time," he deadpanned. Then he signed the mutilated keyboard.

During his 1995 Christmas vacation in Kona Village, Hawaii, Jobs went walking along the beach with his friend Larry Ellison, the irrepressible Oracle chairman. They discussed making a takeover bid for Apple and restoring Jobs as its head. Ellison said he could line up $3 billion in financing: "I will buy Apple, you will get 25% of it right away for being CEO, and we can restore it to its past glory." But Jobs demurred. "I decided I'm not a hostile-takeover kind of guy," he explained. "If they had asked me to come back, it might have been different."

By 1996 Apple's share of the market had fallen to 4% from a high of 16% in the late 1980s. Michael Spindler, the German-born chief of Apple's European operations who had replaced Sculley as CEO in 1993, tried to sell the company to Sun, IBM, and Hewlett-Packard. That failed, and he was ousted in February 1996 and replaced by Gil Amelio, a research engineer who was CEO of National Semiconductor. During his first year the company lost $1 billion, and the stock price, which had been $70 in 1991, fell to $14, even as the tech bubble was pushing other stocks into the stratosphere.

Amelio was not a fan of Jobs. Their first meeting had been in 1994, just after Amelio was elected to the Apple board. Jobs had called him and announced, "I want to come over and see you." Amelio invited him over to his office at National Semiconductor, and he later recalled watching through the glass wall of his office as Jobs arrived. He looked "rather like a boxer, aggressive and elusively graceful, or like an elegant jungle cat ready to spring at its prey." After a few minutes of pleasantries — far more than Jobs usually engaged in — he abruptly announced the reason for his visit. He wanted Amelio to help him return to Apple as the CEO. "There's only one person who can rally the Apple troops," Jobs said, "only one person who can straighten out the company." The Macintosh era had passed, Jobs argued, and it was now time for Apple to create something new that was just as innovative.

"If the Mac is dead, what's going to replace it?" Amelio asked. Jobs's reply didn't impress him. "Steve didn't seem to have a clear answer," Amelio later said. "He seemed to have a set of one-liners." Amelio felt he was witnessing Jobs's reality distortion field and was proud to be immune to it. He shooed Jobs unceremoniously out of his office.

By the summer of 1996 Amelio realized that he had a serious problem. Apple was pinning its hopes on creating a new operating system, called Copland, but Amelio had discovered soon after becoming CEO that it was a bloated piece of vaporware that would not solve Apple's needs for better networking and memory protection, nor would it be ready to ship as scheduled in 1997. He publicly promised that he would quickly find an alternative. His problem was that he didn't have one.

So Apple needed a partner, one that could make a stable operating system, preferably one that was UNIX-like and had an object-oriented application layer. There was one company that could obviously supply such software — NeXT — but it would take a while for Apple to focus on it.

Apple first homed in on a company that had been started by Jean-Louis Gassée, called Be. Gassée began negotiating the sale of Be to Apple, but in August 1996 he overplayed his hand at a meeting with Amelio in Hawaii. He said he wanted to bring his fifty-person team to Apple, and he asked for 15% of the company, worth about $500 million. Amelio was stunned. Apple calculated that Be was worth about $50 million. After a few offers and counteroffers, Gassée refused to budge from demanding at least $275 million. He thought that Apple had no alternatives. It got back to Amelio that Gassée said, "I've got them by the balls, and I'm going to squeeze until it hurts." This did not please Amelio.

Apple's chief technology officer, Ellen Hancock, argued for going with Sun's UNIX-based Solaris operating system, even though it did not yet have a friendly user interface. Amelio began to favor using, of all things, Microsoft's Windows NT, which he felt could be rejiggered on the surface to look and feel just like a Mac while being compatible with the wide range of software available to Windows users. Bill Gates, eager to make a deal, began personally calling Amelio.

There was, of course, one other option. Two years earlier *Macworld* magazine columnist (and former Apple software evangelist) Guy Kawasaki had published a parody press release joking that Apple was buying NeXT and making Jobs its CEO. In the spoof Mike Markkula asked Jobs, "Do you want to spend the rest of your life selling UNIX with a sugarcoating, or change the world?" Jobs responded, "Because I'm now a father, I needed a steadier source of income." The release noted that "because of his experience at Next, he is expected to bring a newfound sense of humility back to Apple." It also quoted Bill Gates as saying there would now be more innovations from Jobs that Microsoft could copy. Everything in the press release was meant as a joke, of course. But reality has an odd habit of catching up with satire.

SLOUCHING TOWARD CUPERTINO

"Does anyone know Steve well enough to call him on this?" Amelio asked his staff. Because his encounter with Jobs two years earlier had ended badly, Amelio didn't want to make the call himself. But as it turned out, he didn't need to. Apple was already getting incoming pings from NeXT. A midlevel product marketer at NeXT, Garrett Rice, had simply picked up the phone and, without consulting Jobs, called Ellen Hancock to see if she might be interested in taking a look at its software. She sent someone to meet with him.

By Thanksgiving of 1996 the two companies had begun midlevel talks, and Jobs picked up the phone to call Amelio directly. "I'm on my way to Japan, but I'll be back in a week and I'd like to see you as soon as I return," he said. "Don't make any decision until we can get together." Amelio, despite his earlier experience with Jobs, was thrilled to hear from him and entranced by the possibility of working with him. "For me, the phone call with Steve was like inhaling the flavors of a great bottle of vintage wine," he recalled. He gave his assurance he would make no deal with Be or anyone else before they got together.

For Jobs, the contest against Be was both professional and personal. NeXT was failing, and the prospect of being bought by Apple was a tantalizing lifeline. In addition, Jobs held grudges, sometimes passionately, and Gassée was near the top of his list, despite the fact that they had seemed to reconcile when Jobs was at NeXT. "Gassée is one of the few people in my life I would say is truly horrible," Jobs later insisted, unfairly. "He knifed me in the back in 1985." Sculley, to his credit, had at least been gentlemanly enough to knife Jobs in the front.

On December 2, 1996, Steve Jobs set foot on Apple's Cupertino campus for the first time since his ouster eleven years earlier. In the executive conference room, he met Amelio and Hancock to make the pitch for NeXT. Once again he was scribbling on the whiteboard there, this time giving his lecture about the four waves of computer systems that had culminated, at least in his telling, with the launch of NeXT. He was at his most seductive, despite the fact that he was speaking to two people he didn't respect. He was particularly adroit at feigning modesty. "It's probably a totally crazy idea," he said, but if they found it appealing, "I'll structure any kind of deal you want — license the software, sell you the company, whatever." He was, in fact, eager to sell everything, and he pushed that approach. "When you take a close look, you'll decide you want more than my software," he told them. "You'll want to buy the whole company and take all the people."

A few weeks later Jobs and his family went to Hawaii for Christmas vacation. Larry Ellison was also there, as he had been the year before. "You know, Larry, I think I've found a way for me to get back into Apple and get control of it without you having to buy it," Jobs said as they walked along the shore. Ellison recalled, "He explained his strategy, which was getting Apple to buy NeXT, then he would go on the board and be one step away from being CEO." Ellison thought that Jobs was missing a key point. "But Steve, there's one thing I don't understand," he said. "If we don't buy the company, how can we make any money?" It was a reminder of how different their desires were. Jobs put his hand on Ellison's left shoulder, pulled him so close that their noses almost touched, and said, "Larry, this is why it's really important that I'm your friend. You don't need any more money."

Ellison recalled that his own answer was almost a whine: "Well, I may not need the money, but why should some fund manager at Fidelity get the money? Why should someone else get it? Why shouldn't it be us?"

"I think if I went back to Apple, and I didn't own any of Apple, and you didn't own any of Apple, I'd have the moral high ground," Jobs replied.

"Steve, that's really expensive real estate, this moral high ground," said Ellison. "Look, Steve, you're my best friend, and Apple is your company. I'll do whatever you want." Although Jobs later said that he was not plotting to take over Apple at the time, Ellison thought it was inevitable. "Anyone who spent more than a half hour with Amelio would realize that he couldn't do anything but self-destruct," he later said.

The big bakeoff between NeXT and Be was held at the Garden Court Hotel in Palo Alto on December 10, in front of Amelio, Hancock, and six other Apple executives. NeXT went first, with Avie Tevanian demonstrating the software while Jobs displayed his hypnotizing salesmanship. They showed how the software could play four video clips on the screen at once, create multimedia, and link to the Internet. "Steve's sales pitch on the NeXT operating system was dazzling," according to Amelio. "He praised the virtues and strengths as though he were describing a performance of Olivier as Macbeth."

Gassée came in afterward, but he acted as if he had the deal in his hand. He provided no new presentation. He simply said that the Apple team knew the capabilities of the Be OS and asked if they had any further questions. It was a short session. While Gassée was presenting, Jobs and Tevanian walked the streets of Palo Alto. After a while they bumped into one of the Apple executives who had been at the meetings. "You're going to win this," he told them.

Tevanian later said that this was no surprise: "We had better technology, we had a solution that was complete, and we had Steve." Amelio knew that bringing Jobs back into the fold would be a double-edged sword, but the same was true of bringing

Gassée back. Larry Tesler, one of the Macintosh veterans from the old days, recommended to Amelio that he choose NeXT, but added, "Whatever company you choose, you'll get someone who will take your job away, Steve or Jean-Louis."

Amelio opted for Jobs. He called Jobs to say that he planned to propose to the Apple board that he be authorized to negotiate a purchase of NeXT. Would he like to be at the meeting? Jobs said he would. When he walked in, there was an emotional moment when he saw Mike Markkula. They had not spoken since Markkula, once his mentor and father figure, had sided with Sculley there back in 1985. Jobs walked over and shook his hand.

Jobs invited Amelio to come to his house in Palo Alto so they could negotiate in a friendly setting. When Amelio arrived in his classic 1973 Mercedes, Jobs was impressed; he liked the car. In the kitchen, which had finally been renovated, Jobs put a kettle on for tea, and then they sat at the wooden table in front of the open-hearth pizza oven. The financial part of the negotiations went smoothly; Jobs was eager not to make Gassée's mistake of overreaching. He suggested that Apple pay $12 a share for NeXT. That would amount to about $500 million. Amelio said that was too high. He countered with $10 a share, or just over $400 million. Unlike Be, NeXT had an actual product, real revenues, and a great team, but Jobs was nevertheless pleasantly surprised at that counteroffer. He accepted immediately.

One sticking point was that Jobs wanted his payout to be in cash. Amelio insisted that he needed to "have skin in the game" and take the payout in stock that he would agree to hold for at least a year. Jobs resisted. Finally, they compromised: Jobs would take $120 million in cash and $37 million in stock, and he pledged to hold the stock for at least six months.

As usual Jobs wanted to have some of their conversation while taking a walk. While they ambled around Palo Alto, he made a pitch to be put on Apple's board. Amelio tried to deflect it, saying there was too much history to do something like that too quickly. "Gil, that really hurts," Jobs said. "This was my company. I've been left out since that horrible day with Sculley." Amelio said he understood, but he was not sure what the board would want. When he was about to begin his negotiations with Jobs, he had made a mental note to "move ahead with logic as my drill sergeant" and "sidestep the charisma." But during the walk he, like so many others, was caught in Jobs's force field. "I was hooked in by Steve's energy and enthusiasm," he recalled.

After circling the long blocks a couple of times, they returned to the house just as Laurene and the kids were arriving home. They all celebrated the easy negotiations, then Amelio rode off in his Mercedes. "He made me feel like a lifelong friend," Amelio recalled. Jobs indeed had a way of doing that. Later, after Jobs had engineered his ouster, Amelio would look back on Jobs's friendliness that day and note wistfully, "As I would painfully discover, it was merely one facet of an extremely complex personality."

After informing Gassée that Apple was buying NeXT, Amelio had what turned out to be an even

441

more uncomfortable task: telling Bill Gates. "He went into orbit," Amelio recalled. Gates found it ridiculous, but perhaps not surprising, that Jobs had pulled off this coup. "Do you really think Steve Jobs has anything there?" Gates asked Amelio. "I know his technology, it's nothing but a warmed-over UNIX, and you'll never be able to make it work on your machines." Gates, like Jobs, had a way of working himself up, and he did so now: "Don't you understand that Steve doesn't know anything about technology? He's just a super salesman. I can't believe you're making such a stupid decision. . . . He doesn't know anything about engineering, and 99% of what he says and thinks is wrong. What the hell are you buying that garbage for?"

Years later, when I raised it with him, Gates did not recall being that upset. The purchase of NeXT, he argued, did not really give Apple a new operating system. "Amelio paid a lot for NeXT, and let's be frank, the NeXT OS was never really used." Instead the purchase ended up bringing in Avie Tevanian, who could help the existing Apple operating system evolve so that it eventually incorporated the kernel of the NeXT technology. Gates knew that the deal was destined to bring Jobs back to power. "But that was a twist of fate," he said. "What they ended up buying was a guy who most people would not have predicted would be a great CEO, because he didn't have much experience at it, but he was a brilliant guy with great design taste and great engineering taste. He suppressed his craziness enough to get himself appointed interim CEO."

■ ■ ■ ■

Despite what both Ellison and Gates believed, Jobs had deeply conflicted feelings about whether he wanted to return to an active role at Apple, at least while Amelio was there. A few days before the NeXT purchase was due to be announced, Amelio asked Jobs to rejoin Apple full-time and take charge of operating system development. Jobs, however, kept deflecting Amelio's request.

Finally, on the day that he was scheduled to make the big announcement, Amelio called Jobs in. He needed an answer. "Steve, do you just want to take your money and leave?" Amelio asked. "It's okay if that's what you want." Jobs did not answer; he just stared. "Do you want to be on the payroll? An advisor?" Again Jobs stayed silent. Amelio went out and grabbed Jobs's lawyer, Larry Sonsini, and asked what he thought Jobs wanted. "Beats me," Sonsini said. So Amelio went back behind closed doors with Jobs and gave it one more try. "Steve, what's on your mind? What are you feeling? Please, I need a decision now."

"I didn't get any sleep last night," Jobs replied.

"Why? What's the problem?"

"I was thinking about all the things that need to be done and about the deal we're making, and it's all running together for me. I'm really tired now and not thinking clearly. I just don't want to be asked any more questions."

Amelio said that wasn't possible. He needed to say something.

Finally Jobs answered, "Look, if you have to tell them something, just say advisor to the chairman."

And that is what Amelio did.

The announcement was made that evening — December 20, 1996 — in front of 250 cheering employees at Apple headquarters. Amelio did as Jobs had requested and described his new role as merely that of a part-time advisor. Instead of appearing from the wings of the stage, Jobs walked in from the rear of the auditorium and ambled down the aisle. Amelio had told the gathering that Jobs would be too tired to say anything, but by then he had been energized by the applause. "I'm very excited," Jobs said. "I'm looking forward to get to reknow some old colleagues." Louise Kehoe of the *Financial Times* came up to the stage afterward and asked Jobs, sounding almost accusatory, whether he was going to end up taking over Apple. "Oh no, Louise," he said. "There are a lot of other things going on in my life now. I have a family. I am involved at Pixar. My time is limited, but I hope I can share some ideas."

The next day Jobs drove to Pixar. He had fallen increasingly in love with the place, and he wanted to let the crew there know he was still going to be president and deeply involved. But the Pixar people were happy to see him go back to Apple part-time; a little less of Jobs's focus would be a good thing. He was useful when there were big negotiations, but he could be dangerous when he had too much time on his hands. When he arrived at Pixar that day, he went to Lasseter's office and explained that even just being an advisor at Apple would take up a lot of his time. He said he wanted Lasseter's blessing. "I keep thinking about all the time away from my family this will cause, and the time away from the other family at Pixar," Jobs said. "But the only reason I want to do it is that the world will be a better place with Apple in it."

Lasseter smiled gently. "You have my blessing," he said.

CHAPTER TWENTY-FOUR: THE RESTORATION

THE LOSER NOW WILL BE LATER TO WIN

HOVERING BACKSTAGE

"It's rare that you see an artist in his thirties or forties able to really contribute something amazing," Jobs declared as he was about to turn thirty.

That held true for Jobs in his thirties, during the decade that began with his ouster from Apple in 1985. But after turning forty in 1995, he flourished. *Toy Story* was released that year, and the following year Apple's purchase of NeXT offered him reentry into the company he had founded. In returning to Apple, Jobs would show that even people over forty could be great innovators. Having transformed personal computers in his twenties, he would now help to do the same for music players, the recording industry's business model, mobile phones, apps, tablet computers, books, and journalism.

He had told Larry Ellison that his return strategy was to sell NeXT to Apple, get appointed to the board, and be there ready when CEO Gil Amelio stumbled. Ellison may have been baffled when Jobs insisted that he was not motivated by money, but it was partly true. He had neither Ellison's conspicuous consumption needs nor Gates's philanthropic

446

impulses nor the competitive urge to see how high on the *Forbes* list he could get. Instead his ego needs and personal drives led him to seek fulfillment by creating a legacy that would awe people. A dual legacy, actually: building innovative products and building a lasting company. He wanted to be in the pantheon with, indeed a notch above, people like Edwin Land, Bill Hewlett, and David Packard. And the best way to achieve all this was to return to Apple and reclaim his kingdom.

And yet when the cup of power neared his lips, he became strangely hesitant, reluctant, perhaps coy.

He returned to Apple officially in January 1997 as a part-time advisor, as he had told Amelio he would. He began to assert himself in some personnel areas, especially in protecting his people who had made the transition from NeXT. But in most other ways he was unusually passive. The decision not to ask him to join the board offended him, and he felt demeaned by the suggestion that he run the company's operating system division. Amelio was thus able to create a situation in which Jobs was both inside the tent and outside the tent, which was not a prescription for tranquillity. Jobs later recalled:

> Gil didn't want me around. And I thought he was a bozo. I knew that before I sold him the company. I thought I was just going to be trotted out now and then for events like Macworld, mainly for show. That was fine, because I was working at Pixar. I rented an office in downtown Palo Alto where I could work a few days a week, and

447

I drove up to Pixar for one or two days. It was a nice life. I could slow down, spend time with my family.

Jobs was, in fact, trotted out for Macworld right at the beginning of January, and this reaffirmed his opinion that Amelio was a bozo. Close to four thousand of the faithful fought for seats in the ballroom of the San Francisco Marriott to hear Amelio's keynote address. He was introduced by the actor Jeff Goldblum. "I play an expert in chaos theory in *The Lost World: Jurassic Park,*" he said. "I figure that will qualify me to speak at an Apple event." He then turned it over to Amelio, who came onstage wearing a flashy sports jacket and a banded-collar shirt buttoned tight at the neck, "looking like a Vegas comic," the *Wall Street Journal* reporter Jim Carlton noted, or in the words of the technology writer Michael Malone, "looking exactly like your newly divorced uncle on his first date."

The bigger problem was that Amelio had gone on vacation, gotten into a nasty tussle with his speechwriters, and refused to rehearse. When Jobs arrived backstage, he was upset by the chaos, and he seethed as Amelio stood on the podium bumbling through a disjointed and endless presentation. Amelio was unfamiliar with the talking points that popped up on his teleprompter and soon was trying to wing his presentation. Repeatedly he lost his train of thought. After more than an hour, the audience was aghast. There were a few welcome breaks, such as when he brought out the singer Peter Gabriel to demonstrate a new music program. He also pointed out Muhammad

Ali in the first row; the champ was supposed to come onstage to promote a website about Parkinson's disease, but Amelio never invited him up or explained why he was there.

Amelio rambled for more than two hours before he finally called onstage the person everyone was waiting to cheer. "Jobs, exuding confidence, style, and sheer magnetism, was the antithesis of the fumbling Amelio as he strode onstage," Carlton wrote. "The return of Elvis would not have provoked a bigger sensation." The crowd jumped to its feet and gave him a raucous ovation for more than a minute. The wilderness decade was over. Finally Jobs waved for silence and cut to the heart of the challenge. "We've got to get the spark back," he said. "The Mac didn't progress much in ten years. So Windows caught up. So we have to come up with an OS that's even better."

Jobs's pep talk could have been a redeeming finale to Amelio's frightening performance. Unfortunately Amelio came back onstage and resumed his ramblings for another hour. Finally, more than three hours after the show began, Amelio brought it to a close by calling Jobs back onstage and then, in a surprise, bringing up Steve Wozniak as well. Again there was pandemonium. But Jobs was clearly annoyed. He avoided engaging in a triumphant trio scene, arms in the air. Instead he slowly edged offstage. "He ruthlessly ruined the closing moment I had planned," Amelio later complained. "His own feelings were more important than good press for Apple." It was only seven days into the new year for Apple, and already it was clear that the center would not hold.

449

Jobs immediately put people he trusted into the top ranks at Apple. "I wanted to make sure the really good people who came in from NeXT didn't get knifed in the back by the less competent people who were then in senior jobs at Apple," he recalled. Ellen Hancock, who had favored choosing Sun's Solaris over NeXT, was on the top of his bozo list, especially when she continued to want to use the kernel of Solaris in the new Apple operating system. In response to a reporter's question about the role Jobs would play in making that decision, she answered curtly, "None." She was wrong. Jobs's first move was to make sure that two of his friends from NeXT took over her duties.

To head software engineering, he tapped his buddy Avie Tevanian. To run the hardware side, he called on Jon Rubinstein, who had done the same at NeXT back when it had a hardware division. Rubinstein was vacationing on the Isle of Skye when Jobs called him. "Apple needs some help," he said. "Do you want to come aboard?" Rubinstein did. He got back in time to attend Macworld and see Amelio bomb onstage. Things were worse than he expected. He and Tevanian would exchange glances at meetings as if they had stumbled into an insane asylum, with people making deluded assertions while Amelio sat at the end of the table in a seeming stupor.

Jobs did not come into the office regularly, but he was on the phone to Amelio often. Once he had succeeded in making sure that Tevanian, Rubinstein, and others he trusted were given top positions, he turned his focus onto the sprawling product line. One of his pet peeves was Newton, the handheld personal digital assistant that boasted handwriting recognition capability. It was not quite as bad as the jokes and *Doonesbury* comic strip made it seem, but Jobs hated it. He disdained the idea of having a stylus or pen for writing on a screen. "God gave us ten styluses," he would say, waving his fingers. "Let's not invent another." In addition, he viewed Newton as John Sculley's one major innovation, his pet project. That alone doomed it in Jobs's eyes.

"You ought to kill Newton," he told Amelio one day by phone.

It was a suggestion out of the blue, and Amelio pushed back. "What do you mean, kill it?" he said. "Steve, do you have any idea how expensive that would be?"

"Shut it down, write it off, get rid of it," said Jobs. "It doesn't matter what it costs. People will cheer you if you got rid of it."

"I've looked into Newton and it's going to be a moneymaker," Amelio declared. "I don't support getting rid of it." By May, however, he announced plans to spin off the Newton division, the beginning of its yearlong stutter-step march to the grave.

Tevanian and Rubinstein would come by Jobs's house to keep him informed, and soon much of Silicon Valley knew that Jobs was quietly wresting power from Amelio. It was not so much a Machiavellian power play as it was Jobs being Jobs. Wanting control was ingrained in his nature. Louise Kehoe, the *Financial Times* reporter who had foreseen this when she questioned Jobs and Amelio at the December announcement, was the first with the story. "Mr. Jobs has become the power behind the throne," she reported at the end of February. "He is said to be directing decisions on which parts of Apple's operations should be cut. Mr. Jobs has urged a number of former Apple colleagues to return to the company, hinting strongly that he plans to take charge, they said. According to one of Mr. Jobs' confidantes, he has decided that Mr. Amelio and his appointees are unlikely to succeed in reviving Apple, and he is intent upon replacing them to ensure the survival of 'his company.' "

That month Amelio had to face the annual stockholders meeting and explain why the results for the final quarter of 1996 showed a 30% plummet in sales from the year before. Shareholders lined up at the microphones to vent their anger. Amelio was clueless about how poorly he handled the meeting. "The presentation was regarded as one of the best I had ever given," he later wrote. But Ed Woolard, the former CEO of DuPont who was now the chair of the Apple board (Markkula had been demoted to vice chair), was appalled. "This is a disaster," his wife whispered to him in the midst of the session. Woolard agreed. "Gil came dressed real cool, but he looked and sounded silly," he recalled. "He couldn't answer the questions, didn't know what he was talking about, and didn't inspire any confidence."

Woolard picked up the phone and called Jobs, whom he'd never met. The pretext was to invite him to Delaware to speak to DuPont executives. Jobs declined, but as Woolard recalled, "the request was a ruse in order to talk to him about

452

Gil." He steered the phone call in that direction and asked Jobs point-blank what his impression of Amelio was. Woolard remembers Jobs being somewhat circumspect, saying that Amelio was not in the right job. Jobs recalled being more blunt:

> I thought to myself, I either tell him the truth, that Gil is a bozo, or I lie by omission. He's on the board of Apple, I have a duty to tell him what I think; on the other hand, if I tell him, he will tell Gil, in which case Gil will never listen to me again, and he'll fuck the people I brought into Apple. All of this took place in my head in less than thirty seconds. I finally decided that I owed this guy the truth. I cared deeply about Apple. So I just let him have it. I said this guy is the worst CEO I've ever seen, I think if you needed a license to be a CEO he wouldn't get one. When I hung up the phone, I thought, I probably just did a really stupid thing.

That spring Larry Ellison saw Amelio at a party and introduced him to the technology journalist Gina Smith, who asked how Apple was doing. "You know, Gina, Apple is like a ship," Amelio answered. "That ship is loaded with treasure, but there's a hole in the ship. And my job is to get everyone to row in the same direction." Smith looked perplexed and asked, "Yeah, but what about the hole?" From then on, Ellison and Jobs joked about the parable of the ship. "When Larry relayed this story to me, we were in this sushi place, and I literally fell off my chair laughing," Jobs recalled. "He was just such a buffoon, and he took himself so seriously. He insisted that everyone call him Dr.

Amelio. That's always a warning sign."

Brent Schlender, *Fortune*'s well-sourced technology reporter, knew Jobs and was familiar with his thinking, and in March he came out with a story detailing the mess. "Apple Computer, Silicon Valley's paragon of dysfunctional management and fumbled techno-dreams, is back in crisis mode, scrambling lugubriously in slow motion to deal with imploding sales, a floundering technology strategy, and a hemorrhaging brand name," he wrote. "To the Machiavellian eye, it looks as if Jobs, despite the lure of Hollywood — lately he has been overseeing Pixar, maker of *Toy Story* and other computer-animated films — might be scheming to take over Apple."

Once again Ellison publicly floated the idea of doing a hostile takeover and installing his "best friend" Jobs as CEO. "Steve's the only one who can save Apple," he told reporters. "I'm ready to help him the minute he says the word." Like the third time the boy cried wolf, Ellison's latest takeover musings didn't get much notice, so later in the month he told Dan Gillmore of the *San Jose Mercury News* that he was forming an investor group to raise $1 billion to buy a majority stake in Apple. (The company's market value was about $2.3 billion.) The day the story came out, Apple stock shot up 11% in heavy trading. To add to the frivolity, Ellison set up an email address, savapple@us.oracle.com, asking the general public to vote on whether he should go ahead with it.

Jobs was somewhat amused by Ellison's self-appointed role. "Larry brings this up now and then," he told a reporter. "I try to explain my role at Apple is to be an advisor." Amelio, however, was livid. He called Ellison to dress him down, but Ellison wouldn't take the call. So Amelio called Jobs, whose response was equivocal but also partly genuine. "I really don't understand what is going on," he told Amelio. "I think all this is crazy." Then he added a reassurance that was not at all genuine: "You and I have a good relationship." Jobs could have ended the speculation by releasing a statement rejecting Ellison's idea, but much to Amelio's annoyance, he didn't. He remained aloof, which served both his interests and his nature.

By then the press had turned against Amelio. *Business Week* ran a cover asking "Is Apple Mincemeat?"; *Red Herring* ran an editorial headlined "Gil Amelio, Please Resign"; and *Wired* ran a cover that showed the Apple logo crucified as a sacred heart with a crown of thorns and the headline "Pray." Mike Barnicle of the *Boston Globe,* railing against years of Apple mismanagement, wrote, "How can these nitwits still draw a paycheck when they took the only computer that didn't frighten people and turned it into the technological equivalent of the 1997 Red Sox bullpen?"

When Jobs and Amelio had signed the contract in February, Jobs began hopping around exuberantly and declared, "You and I need to go out and have a great bottle of wine to celebrate!" Amelio offered to bring wine from his cellar and suggested that they invite their wives. It took until June before they settled on a date, and despite the rising tensions they were able to have a good time. The food and wine were as mismatched as the diners; Amelio brought a bottle of 1964 Cheval Blanc and a Montrachet that each cost about $300; Jobs chose a vegetarian restaurant in Redwood City where the food bill totaled $72. Amelio's wife remarked afterward, "He's such a charmer, and his wife is too."

Jobs could seduce and charm people at will, and he liked to do so. People such as Amelio and Sculley allowed themselves to believe that because Jobs was charming them, it meant that he liked and respected them. It was an impression that he sometimes fostered by dishing out insincere flattery to those hungry for it. But Jobs could be charming to people he hated just as easily as he could be insulting to people he liked. Amelio didn't see this because, like Sculley, he was so eager for Jobs's affection. Indeed the words he used to describe his yearning for a good relationship with Jobs are almost the same as those used by Sculley. "When I was wrestling with a problem, I would walk through the issue with him," Amelio recalled. "Nine times out of ten we would agree." Somehow he willed himself to believe that Jobs really respected him: "I was in awe over the way Steve's mind approached problems, and had the feeling we were building a mutually trusting relationship."

Amelio's disillusionment came a few days after their dinner. During their negotiations, he had insisted that Jobs hold the Apple stock he got for at least six months, and preferably longer. That six months ended in June. When a block of 1.5 million shares was sold, Amelio called Jobs. "I'm telling people that the shares sold were not yours," he said. "Remember, you and I had an understanding that you wouldn't sell any without advising us first."

"That's right," Jobs replied. Amelio took that response to mean that Jobs had not sold his shares, and he issued a statement saying so. But when the next SEC filing came out, it revealed that Jobs had indeed sold the shares. "Dammit, Steve, I asked you point-blank about these shares and you denied it was you." Jobs told Amelio that he had sold in a "fit of depression" about where Apple was going and he didn't want to admit it because he was "a little embarrassed." When I asked him about it years later, he simply said, "I didn't feel I needed to tell Gil."

Why did Jobs mislead Amelio about selling the shares? One reason is simple: Jobs sometimes avoided the truth. Helmut Sonnenfeldt once said of Henry Kissinger, "He lies not because it's in his interest, he lies because it's in his nature." It was in Jobs's nature to mislead or be secretive when he felt it was warranted. But he also indulged in being brutally honest at times, telling the truths that most of us sugarcoat or suppress. Both the dissembling and the truth-telling were simply different aspects of his Nietzschean attitude that ordinary rules didn't apply to him.

EXIT, PURSUED BY A BEAR

Jobs had refused to quash Larry Ellison's takeover talk, and he had secretly sold his shares and been misleading about it. So Amelio finally became convinced that Jobs was gunning for him. "I finally absorbed the fact that I had been too willing and too eager to believe he was on my team,"

Amelio recalled. "Steve's plans to manipulate my termination were charging forward."

Jobs was indeed bad-mouthing Amelio at every opportunity. He couldn't help himself. But there was a more important factor in turning the board against Amelio. Fred Anderson, the chief financial officer, saw it as his fiduciary duty to keep Ed Woolard and the board informed of Apple's dire situation. "Fred was the guy telling me that cash was draining, people were leaving, and more key players were thinking of it," said Woolard. "He made it clear the ship was going to hit the sand soon, and even he was thinking of leaving." That added to the worries Woolard already had from watching Amelio bumble the shareholders meeting.

At an executive session of the board in June, with Amelio out of the room, Woolard described to current directors how he calculated their odds. "If we stay with Gil as CEO, I think there's only a 10% chance we will avoid bankruptcy," he said. "If we fire him and convince Steve to come take over, we have a 60% chance of surviving. If we fire Gil, don't get Steve back, and have to search for a new CEO, then we have a 40% chance of surviving." The board gave him authority to ask Jobs to return.

Woolard and his wife flew to London, where they were planning to watch the Wimbledon tennis matches. He saw some of the tennis during the day, but spent his evenings in his suite at the Inn on the Park calling people back in America, where it was daytime. By the end of his stay, his telephone bill was $2,000.

First, he called Jobs. The board was going to fire Amelio, he said, and it wanted Jobs to come back as CEO. Jobs had been aggressive in deriding Amelio and pushing his own ideas about where to take Apple. But suddenly, when offered the cup, he became coy. "I will help," he replied.

"As CEO?" Woolard asked.

Jobs said no. Woolard pushed hard for him to become at least the acting CEO. Again Jobs demurred. "I will be an advisor," he said. "Unpaid." He also agreed to become a board member — that was something he had yearned for — but declined to be the board chairman. "That's all I can give now," he said. After rumors began circulating, he emailed a memo to Pixar employees assuring them that he was not abandoning them. "I got a call from Apple's board of directors three weeks ago asking me to return to Apple as their CEO," he wrote. "I declined. They then asked me to become chairman, and I again declined. So don't worry — the crazy rumors are just that. I have no plans to leave Pixar. You're stuck with me."

Why did Jobs not seize the reins? Why was he reluctant to grab the job that for two decades he had seemed to desire? When I asked him, he said:

We'd just taken Pixar public, and I was happy being CEO there. I never knew of anyone who served as CEO of two public companies, even temporarily, and I wasn't even sure it was legal. I didn't know what I wanted to do. I was enjoying spending more time with my family. I was torn. I knew Apple was a mess, so I wondered: Do I want to give up this nice lifestyle that I have? What are all the Pixar shareholders going to think? I talked to people I respected. I finally called Andy Grove at about eight one Saturday morning — too early. I gave him the pros and the cons, and in the middle he stopped me and said, "Steve, I don't give a shit about Apple." I was stunned. It was then I realized that I *do* give a shit about Apple — I started it and it is a good thing to have in the world. That was when I decided to go back on a temporary basis to help them hire a CEO.

The claim that he was enjoying spending more time with his family was not convincing. He was never destined to win a Father of the Year trophy, even when he had spare time on his hands. He was getting better at paying heed to his children, especially Reed, but his primary focus was on his work. He was frequently aloof from his two younger daughters, estranged again from Lisa, and often prickly as a husband.

So what was the real reason for his hesitancy in taking over at Apple? For all of his willfulness and insatiable desire to control things, Jobs was indecisive and reticent when he felt unsure about something. He craved perfection, and he was not always good at figuring out how to settle for something less. He did not like to wrestle with complexity or make accommodations. This was true in products, design, and furnishings for the house. It was also true when it came to personal commitments. If he knew for sure a course of action was right, he was unstoppable. But if he had doubts, he sometimes withdrew, preferring not to think about things that did not perfectly suit him. As happened when

460

Amelio had asked him what role he wanted to play, Jobs would go silent and ignore situations that made him uncomfortable.

This attitude arose partly out of his tendency to see the world in binary terms. A person was either a hero or a bozo, a product was either amazing or shit. But he could be stymied by things that were more complex, shaded, or nuanced: getting married, buying the right sofa, committing to run a company. In addition, he didn't want to be set up for failure. "I think Steve wanted to assess whether Apple could be saved," Fred Anderson said.

Woolard and the board decided to go ahead and fire Amelio, even though Jobs was not yet forthcoming about how active a role he would play as an advisor. Amelio was about to go on a picnic with his wife, children, and grandchildren when the call came from Woolard in London. "We need you to step down," Woolard said simply. Amelio replied that it was not a good time to discuss this, but Woolard felt he had to persist. "We are going to announce that we're replacing you."

Amelio resisted. "Remember, Ed, I told the board it was going to take three years to get this company back on its feet again," he said. "I'm not even halfway through."

"The board is at the place where we don't want to discuss it further," Woolard replied. Amelio asked who knew about the decision, and Woolard told him the truth: the rest of the board plus Jobs. "Steve was one of the people we talked to about this," Woolard said. "His view is that you're a really nice guy, but you don't know much about the computer industry."

461

"Why in the world would you involve Steve in a decision like this?" Amelio replied, getting angry. "Steve is not even a member of the board of directors, so what the hell is he doing in any of this conversation?" But Woolard didn't back down, and Amelio hung up to carry on with the family picnic before telling his wife.

At times Jobs displayed a strange mixture of prickliness and neediness. He usually didn't care one iota what people thought of him; he could cut people off and never care to speak to them again. Yet sometimes he also felt a compulsion to explain himself. So that evening Amelio received, to his surprise, a phone call from Jobs. "Gee, Gil, I just wanted you to know, I talked to Ed today about this thing and I really feel bad about it," he said. "I want you to know that I had absolutely nothing to do with this turn of events, it was a decision the board made, but they had asked me for advice and counsel." He told Amelio he respected him for having "the highest integrity of anyone I've ever met," and went on to give some unsolicited advice. "Take six months off," Jobs told him. "When I got thrown out of Apple, I immediately went back to work, and I regretted it." He offered to be a sounding board if Amelio ever wanted more advice.

Amelio was stunned but managed to mumble a few words of thanks. He turned to his wife and recounted what Jobs said. "In ways, I still like the man, but I don't believe him," he told her.

"I was totally taken in by Steve," she said, "and I really feel like an idiot."

"Join the crowd," her husband replied.

Steve Wozniak, who was himself now an informal

advisor to the company, was thrilled that Jobs was coming back. (He forgave easily.) "It was just what we needed," he said, "because whatever you think of Steve, he knows how to get the magic back." Nor did Jobs's triumph over Amelio surprise him. As he told *Wired* shortly after it happened, "Gil Amelio meets Steve Jobs, game over."

That Monday Apple's top employees were summoned to the auditorium. Amelio came in looking calm and relaxed. "Well, I'm sad to report that it's time for me to move on," he said. Fred Anderson, who had agreed to be interim CEO, spoke next, and he made it clear that he would be taking his cues from Jobs. Then, exactly twelve years since he had lost power in a July 4 weekend struggle, Jobs walked back onstage at Apple.

It immediately became clear that, whether or not he wanted to admit it publicly (or even to himself), Jobs was going to take control and not be a mere advisor. As soon as he came onstage that day — wearing shorts, sneakers, and a black turtleneck — he got to work reinvigorating his beloved institution. "Okay, tell me what's wrong with this place," he said. There were some murmurings, but Jobs cut them off. "It's the products!" he answered. "So what's wrong with the products?" Again there were a few attempts at an answer, until Jobs broke in to hand down the correct answer. "The products *suck!*" he shouted. "There's no sex in them anymore!"

Woolard was able to coax Jobs to agree that his role as an advisor would be a very active one. Jobs approved a statement saying that he had "agreed to step up my involvement with Apple for up to 90 days, helping them until they hire a new CEO." The clever formulation that Woolard used in his statement was that Jobs was coming back "as an advisor leading the team."

Jobs took a small office next to the boardroom on the executive floor, conspicuously eschewing Amelio's big corner office. He got involved in all aspects of the business: product design, where to cut, supplier negotiations, and advertising agency review. He believed that he had to stop the hemorrhaging of top Apple employees, and to do so he wanted to reprice their stock options. Apple stock had dropped so low that the options had become worthless. Jobs wanted to lower the exercise price, so they would be valuable again. At the time, that was legally permissible, but it was not considered good corporate practice. On his first Thursday back at Apple, Jobs called for a telephonic board meeting and outlined the problem. The directors balked. They asked for time to do a legal and financial study of what the change would mean. "It has to be done fast," Jobs told them. "We're losing good people."

Even his supporter Ed Woolard, who headed the compensation committee, objected. "At DuPont we never did such a thing," he said.

"You brought me here to fix this thing, and people are the key," Jobs argued. When the board proposed a study that could take two months, Jobs exploded: "Are you nuts?!?" He paused for a long moment of silence, then continued. "Guys, if you don't want to do this, I'm not coming back on Monday. Because I've got thousands of key decisions to make that are far more difficult than this, and if you can't throw your support behind this kind of decision, I will fail. So if you can't do this, I'm out of here, and you can blame it on me, you can say, 'Steve wasn't up for the job.' "

The next day, after consulting with the board, Woolard called Jobs back. "We're going to approve this," he said. "But some of the board members don't like it. We feel like you've put a gun to our head." The options for the top team (Jobs had none) were reset at $13.25, which was the price of the stock the day Amelio was ousted.

Instead of declaring victory and thanking the board, Jobs continued to seethe at having to answer to a board he didn't respect. "Stop the train, this isn't going to work," he told Woolard. "This company is in shambles, and I don't have time to wet-nurse the board. So I need all of you to resign. Or else I'm going to resign and not come back on Monday." The one person who could stay, he said, was Woolard.

Most members of the board were aghast. Jobs was still refusing to commit himself to coming back full-time or being anything more than an advisor, yet he felt he had the power to force them to leave. The hard truth, however, was that he did have that power over them. They could not afford for him to storm off in a fury, nor was the prospect of remaining an Apple board member very enticing by then. "After all they'd been through, most were glad to be let off," Woolard recalled.

Once again the board acquiesced. It made only one request: Would he permit one other director to stay, in addition to Woolard? It would help the optics. Jobs assented. "They were an awful board, a terrible board," he later said. "I agreed they could keep Ed Woolard and a guy named Gareth Chang, who turned out to be a zero. He wasn't terrible, just a zero. Woolard, on the other hand, was one of the best board members I've ever seen. He was a prince, one of the most supportive and wise people I've ever met."

Among those being asked to resign was Mike Markkula, who in 1976, as a young venture capitalist, had visited the Jobs garage, fallen in love with the nascent computer on the workbench, guaranteed a $250,000 line of credit, and become the third partner and one-third owner of the new company. Over the subsequent two decades, he was the one constant on the board, ushering in and out a variety of CEOs. He had supported Jobs at times but also clashed with him, most notably when he sided with Sculley in the showdowns of 1985. With Jobs returning, he knew that it was time for him to leave.

Jobs could be cutting and cold, especially toward people who crossed him, but he could also be sentimental about those who had been with him from the early days. Wozniak fell into that favored category, of course, even though they had drifted apart; so did Andy Hertzfeld and a few others from the Macintosh team. In the end, Mike Markkula did as well. "I felt deeply betrayed by him, but he was like a father and I always cared about him," Jobs later recalled. So when the time came to ask him to resign from the Apple board, Jobs drove to Markkula's chateau-like mansion in the Woodside hills to do it personally. As usual, he asked to take a walk, and they strolled the grounds to a redwood

466

grove with a picnic table. "He told me he wanted a new board because he wanted to start fresh," Markkula said. "He was worried that I might take it poorly, and he was relieved when I didn't."

They spent the rest of the time talking about where Apple should focus in the future. Jobs's ambition was to build a company that would endure, and he asked Markkula what the formula for that would be. Markkula replied that lasting companies know how to reinvent themselves. Hewlett-Packard had done that repeatedly; it started as an instrument company, then became a calculator company, then a computer company. "Apple has been sidelined by Microsoft in the PC business," Markkula said. "You've got to reinvent the company to do some other thing, like other consumer products or devices. You've got to be like a butterfly and have a metamorphosis." Jobs didn't say much, but he agreed.

The old board met in late July to ratify the transition. Woolard, who was as genteel as Jobs was prickly, was mildly taken aback when Jobs appeared dressed in jeans and sneakers, and he worried that Jobs might start berating the veteran board members for screwing up. But Jobs merely offered a pleasant "Hi, everyone." They got down to the business of voting to accept the resignations, elect Jobs to the board, and empower Woolard and Jobs to find new board members.

Jobs's first recruit was, not surprisingly, Larry Ellison. He said he would be pleased to join, but he hated attending meetings. Jobs said it would be fine if he came to only half of them. (After a while Ellison was coming to only a third of the meetings.

467

Jobs took a picture of him that had appeared on the cover of *Business Week* and had it blown up to life size and pasted on a cardboard cutout to put in his chair.)

Jobs also brought in Bill Campbell, who had run marketing at Apple in the early 1980s and been caught in the middle of the Sculley-Jobs clash. Campbell had ended up sticking with Sculley, but he had grown to dislike him so much that Jobs forgave him. Now he was the CEO of Intuit and a walking buddy of Jobs. "We were sitting out in the back of his house," recalled Campbell, who lived only five blocks from Jobs in Palo Alto, "and he said he was going back to Apple and wanted me on the board. I said, 'Holy shit, of course I will do that.' " Campbell had been a football coach at Columbia, and his great talent, Jobs said, was to "get A performances out of B players." At Apple, Jobs told him, he would get to work with A players.

Woolard helped bring in Jerry York, who had been the chief financial officer at Chrysler and then IBM. Others were considered and then rejected by Jobs, including Meg Whitman, who was then the manager of Hasbro's Playskool division and had been a strategic planner at Disney. (In 1998 she became CEO of eBay, and she later ran unsuccessfully for governor of California.) Over the years Jobs would bring in some strong leaders to serve on the Apple board, including Al Gore, Eric Schmidt of Google, Art Levinson of Genentech, Mickey Drexler of the Gap and J. Crew, and Andrea Jung of Avon. But he always made sure they were loyal, sometimes loyal to a fault. Despite their stature, they seemed at times awed or intimidated by Jobs, and they were eager to keep him happy.

At one point he invited Arthur Levitt, the former SEC chairman, to become a board member. Levitt, who bought his first Macintosh in 1984 and was proudly "addicted" to Apple computers, was thrilled. He was excited to visit Cupertino, where he discussed the role with Jobs. But then Jobs read a speech Levitt had given about corporate governance, which argued that boards should play a strong and independent role, and he telephoned to withdraw the invitation. "Arthur, I don't think you'd be happy on our board, and I think it best if we not invite you," Levitt said Jobs told him. "Frankly, I think some of the issues you raised, while appropriate for some companies, really don't apply to Apple's culture." Levitt later wrote, "I was floored. . . . It's plain to me that Apple's board is not designed to act independently of the CEO."

MACWORLD BOSTON, AUGUST 1997

The staff memo announcing the repricing of Apple's stock options was signed "Steve and the executive team," and it soon became public that he was running all of the company's product review meetings. These and other indications that Jobs was now deeply engaged at Apple helped push the stock up from about $13 to $20 during July. It also created a frisson of excitement as the Apple faithful gathered for the August 1997 Macworld in Boston. More than five thousand showed up hours in advance to cram into the Castle convention hall of the Park Plaza hotel for Jobs's keynote speech. They came to see their returning hero — and to find out whether he was really ready to lead them again.

Huge cheers erupted when a picture of Jobs from 1984 was flashed on the overhead screen. "Steve! Steve! Steve!" the crowd started to chant, even as he was still being introduced. When he finally strode onstage — wearing a black vest, collarless white shirt, jeans, and an impish smile — the screams and flashbulbs rivaled those for any rock star. At first he punctured the excitement by reminding them of where he officially worked. "I'm Steve Jobs, the chairman and CEO of Pixar," he introduced himself, flashing a slide onscreen with that title. Then he explained his role at Apple. "I, like a lot of other people, are pulling together to help Apple get healthy again."

But as Jobs paced back and forth across the stage, changing the overhead slides with a clicker in his hand, it was clear that he was now in charge at Apple — and was likely to remain so. He delivered a carefully crafted presentation, using no notes, on why Apple's sales had fallen by 30% over the previous two years. "There are a lot of great people at Apple, but they're doing the wrong things because the plan has been wrong," he said. "I've found people who can't wait to fall into line behind a good strategy, but there just hasn't been one." The crowd again erupted in yelps, whistles, and cheers.

As he spoke, his passion poured forth with increasing intensity, and he began saying "we" and "I" — rather than "they" — when referring to what Apple would be doing. "I think you still have to think differently to buy an Apple computer," he said. "The people who buy them do think different. They are the creative spirits in this world, and they're out to change the world. *We* make tools for those kinds of people." When he stressed the word "we" in that sentence, he cupped his hands and tapped his fingers on his chest. And then, in his final peroration, he continued to stress the word "we" as he talked about Apple's future. "We too are going to think differently and serve the people who have been buying our products from the beginning. Because a lot of people think they're crazy, but in that craziness we see genius." During the prolonged standing ovation, people looked at each other in awe, and a few wiped tears from their eyes. Jobs had made it very clear that he and the "we" of Apple were one.

THE MICROSOFT PACT

The climax of Jobs's August 1997 Macworld appearance was a bombshell announcement, one that made the cover of both *Time* and *Newsweek*. Near the end of his speech, he paused for a sip of water and began to talk in more subdued tones. "Apple lives in an ecosystem," he said. "It needs help from other partners. Relationships that are destructive don't help anybody in this industry." For dramatic effect, he paused again, and then explained: "I'd like to announce one of our first new partnerships today, a very meaningful one, and that is one with Microsoft." The Microsoft and Apple logos appeared together on the screen as people gasped.

Apple and Microsoft had been at war for a decade over a variety of copyright and patent issues, most notably whether Microsoft had stolen the look and feel of Apple's graphical user interface.

Just as Jobs was being eased out of Apple in 1985, John Sculley had struck a surrender deal: Microsoft could license the Apple GUI for Windows 1.0, and in return it would make Excel exclusive to the Mac for up to two years. In 1988, after Microsoft came out with Windows 2.0, Apple sued. Sculley contended that the 1985 deal did not apply to Windows 2.0 and that further refinements to Windows (such as copying Bill Atkinson's trick of "clipping" overlapping windows) had made the infringement more blatant. By 1997 Apple had lost the case and various appeals, but remnants of the litigation and threats of new suits lingered. In addition, President Clinton's Justice Department was preparing a massive antitrust case against Microsoft. Jobs invited the lead prosecutor, Joel Klein, to Palo Alto. Don't worry about extracting a huge remedy against Microsoft, Jobs told him over coffee. Instead simply keep them tied up in litigation. That would allow Apple the opportunity, Jobs explained, to "make an end run" around Microsoft and start offering competing products.

Under Amelio, the showdown had become explosive. Microsoft refused to commit to developing Word and Excel for future Macintosh operating systems, which could have destroyed Apple. In defense of Bill Gates, he was not simply being vindictive. It was understandable that he was reluctant to commit to developing for a future Macintosh operating system when no one, including the ever-changing leadership at Apple, seemed to know what that new operating system would be. Right after Apple bought NeXT, Amelio and Jobs flew together to visit Microsoft, but Gates had trouble

472

figuring out which of them was in charge. A few days later he called Jobs privately. "Hey, what the fuck, am I supposed to put my applications on the NeXT OS?" Gates asked. Jobs responded by "making smart-ass remarks about Gil," Gates recalled, and suggesting that the situation would soon be clarified.

When the leadership issue was partly resolved by Amelio's ouster, one of Jobs's first phone calls was to Gates. Jobs recalled:

I called up Bill and said, "I'm going to turn this thing around." Bill always had a soft spot for Apple. We got him into the application software business. The first Microsoft apps were Excel and Word for the Mac. So I called him and said, "I need help." Microsoft was walking over Apple's patents. I said, "If we kept up our lawsuits, a few years from now we could win a billion-dollar patent suit. You know it, and I know it. But Apple's not going to survive that long if we're at war. I know that. So let's figure out how to settle this right away. All I need is a commitment that Microsoft will keep developing for the Mac and an investment by Microsoft in Apple so it has a stake in our success."

When I recounted to him what Jobs said, Gates agreed it was accurate. "We had a group of people who liked working on the Mac stuff, and we liked the Mac," Gates recalled. He had been negotiating with Amelio for six months, and the proposals kept getting longer and more complicated. "So Steve comes in and says, 'Hey, that deal is too complicated. What I want is a simple deal. I want the commitment and I want an investment.' And so we put that together in just four weeks."

Gates and his chief financial officer, Greg Maffei, made the trip to Palo Alto to work out the framework for a deal, and then Maffei returned alone the following Sunday to work on the details. When he arrived at Jobs's home, Jobs grabbed two bottles of water out of the refrigerator and took Maffei for a walk around the Palo Alto neighborhood. Both men wore shorts, and Jobs walked barefoot. As they sat in front of a Baptist church, Jobs cut to the core issues. "These are the things we care about," he said. "A commitment to make software for the Mac and an investment."

Although the negotiations went quickly, the final details were not finished until hours before Jobs's Macworld speech in Boston. He was rehearsing at the Park Plaza Castle when his cell phone rang. "Hi, Bill," he said as his words echoed through the old hall. Then he walked to a corner and spoke in a soft tone so others couldn't hear. The call lasted an hour. Finally, the remaining deal points were resolved. "Bill, thank you for your support of this company," Jobs said as he crouched in his shorts. "I think the world's a better place for it."

During his Macworld keynote address, Jobs walked through the details of the Microsoft deal. At first there were groans and hisses from the faithful. Particularly galling was Jobs's announcement that, as part of the peace pact, "Apple has decided to make Internet Explorer its default browser on the Macintosh." The audience erupted in boos, and Jobs quickly added, "Since we believe

474

in choice, we're going to be shipping other Internet browsers, as well, and the user can, of course, change their default should they choose to." There were some laughs and scattered applause. The audience was beginning to come around, especially when he announced that Microsoft would be investing $150 million in Apple and getting nonvoting shares.

But the mellower mood was shattered for a moment when Jobs made one of the few visual and public relations gaffes of his onstage career. "I happen to have a special guest with me today via satellite downlink," he said, and suddenly Bill Gates's face appeared on the huge screen looming over Jobs and the auditorium. There was a thin smile on Gates's face that flirted with being a smirk. The audience gasped in horror, followed by some boos and catcalls. The scene was such a brutal echo of the 1984 Big Brother ad that you half expected (and hoped?) that an athletic woman would suddenly come running down the aisle and vaporize the screenshot with a well-thrown hammer.

But it was all for real, and Gates, unaware of the jeering, began speaking on the satellite link from Microsoft headquarters. "Some of the most exciting work that I've done in my career has been the work that I've done with Steve on the Macintosh," he intoned in his high-pitched singsong. As he went on to tout the new version of Microsoft Office that was being made for the Macintosh, the audience quieted down and then slowly seemed to accept the new world order. Gates even was able to rouse some applause when he said that the new Mac versions of Word and Excel would be "in many ways more advanced than what we've done on the Windows platform."

Jobs realized that the image of Gates looming over him and the audience was a mistake. "I wanted him to come to Boston," Jobs later said. "That was my worst and stupidest staging event ever. It was bad because it made me look small, and Apple look small, and as if everything was in Bill's hands." Gates likewise was embarrassed when he saw the videotape of the event. "I didn't know that my face was going to be blown up to looming proportions," he said.

Jobs tried to reassure the audience with an impromptu sermon. "If we want to move forward and see Apple healthy again, we have to let go of a few things here," he told the audience. "We have to let go of this notion that for Apple to win Microsoft has to lose. . . . I think if we want Microsoft Office on the Mac, we better treat the company that puts it out with a little bit of gratitude."

The Microsoft announcement, along with Jobs's passionate reengagement with the company, provided a much-needed jolt for Apple. By the end of the day, its stock had skyrocketed $6.56, or 33%, to close at $26.31, twice the price of the day Amelio resigned. The one-day jump added $830 million to Apple's stock market capitalization. The company was back from the edge of the grave.

CHAPTER TWENTY-FIVE: THINK DIFFERENT

JOBS AS iCEO

HERE'S TO THE CRAZY ONES

Lee Clow, the creative director at Chiat/Day who had done the great "1984" ad for the launch of the Macintosh, was driving in Los Angeles in early July 1997 when his car phone rang. It was Jobs. "Hi, Lee, this is Steve," he said. "Guess what? Amelio just resigned. Can you come up here?"

Apple was going through a review to select a new agency, and Jobs was not impressed by what he had seen. So he wanted Clow and his firm, by then called TBWA\Chiat\Day, to compete for the business. "We have to prove that Apple is still alive," Jobs said, "and that it still stands for something special."

Clow said that he didn't pitch for accounts. "You know our work," he said. But Jobs begged him. It would be hard to reject all the others that were making pitches, including BBDO and Arnold Worldwide, and bring back "an old crony," as Jobs put it. Clow agreed to fly up to Cupertino with something they could show. Recounting the scene years later, Jobs started to cry.

This chokes me up, this really chokes me up.
It was so clear that Lee loved Apple so much.

Here was the best guy in advertising. And he hadn't pitched in ten years. Yet here he was, and he was pitching his heart out, because he loved Apple as much as we did. He and his team had come up with this brilliant idea, "Think Different." And it was ten times better than anything the other agencies showed. It choked me up, and it still makes me cry to think about it, both the fact that Lee cared so much and also how brilliant his "Think Different" idea was. Every once in a while, I find myself in the presence of purity — purity of spirit and love — and I always cry. It always just reaches in and grabs me. That was one of those moments. There was a purity about that I will never forget. I cried in my office as he was showing me the idea, and I still cry when I think about it.

Jobs and Clow agreed that Apple was one of the great brands of the world, probably in the top five based on emotional appeal, but they needed to remind folks what was distinctive about it. So they wanted a brand image campaign, not a set of advertisements featuring products. It was designed to celebrate not what the computers could do, but what creative people could do with the computers. "This wasn't about processor speed or memory," Jobs recalled. "It was about creativity." It was directed not only at potential customers, but also at Apple's own employees: "We at Apple had forgotten who we were. One way to remember who you are is to remember who your heroes are. That was the genesis of that campaign."

Clow and his team tried a variety of approaches

that praised the "crazy ones" who "think different." They did one video with the Seal song "Crazy" ("We're never gonna survive unless we get a little crazy"), but couldn't get the rights to it. Then they tried versions using a recording of Robert Frost reading "The Road Not Taken" and of Robin Williams's speeches from *Dead Poets Society*. Eventually they decided they needed to write their own text; their draft began, "Here's to the crazy ones."

Jobs was as demanding as ever. When Clow's team flew up with a version of the text, he exploded at the young copywriter. "This is shit!" he yelled. "It's advertising agency shit and I hate it." It was the first time the young copywriter had met Jobs, and he stood there mute. He never went back. But those who could stand up to Jobs, including Clow and his teammates Ken Segall and Craig Tanimoto, were able to work with him to create a tone poem that he liked. In its original sixty-second version it read:

Here's to the crazy ones. The misfits. The rebels. The troublemakers. The round pegs in the square holes. The ones who see things differently. They're not fond of rules. And they have no respect for the status quo. You can quote them, disagree with them, glorify or vilify them. About the only thing you can't do is ignore them. Because they change things. They push the human race forward. And while some may see them as the crazy ones, we see genius. Because the people who are crazy enough to think they can change the world are the ones who do.

Jobs, who could identify with each of those sentiments, wrote some of the lines himself, including "They push the human race forward." By the time of the Boston Macworld in early August, they had produced a rough version. They agreed it was not ready, but Jobs used the concepts, and the "think different" phrase, in his keynote speech there. "There's a germ of a brilliant idea there," he said at the time. "Apple is about people who think outside the box, who want to use computers to help them change the world."

They debated the grammatical issue: If "different" was supposed to modify the verb "think," it should be an adverb, as in "think differently." But Jobs insisted that he wanted "different" to be used as a noun, as in "think victory" or "think beauty." Also, it echoed colloquial use, as in "think big." Jobs later explained, "We discussed whether it was correct before we ran it. It's grammatical, if you think about what we're trying to say. It's not think *the same,* it's think *different.* Think a little different, think a lot different, think different. 'Think *differently*' wouldn't hit the meaning for me."

In order to evoke the spirit of *Dead Poets Society,* Clow and Jobs wanted to get Robin Williams to read the narration. His agent said that Williams didn't do ads, so Jobs tried to call him directly. He got through to Williams's wife, who would not let him talk to the actor because she knew how persuasive he could be. They also considered Maya Angelou and Tom Hanks. At a fund-raising dinner featuring Bill Clinton that fall, Jobs pulled the president aside and asked him to telephone Hanks to talk him into it, but the president pocket-vetoed the request. They ended up with Richard Drey-fuss, who was a dedicated Apple fan.

In addition to the television commercials, they created one of the most memorable print campaigns in history. Each ad featured a black-and-white portrait of an iconic historical figure with just the Apple logo and the words "Think Different" in the corner. Making it particularly engaging was that the faces were not captioned. Some of them — Einstein, Gandhi, Lennon, Dylan, Picasso, Edison, Chaplin, King — were easy to identify. But others caused people to pause, puzzle, and maybe ask a friend to put a name to the face: Martha Graham, Ansel Adams, Richard Feynman, Maria Callas, Frank Lloyd Wright, James Watson, Amelia Earhart.

Most were Jobs's personal heroes. They tended to be creative people who had taken risks, defied failure, and bet their career on doing things in a different way. A photography buff, he became involved in making sure they had the perfect iconic portraits. "This is not the right picture of Gandhi," he erupted to Clow at one point. Clow explained that the famous Margaret Bourke-White photograph of Gandhi at the spinning wheel was owned by Time-Life Pictures and was not available for commercial use. So Jobs called Norman Pearlstine, the editor in chief of Time Inc., and badgered him into making an exception. He called Eunice Shriver to convince her family to release a picture that he loved, of her brother Bobby Kennedy touring Appalachia, and he talked to Jim Henson's children personally to get the right shot of the late Muppeteer.

He likewise called Yoko Ono for a picture of her late husband, John Lennon. She sent him one, but it was not Jobs's favorite. "Before it ran, I was in New York, and I went to this small Japanese restaurant that I love, and let her know I would be there," he recalled. When he arrived, she came over to his table. "This is a better one," she said, handing him an envelope. "I thought I would see you, so I had this with me." It was the classic photo of her and John in bed together, holding flowers, and it was the one that Apple ended up using. "I can see why John fell in love with her," Jobs recalled.

The narration by Richard Dreyfuss worked well, but Lee Clow had another idea. What if Jobs did the voice-over himself? "You really believe this," Clow told him. "You should do it." So Jobs sat in a studio, did a few takes, and soon produced a voice track that everyone liked. The idea was that, if they used it, they would not tell people who was speaking the words, just as they didn't caption the iconic pictures. Eventually people would figure out it was Jobs. "This will be really powerful to have it in your voice," Clow argued. "It will be a way to reclaim the brand."

Jobs couldn't decide whether to use the version with his voice or to stick with Dreyfuss. Finally, the night came when they had to ship the ad; it was due to air, appropriately enough, on the television premiere of *Toy Story*. As was often the case, Jobs did not like to be forced to make a decision. He told Clow to ship both versions; this would give him until the morning to decide. When morning came, Jobs called and told them to use the Dreyfuss version. "If we use my voice, when people find out they will say it's about me," he told Clow. "It's not. It's about Apple."

Ever since he left the apple commune, Jobs had defined himself, and by extension Apple, as a child of the counterculture. In ads such as "Think Different" and "1984," he positioned the Apple brand so that it reaffirmed his own rebel streak, even after he became a billionaire, and it allowed other baby boomers and their kids to do the same. "From when I first met him as a young guy, he's had the greatest intuition of the impact he wants his brand to have on people," said Clow.

Very few other companies or corporate leaders — perhaps none — could have gotten away with the brilliant audacity of associating their brand with Gandhi, Einstein, Picasso, and the Dalai Lama. Jobs was able to encourage people to define themselves as anticorporate, creative, innovative rebels simply by the computer they used. "Steve created the only lifestyle brand in the tech industry," Larry Ellison said. "There are cars people are proud to have — Porsche, Ferrari, Prius — because what I drive says something about me. People feel the same way about an Apple product."

Starting with the "Think Different" campaign, and continuing through the rest of his years at Apple, Jobs held a freewheeling three-hour meeting every Wednesday afternoon with his top agency, marketing, and communications people to kick around messaging strategy. "There's not a CEO on the planet who deals with marketing the way Steve does," said Clow. "Every Wednesday he approves each new commercial, print ad, and billboard." At the end of the meeting, he would often take Clow and his two agency colleagues, Duncan Milner and James Vincent, to Apple's closely guarded design studio to see what products were in the works. "He gets very passionate and emotional when he shows us what's in development," said Vincent. By sharing with his marketing gurus his passion for the products as they were being created, he was able to ensure that almost every ad they produced was infused with his emotion.

iCEO

As he was finishing work on the "Think Different" ad, Jobs did some different thinking of his own. He decided that he would officially take over running the company, at least on a temporary basis. He had been the de facto leader since Amelio's ouster ten weeks earlier, but only as an advisor. Fred Anderson had the titular role of interim CEO. On September 16, 1997, Jobs announced that he would take over that title, which inevitably got abbreviated as iCEO. His commitment was tentative: He took no salary and signed no contract. But he was not tentative in his actions. He was in charge, and he did not rule by consensus.

That week he gathered his top managers and staff in the Apple auditorium for a rally, followed by a picnic featuring beer and vegan food, to celebrate his new role and the company's new ads. He was wearing shorts, walking around the campus barefoot, and had a stubble of beard. "I've been back about ten weeks, working really hard," he said, looking tired but deeply determined. "What we're trying to do is not highfa-lutin. We're trying to get back to the basics of great products, great marketing, and great dis-tribution. Apple has drifted away from doing the basics really well."

For a few more weeks Jobs and the board kept looking for a permanent CEO. Various names surfaced — George M. C. Fisher of Kodak, Sam Palmisano at IBM, Ed Zander at Sun Microsys-tems — but most of the candidates were under-standably reluctant to consider becoming CEO if Jobs was going to remain an active board member. The *San Francisco Chronicle* reported that Zander declined to be considered because he "didn't want Steve looking over his shoulder, second-guessing him on every decision." At one point Jobs and El-lison pulled a prank on a clueless computer con-sultant who was campaigning for the job; they sent him an email saying that he had been selected, which caused both amusement and embarrass-ment when stories appeared in the papers that they were just toying with him.

By December it had become clear that Jobs's iCEO status had evolved from *interim* to *indefinite*. As Jobs continued to run the company, the board quietly deactivated its search. "I went back to Apple and tried to hire a CEO, with the help of a recruiting agency, for almost four months," he re-called. "But they didn't produce the right people. That's why I finally stayed. Apple was in no shape to attract anybody good."

The problem Jobs faced was that running two companies was brutal. Looking back on it, he traced his health problems back to those days:

It was rough, really rough, the worst time in my life. I had a young family. I had Pixar. I would go to work at 7 a.m. and I'd get back at 9 at night, and the kids would be in bed. And I couldn't speak, I literally couldn't, I was so exhausted. I couldn't speak to Laurene. All I could do was watch a half hour of TV and vegetate. It got close to killing me. I was driving up to Pixar and down to Apple in a black Porsche convertible, and I started to get kidney stones. I would rush to the hospital and the hospital would give me a shot of Demerol in the butt and eventually I would pass it.

Despite the grueling schedule, the more that Jobs immersed himself in Apple, the more he realized that he would not be able to walk away. When Michael Dell was asked at a computer trade show in October 1997 what he would do if he were Steve Jobs and taking over Apple, he replied, "I'd shut it down and give the money back to the shareholders." Jobs fired off an email to Dell. "CEOs are supposed to have class," it said. "I can see that isn't an opinion you hold." Jobs liked to stoke up rivalries as a way to rally his team — he had done so with IBM and Microsoft — and he did so with Dell. When he called together his managers to institute a build-to-order system for manufacturing and distribution, Jobs used as a backdrop a blown-up picture of Michael Dell with a target on his face. "We're coming after you, buddy," he said to cheers from his troops.

One of his motivating passions was to build a lasting company. At age twelve, when he got a summer job at Hewlett-Packard, he learned that a properly run company could spawn innovation far more than any single creative individual. "I discovered that the best innovation is sometimes the company, the way you organize a company," he recalled. "The whole notion of how you build a company is fascinating. When I got the chance to come back to Apple, I realized that I would be useless without the company, and that's why I decided to stay and rebuild it."

KILLING THE CLONES

One of the great debates about Apple was whether it should have licensed its operating system more aggressively to other computer makers, the way Microsoft licensed Windows. Wozniak had favored that approach from the beginning. "We had the most beautiful operating system," he said, "but to get it you had to buy our hardware at twice the price. That was a mistake. What we should have done was calculate an appropriate price to license the operating system." Alan Kay, the star of Xerox PARC who came to Apple as a fellow in 1984, also fought hard for licensing the Mac OS software. "Software people are always multiplatform, because you want to run on everything," he recalled. "And that was a huge battle, probably the largest battle I lost at Apple."

Bill Gates, who was building a fortune by licensing Microsoft's operating system, had urged Apple to do the same in 1985, just as Jobs was being eased out. Gates believed that, even if Apple took away some of Microsoft's operating system customers, Microsoft could make money by creating versions of its applications software, such as Word and Excel, for the users of the Macintosh and its clones. "I was trying to do everything to get them to be a strong licensor," he recalled. He sent a formal memo to Sculley making the case. "The industry has reached the point where it is now impossible for Apple to create a standard out of their innovative technology without support from, and the resulting credibility of, other personal computer manufacturers," he argued. "Apple should license Macintosh technology to 3–5 significant manufacturers for the development of 'Mac Compatibles.'" Gates got no reply, so he wrote a second memo suggesting some companies that would be good at cloning the Mac, and he added, "I want to help in any way I can with the licensing. Please give me a call."

Apple resisted licensing out the Macintosh operating system until 1994, when CEO Michael Spindler allowed two small companies, Power Computing and Radius, to make Macintosh clones. When Gil Amelio took over in 1996, he added Motorola to the list. It turned out to be a dubious business strategy: Apple got an $80 licensing fee for each computer sold, but instead of expanding the market, the cloners cannibalized the sales of Apple's own high-end computers, on which it made up to $500 in profit.

Jobs's objections to the cloning program were not just economic, however. He had an inbred aversion to it. One of his core principles was that hardware and software should be tightly integrated. He loved to control all aspects of his life, and the only way to do that with computers was to take responsibility for the user experience from end to end.

So upon his return to Apple he made killing the Macintosh clones a priority. When a new version of the Mac operating system shipped in July 1997, weeks after he had helped oust Amelio, Jobs did not allow the clone makers to upgrade to it. The head of Power Computing, Stephen "King" Kahng, organized pro-cloning protests when Jobs appeared at Boston Macworld that August and publicly warned that the Macintosh OS would die if Jobs declined to keep licensing it out. "If the platform goes closed, it is over," Kahng said. "Total destruction. Closed is the kiss of death."

Jobs disagreed. He telephoned Ed Woolard to say he was getting Apple out of the licensing business. The board acquiesced, and in September he reached a deal to pay Power Computing $100 million to relinquish its license and give Apple access to its database of customers. He soon terminated the licenses of the other cloners as well. "It was the dumbest thing in the world to let companies making crappier hardware use our operating system and cut into our sales," he later said.

PRODUCT LINE REVIEW

One of Jobs's great strengths was knowing how to focus. "Deciding what *not* to do is as important as deciding what to do," he said. "That's true for companies, and it's true for products."

He went to work applying this principle as soon as he returned to Apple. One day he was walking the halls and ran into a young Wharton School graduate who had been Amelio's assistant and who said he was wrapping up his work. "Well, good, because I need someone to do grunt work," Jobs told him. His new role was to take notes as Jobs met with the dozens of product teams at Apple, asked them to explain what they were doing, and forced them to justify going ahead with their products or projects.

He also enlisted a friend, Phil Schiller, who had worked at Apple but was then at the graphics software company Macromedia. "Steve would summon the teams into the boardroom, which seats twenty, and they would come with thirty people and try to show PowerPoints, which Steve didn't want to see," Schiller recalled. One of the first things Jobs did during the product review process was ban PowerPoints. "I hate the way people use slide presentations instead of thinking," Jobs later recalled. "People would confront a problem by creating a presentation. I wanted them to engage, to hash things out at the table, rather than show a bunch of slides. People who know what they're talking about don't need PowerPoint."

The product review revealed how unfocused Apple had become. The company was churning out multiple versions of each product because of bureaucratic momentum and to satisfy the whims of retailers. "It was insanity," Schiller recalled. "Tons of products, most of them crap, done by deluded teams." Apple had a dozen versions of the Macintosh, each with a different confusing number, ranging from 1400 to 9600. "I had people explaining this to me for three weeks," Jobs said. "I couldn't figure it out." He finally began asking simple questions, like, "Which ones do I tell my friends to buy?"

When he couldn't get simple answers, he began

slashing away at models and products. Soon he had cut 70% of them. "You are bright people," he told one group. "You shouldn't be wasting your time on such crappy products." Many of the engineers were infuriated at his slash-and-burn tactics, which resulted in massive layoffs. But Jobs later claimed that the good engineers, including some whose projects were killed, were appreciative. He told one staff meeting in September 1997, "I came out of the meeting with people who had just gotten their products canceled and they were three feet off the ground with excitement because they finally understood where in the heck we were going."

After a few weeks Jobs finally had enough. "Stop!" he shouted at one big product strategy session. "This is crazy." He grabbed a magic marker, padded to a whiteboard, and drew a horizontal and vertical line to make a four-squared chart. "Here's what we need," he continued. Atop the two columns he wrote "Consumer" and "Pro"; he labeled the two rows "Desktop" and "Portable." Their job, he said, was to make four great products, one for each quadrant. "The room was in dumb silence," Schiller recalled.

There was also a stunned silence when Jobs presented the plan to the September meeting of the Apple board. "Gil had been urging us to approve more and more products every meeting," Woolard recalled. "He kept saying we need more products. Steve came in and said we needed fewer. He drew a matrix with four quadrants and said that this was where we should focus." At first the board pushed back. It was a risk, Jobs was told. "I can make it work," he replied. The board never voted on the new strategy. Jobs was in charge, and he forged ahead.

The result was that the Apple engineers and managers suddenly became sharply focused on just four areas. For the professional desktop quadrant, they would work on making the Power Macintosh G3. For the professional portable, there would be the PowerBook G3. For the consumer desktop, work would begin on what became the iMac. And for the consumer portable, they would focus on what would become the iBook. The "i," Jobs later explained, was to emphasize that the devices would be seamlessly integrated with the Internet.

Apple's sharper focus meant getting the company out of other businesses, such as printers and servers. In 1997 Apple was selling StyleWriter color printers that were basically a version of the Hewlett-Packard DeskJet. HP made most of its money by selling the ink cartridges. "I don't understand," Jobs said at the product review meeting. "You're going to ship a million and not make money on these? This is nuts." He left the room and called the head of HP. Let's tear up our arrangement, Jobs proposed, and we will get out of the printer business and just let you do it. Then he came back to the boardroom and announced the decision. "Steve looked at the situation and instantly knew we needed to get outside of the box," Schiller recalled.

The most visible decision he made was to kill, once and for all, the Newton, the personal digital assistant with the almost-good handwriting-recognition system. Jobs hated it because it was Sculley's pet project, because it didn't work perfectly, and because he had an aversion to stylus devices. He had tried to get Amelio to kill it early in 1997 and succeeded only in convincing him to try to spin off the division. By late 1997, when Jobs did his product reviews, it was still around. He later described his thinking:

> If Apple had been in a less precarious situation, I would have drilled down myself to figure out how to make it work. I didn't trust the people running it. My gut was that there was some really good technology, but it was fucked up by mismanagement. By shutting it down, I freed up some good engineers who could work on new mobile devices. And eventually we got it right when we moved on to iPhones and the iPad.

This ability to focus saved Apple. In his first year back, Jobs laid off more than three thousand people, which salvaged the company's balance sheet. For the fiscal year that ended when Jobs became interim CEO in September 1997, Apple lost $1.04 billion. "We were less than ninety days from being insolvent," he recalled. At the January 1998 San Francisco Macworld, Jobs took the stage where Amelio had bombed a year earlier. He sported a full beard and a leather jacket as he touted the new product strategy. And for the first time he ended the presentation with a phrase that he would make his signature coda: "Oh, and one more thing . . ." This time the "one more thing" was "Think Profit." When he said those words, the crowd erupted in applause. After two years

493

of staggering losses, Apple had enjoyed a profitable quarter, making $45 million. For the full fis-
cal year of 1998, it would turn in a $309 million
profit. Jobs was back, and so was Apple.

CHAPTER TWENTY-SIX:
DESIGN PRINCIPLES
THE STUDIO OF JOBS AND IVE

JONY IVE

When Jobs gathered his top management for a pep talk just after he became iCEO in September 1997, sitting in the audience was a sensitive and passionate thirty-year-old Brit who was head of the company's design team. Jonathan Ive, known to all as Jony, was planning to quit. He was sick of the company's focus on profit maximization rather than product design. Jobs's talk led him to reconsider. "I remember very clearly Steve announcing that our goal is not just to make money but to make great products," Ive recalled. "The decisions you make based on that philosophy are fundamentally different from the ones we had been making at Apple." Ive and Jobs would soon forge a bond that would lead to the greatest industrial design collaboration of their era.

Ive grew up in Chingford, a town on the northeast edge of London. His father was a silversmith who taught at the local college. "He's a fantastic craftsman," Ive recalled. "His Christmas gift to me would be one day of his time in his college workshop, during the Christmas break when no one else was there, helping me make whatever I

495

dreamed up." The only condition was that Jony had to draw by hand what they planned to make. "I always understood the beauty of things made by hand. I came to realize that what was really important was the care that was put into it. What I really despise is when I sense some carelessness in a product."

Ive enrolled in Newcastle Polytechnic and spent his spare time and summers working at a design consultancy. One of his creations was a pen with a little ball on top that was fun to fiddle with. It helped give the owner a playful emotional connection to the pen. For his thesis he designed a microphone and earpiece — in purest white plastic — to communicate with hearing-impaired kids. His flat was filled with foam models he had made to help him perfect the design. He also designed an ATM machine and a curved phone, both of which won awards from the Royal Society of Arts. Unlike some designers, he didn't just make beautiful sketches; he also focused on how the engineering and inner components would work. He had an epiphany in college when he was able to design on a Macintosh. "I discovered the Mac and felt I had a connection with the people who were making this product," he recalled. "I suddenly understood what a company was, or was supposed to be."

After graduation Ive helped to build a design firm in London, Tangerine, which got a consulting contract with Apple. In 1992 he moved to Cupertino to take a job in the Apple design department. He became the head of the department in 1996, the year before Jobs returned, but wasn't happy. Amelio had little appreciation for design. "There wasn't that feeling of putting care into a product, because we were trying to maximize the money we made," Ive said. "All they wanted from us designers was a model of what something was supposed to look like on the outside, and then engineers would make it as cheap as possible. I was about to quit."

When Jobs took over and gave his pep talk, Ive decided to stick around. But Jobs at first looked around for a world-class designer from the outside. He talked to Richard Sapper, who designed the IBM ThinkPad, and Giorgetto Giugiaro, who designed the Ferrari 250 and the Maserati Ghibli. But then he took a tour of Apple's design studio and bonded with the affable, eager, and very earnest Ive. "We discussed approaches to forms and materials," Ive recalled. "We were on the same wavelength. I suddenly understood why I loved the company."

Ive reported, at least initially, to Jon Rubinstein, whom Jobs had brought in to head the hardware division, but he developed a direct and unusually strong relationship with Jobs. They began to have lunch together regularly, and Jobs would end his day by dropping by Ive's design studio for a chat. "Jony had a special status," said Laurene Powell. "He would come by our house, and our families became close. Steve is never intentionally wounding to him. Most people in Steve's life are replaceable. But not Jony."

Jobs described to me his respect for Ive:

The difference that Jony has made, not only at Apple but in the world, is huge. He is a wickedly intelligent person in all ways. He understands business concepts, marketing concepts. He picks stuff up just like that, click. He understands what we do at our core better than anyone. If I had a spiritual partner at Apple, it's Jony. Jony and I think up most of the products together and then pull others in and say, "Hey, what do you think about this?" He gets the big picture as well as the most infinitesimal details about each product. And he understands that Apple is a product company. He's not just a designer. That's why he works directly for me. He has more operational power than anyone else at Apple except me. There's no one who can tell him what to do, or to butt out. That's the way I set it up.

Like most designers, Ive enjoyed analyzing the philosophy and the step-by-step thinking that went into a particular design. For Jobs, the process was more intuitive. He would point to models and sketches he liked and dump on the ones he didn't. Ive would then take the cues and develop the concepts Jobs blessed.

Ive was a fan of the German industrial designer Dieter Rams, who worked for the electronics firm Braun. Rams preached the gospel of "Less but better," *Weniger aber besser,* and likewise Jobs and Ive wrestled with each new design to see how much they could simplify it. Ever since Apple's first brochure proclaimed "Simplicity is the ultimate sophistication," Jobs had aimed for the simplicity that comes from conquering complexities, not ignoring them. "It takes a lot of hard work," he said, "to make something simple, to truly understand the underlying challenges and come up with elegant solutions."

In Ive, Jobs met his soul mate in the quest for true rather than surface simplicity. Sitting in his design studio, Ive described his philosophy:

> Why do we assume that simple is good? Because with physical products, we have to feel we can dominate them. As you bring order to complexity, you find a way to make the product defer to you. Simplicity isn't just a visual style. It's not just minimalism or the absence of clutter. It involves digging through the depth of the complexity. To be truly simple, you have to go really deep. For example, to have no screws on something, you can end up having a product that is so convoluted and so complex. The better way is to go deeper with the simplicity, to understand everything about it and how it's manufactured. You have to deeply understand the essence of a product in order to be able to get rid of the parts that are not essential.

That was the fundamental principle Jobs and Ive shared. Design was not just about what a product looked like on the surface. It had to reflect the product's essence. "In most people's vocabularies, design means veneer," Jobs told *Fortune* shortly after retaking the reins at Apple. "But to me, nothing could be further from the meaning of design. Design is the fundamental soul of a man-made creation that ends up expressing itself in successive outer layers."

As a result, the process of designing a product

at Apple was integrally related to how it would be engineered and manufactured. Ive described one of Apple's Power Macs. "We wanted to get rid of anything other than what was absolutely essential," he said. "To do so required total collaboration between the designers, the product developers, the engineers, and the manufacturing team. We kept going back to the beginning, again and again. Do we need that part? Can we get it to perform the function of the other four parts?"

The connection between the design of a product, its essence, and its manufacturing was illustrated for Jobs and Ive when they were traveling in France and went into a kitchen supply store. Ive picked up a knife he admired, but then put it down in disappointment. Jobs did the same. "We both noticed a tiny bit of glue between the handle and the blade," Ive recalled. They talked about how the knife's good design had been ruined by the way it was manufactured. "We don't like to think of our knives as being glued together," Ive said. "Steve and I care about things like that, which ruin the purity and detract from the essence of something like a utensil, and we think alike about how products should be made to look pure and seamless."

At most other companies, engineering tends to drive design. The engineers set forth their specifications and requirements, and the designers then come up with cases and shells that will accommodate them. For Jobs, the process tended to work the other way. In the early days of Apple, Jobs had approved the design of the case of the original Macintosh, and the engineers had to make their boards and components fit.

After he was forced out, the process at Apple reverted to being engineer-driven. "Before Steve came back, engineers would say 'Here are the guts' — processor, hard drive — and then it would go to the designers to put it in a box," said Apple's marketing chief Phil Schiller. "When you do it that way, you come up with awful products." But when Jobs returned and forged his bond with Ive, the balance was again tilted toward the designers. "Steve kept impressing on us that the design was integral to what would make us great," said Schiller. "Design once again dictated the engineering, not just vice versa."

On occasion this could backfire, such as when Jobs and Ive insisted on using a solid piece of brushed aluminum for the edge of the iPhone 4 even when the engineers worried that it would compromise the antenna. But usually the distinctiveness of its designs — for the iMac, the iPod, the iPhone, and the iPad — would set Apple apart and lead to its triumphs in the years after Jobs returned.

INSIDE THE STUDIO

The design studio where Jony Ive reigns, on the ground floor of Two Infinite Loop on the Apple campus, is shielded by tinted windows and a heavy clad, locked door. Just inside is a glass-booth reception desk where two assistants guard access. Even high-level Apple employees are not allowed in without special permission. Most of my interviews with Jony Ive for this book were held elsewhere, but one day in 2010 he arranged for me to spend an afternoon touring the studio and talking about how he and Jobs collaborate there.

To the left of the entrance is a bullpen of desks with young designers; to the right is the cavernous main room with six long steel tables for displaying and playing with works in progress. Beyond the main room is a computer-aided design studio, filled with workstations, that leads to a room with molding machines to turn what's on the screens into foam models. Beyond that is a robot-controlled spray-painting chamber to make the models look real. The look is sparse and industrial, with metallic gray décor. Leaves from the trees outside cast moving patterns of light and shadows on the tinted windows. Techno and jazz play in the background.

Almost every day when Jobs was healthy and in the office, he would have lunch with Ive and then wander by the studio in the afternoon. As he entered, he could survey the tables and see the products in the pipeline, sense how they fit into Apple's strategy, and inspect with his fingertips the evolving design of each. Usually it was just the two of them alone, while the other designers glanced up from their work but kept a respectful distance. If Jobs had a specific issue, he might call over the head of mechanical design or another of Ive's deputies. If something excited him or sparked some thoughts about corporate strategy, he might ask the chief operating officer Tim Cook or the marketing head Phil Schiller to come over and join them. Ive described the usual process:

This great room is the one place in the company where you can look around and see everything

502

we have in the works. When Steve comes in, he will sit at one of these tables. If we're working on a new iPhone, for example, he might grab a stool and start playing with different models and feeling them in his hands, remarking on which ones he likes best. Then he will graze by the other tables, just him and me, to see where all the other products are heading. He can get a sense of the sweep of the whole company, the iPhone and iPad, the iMac and laptop and everything we're considering. That helps him see where the company is spending its energy and how things connect. And he can ask, "Does doing this make sense, because over here is where we are growing a lot?" or questions like that. He gets to see things in relationship to each other, which is pretty hard to do in a big company. Looking at the models on these tables, he can see the future for the next three years.

Much of the design process is a conversation, a back-and-forth as we walk around the tables and play with the models. He doesn't like to read complex drawings. He wants to see and feel a model. He's right. I get surprised when we make a model and then realize it's rubbish, even though based on the CAD [computer-aided design] renderings it looked great.

He loves coming in here because it's calm and gentle. It's a paradise if you're a visual person. There are no formal design reviews, so there are no huge decision points. Instead, we can make the decisions fluid. Since we iterate every day and never have dumb-ass presentations, we don't run into major disagreements.

On this day Ive was overseeing the creation of a new European power plug and connector for the Macintosh. Dozens of foam models, each with the tiniest variation, have been cast and painted for inspection. Some would find it odd that the head of design would fret over something like this, but Jobs got involved as well. Ever since he had a special power supply made for the Apple II, Jobs has cared about not only the engineering but also the design of such parts. His name is listed on the patent for the white power brick used by the MacBook as well as its magnetic connector with its satisfying click. In fact he is listed as one of the inventors for 212 different Apple patents in the United States as of the beginning of 2011.

Ive and Jobs have even obsessed over, and patented, the packaging for various Apple products. U.S. patent D558572, for example, granted on January 1, 2008, is for the iPod Nano box, with four drawings showing how the device is nestled in a cradle when the box is opened. Patent D596485, issued on July 21, 2009, is for the iPhone packaging, with its sturdy lid and little glossy plastic tray inside.

Early on, Mike Markkula had taught Jobs to "impute" — to understand that people *do* judge a book by its cover — and therefore to make sure all the trappings and packaging of Apple signaled that there was a beautiful gem inside. Whether it's an iPod Mini or a MacBook Pro, Apple customers know the feeling of opening up the well-crafted box and finding the product nestled in an inviting fashion. "Steve and I spend a lot of time on the packaging," said Ive. "I love the process of unpacking something. You design a ritual of unpacking to make the product feel special. Packaging can be theater, it can create a story."

Ive, who has the sensitive temperament of an artist, at times got upset with Jobs for taking too much credit, a habit that has bothered other colleagues over the years. His personal feelings for Jobs were so intense that at times he got easily bruised. "He will go through a process of looking at my ideas and say, 'That's no good. That's not very good. I like that one,' " Ive said. "And later I will be sitting in the audience and he will be talking about it as if it was his idea. I pay maniacal attention to where an idea comes from, and I even keep notebooks filled with my ideas. So it hurts when he takes credit for one of my designs." Ive also has bristled when outsiders portrayed Jobs as the only ideas guy at Apple. "That makes us vulnerable as a company," Ive said earnestly, his voice soft. But then he paused to recognize the role Jobs in fact played. "In so many other companies, ideas and great design get lost in the process," he said. "The ideas that come from me and my team would have been completely irrelevant, nowhere, if Steve hadn't been here to push us, work with us, and drive through all the resistance to turn our ideas into products."

Chapter Twenty-Seven:
The iMac
HELLO (AGAIN)

Back to the Future

The first great design triumph to come from the Jobs-Ive collaboration was the iMac, a desktop computer aimed at the home consumer market that was introduced in May 1998. Jobs had certain specifications. It should be an all-in-one product, with keyboard and monitor and computer ready to use right out of the box. It should have a distinctive design that made a brand statement. And it should sell for $1,200 or so. (Apple had no computer selling for less than $2,000 at the time.) "He told us to go back to the roots of the original 1984 Macintosh, an all-in-one consumer appliance," recalled Schiller. "That meant design and engineering had to work together."

The initial plan was to build a "network computer," a concept championed by Oracle's Larry Ellison, which was an inexpensive terminal without a hard drive that would mainly be used to connect to the Internet and other networks. But Apple's chief financial officer Fred Anderson led the push to make the product more robust by adding a disk drive so it could become a full-fledged desktop computer for the home. Jobs eventually agreed.

506

Jon Rubinstein, who was in charge of hardware, adapted the microprocessor and guts of the Power-Mac G3, Apple's high-end professional computer, for use in the proposed new machine. It would have a hard drive and a tray for compact disks, but in a rather bold move, Jobs and Rubinstein decided not to include the usual floppy disk drive. Jobs quoted the hockey star Wayne Gretzky's maxim, "Skate where the puck's going, not where it's been." He was a bit ahead of his time, but eventually most computers eliminated floppy disks.

Ive and his top deputy, Danny Coster, began to sketch out futuristic designs. Jobs brusquely rejected the dozen foam models they initially produced, but Ive knew how to guide him gently. Ive agreed that none of them was quite right, but he pointed out one that had promise. It was curved, playful looking, and did not seem like an unmovable slab rooted to the table. "It has a sense that it's just arrived on your desktop or it's just about to hop off and go somewhere," he told Jobs.

By the next showing Ive had refined the playful model. This time Jobs, with his binary view of the world, raved that he loved it. He took the foam prototype and began carrying it around the headquarters with him, showing it in confidence to trusted lieutenants and board members. In its ads Apple was celebrating the glories of being able to think different, yet until now nothing had been proposed that was much different from existing computers. Finally, Jobs had something new.

The plastic casing that Ive and Coster proposed was sea-green blue, later named bondi blue after the color of the water at a beach in Australia, and it was translucent so that you could see through to the inside of the machine. "We were trying to convey a sense of the computer being changeable based on your needs, to be like a chameleon," said Ive. "That's why we liked the translucency. You could have color but it felt so unstatic. And it came across as cheeky."

Both metaphorically and in reality, the translucency connected the inner engineering of the computer to the outer design. Jobs had always insisted that the rows of chips on the circuit boards look neat, even though they would never be seen. Now they would be seen. The casing would make visible the care that had gone into making all components of the computer and fitting them together. The playful design would convey simplicity while also revealing the depths that true simplicity entails.

Even the simplicity of the plastic shell itself involved great complexity. Ive and his team worked with Apple's Korean manufacturers to perfect the process of making the cases, and they even went to a jelly bean factory to study how to make translucent colors look enticing. The cost of each case was more than $60 per unit, three times that of a regular computer case. Other companies would probably have demanded presentations and studies to show whether the translucent case would increase sales enough to justify the extra cost. Jobs asked for no such analysis.

Topping off the design was the handle nestled into the iMac. It was more playful and semiotic than it was functional. This was a desktop computer; not many people were really going to carry it around. But as Ive later explained:

Back then, people weren't comfortable with technology. If you're scared of something, then you won't touch it. I could see my mum being scared to touch it. So I thought, if there's this handle on it, it makes a relationship possible. It's approachable. It's intuitive. It gives you permission to touch. It gives a sense of its deference to you. Unfortunately, manufacturing a recessed handle costs a lot of money. At the old Apple, I would have lost the argument. What was really great about Steve is that he saw it and said, "That's cool!" I didn't explain all the thinking, but he intuitively got it. He just knew that it was part of the iMac's friendliness and playfulness.

Jobs had to fend off the objections of the manufacturing engineers, supported by Rubinstein, who tended to raise practical cost considerations when faced with Ive's aesthetic desires and various design whims. "When we took it to the engineers," Jobs said, "they came up with thirty-eight reasons they couldn't do it. And I said, 'No, no, we're doing this.' And they said, 'Well, why?' And I said, 'Because I'm the CEO, and I think it can be done.' And so they kind of grudgingly did it."

Jobs asked Lee Clow and Ken Segall and others from the TBWA\Chiat\Day ad team to fly up to see what he had in the works. He brought them into the guarded design studio and dramatically unveiled Ive's translucent teardrop-shaped design, which looked like something from *The Jetsons,* the animated TV show set in the future. For a moment they were taken aback. "We were pretty shocked, but we couldn't be frank," Segall recalled. "We

509

were really thinking, 'Jesus, do they know what they are doing?' It was so radical." Jobs asked them to suggest names. Segall came back with five options, one of them "iMac." Jobs didn't like any of them at first, so Segall came up with another list a week later, but he said that the agency still preferred "iMac." Jobs replied, "I don't hate it this week, but I still don't like it." He tried silk-screening it on some of the prototypes, and the name grew on him. And thus it became the iMac.

As the deadline for completing the iMac drew near, Jobs's legendary temper reappeared in force, especially when he was confronting manufacturing issues. At one product review meeting, he learned that the process was going slowly. "He did one of his displays of awesome fury, and the fury was absolutely pure," recalled Ive. He went around the table assailing everyone, starting with Rubinstein. "You know we're trying to save the company here," he shouted, "and you guys are screwing it up!"

Like the original Macintosh team, the iMac crew staggered to completion just in time for the big announcement. But not before Jobs had one last explosion. When it came time to rehearse for the launch presentation, Rubinstein cobbled together two working prototypes. Jobs had not seen the final product before, and when he looked at it onstage he saw a button on the front, under the display. He pushed it and the CD tray opened. "What the fuck is this?!?" he asked, though not as politely. "None of us said anything," Schiller recalled, "because he obviously knew what a CD tray was." So Jobs continued to rail. It was supposed to have a clean CD slot, he insisted, referring to the elegant slot drives that were already to be found in upscale cars. "Steve, this is exactly the drive I showed you when we talked about the components," Rubinstein explained. "No, there was never a tray, just a slot," Jobs insisted. Rubinstein didn't back down. Jobs's fury didn't abate. "I almost started crying, because it was too late to do anything about it," Jobs later recalled.

They suspended the rehearsal, and for a while it seemed as if Jobs might cancel the entire product launch. "Ruby looked at me as if to say, 'Am I crazy?' " Schiller recalled. "It was my first product launch with Steve and the first time I saw his mind-set of 'If it's not right we're not launching it.' " Finally, they agreed to replace the tray with a slot drive for the next version of the iMac. "I'm only going to go ahead with the launch if you promise we're going to go to slot mode as soon as possible," Jobs said tearfully.

There was also a problem with the video he planned to show. In it, Jony Ive is shown describing his design thinking and asking, "What computer would the Jetsons have had? It was like, the future yesterday." At that moment there was a two-second snippet from the cartoon show, showing Jane Jetson looking at a video screen, followed by another two-second clip of the Jetsons giggling by a Christmas tree. At a rehearsal a production assistant told Jobs they would have to remove the clips because Hanna-Barbera had not given permission to use them. "Keep it in," Jobs barked at him. The assistant explained that there were rules against that. "I don't care," Jobs said. "We're using it." The clip stayed in.

Lee Clow was preparing a series of colorful magazine ads, and when he sent Jobs the page proofs he got an outraged phone call in response. The blue in the ad, Jobs insisted, was different from that of the iMac. "You guys don't know what you're doing!" Jobs shouted. "I'm going to get someone else to do the ads, because this is fucked up." Clow argued back. Compare them, he said. Jobs, who was not in the office, insisted he was right and continued to shout. Eventually Clow got him to sit down with the original photographs. "I finally proved to him that the blue was the blue was the blue." Years later, on a Steve Jobs discussion board on the website Gawker, the following tale appeared from someone who had worked at the Whole Foods store in Palo Alto a few blocks from Jobs's home: "I was shagging carts one afternoon when I saw this silver Mercedes parked in a handicapped spot. Steve Jobs was inside screaming at his car phone. This was right before the first iMac was unveiled and I'm pretty sure I could make out, 'Not. Fucking. Blue. Enough!!!' "

As always, Jobs was compulsive in preparing for the dramatic unveiling. Having stopped one rehearsal because he was angry about the CD drive tray, he stretched out the other rehearsals to make sure the show would be stellar. He repeatedly went over the climactic moment when he would walk across the stage and proclaim, "Say hello to the new iMac." He wanted the lighting to be perfect so that the translucence of the new machine would be vivid. But after a few run-throughs he was still unsatisfied, an echo of his obsession with stage lighting that Sculley had witnessed at the rehears-als for the original 1984 Macintosh launch. He or-dered the lights to be brighter and come on earlier, but that still didn't please him. So he jogged down the auditorium aisle and slouched into a center seat, draping his legs over the seat in front. "Let's keep doing it till we get it right, okay?" he said. They made another attempt. "No, no," Jobs com-plained. "This isn't working at all." The next time, the lights were bright enough, but they came on too late. "I'm getting tired of asking about this," Jobs growled. Finally, the iMac shone just right. "Oh! Right there! That's great!" Jobs yelled.

A year earlier Jobs had ousted Mike Markkula, his early mentor and partner, from the board. But he was so proud of what he had wrought with the new iMac, and so sentimental about its connection to the original Macintosh, that he invited Mark-kula to Cupertino for a private preview. Mark-kula was impressed. His only objection was to the new mouse that Ive had designed. It looked like a hockey puck, Markkula said, and people would hate it. Jobs disagreed, but Markkula was right. Otherwise the machine had turned out to be, as had its predecessor, insanely great.

THE LAUNCH, MAY 6, 1998

With the launch of the original Macintosh in 1984, Jobs had created a new kind of theater: the product debut as an epochal event, climaxed by a let-there-be-light moment in which the skies part, a light shines down, the angels sing, and a chorus of the chosen faithful sings "Hallelujah." For the grand unveiling of the product that he hoped would save

513

Apple and again transform personal computing, Jobs symbolically chose the Flint Auditorium of De Anza Community College in Cupertino, the same venue he had used in 1984. He would be pulling out all the stops in order to dispel doubts, rally the troops, enlist support in the developers' community, and jump-start the marketing of the new machine. But he was also doing it because he enjoyed playing impresario. Putting on a great show piqued his passions in the same way as putting out a great product.

Displaying his sentimental side, he began with a graceful shout-out to three people he had invited to be up front in the audience. He had become estranged from all of them, but now he wanted them rejoined. "I started the company with Steve Wozniak in my parents' garage, and Steve is here today," he said, pointing him out and prompting applause. "We were joined by Mike Markkula and soon after that our first president, Mike Scott," he continued. "Both of those folks are in the audience today. And none of us would be here without these three guys." His eyes misted for a moment as the applause again built. Also in the audience were Andy Hertzfeld and most of the original Mac team. Jobs gave them a smile. He believed he was about to do them proud.

After showing the grid of Apple's new product strategy and going through some slides about the new computer's performance, he was ready to unveil his new baby. "This is what computers look like today," he said as a picture of a beige set of boxy components and monitor was projected on the big screen behind him. "And I'd like to take

514

the privilege of showing you what they are going to look like from today on." He pulled the cloth from the table at center stage to reveal the new iMac, which gleamed and sparkled as the lights came up on cue. He pressed the mouse, and as at the launch of the original Macintosh, the screen flashed with fast-paced images of all the wondrous things the computer could do. At the end, the word "hello" appeared in the same playful script that had adorned the 1984 Macintosh, this time with the word "again" below it in parentheses: *Hello (again)*. There was thunderous applause. Jobs stood back and proudly gazed at his new Macintosh. "It looks like it's from another planet," he said, as the audience laughed. "A good planet. A planet with better designers."

Once again Jobs had produced an iconic new product, this one a harbinger of a new millennium. It fulfilled the promise of "Think Different." Instead of beige boxes and monitors with a welter of cables and a bulky setup manual, here was a friendly and spunky appliance, smooth to the touch and as pleasing to the eye as a robin's egg. You could grab its cute little handle and lift it out of the elegant white box and plug it right into a wall socket. People who had been afraid of computers now wanted one, and they wanted to put it in a room where others could admire and perhaps covet it. "A piece of hardware that blends sci-fi shimmer with the kitsch whimsy of a cocktail umbrella," Steven Levy wrote in *Newsweek,* "it is not only the coolest-looking computer introduced in years, but a chest-thumping statement that Silicon Valley's original dream company is no longer som-

515

nambulant." *Forbes* called it "an industry-altering success," and John Sculley later came out of exile to gush, "He has implemented the same simple strategy that made Apple so successful 15 years ago: make hit products and promote them with terrific marketing."

Carping was heard from only one familiar corner. As the iMac garnered kudos, Bill Gates assured a gathering of financial analysts visiting Microsoft that this would be a passing fad. "The one thing Apple's providing now is leadership in colors," Gates said as he pointed to a Windows-based PC that he jokingly had painted red. "It won't take long for us to catch up with that, I don't think." Jobs was furious, and he told a reporter that Gates, the man he had publicly decried for being completely devoid of taste, was clueless about what made the iMac so much more appealing than other computers. "The thing that our competitors are missing is that they think it's about fashion, and they think it's about surface appearance," he said. "They say, We'll slap a little color on this piece of junk computer, and we'll have one, too."

The iMac went on sale in August 1998 for $1,299. It sold 278,000 units in its first six weeks, and would sell 800,000 by the end of the year, making it the fastest-selling computer in Apple history. Most notably, 32% of the sales went to people who were buying a computer for the first time, and another 12% to people who had been using Windows machines.

Ive soon came up with four new juicy-looking colors, in addition to bondi blue, for the iMacs. Offering the same computer in five colors would of course create huge challenges for manufacturing, inventory, and distribution. At most companies, including even the old Apple, there would have been studies and meetings to look at the costs and benefits. But when Jobs looked at the new colors, he got totally psyched and summoned other executives over to the design studio. "We're going to do all sorts of colors!" he told them excitedly. When they left, Ive looked at his team in amazement. "In most places that decision would have taken months," Ive recalled. "Steve did it in a half hour."

There was one other important refinement that Jobs wanted for the iMac: getting rid of that detested CD tray. "I'd seen a slot-load drive on a very high-end Sony stereo," he said, "so I went to the drive manufacturers and got them to do a slot-load drive for us for the version of the iMac we did nine months later." Rubinstein tried to argue him out of the change. He predicted that new drives would come along that could burn music onto CDs rather than merely play them, and they would be available in tray form before they were made to work in slots. "If you go to slots, you will always be behind on the technology," Rubinstein argued.

"I don't care, that's what I want," Jobs snapped back. They were having lunch at a sushi bar in San Francisco, and Jobs insisted that they continue the conversation over a walk. "I want you to do the slot-load drive for me as a personal favor," Jobs asked. Rubinstein agreed, of course, but he turned out to be right. Panasonic came out with a CD drive that could rip and burn music, and it was available first for computers that had old-fashioned tray loaders. The effects of this would ripple over the next few years: It would cause Apple to be slow in catering to users who wanted to rip and burn their own music, but that would then force Apple to be imaginative and bold in finding a way to leapfrog over its competitors when Jobs finally realized that he had to get into the music market.

CHAPTER TWENTY-EIGHT: CEO
STILL CRAZY AFTER ALL THESE YEARS

TIM COOK

When Steve Jobs returned to Apple and produced the "Think Different" ads and the iMac in his first year, it confirmed what most people already knew: that he could be creative and a visionary. He had shown that during his first round at Apple. What was less clear was whether he could run a company. He had definitely *not* shown that during his first round.

Jobs threw himself into the task with a detail-oriented realism that astonished those who were used to his fantasy that the rules of this universe need not apply to him. "He became a manager, which is different from being an executive or visionary, and that pleasantly surprised me," recalled Ed Woolard, the board chair who lured him back.

His management mantra was "Focus." He eliminated excess product lines and cut extraneous features in the new operating system software that Apple was developing. He let go of his control-freak desire to manufacture products in his own factories and instead outsourced the making of everything from the circuit boards to the finished computers. And he enforced on Apple's suppliers a

519

rigorous discipline. When he took over, Apple had more than two months' worth of inventory sitting in warehouses, more than any other tech company. Like eggs and milk, computers have a short shelf life, so this amounted to at least a $500 million hit to profits. By early 1998 he had halved that to a month.

Jobs's successes came at a cost, since velvety diplomacy was still not part of his repertoire. When he decided that a division of Airborne Express wasn't delivering spare parts quickly enough, he ordered an Apple manager to break the contract. When the manager protested that doing so could lead to a lawsuit, Jobs replied, "Just tell them if they fuck with us, they'll never get another fucking dime from this company, ever." The manager quit, there was a lawsuit, and it took a year to resolve. "My stock options would be worth $10 million had I stayed," the manager said, "but I knew I couldn't have stood it — and he'd have fired me anyway." The new distributor was ordered to cut inventory 75%, and did. "Under Steve Jobs, there's zero tolerance for not performing," its CEO said. At another point, when VLSI Technology was having trouble delivering enough chips on time, Jobs stormed into a meeting and started shouting that they were "fucking dickless assholes." The company ended up getting the chips to Apple on time, and its executives made jackets that boasted on the back, "Team FDA."

After three months of working under Jobs, Apple's head of operations decided he could not bear the pressure, and he quit. For almost a year Jobs ran operations himself, because all the prospects he interviewed "seemed like they were old-wave manufacturing people," he recalled. He wanted someone who could build just-in-time factories and supply chains, as Michael Dell had done. Then, in 1998, he met Tim Cook, a courtly thirty-seven-year-old procurement and supply chain manager at Compaq Computers, who not only would become his operations manager but would grow into an indispensable backstage partner in running Apple. As Jobs recalled:

Tim Cook came out of procurement, which is just the right background for what we needed. I realized that he and I saw things exactly the same way. I had visited a lot of just-in-time factories in Japan, and I'd built one for the Mac and at NeXT. I knew what I wanted, and I met Tim, and he wanted the same thing. So we started to work together, and before long I trusted him to know exactly what to do. He had the same vision I did, and we could interact at a high strategic level, and I could just forget about a lot of things unless he came and pinged me.

Cook, the son of a shipyard worker, was raised in Robertsdale, Alabama, a small town between Mobile and Pensacola a half hour from the Gulf Coast. He majored in industrial engineering at Auburn, got a business degree at Duke, and for the next twelve years worked for IBM in the Research Triangle of North Carolina. When Jobs interviewed him, he had recently taken a job at Compaq. He had always been a very logical engineer, and Compaq then seemed a more sensible career option, but he was snared by Jobs's aura. "Five minutes into my initial interview with Steve, I wanted to throw caution and logic to the wind and join Apple," he later said. "My intuition told me that joining Apple would be a once-in-a-lifetime opportunity to work for a creative genius." And so he did. "Engineers are taught to make a decision analytically, but there are times when relying on gut or intuition is most indispensable."

At Apple his role became implementing Jobs's intuition, which he accomplished with a quiet diligence. Never married, he threw himself into his work. He was up most days at 4:30 sending emails, then spent an hour at the gym, and was at his desk shortly after 6. He scheduled Sunday evening conference calls to prepare for each week ahead. In a company that was led by a CEO prone to tantrums and withering blasts, Cook commanded situations with a calm demeanor, a soothing Alabama accent, and silent stares. "Though he's capable of mirth, Cook's default facial expression is a frown, and his humor is of the dry variety," Adam Lashinsky wrote in *Fortune.* "In meetings he's known for long, uncomfortable pauses, when all you hear is the sound of his tearing the wrapper off the energy bars he constantly eats."

At a meeting early in his tenure, Cook was told of a problem with one of Apple's Chinese suppliers. "This is really bad," he said. "Someone should be in China driving this." Thirty minutes later he looked at an operations executive sitting at the table and unemotionally asked, "Why are you still here?" The executive stood up, drove directly to the San Francisco airport, and bought a ticket to

China. He became one of Cook's top deputies.

Cook reduced the number of Apple's key suppliers from a hundred to twenty-four, forced them to cut better deals to keep the business, convinced many to locate next to Apple's plants, and closed ten of the company's nineteen warehouses. By reducing the places where inventory could pile up, he reduced inventory. Jobs had cut inventory from two months' worth of product down to one by early 1998. By September of that year, Cook had gotten it down to six days. By the following September, it was down to an amazing two days' worth. In addition, he cut the production process for making an Apple computer from four months to two. All of this not only saved money, it also allowed each new computer to have the very latest components available.

Mock Turtlenecks and Teamwork

On a trip to Japan in the early 1980s, Jobs asked Sony's chairman, Akio Morita, why everyone in his company's factories wore uniforms. "He looked very ashamed and told me that after the war, no one had any clothes, and companies like Sony had to give their workers something to wear each day," Jobs recalled. Over the years the uniforms developed their own signature style, especially at companies such as Sony, and it became a way of bonding workers to the company. "I decided that I wanted that type of bonding for Apple," Jobs recalled.

Sony, with its appreciation for style, had gotten the famous designer Issey Miyake to create one of its uniforms. It was a jacket made of ripstop nylon with sleeves that could unzip to make it a vest. "So I called Issey and asked him to design a vest for Apple," Jobs recalled. "I came back with some samples and told everyone it would be great if we would all wear these vests. Oh man, did I get booed off the stage. Everybody hated the idea."

In the process, however, he became friends with Miyake and would visit him regularly. He also came to like the idea of having a uniform for himself, because of both its daily convenience (the rationale he claimed) and its ability to convey a signature style. "So I asked Issey to make me some of his black turtlenecks that I liked, and he made me like a hundred of them." Jobs noticed my surprise when he told this story, so he gestured to them stacked up in the closet. "That's what I wear," he said. "I have enough to last for the rest of my life."

Despite his autocratic nature — he never worshipped at the altar of consensus — Jobs worked hard to foster a culture of collaboration at Apple. Many companies pride themselves on having few meetings. Jobs had many: an executive staff session every Monday, a marketing strategy session all Wednesday afternoon, and endless product review sessions. Still allergic to PowerPoints and formal presentations, he insisted that the people around the table hash out issues from various vantages and the perspectives of different departments.

Because he believed that Apple's great advantage was its integration of the whole widget — from design to hardware to software to content — he wanted all departments at the company to work together in parallel. The phrases he used were "deep collaboration" and "concurrent engineering." Instead of a development process in which a product would be passed sequentially from engineering to design to manufacturing to marketing and distribution, these various departments collaborated simultaneously. "Our method was to develop integrated products, and that meant our process had to be integrated and collaborative," Jobs said.

This approach also applied to key hires. He would have candidates meet the top leaders — Cook, Tevanian, Schiller, Rubinstein, Ive — rather than just the managers of the department where they wanted to work. "Then we all get together without the person and talk about whether they'll fit in," Jobs said. His goal was to be vigilant against "the bozo explosion" that leads to a company's being larded with second-rate talent:

> For most things in life, the range between best and average is 30% or so. The best airplane flight, the best meal, they may be 30% better than your average one. What I saw with Woz was somebody who was fifty times better than the average engineer. He could have meetings in his head. The Mac team was an attempt to build a whole team like that, A players. People said they wouldn't get along, they'd hate working with each other. But I realized that A players like to work with A players, they just didn't like working with C players. At Pixar, it was a whole company of A players. When I got back to Apple, that's what I decided to try to do. You need to have a collaborative hiring process. When we hire someone, even if they're going to be in marketing, I will have them talk to the design folks and the engineers.

My role model was J. Robert Oppenheimer. I read about the type of people he sought for the atom bomb project. I wasn't nearly as good as he was, but that's what I aspired to do.

The process could be intimidating, but Jobs had an eye for talent. When they were looking for people to design the graphical interface for Apple's new operating system, Jobs got an email from a young man and invited him in. The applicant was nervous, and the meeting did not go well. Later that day Jobs bumped into him, dejected, sitting in the lobby. The guy asked if he could just show him one of his ideas, so Jobs looked over his shoulder and saw a little demo, using Adobe Director, of a way to fit more icons in the dock at the bottom of a screen. When the guy moved the cursor over the icons crammed into the dock, the cursor mimicked a magnifying glass and made each icon balloon bigger. "I said, 'My God,' and hired him on the spot," Jobs recalled. The feature became a lovable part of Mac OSX, and the designer went on to design such things as inertial scrolling for multi-touch screens (the delightful feature that makes the screen keep gliding for a moment after you've finished swiping).

Jobs's experiences at NeXT had matured him, but they had not mellowed him much. He still had no license plate on his Mercedes, and he still parked in the handicapped spaces next to the front door, sometimes straddling two slots. It became a running gag. Employees made signs saying, "Park Different," and someone painted over the handicapped wheelchair symbol with a Mercedes logo.

People were allowed, even encouraged, to challenge him, and sometimes he would respect them for it. But you had to be prepared for him to attack you, even bite your head off, as he processed your ideas. "You never win an argument with him at the time, but sometimes you eventually win," said James Vincent, the creative young adman who worked with Lee Clow. "You propose something and he declares, 'That's a stupid idea,' and later he comes back and says, 'Here's what we're going to do.' And you want to say, 'That's what I told you two weeks ago and you said that's a stupid idea.' But you can't do that. Instead you say, 'That's a great idea, let's do that.' "

People also had to put up with Jobs's occasional irrational or incorrect assertions. To both family and colleagues, he was apt to declare, with great conviction, some scientific or historical fact that had scant relationship to reality. "There can be something he knows absolutely nothing about, and because of his crazy style and utter conviction, he can convince people that he knows what he's talking about," said Ive, who described the trait as weirdly endearing. Yet with his eye for detail, Jobs sometimes correctly pounced on tiny things others had missed. Lee Clow recalled showing Jobs a cut of a commercial, making some minor changes he requested, and then being assaulted with a tirade about how the ad had been completely destroyed. "He discovered we had cut two extra frames, something so fleeting it was nearly impossible to notice," said Clow. "But he wanted to be sure that an image hit at the exact moment as a beat of the music, and he was totally right."

From iCEO to CEO

Ed Woolard, his mentor on the Apple board, pressed Jobs for more than two years to drop the *interim* in front of his CEO title. Not only was Jobs refusing to commit himself, but he was baffling everyone by taking only $1 a year in pay and no stock options. "I make 50 cents for showing up," he liked to joke, "and the other 50 cents is based on performance." Since his return in July 1997, Apple stock had gone from just under $14 to just over $102 at the peak of the Internet bubble at the beginning of 2000. Woolard had begged him to take at least a modest stock grant back in 1997, but Jobs had declined, saying, "I don't want the people I work with at Apple to think I am coming back to get rich." Had he accepted that modest grant, it would have been worth $400 million. Instead he made $2.50 during that period.

The main reason he clung to his *interim* designation was a sense of uncertainty about Apple's future. But as 2000 approached, it was clear that Apple had rebounded, and it was because of him. He took a long walk with Laurene and discussed what to most people by now seemed a formality but to him was still a big deal. If he dropped the *interim* designation, Apple could be the base for all the things he envisioned, including the possibility of getting Apple into products beyond computers. He decided to do so.

Woolard was thrilled, and he suggested that the board was willing to give him a massive stock grant. "Let me be straight with you," Jobs replied. "What I'd rather have is an airplane. We just had a third kid. I don't like flying commercial. I like to take my family to Hawaii. When I go east, I'd like to have pilots I know." He was never the type of person who could display grace and patience in a commercial airplane or terminal, even before the days of the TSA. Board member Larry Ellison, whose plane Jobs sometimes used (Apple paid $102,000 to Ellison in 1999 for Jobs's use of it), had no qualms. "Given what he's accomplished, we should give him five airplanes!" Ellison argued. He later said, "It was the perfect thank-you gift for Steve, who had saved Apple and gotten nothing in return."

So Woolard happily granted Jobs's wish, with a Gulfstream V, and also offered him fourteen million stock options. Jobs gave an unexpected response. He wanted more: twenty million options. Woolard was baffled and upset. The board had authority from the stockholders to give out only fourteen million. "You said you didn't want any, and we gave you a plane, which you did want," Woolard said.

"I hadn't been insisting on options before," Jobs replied, "but you suggested it could be up to 5% of the company in options, and that's what I now want." It was an awkward tiff in what should have been a celebratory period. In the end, a complex solution was worked out that granted him ten million shares in January 2000 that were valued at the current price but timed to vest as if granted in 1997, plus another grant due in 2001. Making matters worse, the stock fell with the burst of the Internet bubble. Jobs never exercised the options, and at the end of 2001 he asked that they be replaced by a new grant with a lower strike price.

529

The wrestling over options would come back to haunt the company.

Even if he didn't profit from the options, at least he got to enjoy the airplane. Not surprisingly he fretted over how the interior would be designed. It took him more than a year. He used Ellison's plane as a starting point and hired his designer. Pretty soon he was driving her crazy. For example, Ellison's had a door between cabins with an open button and a close button. Jobs insisted that his have a single button that toggled. He didn't like the polished stainless steel of the buttons, so he had them replaced with brushed metal ones. But in the end he got the plane he wanted, and he loved it. "I look at his airplane and mine, and everything he changed was better," said Ellison.

At the January 2000 Macworld in San Francisco, Jobs rolled out the new Macintosh operating system, OSX, which used some of the software that Apple had bought from NeXT three years earlier. It was fitting, and not entirely coincidental, that he was willing to incorporate himself back at Apple at the same moment as the NeXT OS was incorporated into Apple's. Avie Tevanian had taken the UNIX-related Mach kernel of the NeXT operating system and turned it into the Mac OS kernel, known as Darwin. It offered protected memory, advanced networking, and preemptive multitasking. It was precisely what the Macintosh needed, and it would be the foundation of the Mac OS henceforth. Some critics, including Bill Gates, noted that Apple ended up not adopting the entire NeXT operating system. There's some truth to that, because Apple decided not to leap into a completely new system but instead to evolve the existing one. Application software written for the old Macintosh system was generally compatible with or easy to port to the new one, and a Mac user who upgraded would notice a lot of new features but not a whole new interface.

The fans at Macworld received the news with enthusiasm, of course, and they especially cheered when Jobs showed off the dock and how the icons in it could be magnified by passing the cursor over them. But the biggest applause came for the announcement he reserved for his "Oh, and one more thing" coda. He spoke about his duties at both Pixar and Apple, and said that he had become comfortable that the situation could work. "So I am pleased to announce today that I'm going to drop the interim title," he said with a big smile. The crowd jumped to its feet, screaming as if the Beatles had reunited. Jobs bit his lip, adjusted his wire rims, and put on a graceful show of humility. "You guys are making me feel funny now. I get to come to work every day and work with the most talented people on the planet, at Apple and Pixar. But these jobs are team sports. I accept your thanks on behalf of everybody at Apple."

CHAPTER TWENTY-NINE: APPLE STORES

GENIUS BARS AND SIENA SANDSTONE

THE CUSTOMER EXPERIENCE

Jobs hated to cede control of anything, especially when it might affect the customer experience. But he faced a problem. There was one part of the process he didn't control: the experience of buying an Apple product in a store.

The days of the Byte Shop were over. Industry sales were shifting from local computer specialty shops to megachains and big box stores, where most clerks had neither the knowledge nor the incentive to explain the distinctive nature of Apple products. "All that the salesman cared about was a $50 spiff," Jobs said. Other computers were pretty generic, but Apple's had innovative features and a higher price tag. He didn't want an iMac to sit on a shelf between a Dell and a Compaq while an uninformed clerk recited the specs of each. "Unless we could find ways to get our message to customers at the store, we were screwed."

In great secrecy, Jobs began in late 1999 to interview executives who might be able to develop a string of Apple retail stores. One of the candidates had a passion for design and the boyish enthusiasm of a natural-born retailer: Ron Johnson, the vice president for merchandising at Target, who was responsible for launching distinctive-looking products, such as a teakettle designed by Michael Graves. "Steve is very easy to talk to," said Johnson in recalling their first meeting. "All of a sudden there's a torn pair of jeans and turtleneck, and he's off and running about why he needed great stores. If Apple is going to succeed, he told me, we're going to win on innovation. And you can't win on innovation unless you have a way to communicate to customers."

When Johnson came back in January 2000 to be interviewed again, Jobs suggested that they take a walk. They went to the sprawling 140-store Stanford Shopping Mall at 8:30 a.m. The stores weren't open yet, so they walked up and down the entire mall repeatedly and discussed how it was organized, what role the big department stores played relative to the other stores, and why certain specialty shops were successful.

They were still walking and talking when the stores opened at 10, and they went into Eddie Bauer. It had an entrance off the mall and another off the parking lot. Jobs decided that Apple stores should have only one entrance, which would make it easier to control the experience. And the Eddie Bauer store, they agreed, was too long and narrow. It was important that customers intuitively grasp the layout of a store as soon as they entered.

There were no tech stores in the mall, and Johnson explained why: The conventional wisdom was that a consumer, when making a major and infrequent purchase such as a computer, would be willing to drive to a less convenient location, where the rent would be cheaper. Jobs disagreed. Apple stores should be in malls and on Main Streets — in areas with a lot of foot traffic, no matter how expensive. "We may not be able to get them to drive ten miles to check out our products, but we can get them to walk ten feet," he said. The Windows users, in particular, had to be ambushed: "If they're passing by, they will drop in out of curiosity, if we make it inviting enough, and once we get a chance to show them what we have, we will win."

Johnson said that the size of a store signaled the importance of the brand. "Is Apple as big of a brand as the Gap?" he asked. Jobs said it was much bigger. Johnson replied that its stores should therefore be bigger. "Otherwise you won't be relevant." Jobs described Mike Markkula's maxim that a good company must "impute" — it must convey its values and importance in everything it does, from packaging to marketing. Johnson loved it. It definitely applied to a company's stores. "The store will become the most powerful physical expression of the brand," he predicted. He said that when he was young he had gone to the wood-paneled, art-filled mansion-like store that Ralph Lauren had created at Seventy-second and Madison in Manhattan. "Whenever I buy a polo shirt, I think of that mansion, which was a physical expression of Ralph's ideals," Johnson said. "Mickey Drexler did that with the Gap. You couldn't think of a Gap product without thinking of the great Gap store with the clean space and wood floors and white walls and folded merchandise."

When they finished, they drove to Apple and sat in a conference room playing with the company's products. There weren't many, not enough to fill the shelves of a conventional store, but that was an advantage. The type of store they would build, they decided, would benefit from having few products. It would be minimalist and airy and offer a lot of places for people to try out things. "Most people don't know Apple products," Johnson said. "They think of Apple as a cult. You want to move from a cult to something cool, and having an awesome store where people can try things will help that." The stores would impute the ethos of Apple products: playful, easy, creative, and on the bright side of the line between hip and intimidating.

THE PROTOTYPE

When Jobs finally presented the idea, the board was not thrilled. Gateway Computers was going down in flames after opening suburban stores, and Jobs's argument that his would do better because they would be in more expensive locations was not, on its face, reassuring. "Think different" and "Here's to the crazy ones" made for good advertising slogans, but the board was hesitant to make them guidelines for corporate strategy. "I'm scratching my head and thinking this is crazy," recalled Art Levinson, the CEO of Genentech who joined the Apple board in 2000. "We are a small company, a marginal player. I said that I'm not sure I can support something like this." Ed Woolard was also dubious. "Gateway has tried this and failed, while Dell is selling direct to consumers without stores and succeeding," he argued. Jobs was not appreciative of too much pushback from the board. The last time that happened, he had replaced most of

535

the members. This time, for personal reasons as well as being tired of playing tug-of-war with Jobs, Woolard decided to step down. But before he did, the board approved a trial run of four Apple stores.

Jobs did have one supporter on the board. In 1999 he had recruited the Bronx-born retailing prince Millard "Mickey" Drexler, who as CEO of Gap had transformed a sleepy chain into an icon of American casual culture. He was one of the few people in the world who were as successful and savvy as Jobs on matters of design, image, and consumer yearnings. In addition, he had insisted on end-to-end control: Gap stores sold only Gap products, and Gap products were sold almost exclusively in Gap stores. "I left the department store business because I couldn't stand not controlling my own product, from how it's manufactured to how it's sold," Drexler said. "Steve is just that way, which is why I think he recruited me."

Drexler gave Jobs a piece of advice: Secretly build a prototype of the store near the Apple campus, furnish it completely, and then hang out there until you feel comfortable with it. So Johnson and Jobs rented a vacant warehouse in Cupertino. Every Tuesday for six months, they convened an all-morning brainstorming session there, refining their retailing philosophy as they walked the space. It was the store equivalent of Ive's design studio, a haven where Jobs, with his visual approach, could come up with innovations by touching and seeing the options as they evolved. "I loved to wander over there on my own, just checking it out," Jobs recalled.

Sometimes he made Drexler, Larry Ellison, and

other trusted friends come look. "On too many weekends, when he wasn't making me watch new scenes from *Toy Story,* he made me go to the warehouse and look at the mockups for the store," Ellison said. "He was obsessed by every detail of the aesthetic and the service experience. It got to the point where I said, 'Steve I'm not coming to see you if you're going to make me go to the store again.' "

Ellison's company, Oracle, was developing software for the handheld checkout system, which avoided having a cash register counter. On each visit Jobs prodded Ellison to figure out ways to streamline the process by eliminating some unnecessary step, such as handing over the credit card or printing a receipt. "If you look at the stores and the products, you will see Steve's obsession with beauty as simplicity — this Bauhaus aesthetic and wonderful minimalism, which goes all the way to the checkout process in the stores," said Ellison. "It means the absolute minimum number of steps. Steve gave us the exact, explicit recipe for how he wanted the checkout to work."

When Drexler came to see the prototype, he had some criticisms: "I thought the space was too chopped up and not clean enough. There were too many distracting architectural features and colors." He emphasized that a customer should be able to walk into a retail space and, with one sweep of the eye, understand the flow. Jobs agreed that simplicity and lack of distractions were keys to a great store, as they were to a product. "After that, he nailed it," said Drexler. "The vision he had was complete control of the entire experience of his product, from how it was designed and made to how it was sold."

In October 2000, near what he thought was the end of the process, Johnson woke up in the middle of a night before one of the Tuesday meetings with a painful thought: They had gotten something fundamentally wrong. They were organizing the store around each of Apple's main product lines, with areas for the PowerMac, iMac, iBook, and PowerBook. But Jobs had begun developing a new concept: the computer as a hub for all your digital activity. In other words, your computer might handle video and pictures from your cameras, and perhaps someday your music player and songs, or your books and magazines. Johnson's predawn brainstorm was that the stores should organize displays not just around the company's four lines of computers, but also around things people might want to do. "For example, I thought there should be a movie bay where we'd have various Macs and PowerBooks running iMovie and showing how you can import from your video camera and edit."

Johnson arrived at Jobs's office early that Tuesday and told him about his sudden insight that they needed to reconfigure the stores. He had heard tales of his boss's intemperate tongue, but he had not yet felt its lash — until now. Jobs erupted. "Do you know what a big change this is?" he yelled. "I've worked my ass off on this store for six months, and now you want to change everything!" Jobs suddenly got quiet. "I'm tired. I don't know if I can design another store from scratch."

Johnson was speechless, and Jobs made sure he remained so. On the ride to the prototype store, where people had gathered for the Tuesday meeting, he told Johnson not to say a word, either to him or to the other members of the team. So the seven-minute drive proceeded in silence. When they arrived, Jobs had finished processing the information. "I knew Ron was right," he recalled. So to Johnson's surprise, Jobs opened the meeting by saying, "Ron thinks we've got it all wrong. He thinks it should be organized not around products but instead around what people do." There was a pause, then Jobs continued. "And you know, he's right." He said they would redo the layout, even though it would likely delay the planned January rollout by three or four months. "We've only got one chance to get it right."

Jobs liked to tell the story — and he did so to his team that day — about how everything that he had done correctly had required a moment when he hit the rewind button. In each case he had to rework something that he discovered was not perfect. He talked about doing it on *Toy Story,* when the character of Woody had evolved into being a jerk, and on a couple of occasions with the original Macintosh. "If something isn't right, you can't just ignore it and say you'll fix it later," he said. "That's what other companies do."

When the revised prototype was finally completed in January 2001, Jobs allowed the board to see it for the first time. He explained the theories behind the design by sketching on a whiteboard; then he loaded board members into a van for the two-mile trip. When they saw what Jobs and Johnson had built, they unanimously approved going ahead. It would, the board agreed, take the relationship between retailing and brand image to a new level. It would also ensure that consumers did not see Apple computers as merely a commodity product like Dell or Compaq.

Most outside experts disagreed. "Maybe it's time Steve Jobs stopped thinking quite so differently," *Business Week* wrote in a story headlined "Sorry Steve, Here's Why Apple Stores Won't Work." Apple's former chief financial officer, Joseph Graziano, was quoted as saying, "Apple's problem is it still believes the way to grow is serving caviar in a world that seems pretty content with cheese and crackers." And the retail consultant David Goldstein declared, "I give them two years before they're turning out the lights on a very painful and expensive mistake."

WOOD, STONE, STEEL, GLASS

On May 19, 2001, the first Apple store opened in Tyson's Corner, Virginia, with gleaming white counters, bleached wood floors, and a huge "Think Different" poster of John and Yoko in bed. The skeptics were wrong. Gateway stores had been averaging 250 visitors a week. By 2004 Apple stores were averaging 5,400 per week. That year the stores had $1.2 billion in revenue, setting a record in the retail industry for reaching the billion-dollar milestone. Sales in each store were tabulated every four minutes by Ellison's software, giving instant information on how to integrate manufacturing, supply, and sales channels.

As the stores flourished, Jobs stayed involved in every aspect. Lee Clow recalled, "In one of our marketing meetings just as the stores were opening, Steve made us spend a half hour deciding what hue of gray the restroom signs should be." The architectural firm of Bohlin Cywinski Jackson designed the signature stores, but Jobs made all of the major decisions.

Jobs particularly focused on the staircases, which echoed the one he had built at NeXT. When he visited a store as it was being constructed, he invariably suggested changes to the staircase. His name is listed as the lead inventor on two patent applications on the staircases, one for the see-through look that features all-glass treads and glass supports melded together with titanium, the other for the engineering system that uses a monolithic unit of glass containing multiple glass sheets laminated together for supporting loads.

In 1985, as he was being ousted from his first tour at Apple, he had visited Italy and been impressed by the gray stone of Florence's sidewalks. In 2002, when he came to the conclusion that the light wood floors in the stores were beginning to look somewhat pedestrian — a concern that it's hard to imagine bedeviling someone like Microsoft CEO Steve Ballmer — Jobs wanted to use that stone instead. Some of his colleagues pushed to replicate the color and texture using concrete, which would have been ten times cheaper, but Jobs insisted that it had to be authentic. The gray-blue Pietra Serena sandstone, which has a fine-grained texture, comes from a family-owned quarry, Il Casone, in Firenzuola outside of Florence. "We select only 3% of what comes out of the mountain, because it has to have the right shading and veining and purity," said Johnson. "Steve felt very strongly that we had to get the color right and it had to be a material with high integrity." So designers in Florence picked out just the right quarried stone, oversaw cutting it into the proper tiles, and made sure each tile was marked with a sticker to ensure that it was laid out next to its companion tiles. "Knowing that it's the same stone that Florence uses for its sidewalks assures you that it can stand the test of time," said Johnson.

Another notable feature of the stores was the Genius Bar. Johnson came up with the idea on a two-day retreat with his team. He had asked them all to describe the best service they'd ever enjoyed. Almost everyone mentioned some nice experience at a Four Seasons or Ritz-Carlton hotel. So Johnson sent his first five store managers through the Ritz-Carlton training program and came up with the idea of replicating something between a concierge desk and a bar. "What if we staffed the bar with the smartest Mac people," he said to Jobs. "We could call it the Genius Bar."

Jobs called the idea crazy. He even objected to the name. "You can't call them geniuses," he said. "They're geeks. They don't have the people skills to deliver on something called the genius bar." Johnson thought he had lost, but the next day he ran into Apple's general counsel, who said, "By the way, Steve just told me to trademark the name 'genius bar.' "

Many of Jobs's passions came together for Manhattan's Fifth Avenue store, which opened in 2006: a cube, a signature staircase, glass, and making a maximum statement through minimalism. "It was really Steve's store," said Johnson. Open 24/7, it vindicated the strategy of finding signature high-traffic locations by attracting fifty thousand visitors a week during its first year. (Remember Gateway's draw: 250 visitors a week.) "This store grosses more per square foot than any store in the world," Jobs proudly noted in 2010. "It also grosses more in total — absolute dollars, not just per square foot — than any store in New York. That includes Saks and Bloomingdale's."

Jobs was able to drum up excitement for store openings with the same flair he used for product releases. People began to travel to store openings and spend the night outside so they could be among the first in. "My then 14-year-old son suggested my first overnighter at Palo Alto, and the experience turned into an interesting social event," wrote Gary Allen, who started a website that caters to Apple store fans. "He and I have done several overnighters, including five in other countries, and have met so many great people."

In July 2011, a decade after the first ones opened, there were 326 Apple stores. The biggest was in London's Covent Garden, the tallest in Tokyo's Ginza. The average annual revenue per store was $34 million, and the total net sales in fiscal 2010 were $9.8 billion. But the stores did even more. They directly accounted for only 15% of Apple's revenue, but by creating buzz and brand awareness they indirectly helped boost everything the company did.

Even as he was fighting the effects of cancer in 2011, Jobs spent time envisioning future store projects, such as the one he wanted to build in New York City's Grand Central Terminal. One afternoon he showed me a picture of the Fifth Avenue store and pointed to the eighteen pieces of glass on each side. "This was state of the art in glass technology at the time," he said. "We had to build our own autoclaves to make the glass." Then he pulled out a drawing in which the eighteen panes were replaced by four huge panes. That is what he wanted to do next, he said. Once again, it was a challenge at the intersection of aesthetics and technology. "If we wanted to do it with our current technology, we would have to make the cube a foot shorter," he said. "And I didn't want to do that. So we have to build some new autoclaves in China."

Ron Johnson was not thrilled by the idea. He thought the eighteen panes actually looked better than four panes would. "The proportions we have today work magically with the colonnade of the GM Building," he said. "It glitters like a jewel box. I think if we get the glass too transparent, it will almost go away to a fault." He debated the point with Jobs, but to no avail. "When technology enables something new, he wants to take advantage of that," said Johnson. "Plus, for Steve, less is always more, simpler is always better. Therefore, if you can build a glass box with fewer elements, it's better, it's simpler, and it's at the forefront of technology. That's where Steve likes to be, in both his products and his stores."

CHAPTER THIRTY:
THE DIGITAL HUB
FROM iTUNES TO THE iPOD

CONNECTING THE DOTS

Once a year Jobs took his most valuable employees on a retreat, which he called "The Top 100." They were picked based on a simple guideline: the people you would bring if you could take only a hundred people with you on a lifeboat to your next company. At the end of each retreat, Jobs would stand in front of a whiteboard (he loved whiteboards because they gave him complete control of a situation and they engendered focus) and ask, "What are the ten things we should be doing next?" People would fight to get their suggestions on the list. Jobs would write them down, and then cross off the ones he decreed dumb. After much jockeying, the group would come up with a list of ten. Then Jobs would slash the bottom seven and announce, "We can only do three."

By 2001 Apple had revived its personal computer offerings. It was now time to think different. A set of new possibilities topped the what-next list on his whiteboard that year.

At the time, a pall had descended on the digital realm. The dot-com bubble had burst, and the NASDAQ had fallen more than 50% from its peak.

Only three tech companies had ads during the January 2001 Super Bowl, compared to seventeen the year before. But the sense of deflation went deeper. For the twenty-five years since Jobs and Wozniak had founded Apple, the personal computer had been the centerpiece of the digital revolution. Now experts were predicting that its central role was ending. It had "matured into something boring," wrote the *Wall Street Journal*'s Walt Mossberg. Jeff Weitzen, the CEO of Gateway, proclaimed, "We're clearly migrating away from the PC as the centerpiece."

It was at that moment that Jobs launched a new grand strategy that would transform Apple — and with it the entire technology industry. The personal computer, instead of edging toward the sidelines, would become a "digital hub" that coordinated a variety of devices, from music players to video recorders to cameras. You'd link and sync all these devices with your computer, and it would manage your music, pictures, video, text, and all aspects of what Jobs dubbed your "digital lifestyle." Apple would no longer be just a computer company — indeed it would drop that word from its name — but the Macintosh would be reinvigorated by becoming the hub for an astounding array of new gadgets, including the iPod and iPhone and iPad.

When he was turning thirty, Jobs had used a metaphor about record albums. He was musing about why folks over thirty develop rigid thought patterns and tend to be less innovative. "People get stuck in those patterns, just like grooves in a record, and they never get out of them," he said.

At age forty-five, Jobs was now about to get out of his groove.

Jobs's vision that your computer could become your digital hub went back to a technology called FireWire, which Apple developed in the early 1990s. It was a high-speed serial port that moved digital files such as video from one device to another. Japanese camcorder makers adopted it, and Jobs decided to include it on the updated versions of the iMac that came out in October 1999. He began to see that FireWire could be part of a system that moved video from cameras onto a computer, where it could be edited and distributed.

To make this work, the iMac needed to have great video editing software. So Jobs went to his old friends at Adobe, the digital graphics company, and asked them to make a new Mac version of Adobe Premiere, which was popular on Windows computers. Adobe's executives stunned Jobs by flatly turning him down. The Macintosh, they said, had too few users to make it worthwhile. Jobs was furious and felt betrayed. "I put Adobe on the map, and they screwed me," he later claimed. Adobe made matters even worse when it also didn't write its other popular programs, such as Photoshop, for the Mac OSX, even though the Macintosh was popular among designers and other creative people who used those applications.

Jobs never forgave Adobe, and a decade later he got into a public war with the company by not permitting Adobe Flash to run on the iPad. He took away a valuable lesson that reinforced his

547

desire for end-to-end control of all key elements of a system: "My primary insight when we were screwed by Adobe in 1999 was that we shouldn't get into any business where we didn't control both the hardware and the software, otherwise we'd get our head handed to us."

So starting in 1999 Apple began to produce application software for the Mac, with a focus on people at the intersection of art and technology. These included Final Cut Pro, for editing digital video; iMovie, which was a simpler consumer version; iDVD, for burning video or music onto a disc; iPhoto, to compete with Adobe Photoshop; GarageBand, for creating and mixing music; iTunes, for managing your songs; and the iTunes Store, for buying songs.

The idea of the digital hub quickly came into focus. "I first understood this with the camcorder," Jobs said. "Using iMovie makes your camcorder ten times more valuable." Instead of having hundreds of hours of raw footage you would never really sit through, you could edit it on your computer, make elegant dissolves, add music, and roll credits, listing yourself as executive producer. It allowed people to be creative, to express themselves, to make something emotional. "That's when it hit me that the personal computer was going to morph into something else."

Jobs had another insight: If the computer served as the hub, it would allow the portable devices to become simpler. A lot of the functions that the devices tried to do, such as editing the video or pictures, they did poorly because they had small screens and could not easily accommodate menus filled with lots of functions. Computers could handle that more easily.

And one more thing . . . What Jobs also saw was that this worked best when everything — the device, computer, software, applications, FireWire — was all tightly integrated. "I became even more of a believer in providing end-to-end solutions," he recalled.

The beauty of this realization was that there was only one company that was well-positioned to provide such an integrated approach. Microsoft wrote software, Dell and Compaq made hardware, Sony produced a lot of digital devices, Adobe developed a lot of applications. But only Apple did all of these things. "We're the only company that owns the whole widget — the hardware, the software and the operating system," he explained to *Time.* "We can take full responsibility for the user experience. We can do things that the other guys can't do."

Apple's first integrated foray into the digital hub strategy was video. With FireWire, you could get your video onto your Mac, and with iMovie you could edit it into a masterpiece. Then what? You'd want to burn some DVDs so you and your friends could watch it on a TV. "So we spent a lot of time working with the drive manufacturers to get a consumer drive that could burn a DVD," he said. "We were the first to ever ship that." As usual Jobs focused on making the product as simple as possible for the user, and this was the key to its success. Mike Evangelist, who worked at Apple on software design, recalled demonstrating to Jobs an early version of the interface. After looking at a bunch of screenshots, Jobs jumped up, grabbed a marker, and drew a simple rectangle on a whiteboard. "Here's the new application," he said. "It's got one window. You drag your video into the window. Then you click the button that says 'Burn.' That's it. That's what we're going to make." Evangelist was dumbfounded, but it led to the simplicity of what became iDVD. Jobs even helped design the "Burn" button icon.

Jobs knew digital photography was also about to explode, so Apple developed ways to make the computer the hub of your photos. But for the first year at least, he took his eye off one really big opportunity. HP and a few others were producing a drive that burned music CDs, but Jobs decreed that Apple should focus on video rather than music. In addition, his angry insistence that the iMac get rid of its tray disk drive and use instead a more elegant slot drive meant that it could not include the first CD burners, which were initially made for the tray format. "We kind of missed the boat on that," he recalled. "So we needed to catch up real fast."

The mark of an innovative company is not only that it comes up with new ideas first, but also that it knows how to leapfrog when it finds itself behind.

iTUNES

It didn't take Jobs long to realize that music was going to be huge. By 2000 people were ripping music onto their computers from CDs, or downloading it from file-sharing services such as Napster, and burning playlists onto their own blank disks. That year the number of blank CDs sold in the United States was 320 million. There were only 281 million people in the country. That meant some people were *really* into burning CDs, and Apple wasn't catering to them. "I felt like a dope," he told *Fortune.* "I thought we had missed it. We had to work hard to catch up."

Jobs added a CD burner to the iMac, but that wasn't enough. His goal was to make it simple to transfer music from a CD, manage it on your computer, and then burn playlists. Other companies were already making music-management applications, but they were clunky and complex. One of Jobs's talents was spotting markets that were filled with second-rate products. He looked at the music apps that were available — including Real Jukebox, Windows Media Player, and one that HP was including with its CD burner — and came to a conclusion: "They were so complicated that only a genius could figure out half of their features."

That is when Bill Kincaid came in. A former Apple software engineer, he was driving to a track in Willows, California, to race his Formula Ford sports car while (a bit incongruously) listening to National Public Radio. He heard a report about a portable music player called the Rio that played a digital song format called MP3. He perked up when the reporter said something like, "Don't get excited, Mac users, because it won't work with Macs." Kincaid said to himself, "Ha! I can fix that!"

To help him write a Rio manager for the Mac, he called his friends Jeff Robbin and Dave Heller, also former Apple software engineers. Their product, known as SoundJam, offered Mac users an interface for the Rio and software for managing the songs on their computer. In July 2000, when Jobs was pushing his team to come up with music-management software, Apple swooped in and bought SoundJam, bringing its founders back into the Apple fold. (All three stayed with the company, and Robbin continued to run the music software development team for the next decade. Jobs considered Robbin so valuable he once allowed a *Time* reporter to meet him only after extracting the promise that the reporter would not print his last name.)

Jobs personally worked with them to transform SoundJam into an Apple product. It was laden with all sorts of features, and consequently a lot of complex screens. Jobs pushed them to make it simpler and more fun. Instead of an interface that made you specify whether you were searching for an artist, song, or album, Jobs insisted on a simple box where you could type in anything you wanted. From iMovie the team adopted the sleek brushed-metal look and also a name. They dubbed it iTunes.

Jobs unveiled iTunes at the January 2001 Macworld as part of the digital hub strategy. It would be free to all Mac users, he announced. "Join the music revolution with iTunes, and make your music devices ten times more valuable," he concluded to great applause. As his advertising slogan would later put it: *Rip. Mix. Burn.*

That afternoon Jobs happened to be meeting with John Markoff of the *New York Times*. The interview was going badly, but at the end Jobs sat down at his Mac and showed off iTunes. "It reminds me of my youth," he said as the psychedelic patterns danced on the screen. That led him to reminisce about dropping acid. Taking LSD was one of the two or three most important things he'd done in his life, Jobs told Markoff. People who had never taken acid would never fully understand him.

THE iPOD

The next step for the digital hub strategy was to make a portable music player. Jobs realized that Apple had the opportunity to design such a device in tandem with the iTunes software, allowing it to be simpler. Complex tasks could be handled on the computer, easy ones on the device. Thus was born the iPod, the device that would begin the transformation of Apple from being a computer maker into being the world's most valuable company.

Jobs had a special passion for the project because he loved music. The music players that were already on the market, he told his colleagues, "truly sucked." Phil Schiller, Jon Rubinstein, and the rest of the team agreed. As they were building iTunes, they spent time with the Rio and other players while merrily trashing them. "We would sit around and say, 'These things really stink,' " Schiller recalled. "They held about sixteen songs, and you couldn't figure out how to use them."

Jobs began pushing for a portable music player in the fall of 2000, but Rubinstein responded that the necessary components were not available yet. He asked Jobs to wait. After a few months Rubinstein was able to score a suitable small LCD screen and rechargeable lithium-polymer battery.

The tougher challenge was finding a disk drive that was small enough but had ample memory to make a great music player. Then, in February 2001, he took one of his regular trips to Japan to visit Apple's suppliers.

At the end of a routine meeting with Toshiba, the engineers mentioned a new product they had in the lab that would be ready by that June. It was a tiny, 1.8-inch drive (the size of a silver dollar) that would hold five gigabytes of storage (about a thousand songs), and they were not sure what to do with it. When the Toshiba engineers showed it to Rubinstein, he knew immediately what it could be used for. A thousand songs in his pocket! Perfect. But he kept a poker face. Jobs was also in Japan, giving the keynote speech at the Tokyo Macworld conference. They met that night at the Hotel Okura, where Jobs was staying. "I know how to do it now," Rubinstein told him. "All I need is a $10 million check." Jobs immediately authorized it. So Rubinstein started negotiating with Toshiba to have exclusive rights to every one of the disks it could make, and he began to look around for someone who could lead the development team.

Tony Fadell was a brash entrepreneurial programmer with a cyberpunk look and an engaging smile who had started three companies while still at the University of Michigan. He had gone to work at the handheld device maker General Magic (where he met Apple refugees Andy Hertzfeld and Bill Atkinson), and then spent some awkward time at Philips Electronics, where he bucked the staid culture with his short bleached hair and rebellious style. He had come up with some ideas for creating a better digital music player, which he had shopped around unsuccessfully to RealNetworks, Sony, and Philips. One day he was in Colorado, skiing with an uncle, and his cell phone rang while he was riding on the chairlift. It was Rubinstein, who told him that Apple was looking for someone who could work on a "small electronic device." Fadell, not lacking in confidence, boasted that he was a wizard at making such devices. Rubinstein invited him to Cupertino.

Fadell assumed that he was being hired to work on a personal digital assistant, some successor to the Newton. But when he met with Rubinstein, the topic quickly turned to iTunes, which had been out for three months. "We've been trying to hook up the existing MP3 players to iTunes and they've been horrible, absolutely horrible," Rubinstein told him. "We think we should make our own version."

Fadell was thrilled. "I was passionate about music. I was trying to do some of that at RealNetworks, and I was pitching an MP3 player to Palm." He agreed to come aboard, at least as a consultant. After a few weeks Rubinstein insisted that if he was to lead the team, he had to become a full-time Apple employee. But Fadell resisted; he liked his freedom. Rubinstein was furious at what he considered Fadell's whining. "This is one of those life decisions," he told Fadell. "You'll never regret it."

He decided to force Fadell's hand. He gathered a roomful of the twenty or so people who had been assigned to the project. When Fadell walked in, Rubinstein told him, "Tony, we're not doing this project unless you sign on full-time. Are you in or out? You have to decide right now."

Fadell looked Rubinstein in the eye, then turned to the audience and said, "Does this always happen at Apple, that people are put under duress to sign an offer?" He paused for a moment, said yes, and grudgingly shook Rubinstein's hand. "It left some very unsettling feeling between Jon and me for many years," Fadell recalled. Rubinstein agreed: "I don't think he ever forgave me for that."

Fadell and Rubinstein were fated to clash because they both thought that they had fathered the iPod. As Rubinstein saw it, he had been given the mission by Jobs months earlier, found the Toshiba disk drive, and figured out the screen, battery, and other key elements. He had then brought in Fadell to put it together. He and others who resented Fadell's visibility began to refer to him as "Tony Baloney." But from Fadell's perspective, before he came to Apple he had already come up with plans for a great MP3 player, and he had been shopping it around to other companies before he had agreed to come to Apple. The issue of who deserved the most credit for the iPod, or should get the title Podfather, would be fought over the years in interviews, articles, web pages, and even *Wikipedia* entries.

But for the next few months they were too busy to bicker. Jobs wanted the iPod out by Christmas, and this meant having it ready to unveil in October. They looked around for other companies that were designing MP3 players that could serve as the foundation for Apple's work and settled on a small company named PortalPlayer. Fadell told the team there, "This is the project that's going to remold Apple, and ten years from now, it's going to be a music business, not a computer business." He convinced them to sign an exclusive deal, and his group began to modify PortalPlayer's deficiencies, such as its complex interfaces, short battery life, and inability to make a playlist longer than ten songs.

THAT'S IT!

There are certain meetings that are memorable both because they mark a historic moment and because they illuminate the way a leader operates. Such was the case with the gathering in Apple's fourth-floor conference room in April 2001, where Jobs decided on the fundamentals of the iPod. There to hear Fadell present his proposals to Jobs were Rubinstein, Schiller, Ive, Jeff Robbin, and marketing director Stan Ng. Fadell didn't know Jobs, and he was understandably intimidated. "When he walked into the conference room, I sat up and thought, 'Whoa, there's Steve!' I was really on guard, because I'd heard how brutal he could be."

The meeting started with a presentation of the potential market and what other companies were doing. Jobs, as usual, had no patience. "He won't pay attention to a slide deck for more than a minute," Fadell said. When a slide showed other possible players in the market, he waved it away. "Don't worry about Sony," he said. "We know what we're doing, and they don't." After that, they quit showing slides, and instead Jobs peppered the group with questions. Fadell took away a lesson: "Steve prefers to be in the moment, talking things through. He once told me, 'If you need slides, it shows you don't know what you're talking about.' "

Instead Jobs liked to be shown physical objects that he could feel, inspect, and fondle. So Fadell brought three different models to the conference room; Rubinstein had coached him on how to reveal them sequentially so that his preferred choice would be the pièce de résistance. They hid the mockup of that option under a wooden bowl at the center of the table.

Fadell began his show-and-tell by taking the various parts they were using out of a box and spreading them on the table. There were the 1.8-inch drive, LCD screen, boards, and batteries, all labeled with their cost and weight. As he displayed them, they discussed how the prices or sizes might come down over the next year or so. Some of the pieces could be put together, like Lego blocks, to show the options.

Then Fadell began unveiling his models, which were made of Styrofoam with fishing leads inserted to give them the proper weight. The first had a slot for a removable memory card for music. Jobs dismissed it as complicated. The second had dynamic RAM memory, which was cheap but would lose all of the songs if the battery ran out. Jobs was not pleased. Next Fadell put a few of the pieces together to show what a device with the 1.8-inch hard drive would be like. Jobs seemed intrigued. The show climaxed with Fadell lifting the bowl and revealing a fully assembled model of that alternative. "I was hoping to be able to play more with the Lego parts, but Steve settled right on the hard-drive option just the way we had modeled it," Fadell recalled. He was rather stunned by the process. "I was used to being at Philips, where decisions like this would take meeting after meeting, with a lot of PowerPoint presentations and going back for more study."

Next it was Phil Schiller's turn. "Can I bring out my idea now?" he asked. He left the room and returned with a handful of iPod models, all of which had the same device on the front: the soon-to-be-famous trackwheel. "I had been thinking of how you go through a playlist," he recalled. "You can't press a button hundreds of times. Wouldn't it be great if you could have a wheel?" By turning the wheel with your thumb, you could scroll through songs. The longer you kept turning, the faster the scrolling got, so you could zip through hundreds easily. Jobs shouted, "That's it!" He got Fadell and the engineers working on it.

Once the project was launched, Jobs immersed himself in it daily. His main demand was "Simplify!" He would go over each screen of the user interface and apply a rigid test: If he wanted a song or a function, he should be able to get there in three clicks. And the click should be intuitive. If he couldn't figure out how to navigate to something, or if it took more than three clicks, he would be brutal. "There would be times when we'd rack our brains on a user interface problem, and think we'd considered every option, and he would go, 'Did you think of this?' " said Fadell. "And then we'd all go, 'Holy shit.' He'd redefine the problem or approach, and our little problem would go away."

Every night Jobs would be on the phone with ideas. Fadell and the others would call each other up, discuss Jobs's latest suggestion, and conspire on how to nudge him to where they wanted him to go, which worked about half the time. "We would have this swirling thing of Steve's latest idea, and we would all try to stay ahead of it," said Fadell. "Every day there was something like that, whether it was a switch here, or a button color, or a pricing strategy issue. With his style, you needed to work with your peers, watch each other's back."

One key insight Jobs had was that as many functions as possible should be performed using iTunes on your computer rather than on the iPod. As he later recalled:

In order to make the iPod really easy to use — and this took a lot of arguing on my part — we needed to limit what the device itself would do. Instead we put that functionality in iTunes on the computer. For example, we made it so you couldn't make playlists using the device. You made playlists on iTunes, and then you synced with the device. That was controversial. But what made the Rio and other devices so brain-dead was that they were complicated. They had to do things like make playlists, because they weren't integrated with the jukebox software on your computer. So by owning the iTunes software and the iPod device, that allowed us to make the computer and the device work together, and it allowed us to put the complexity in the right place.

The most Zen of all simplicities was Jobs's decree, which astonished his colleagues, that the iPod would not have an on-off switch. It became true of most Apple devices. There was no need for one. Apple's devices would go dormant if they were not being used, and they would wake up when you touched any key. But there was no need for a switch that would go "Click — you're off. Good-bye."

Suddenly everything had fallen into place: a drive that would hold a thousand songs; an interface and scroll wheel that would let you navigate a thousand songs; a FireWire connection that could sync a thousand songs in under ten minutes; and a battery that would last through a thousand songs. "We suddenly were looking at one another and saying, 'This is going to be so cool,' " Jobs recalled. "We knew how cool it was, because we knew how badly we each wanted one personally. And the concept became so beautifully simple: a thousand songs in your pocket." One of the copywriters suggested they call it a "Pod." Jobs was the one who, borrowing from the iMac and iTunes names, modified that to iPod.

THE WHITENESS OF THE WHALE

Jony Ive had been playing with the foam model of the iPod and trying to conceive what the finished product should look like when an idea occurred to him on a morning drive from his San Francisco home to Cupertino. Its face should be pure white, he told his colleague in the car, and it should connect seamlessly to a polished stainless steel back. "Most small consumer products have this disposable feel to them," said Ive. "There is no cultural gravity to them. The thing I'm proudest of about the iPod is that there is something about it that makes it feel significant, not disposable."

The white would be not just white, but *pure* white. "Not only the device, but the headphones and the wires and even the power block," he recalled. "*Pure* white." Others kept arguing that the headphones, of course, should be black, like all headphones. "But Steve got it immediately, and embraced white," said Ive. "There would be a purity to it." The sinuous flow of the white earbud wires helped make the iPod an icon. As Ive described it:

> There was something very significant and non-disposable about it, yet there was also something very quiet and very restrained. It wasn't wagging its tail in your face. It was restrained, but it was also crazy, with those flowing headphones. That's why I like white. White isn't just a neutral color. It is so pure and quiet. Bold and conspicuous and yet so inconspicuous as well.

Lee Clow's advertising team at TBWA\Chiat\Day wanted to celebrate the iconic nature of the iPod and its whiteness rather than create more traditional product-introduction ads that showed off the device's features. James Vincent, a lanky young Brit who had played in a band and worked as a DJ, had recently joined the agency, and he was a natural to help focus Apple's advertising on hip millennial-generation music lovers rather than rebel baby boomers. With the help of the art director Susan Alinsangan, they created a series of billboards and posters for the iPod, and they spread the options on Jobs's conference room table for his inspection. At the far right end they placed the most traditional options, which featured straightforward photos of the iPod on a white background. At the far left end they placed the most graphic and iconic treatments, which showed just a silhouette of someone dancing while listening to an iPod, its white earphone wires waving with the music. "It understood your emotional and intensely personal relationship with the music," Vincent said. He suggested to Duncan Milner, the creative director, that they all stand firmly at the far left end, to see if they could get Jobs to gravitate there. When he walked in, he went immediately to the right, looking at the stark product pictures. "This looks great," he said. "Let's talk about these." Vincent, Milner, and Clow did not budge from the other end. Finally, Jobs looked up, glanced at the iconic treatments, and said, "Oh, I guess you like this stuff." He shook his head. "It doesn't show the product. It doesn't say what it is." Vincent proposed that they use the iconic images but add the tagline, "1,000 songs in your pocket." That would say it all. Jobs glanced back toward the right end of the table, then finally agreed. Not surprisingly he was soon claiming that it was his idea to push for the more iconic ads. "There were some skeptics around who asked, 'How's this going to actually sell an iPod?' " Jobs recalled. "That's when it came in handy to be the CEO, so I could push the idea through."

Jobs realized that there was yet another advantage to the fact that Apple had an integrated system of computer, software, and device. It meant that sales of the iPod would drive sales of the iMac. That, in turn, meant that he could take money that Apple was spending on iMac advertising and shift it to spending on iPod ads — getting a double bang for the buck. A triple bang, actually, because the ads would lend luster and youthfulness to the whole Apple brand. He recalled:

> I had this crazy idea that we could sell just as many Macs by advertising the iPod. In addition, the iPod would position Apple as evoking innovation and youth. So I moved $75 million of advertising money to the iPod, even though the category didn't justify one hundredth of that. That meant that we completely dominated the market for music players. We outspent everybody by a factor of about a hundred.

The television ads showed the iconic silhouettes dancing to songs picked by Jobs, Clow, and Vincent. "Finding the music became our main fun at our weekly marketing meetings," said Clow. "We'd play some edgy cut, Steve would say, 'I hate that,' and James would have to talk him into it." The ads helped popularize many new bands, most notably the Black Eyed Peas; the ad with "Hey Mama" is the classic of the silhouettes genre. When a new ad was about to go into production, Jobs would often have second thoughts, call up Vincent, and insist that he cancel it. "It sounds a bit poppy" or "It sounds a bit trivial," he would say. "Let's call it off." James would get flustered and try to talk him around. "Hold on, it's going to be great," he would argue. Invariably Jobs would relent, the ad would be made, and he would love it.

■ ■ ■

Jobs unveiled the iPod on October 23, 2001, at one of his signature product launch events. "Hint: It's not a Mac," the invitation teased. When it came time to reveal the product, after he described its technical capabilities, Jobs did not do his usual trick of walking over to a table and pulling off a velvet cloth. Instead he said, "I happen to have one right here in my pocket." He reached into his jeans and pulled out the gleaming white device. "This amazing little device holds a thousand songs, and it goes right in my pocket." He slipped it back in and ambled offstage to applause.

Initially there was some skepticism among tech geeks, especially about the $399 price. In the blogosphere, the joke was that iPod stood for "idiots price our devices." However, consumers soon made it a hit. More than that, the iPod became the essence of everything Apple was destined to be: poetry connected to engineering, arts and creativity intersecting with technology, design that's bold and simple. It had an ease of use that came from being an integrated end-to-end system, from computer to FireWire to device to software to content management. When you took an iPod out of the box, it was so beautiful that it seemed to glow, and it made all other music players look as if they had been designed and manufactured in Uzbekistan.

Not since the original Mac had a clarity of product vision so propelled a company into the future. "If anybody was ever wondering why Apple is on the earth, I would hold up this as a good example," Jobs told *Newsweek*'s Steve Levy at the time. Wozniak, who had long been skeptical of integrated systems, began to revise his philosophy. "Wow, it makes sense that Apple was the one to come up with it," Wozniak enthused after the iPod came out. "After all, Apple's whole history is making both the hardware and the software, with the result that the two work better together."

The day that Levy got his press preview of the iPod, he happened to be meeting Bill Gates at a dinner, and he showed it to him. "Have you seen this yet?" Levy asked. Levy noted, "Gates went into a zone that recalls those science fiction films where a space alien, confronted with a novel object, creates some sort of force tunnel between him and the object, allowing him to suck directly into his brain all possible information about it." Gates played with the scroll wheel and pushed every button combination, while his eyes stared fixedly at the screen. "It looks like a great product," he finally said. Then he paused and looked puzzled. "It's only for Macintosh?" he asked.

CHAPTER THIRTY-ONE:
THE iTUNES STORE
I'M THE PIED PIPER

WARNER MUSIC

At the beginning of 2002 Apple faced a challenge. The seamless connection between your iPod, iTunes software, and computer made it easy to manage the music you already owned. But to get new music, you had to venture out of this cozy environment and go buy a CD or download the songs online. The latter endeavor usually meant foraying into the murky domains of file-sharing and piracy services. So Jobs wanted to offer iPod users a way to download songs that was simple, safe, and legal.

The music industry also faced a challenge. It was being plagued by a bestiary of piracy services — Napster, Grokster, Gnutella, Kazaa — that enabled people to get songs for free. Partly as a result, legal sales of CDs were down 9% in 2002.

The executives at the music companies were desperately scrambling, with the elegance of second-graders playing soccer, to agree on a common standard for copy-protecting digital music. Paul Vidich of Warner Music and his corporate colleague Bill Raduchel of AOL Time Warner were working with Sony in that effort, and they hoped to get Apple to be part of their consortium. So a group of them flew to Cupertino in January 2002 to see Jobs.

It was not an easy meeting. Vidich had a cold and was losing his voice, so his deputy, Kevin Gage, began the presentation. Jobs, sitting at the head of the conference table, fidgeted and looked annoyed. After four slides, he waved his hand and broke in. "You have your heads up your asses," he pointed out. Everyone turned to Vidich, who struggled to get his voice working. "You're right," he said after a long pause. "We don't know what to do. You need to help us figure it out." Jobs later recalled being slightly taken aback, and he agreed that Apple would work with the Warner-Sony effort.

If the music companies had been able to agree on a standardized encoding method for protecting music files, then multiple online stores could have proliferated. That would have made it hard for Jobs to create an iTunes Store that allowed Apple to control how online sales were handled. Sony, however, handed Jobs that opportunity when it decided, after the January 2002 Cupertino meeting, to pull out of the talks because it favored its own proprietary format, from which it would get royalties.

"You know Steve, he has his own agenda," Sony's CEO Nobuyuki Idei explained to *Red Herring* editor Tony Perkins. "Although he is a genius, he doesn't share everything with you. This is a difficult person to work with if you are a big company. . . . It is a nightmare." Howard Stringer, then head of Sony North America, added about Jobs: "Trying to get together would frankly be a waste of time."

Instead Sony joined with Universal to create a subscription service called Pressplay. Meanwhile, AOL Time Warner, Bertelsmann, and EMI teamed up with RealNetworks to create Music-Net. Neither would license its songs to the rival service, so each offered only about half the music available. Both were subscription services that allowed customers to stream songs but not keep them, so you lost access to them if your subscription lapsed. They had complicated restrictions and clunky interfaces. Indeed they would earn the dubious distinction of becoming number nine on *PC World*'s list of "the 25 worst tech products of all time." The magazine declared, "The services' stunningly brain-dead features showed that the record companies still didn't get it."

At this point Jobs could have decided simply to indulge piracy. Free music meant more valuable iPods. Yet because he *really* liked music, and the artists who made it, he was opposed to what he saw as the theft of creative products. As he later told me:

> From the earliest days at Apple, I realized that we thrived when we created intellectual property. If people copied or stole our software, we'd be out of business. If it weren't protected, there'd be no incentive for us to make new software or product designs. If protection of intellectual property begins to disappear, creative companies will disappear or never get started. But there's a simpler reason: It's wrong to steal. It hurts other people. And it hurts your own character.

He knew, however, that the best way to stop piracy — in fact the only way — was to offer an alternative that was more attractive than the brain-dead services that music companies were concocting. "We believe that 80% of the people stealing stuff don't want to be, there's just no legal alternative," he told Andy Langer of *Esquire.* "So we said, 'Let's create a legal alternative to this.' Everybody wins. Music companies win. The artists win. Apple wins. And the user wins, because he gets a better service and doesn't have to be a thief."

So Jobs set out to create an "iTunes Store" and to persuade the five top record companies to allow digital versions of their songs to be sold there. "I've never spent so much of my time trying to convince people to do the right thing for themselves," he recalled. Because the companies were worried about the pricing model and unbundling of albums, Jobs pitched that his new service would be only on the Macintosh, a mere 5% of the market. They could try the idea with little risk. "We used our small market share to our advantage by arguing that if the store turned out to be destructive it wouldn't destroy the entire universe," he recalled.

Jobs's proposal was to sell digital songs for 99 cents — a simple and impulsive purchase. The record companies would get 70 cents of that. Jobs insisted that this would be more appealing than the monthly subscription model preferred by the music companies. He believed that people had an emotional connection to the songs they loved. They wanted to *own* "Sympathy for the Devil" and "Shelter from the Storm," not just rent them. As he told Jeff Goodell of *Rolling Stone* at the time, "I think you could make available the Second Coming in a subscription model and it might not be successful."

Jobs also insisted that the iTunes Store would sell individual songs, not just entire albums. That ended up being the biggest cause of conflict with the record companies, which made money by putting out albums that had two or three great songs and a dozen or so fillers; to get the song they wanted, consumers had to buy the whole album. Some musicians objected on artistic grounds to Jobs's plan to disaggregate albums. "There's a flow to a good album," said Trent Reznor of Nine Inch Nails. "The songs support each other. That's the way I like to make music." But the objections were moot. "Piracy and online downloads had already deconstructed the album," recalled Jobs. "You couldn't compete with piracy unless you sold the songs individually."

At the heart of the problem was a chasm between the people who loved technology and those who loved artistry. Jobs loved both, as he had demonstrated at Pixar and Apple, and he was thus positioned to bridge the gap. He later explained:

When I went to Pixar, I became aware of a great divide. Tech companies don't understand creativity. They don't appreciate *intuitive* thinking, like the ability of an A&R guy at a music label to listen to a hundred artists and have a feel for which five might be successful. And they think that creative people just sit around on couches all day and are undisciplined, because they've not seen how driven and disciplined the creative folks at places

571

like Pixar are. On the other hand, music companies are completely clueless about technology. They think they can just go out and hire a few tech folks. But that would be like Apple trying to hire people to produce music. We'd get second-rate A&R people, just like the music companies ended up with second-rate tech people. I'm one of the few people who understands how producing technology requires intuition and creativity, and how producing something artistic takes real discipline.

Jobs had a long relationship with Barry Schuler, the CEO of the AOL unit of Time Warner, and began to pick his brain about how to get the music labels into the proposed iTunes Store. "Piracy is flipping everyone's circuit breakers," Schuler told him. "You should use the argument that because you have an integrated end-to-end service, from iPods to the store, you can best protect how the music is used."

One day in March 2002, Schuler got a call from Jobs and decided to conference-in Vidich. Jobs asked Vidich if he would come to Cupertino and bring the head of Warner Music, Roger Ames. This time Jobs was charming. Ames was a sardonic, fun, and clever Brit, a type (such as James Vincent and Jony Ive) that Jobs tended to like. So the Good Steve was on display. At one point early in the meeting, Jobs even played the unusual role of diplomat. Ames and Eddy Cue, who ran iTunes for Apple, got into an argument over why radio in England was not as vibrant as in the United States, and Jobs stepped in, saying, "We know about tech, but we don't know as much about music, so let's not argue."

Ames had just lost a boardroom battle to have his corporation's AOL division improve its own fledgling music download service. "When I did a digital download using AOL, I could never find the song on my shitty computer," he recalled. So when Jobs demonstrated a prototype of the iTunes Store, Ames was impressed. "Yes, yes, that's exactly what we've been waiting for," he said. He agreed that Warner Music would sign up, and he offered to help enlist other music companies.

Jobs flew east to show the service to other Time Warner execs. "He sat in front of a Mac like a kid with a toy," Vidich recalled. "Unlike any other CEO, he was totally engaged with the product." Ames and Jobs began to hammer out the details of the iTunes Store, including the number of times a track could be put on different devices and how the copy-protection system would work. They soon were in agreement and set out to corral other music labels.

HERDING CATS

The key player to enlist was Doug Morris, head of the Universal Music Group. His domain included must-have artists such as U2, Eminem, and Mariah Carey, as well as powerful labels such as Motown and Interscope-Geffen-A&M. Morris was eager to talk. More than any other mogul, he was upset about piracy and fed up with the caliber of the technology people at the music companies. "It was like the Wild West," Morris recalled. "No one was selling digital music, and it was awash with piracy.

Everything we tried at the record companies was a failure. The difference in skill sets between the music folks and technologists is just huge."

As Ames walked with Jobs to Morris's office on Broadway he briefed Jobs on what to say. It worked. What impressed Morris was that Jobs tied everything together in a way that made things easy for the consumer and also safe for the record companies. "Steve did something brilliant," said Morris. "He proposed this complete system: the iTunes Store, the music-management software, the iPod itself. It was so smooth. He had the whole package."

Morris was convinced that Jobs had the technical vision that was lacking at the music companies. "Of course we have to rely on Steve Jobs to do this," he told his own tech vice president, "because we don't have anyone at Universal who knows anything about technology." That did not make Universal's technologists eager to work with Jobs, and Morris had to keep ordering them to surrender their objections and make a deal quickly. They were able to add a few more restrictions to FairPlay, the Apple system of digital rights management, so that a purchased song could not be spread to too many devices. But in general, they went along with the concept of the iTunes Store that Jobs had worked out with Ames and his Warner colleagues.

Morris was so smitten with Jobs that he called Jimmy Iovine, the fast-talking and brash chief of Interscope-Geffen-A&M. Iovine and Morris were best friends who had spoken every day for the past thirty years. "When I met Steve, I thought he was our savior, so I immediately brought Jimmy in to get his impression," Morris recalled.

Jobs could be extraordinarily charming when he wanted to be, and he turned it on when Iovine flew out to Cupertino for a demo. "See how simple it is?" he asked Iovine. "Your tech folks are never going to do this. There's no one at the music companies who can make it simple enough."

Iovine called Morris right away. "This guy is unique!" he said. "You're right. He's got a turnkey solution." They complained about how they had spent two years working with Sony, and it hadn't gone anywhere. "Sony's never going to figure things out," he told Morris. They agreed to quit dealing with Sony and join with Apple instead. "How Sony missed this is completely mind-boggling to me, a historic fuckup," Iovine said. "Steve would fire people if the divisions didn't work together, but Sony's divisions were at war with one another."

Indeed Sony provided a clear counterexample to Apple. It had a consumer electronics division that made sleek products and a music division with beloved artists (including Bob Dylan). But because each division tried to protect its own interests, the company as a whole never got its act together to produce an end-to-end service.

Andy Lack, the new head of Sony music, had the unenviable task of negotiating with Jobs about whether Sony would sell its music in the iTunes Store. The irrepressible and savvy Lack had just come from a distinguished career in television journalism — a producer at CBS News and president of NBC — and he knew how to size people up and keep his sense of humor. He realized that, for Sony, selling its songs in the iTunes Store was both insane and necessary — which seemed to be the case with a lot of decisions in the music business. Apple would make out like a bandit, not just from its cut on song sales, but from driving the sale of iPods. Lack believed that since the music companies would be responsible for the success of the iPod, they should get a royalty from each device sold.

Jobs would agree with Lack in many of their conversations and claim that he wanted to be a true partner with the music companies. "Steve, you've got me if you just give me *something* for every sale of your device," Lack told him in his booming voice. "It's a beautiful device. But our music is helping to sell it. That's what true partnership means to me."

"I'm with you," Jobs replied on more than one occasion. But then he would go to Doug Morris and Roger Ames to lament, in a conspiratorial fashion, that Lack just didn't get it, that he was clueless about the music business, that he wasn't as smart as Morris and Ames. "In classic Steve fashion, he would agree to something, but it would never happen," said Lack. "He would set you up and then pull it off the table. He's pathological, which can be useful in negotiations. And he's a genius."

Lack knew that he could not win his case unless he got support from others in the industry. But Jobs used flattery and the lure of Apple's marketing clout to keep the other record labels in line. "If the industry had stood together, we could have gotten a license fee, giving us the dual revenue stream we desperately needed," Lack said. "We were the ones making the iPod sell, so it would have been equitable." That, of course, was one of the beauties of Jobs's end-to-end strategy: Sales of songs on iTunes would drive iPod sales, which would drive Macintosh sales. What made it all the more infuriating to Lack was that Sony could have done the same, but it never could get its hardware and software and content divisions to row in unison.

Jobs tried hard to seduce Lack. During one visit to New York, he invited Lack to his penthouse at the Four Seasons hotel. Jobs had already ordered a breakfast spread — oatmeal and berries for them both — and was "beyond solicitous," Lack recalled. "But Jack Welch taught me not to fall in love. Morris and Ames could be seduced. They would say, 'You don't get it, you're supposed to fall in love,' and they did. So I ended up isolated in the industry."

Even after Sony agreed to sell its music in the iTunes Store, the relationship remained contentious. Each new round of renewals or changes would bring a showdown. "With Andy, it was mostly about his big ego," Jobs claimed. "He never really understood the music business, and he could never really deliver. I thought he was sometimes a dick." When I told him what Jobs said, Lack responded, "I fought for Sony and the music industry, so I can see why he thought I was a dick."

Corralling the record labels to go along with the iTunes plan was not enough, however. Many of their artists had carve-outs in their contracts that allowed them personally to control the digital distribution of their music or prevent their songs from being unbundled from their albums and sold singly. So Jobs set about cajoling various top musicians, which he found fun but also a lot harder than he expected.

Before the launch of iTunes, Jobs met with almost two dozen major artists, including Bono, Mick Jagger, and Sheryl Crow. "He would call me at home, relentless, at ten at night, to say he still needed to get to Led Zeppelin or Madonna," Ames recalled. "He was determined, and nobody else could have convinced some of these artists."

Perhaps the oddest meeting was when Dr. Dre came to visit Jobs at Apple headquarters. Jobs loved the Beatles and Dylan, but he admitted that the appeal of rap eluded him. Now Jobs needed Eminem and other rappers to agree to be sold in the iTunes Store, so he huddled with Dr. Dre, who was Eminem's mentor. After Jobs showed him the seamless way the iTunes Store would work with the iPod, Dr. Dre proclaimed, "Man, somebody finally got it right."

On the other end of the musical taste spectrum was the trumpeter Wynton Marsalis. He was on a West Coast fund-raising tour for Jazz at Lincoln Center and was meeting with Jobs's wife, Laurene. Jobs insisted that he come over to the house in Palo Alto, and he proceeded to show off iTunes. "What do you want to search for?" he asked Marsalis. Beethoven, the trumpeter replied. "Watch what it can do!" Jobs kept insisting when Marsalis's attention would wander. "See how the interface works." Marsalis later recalled, "I don't care much about computers, and kept telling him so, but he goes on for two hours. He was a man possessed.

After a while, I started looking at him and not the computer, because I was so fascinated with his passion."

Jobs unveiled the iTunes Store on April 28, 2003, at San Francisco's Moscone Center. With hair now closely cropped and receding, and a studied unshaven look, Jobs paced the stage and described how Napster "demonstrated that the Internet was made for music delivery." Its offspring, such as Kazaa, he said, offered songs for free. How do you compete with that? To answer that question, he began by describing the downsides of using these free services. The downloads were unreliable and the quality was often bad. "A lot of these songs are encoded by seven-year-olds, and they don't do a great job." In addition, there were no previews or album art. Then he added, "Worst of all it's stealing. It's best not to mess with karma."

Why had these piracy sites proliferated, then? Because, Jobs said, there was no alternative. The subscription services, such as Pressplay and Music-Net, "treat you like a criminal," he said, showing a slide of an inmate in striped prison garb. Then a slide of Bob Dylan came on the screen. "People want to own the music they love."

After a lot of negotiating with the record companies, he said, "they were willing to do something with us to change the world." The iTunes Store would start with 200,000 tracks, and it would grow each day. By using the store, he said, you can own your songs, burn them on CDs, be assured of the download quality, get a preview of a song before you download it, and use it with your iMovies and iDVDs to "make the soundtrack of your life." The price? Just 99 cents, he said, less than a third of what a Starbucks latte cost. Why was it worth it? Because to get the right song from Kazaa took about fifteen minutes, rather than a minute. By spending an hour of your time to save about four dollars, he calculated, "you're working for under the minimum wage!" And one more thing . . . "With iTunes, it's not stealing anymore. It's good karma."

Clapping the loudest for that line were the heads of the record labels in the front row, including Doug Morris sitting next to Jimmy Iovine, in his usual baseball cap, and the whole crowd from Warner Music. Eddy Cue, who was in charge of the store, predicted that Apple would sell a million songs in six months. Instead the iTunes Store sold a million songs in six *days*. "This will go down in history as a turning point for the music industry," Jobs declared.

MICROSOFT

"We were smoked."

That was the blunt email sent to four colleagues by Jim Allchin, the Microsoft executive in charge of Windows development, at 5 p.m. the day he saw the iTunes Store. It had only one other line: "How did they get the music companies to go along?"

Later that evening a reply came from David Cole, who was running Microsoft's online business group. "When Apple brings this to Windows (I assume they won't make the mistake of not bringing it to Windows), we will really be smoked." He said that the Windows team needed "to bring this kind

580

of solution to market," adding, "That will require focus and goal alignment around an *end-to-end service* which delivers direct user value, something we don't have today." Even though Microsoft had its own Internet service (MSN), it was not used to providing end-to-end service the way Apple was.

Bill Gates himself weighed in at 10:46 that night. His subject line, "Apple's Jobs again," indicated his frustration. "Steve Jobs's ability to focus in on a few things that count, get people who get user interface right, and market things as revolutionary are amazing things," he said. He too expressed surprise that Jobs had been able to convince the music companies to go along with his store. "This is very strange to me. The music companies' own operations offer a service that is truly unfriendly to the user. Somehow they decide to give Apple the ability to do something pretty good."

Gates also found it strange that no one else had created a service that allowed people to buy songs rather than subscribe on a monthly basis. "I am not saying this strangeness means we messed up — at least if we did, so did Real and Pressplay and MusicNet and basically everyone else," he wrote. "Now that Jobs has done it we need to move fast to get something where the user interface and Rights are as good. . . . I think we need some plan to prove that, even though Jobs has us a bit flat footed again, we can move quick and both match and do stuff better." It was an astonishing private admission: Microsoft had again been caught flat-footed, and it would again try to catch up by copying Apple. But like Sony, Microsoft could never make it happen, even after Jobs showed the way.

Instead Apple continued to smoke Microsoft in the way that Cole had predicted: It ported the iTunes software and store to Windows. But that took some internal agonizing. First, Jobs and his team had to decide whether they wanted the iPod to work with Windows computers. Jobs was initially opposed. "By keeping the iPod for Mac only, it was driving the sales of Macs even more than we expected," he recalled. But lined up against him were all four of his top executives: Schiller, Rubinstein, Robbin, and Fadell. It was an argument about what the future of Apple should be. "We felt we should be in the music player business, not just in the Mac business," said Schiller.

Jobs always wanted Apple to create its own unified utopia, a magical walled garden where hardware and software and peripheral devices worked well together to create a great experience, and where the success of one product drove sales of all the companions. Now he was facing pressure to have his hottest new product work with Windows machines, and it went against his nature. "It was a really big argument for months," Jobs recalled, "me against everyone else." At one point he declared that Windows users would get to use iPods "over my dead body." But still his team kept pushing. "This *needs* to get to the PC," said Fadell.

Finally Jobs declared, "Until you can prove to me that it will make business sense, I'm not going to do it." That was actually his way of backing down. If you put aside emotion and dogma, it was easy to prove that it made business sense to allow Windows users to buy iPods. Experts were called in, sales scenarios developed, and everyone concluded this would bring in more profits. "We developed a spreadsheet," said Schiller. "Under all scenarios, there was no amount of cannibalization of Mac sales that would outweigh the sales of iPods." Jobs was sometimes willing to surrender, despite his reputation, but he never won any awards for gracious concession speeches. "Screw it," he said at one meeting where they showed him the analysis. "I'm sick of listening to you assholes. Go do whatever the hell you want."

That left another question: When Apple allowed the iPod to be compatible with Windows machines, should it also create a version of iTunes to serve as the music-management software for those Windows users? As usual, Jobs believed the hardware and software should go together: The user experience depended on the iPod working in complete sync (so to speak) with iTunes software on the computer. Schiller was opposed. "I thought that was crazy, since we don't make Windows software," Schiller recalled. "But Steve kept arguing, 'If we're going to do it, we should do it right.' "

Schiller prevailed at first. Apple decided to allow the iPod to work with Windows by using software from MusicMatch, an outside company. But the software was so clunky that it proved Jobs's point, and Apple embarked on a fast-track effort to produce iTunes for Windows. Jobs recalled:

To make the iPod work on PCs, we initially partnered with another company that had a jukebox, gave them the secret sauce to connect to the

583

iPod, and they did a crappy job. That was the worst of all worlds, because this other company was controlling a big piece of the user experience. So we lived with this crappy outside jukebox for about six months, and then we finally got iTunes written for Windows. In the end, you just don't want someone else to control a big part of the user experience. People may disagree with me, but I am pretty consistent about that.

Porting iTunes to Windows meant going back to all of the music companies — which had made deals to be in iTunes based on the assurance that it would be for only the small universe of Macintosh users — and negotiate again. Sony was especially resistant. Andy Lack thought it another example of Jobs changing the terms after a deal was done. It was. But by then the other labels were happy about how the iTunes Store was working and went along, so Sony was forced to capitulate.

Jobs announced the launch of iTunes for Windows in October 2003. "Here's a feature that people thought we'd never add until this happened," he said, waving his hand at the giant screen behind him. "Hell froze over," proclaimed the slide. The show included iChat appearances and videos from Mick Jagger, Dr. Dre, and Bono. "It's a very cool thing for musicians and music," Bono said of the iPod and iTunes. "That's why I'm here to kiss the corporate ass. I don't kiss everybody's."

Jobs was never prone to understatement. To the cheers of the crowd, he declared, "iTunes for Windows is probably the best Windows app ever written."

Microsoft was not grateful. "They're pursuing the same strategy that they pursued in the PC business, controlling both the hardware and software," Bill Gates told *Business Week.* "We've always done things a little bit differently than Apple in terms of giving people choice." It was not until three years later, in November 2006, that Microsoft was finally able to release its own answer to the iPod. It was called the Zune, and it looked like an iPod, though a bit clunkier. Two years later it had achieved a market share of less than 5%. Jobs was brutal about the cause of the Zune's uninspired design and market weakness:

> The older I get, the more I see how much motivations matter. The Zune was crappy because the people at Microsoft don't really love music or art the way we do. We won because we personally love music. We made the iPod for ourselves, and when you're doing something for yourself, or your best friend or family, you're not going to cheese out. If you don't love something, you're not going to go the extra mile, work the extra weekend, challenge the status quo as much.

MR. TAMBOURINE MAN

Andy Lack's first annual meeting at Sony was in April 2003, the same week that Apple launched the iTunes Store. He had been made head of the music division four months earlier, and had spent much of that time negotiating with Jobs. In fact he arrived in Tokyo directly from Cupertino, carry-

585

ing the latest version of the iPod and a description of the iTunes Store. In front of the two hundred managers gathered, he pulled the iPod out of his pocket. "Here it is," he said as CEO Nobuyuki Idei and Sony's North America head Howard Stringer looked on. "Here's the Walkman killer. There's no mystery meat. The reason you bought a music company is so that you could be the one to make a device like this. You can do better."

But Sony couldn't. It had pioneered portable music with the Walkman, it had a great record company, and it had a long history of making beautiful consumer devices. It had all of the assets to compete with Jobs's strategy of integration of hardware, software, devices, and content sales. Why did it fail? Partly because it was a company, like AOL Time Warner, that was organized into divisions (that word itself was ominous) with their own bottom lines; the goal of achieving synergy in such companies by prodding the divisions to work together was usually elusive.

Jobs did not organize Apple into semiautonomous divisions; he closely controlled all of his teams and pushed them to work as one cohesive and flexible company, with one profit-and-loss bottom line. "We don't have 'divisions' with their own P&L," said Tim Cook. "We run one P&L for the company."

In addition, like many companies, Sony worried about cannibalization. If it built a music player and service that made it easy for people to share digital songs, that might hurt sales of its record division. One of Jobs's business rules was to never be afraid of cannibalizing yourself. "If you don't cannibalize yourself, someone else will," he said. So even though an iPhone might cannibalize sales of an iPod, or an iPad might cannibalize sales of a laptop, that did not deter him.

That July, Sony appointed a veteran of the music industry, Jay Samit, to create its own iTunes-like service, called Sony Connect, which would sell songs online and allow them to play on Sony's portable music devices. "The move was immediately understood as a way to unite the sometimes conflicting electronics and content divisions," the *New York Times* reported. "That internal battle was seen by many as the reason Sony, the inventor of the Walkman and the biggest player in the portable audio market, was being trounced by Apple." Sony Connect launched in May 2004. It lasted just over three years before Sony shut it down.

Microsoft was willing to license its Windows Media software and digital rights format to other companies, just as it had licensed out its operating system in the 1980s. Jobs, on the other hand, would not license out Apple's FairPlay to other device makers; it worked only on an iPod. Nor would he allow other online stores to sell songs for use on iPods. A variety of experts said this would eventually cause Apple to lose market share, as it did in the computer wars of the 1980s. "If Apple continues to rely on a proprietary architecture," the Harvard Business School professor Clayton Christensen told *Wired,* "the iPod will likely become a niche product." (Other than in this case, Christensen was one of the world's most insightful business analysts, and Jobs was deeply influenced by his book

The Innovator's Dilemma.) Bill Gates made the same argument. "There's nothing unique about music," he said. "This story has played out on the PC."

Rob Glaser, the founder of RealNetworks, tried to circumvent Apple's restrictions in July 2004 with a service called Harmony. He had attempted to convince Jobs to license Apple's FairPlay format to Harmony, but when that didn't happen, Glaser just reverse-engineered it and used it with the songs that Harmony sold. Glaser's strategy was that the songs sold by Harmony would play on any device, including an iPod or a Zune or a Rio, and he launched a marketing campaign with the slogan "Freedom of Choice." Jobs was furious and issued a release saying that Apple was "stunned that RealNetworks has adopted the tactics and ethics of a hacker to break into the iPod." RealNetworks responded by launching an Internet petition that demanded "Hey Apple! Don't break my iPod." Jobs kept quiet for a few months, but in October he released a new version of the iPod software that caused songs bought through Harmony to become inoperable. "Steve is a one-of-a-kind guy," Glaser said. "You know that about him when you do business with him."

In the meantime Jobs and his team — Rubinstein, Fadell, Robbin, Ive — were able to keep coming up with new versions of the iPod that extended Apple's lead. The first major revision, announced in January 2004, was the iPod Mini. Far smaller than the original iPod — just the size of a business card — it had less capacity and was about the same price. At one point Jobs decided to kill it, not seeing why anyone would want to pay the same for less. "He doesn't do sports, so he didn't relate to how it would be great on a run or in the gym," said Fadell. In fact the Mini was what truly launched the iPod to market dominance, by eliminating the competition from smaller flash-drive players. In the eighteen months after it was introduced, Apple's market share in the portable music player market shot from 31% to 74%.

The iPod Shuffle, introduced in January 2005, was even more revolutionary. Jobs learned that the shuffle feature on the iPod, which played songs in random order, had become very popular. People liked to be surprised, and they were also too lazy to keep setting up and revising their playlists. Some users even became obsessed with figuring out whether the song selection was truly random, and if so, why their iPod kept coming back to, say, the Neville Brothers. That feature led to the iPod Shuffle. As Rubinstein and Fadell were working on creating a flash player that was small and inexpensive, they kept doing things like making the screen tinier. At one point Jobs came in with a crazy suggestion: Get rid of the screen altogether. "What?!?" Fadell responded. "Just get rid of it," Jobs insisted. Fadell asked how users would navigate the songs. Jobs's insight was that you wouldn't need to navigate; the songs would play randomly. After all, they were songs you had chosen. All that was needed was a button to skip over a song if you weren't in the mood for it. "Embrace uncertainty," the ads read.

As competitors stumbled and Apple continued to innovate, music became a larger part of Apple's business. In January 2007 iPod sales were half of

Apple's revenues. The device also added luster to the Apple brand. But an even bigger success was the iTunes Store. Having sold one million songs in the first six days after it was introduced in April 2003, the store went on to sell seventy million songs in its first year. In February 2006 the store sold its one billionth song when Alex Ostrovsky, sixteen, of West Bloomfield, Michigan, bought Coldplay's "Speed of Sound" and got a congratulatory call from Jobs, bestowing upon him ten iPods, an iMac, and a $10,000 music gift certificate.

The success of the iTunes Store also had a more subtle benefit. By 2011 an important new business had emerged: being the service that people trusted with their online identity and payment information. Along with Amazon, Visa, PayPal, American Express, and a few other services, Apple had built up databases of people who trusted them with their email address and credit card information to facilitate safe and easy shopping. This allowed Apple to sell, for example, a magazine subscription through its online store; when that happened, Apple, not the magazine publisher, would have a direct relationship with the subscriber. As the iTunes Store sold videos, apps, and subscriptions, it built up a database of 225 million active users by June 2011, which positioned Apple for the next age of digital commerce.

CHAPTER THIRTY-TWO: MUSIC MAN

THE SOUND TRACK OF HIS LIFE

ON HIS iPOD

As the iPod phenomenon grew, it spawned a question that was asked of presidential candidates, B-list celebrities, first dates, the queen of England, and just about anyone else with white earbuds: "What's on your iPod?" The parlor game took off when Elisabeth Bumiller wrote a piece in the *New York Times* in early 2005 dissecting the answer that President George W. Bush gave when she asked him that question. "Bush's iPod is heavy on traditional country singers," she reported. "He has selections by Van Morrison, whose 'Brown Eyed Girl' is a Bush favorite, and by John Fogerty, most predictably 'Centerfield.' " She got a *Rolling Stone* editor, Joe Levy, to analyze the selection, and he commented, "One thing that's interesting is that the president likes artists who don't like him."

"Simply handing over your iPod to a friend, your blind date, or the total stranger sitting next to you on the plane opens you up like a book," Steven Levy wrote in *The Perfect Thing*. "All somebody needs to do is scroll through your library on that click wheel, and, musically speaking, you're naked. It's not just what you like — it's *who you are*." So

591

one day, when we were sitting in his living room listening to music, I asked Jobs to let me see his. As we sat there, he flicked through his favorite songs.

Not surprisingly, there were all six volumes of Dylan's bootleg series, including the tracks Jobs had first started worshipping when he and Wozniak were able to score them on reel-to-reel tapes years before the series was officially released. In addition, there were fifteen other Dylan albums, starting with his first, *Bob Dylan* (1962), but going only up to *Oh Mercy* (1989). Jobs had spent a lot of time arguing with Andy Hertzfeld and others that Dylan's subsequent albums, indeed any of his albums after *Blood on the Tracks* (1975), were not as powerful as his early performances. The one exception he made was Dylan's track "Things Have Changed" from the 2000 movie *Wonder Boys.* Notably his iPod did not include *Empire Burlesque* (1985), the album that Hertzfeld had brought him the weekend he was ousted from Apple.

The other great trove on his iPod was the Beatles. He included songs from seven of their albums: *A Hard Day's Night, Abbey Road, Help!, Let It Be, Magical Mystery Tour, Meet the Beatles!* and *Sgt. Pepper's Lonely Hearts Club Band*. The solo albums missed the cut. The Rolling Stones clocked in next, with six albums: *Emotional Rescue, Flashpoint, Jump Back, Some Girls, Sticky Fingers,* and *Tattoo You*. In the case of the Dylan and the Beatles albums, most were included in their entirety. But true to his belief that albums can and should be disaggregated, those of the Stones and most other artists on his iPod included only three or four cuts. His onetime girlfriend Joan Baez was amply represented by selections from four albums, including two different versions of "Love Is Just a Four-Letter Word."

His iPod selections were those of a kid from the seventies with his heart in the sixties. There were Aretha, B. B. King, Buddy Holly, Buffalo Springfield, Don McLean, Donovan, the Doors, Janis Joplin, Jefferson Airplane, Jimi Hendrix, Johnny Cash, John Mellencamp, Simon and Garfunkel, and even The Monkees ("I'm a Believer") and Sam the Sham ("Wooly Bully"). Only about a quarter of the songs were from more contemporary artists, such as 10,000 Maniacs, Alicia Keys, Black Eyed Peas, Coldplay, Dido, Green Day, John Mayer (a friend of both his and Apple), Moby (likewise), U2, Seal, and Talking Heads. As for classical music, there were a few recordings of Bach, including the Brandenburg Concertos, and three albums by Yo-Yo Ma.

Jobs told Sheryl Crow in May 2003 that he was downloading some Eminem tracks, admitting, "He's starting to grow on me." James Vincent subsequently took him to an Eminem concert. Even so, the rapper missed making it onto Jobs's iPod. As Jobs said to Vincent after the concert, "I don't know . . ." He later told me, "I respect Eminem as an artist, but I just don't want to listen to his music, and I can't relate to his values the way I can to Dylan's."

His favorites did not change over the years. When the iPad 2 came out in March 2011, he transferred his favorite music to it. One afternoon we sat in his living room as he scrolled through the songs on his new iPad and, with a mellow nostalgia, tapped on ones he wanted to hear.

We went through the usual Dylan and Beatles favorites, then he became more reflective and tapped on a Gregorian chant, "Spiritus Domini," performed by Benedictine monks. For a minute or so he zoned out, almost in a trance. "That's really beautiful," he murmured. He followed with Bach's Second Brandenburg Concerto and a fugue from *The Well-Tempered Clavier.* Bach, he declared, was his favorite classical composer. He was particularly fond of listening to the contrasts between the two versions of the "Goldberg Variations" that Glenn Gould recorded, the first in 1955 as a twenty-two-year-old little-known pianist and the second in 1981, a year before he died. "They're like night and day," Jobs said after playing them sequentially one afternoon. "The first is an exuberant, young, brilliant piece, played so fast it's a revelation. The later one is so much more spare and stark. You sense a very deep soul who's been through a lot in life. It's deeper and wiser." Jobs was on his third medical leave that afternoon when he played both versions, and I asked which he liked better. "Gould liked the later version much better," he said. "I used to like the earlier, exuberant one. But now I can see where he was coming from."

He then jumped from the sublime to the sixties: Donovan's "Catch the Wind." When he noticed me look askance, he protested, "Donovan did some really good stuff, really." He punched up "Mellow Yellow," and then admitted that perhaps it was not the best example. "It sounded better when we were young."

I asked what music from our childhood actually

held up well these days. He scrolled down the list on his iPad and called up the Grateful Dead's 1969 song "Uncle John's Band." He nodded along with the lyrics: "When life looks like Easy Street, there is danger at your door." For a moment we were back at that tumultuous time when the mellowness of the sixties was ending in discord. "Whoa, oh, what I want to know is, are you kind?"

Then he turned to Joni Mitchell. "She had a kid she put up for adoption," he said. "This song is about her little girl." He tapped on "Little Green," and we listened to the mournful melody and lyrics that describe the feelings of a mother who gives up a child. "So you sign all the papers in the family name / You're sad and you're sorry, but you're not ashamed." I asked whether he still often thought about being put up for adoption. "No, not much," he said. "Not too often."

These days, he said, he thought more about getting older than about his birth. That led him to play Joni Mitchell's greatest song, "Both Sides Now," with its lyrics about being older and wiser: "I've looked at life from both sides now, / From win and lose, and still somehow, / It's life's illusions I recall, / I really don't know life at all." As Glenn Gould had done with Bach's "Goldberg Variations," Mitchell had recorded "Both Sides Now" many years apart, first in 1969 and then in an excruciatingly haunting slow version in 2000. He played the latter. "It's interesting how people age," he noted.

Some people, he added, don't age well even when they are young. I asked who he had in mind. "John Mayer is one of the best guitar players who's ever lived, and I'm just afraid he's blowing it big time," Jobs replied. Jobs liked Mayer and occasionally had him over for dinner in Palo Alto. When he was twenty-seven, Mayer appeared at the January 2004 Macworld, where Jobs introduced Garage-Band, and he became a fixture at the event most years. Jobs punched up Mayer's hit "Gravity." The lyrics are about a guy filled with love who inexplicably dreams of ways to throw it away: "Gravity is working against me, / And gravity wants to bring me down." Jobs shook his head and commented, "I think he's a really good kid underneath, but he's just been out of control."

At the end of the listening session, I asked him a well-worn question: the Beatles or the Stones? "If the vault was on fire and I could grab only one set of master tapes, I would grab the Beatles," he answered. "The hard one would be between the Beatles and Dylan. Somebody else could have replicated the Stones. No one could have been Dylan or the Beatles." As he was ruminating about how fortunate we were to have all of them when we were growing up, his son, then eighteen, came in the room. "Reed doesn't understand," Jobs lamented. Or perhaps he did. He was wearing a Joan Baez T-shirt, with the words "Forever Young" on it.

BOB DYLAN

The only time Jobs can ever recall being tongue-tied was in the presence of Bob Dylan. He was playing near Palo Alto in October 2004, and Jobs was recovering from his first cancer surgery. Dylan was not a gregarious man, not a Bono or a

Bowie. He was never Jobs's friend, nor did he care to be. He did, however, invite Jobs to visit him at his hotel before the concert. Jobs recalled:

> We sat on the patio outside his room and talked for two hours. I was really nervous, because he was one of my heroes. And I was also afraid that he wouldn't be really smart anymore, that he'd be a caricature of himself, like happens to a lot of people. But I was delighted. He was as sharp as a tack. He was everything I'd hoped. He was really open and honest. He was just telling me about his life and about writing his songs. He said, "They just came through me, it wasn't like I was having to compose them. That doesn't happen anymore, I just can't write them that way anymore." Then he paused and said to me with his raspy voice and little smile, "But I still can sing them."

The next time Dylan played nearby, he invited Jobs to drop by his tricked-up tour bus just before the concert. When Dylan asked what his favorite song was, Jobs said "One Too Many Mornings." So Dylan sang it that night. After the concert, as Jobs was walking out the back, the tour bus came by and screeched to a stop. The door flipped open. "So, did you hear my song I sang for you?" Dylan rasped. Then he drove off. When Jobs tells the tale, he does a pretty good impression of Dylan's voice. "He's one of my all-time heroes," Jobs recalled. "My love for him has grown over the years, it's ripened. I can't figure out how he did it when he was so young."

A few months after seeing him in concert, Jobs

came up with a grandiose plan. The iTunes Store should offer a digital "boxed set" of every Dylan song every recorded, more than seven hundred in all, for $199. Jobs would be the curator of Dylan for the digital age. But Andy Lack of Sony, which was Dylan's label, was in no mood to make a deal without some serious concessions regarding iTunes. In addition, Lack felt the price was too low and would cheapen Dylan. "Bob is a national treasure," said Lack, "and Steve wanted him on iTunes at a price that commoditized him." It got to the heart of the problems that Lack and other record executives were having with Jobs: He was getting to set the price points, not them. So Lack said no.

"Okay, then I will call Dylan directly," Jobs said. But it was not the type of thing that Dylan ever dealt with, so it fell to his agent, Jeff Rosen, to sort things out.

"It's a really bad idea," Lack told Rosen, showing him the numbers. "Bob is Steve's hero. He'll sweeten the deal." Lack had both a professional and a personal desire to fend Jobs off, even to yank his chain a bit. So he made an offer to Rosen. "I will write you a check for a million dollars tomorrow if you hold off for the time being." As Lack later explained, it was an advance against future royalties, "one of those accounting things record companies do." Rosen called back forty-five minutes later and accepted. "Andy worked things out with us and asked us not to do it, which we didn't," he recalled. "I think Andy gave us some sort of an advance to hold off doing it."

By 2006, however, Lack had stepped aside as the CEO of what was by then Sony BMG, and Jobs reopened negotiations. He sent Dylan an iPod with all of his songs on it, and he showed Rosen the type of marketing campaign that Apple could mount. In August he announced a grand deal. It allowed Apple to sell the $199 digital boxed set of all the songs Dylan ever recorded, plus the exclusive right to offer Dylan's new album, *Modern Times,* for pre-release orders. "Bob Dylan is one of the most respected poets and musicians of our time, and he is a personal hero of mine," Jobs said at the announcement. The 773-track set included forty-two rarities, such as a 1961 tape of "Wade in the Water" made in a Minnesota hotel, a 1962 version of "Handsome Molly" from a live concert at the Gaslight Café in Greenwich Village, the truly awesome rendition of "Mr. Tambourine Man" from the 1964 Newport Folk Festival (Jobs's favorite), and an acoustic version of "Outlaw Blues" from 1965.

As part of the deal, Dylan appeared in a television ad for the iPod, featuring his new album, *Modern Times*. This was one of the most astonishing cases of flipping the script since Tom Sawyer persuaded his friends to whitewash the fence. In the past, getting celebrities to do an ad required paying them a lot of money. But by 2006 the tables were turned. Major artists *wanted* to appear in iPod ads; the exposure would guarantee success. James Vincent had predicted this a few years earlier, when Jobs said he had contacts with many musicians and could pay them to appear in ads. "No, things are going to soon change," Vincent replied. "Apple is a different kind of brand, and it's cooler than the brand of most artists. We should talk about the opportunity we offer the bands, not pay them."

Lee Clow recalled that there was actually some resistance among the younger staffers at Apple and the ad agency to using Dylan. "They wondered whether he was still cool enough," Clow said. Jobs would hear none of that. He was thrilled to have Dylan.

Jobs became obsessed by every detail of the Dylan commercial. Rosen flew to Cupertino so that they could go through the album and pick the song they wanted to use, which ended up being "Someday Baby." Jobs approved a test video that Clow made using a stand-in for Dylan, which was then shot in Nashville with Dylan himself. But when it came back, Jobs hated it. It wasn't distinctive enough. He wanted a new style. So Clow hired another director, and Rosen was able to convince Dylan to retape the entire commercial. This time it was done with a gently backlit cowboy-hatted Dylan sitting on a stool, strumming and singing, while a hip woman in a newsboy cap dances with her iPod. Jobs loved it.

The ad showed the halo effect of the iPod's marketing: It helped Dylan win a younger audience, just as the iPod had done for Apple computers. Because of the ad, Dylan's album was number one on the *Billboard* chart its first week, topping hot-selling albums by Christina Aguilera and Outkast. It was the first time Dylan had reached the top spot since *Desire* in 1976, thirty years earlier. *Ad Age* headlined Apple's role in propelling Dylan. "The iTunes spot wasn't just a run-of-the-mill celebrity-endorsement deal in which a big brand signs a big check to tap into the equity of a big star," it reported. "This one flipped the formula, with the all-powerful Apple brand giving Mr. Dylan access to younger demographics and helping propel his sales to places they hadn't been since the Ford administration."

THE BEATLES

Among Jobs's prized CDs was a bootleg that contained a dozen or so taped sessions of the Beatles revising "Strawberry Fields Forever." It became the musical score to his philosophy of how to perfect a product. Andy Hertzfeld had found the CD and made a copy of it for Jobs in 1986, though Jobs sometimes told folks that it had come from Yoko Ono. Sitting in the living room of his Palo Alto home one day, Jobs rummaged around in some glass-enclosed bookcases to find it, then put it on while describing what it had taught him:

It's a complex song, and it's fascinating to watch the creative process as they went back and forth and finally created it over a few months. Lennon was always my favorite Beatle. [He laughs as Lennon stops during the first take and makes the band go back and revise a chord.] Did you hear that little detour they took? It didn't work, so they went back and started from where they were. It's so raw in this version. It actually makes them sound like mere mortals. You could actually imagine other people doing this, up to this version. Maybe not writing and conceiving it, but certainly playing it. Yet they just didn't stop. They were such perfectionists they kept it going and going. This made a big impression on me when I was in my thirties. You could just tell how much they worked at this.

They did a bundle of work between each of these recordings. They kept sending it back to make it closer to perfect. [As he listens to the third take, he points out how the instrumentation has gotten more complex.] The way we build stuff at Apple is often this way. Even the number of models we'd make of a new notebook or iPod. We would start off with a version and then begin refining and refining, doing detailed models of the design, or the buttons, or how a function operates. It's a lot of work, but in the end it just gets better, and soon it's like, "Wow, how did they do that?!? Where are the screws?"

It was thus understandable that Jobs was driven to distraction by the fact that the Beatles were not on iTunes.

His struggle with Apple Corps, the Beatles' business holding company, stretched more than three decades, causing too many journalists to use the phrase "long and winding road" in stories about the relationship. It began in 1978, when Apple Computers, soon after its launch, was sued by Apple Corps for trademark infringement, based on the fact that the Beatles' former recording label was called Apple. The suit was settled three years later, when Apple Computers paid Apple Corps $80,000. The settlement had what seemed back then an innocuous stipulation: The Beatles would not produce any computer equipment and Apple would not market any music products.

The Beatles kept their end of the bargain; none

of them ever produced any computers. But Apple ended up wandering into the music business. It got sued again in 1991, when the Mac incorporated the ability to play musical files, then again in 2003, when the iTunes Store was launched. The legal issues were finally resolved in 2007, when Apple made a deal to pay Apple Corps $500 million for all worldwide rights to the name, and then licensed back to the Beatles the right to use Apple Corps for their record and business holdings.

Alas, this did not resolve the issue of getting the Beatles onto iTunes. For that to happen, the Beatles and EMI Music, which held the rights to most of their songs, had to negotiate their own differences over how to handle the digital rights. "The Beatles all want to be on iTunes," Jobs later recalled, "but they and EMI are like an old married couple. They hate each other but can't get divorced. The fact that my favorite band was the last holdout from iTunes was something I very much hoped I would live to resolve." As it turned out, he would.

BONO

Bono, the lead singer of U2, deeply appreciated Apple's marketing muscle. He was confident that his Dublin-based band was still the best in the world, but in 2004 it was trying, after almost thirty years together, to reinvigorate its image. It had produced an exciting new album with a song that the band's lead guitarist, The Edge, declared to be "the mother of all rock tunes." Bono knew he needed to find a way to get it some traction, so he placed a call to Jobs.

"I wanted something specific from Apple," Bono recalled. "We had a song called 'Vertigo' that featured an aggressive guitar riff that I knew would be contagious, but only if people were exposed to it many, many times." He was worried that the era of promoting a song through airplay on the radio was over. So Bono visited Jobs at home in Palo Alto, walked around the garden, and made an unusual pitch. Over the years U2 had spurned offers as high as $23 million to be in commercials. Now he wanted Jobs to use the band in an iPod commercial for free — or at least as part of a mutually beneficial package. "They had never done a commercial before," Jobs later recalled. "But they were getting ripped off by free downloading, they liked what we were doing with iTunes, and they thought we could promote them to a younger audience."

Any other CEO would have jumped into a mosh pit to have U2 in an ad, but Jobs pushed back a bit. Apple didn't feature recognizable people in the iPod ads, just silhouettes. (The Dylan ad had not yet been made.) "You have silhouettes of fans," Bono replied, "so couldn't the next phase be silhouettes of artists?" Jobs said it sounded like an idea worth exploring. Bono left a copy of the unreleased album, *How to Dismantle an Atomic Bomb,* for Jobs to hear. "He was the only person outside the band who had it," Bono said.

A round of meetings ensued. Jobs flew down to talk to Jimmy Iovine, whose Interscope records distributed U2, at his house in the Holmby Hills section of Los Angeles. The Edge was there, along with U2's manager, Paul McGuinness. Another meeting took place in Jobs's kitchen, with Mc-

Guinness writing down the deal points in the back of his diary. U2 would appear in the commercial, and Apple would vigorously promote the album in multiple venues, ranging from billboards to the iTunes homepage. The band would get no direct fee, but it would get royalties from the sale of a special U2 edition of the iPod. Bono believed, like Lack, that the musicians should get a royalty on each iPod sold, and this was his small attempt to assert the principle in a limited way for his band. "Bono and I asked Steve to make us a black one," Iovine recalled. "We weren't just doing a commercial sponsorship, we were making a co-branding deal."

"We wanted our own iPod, something distinct from the regular white ones," Bono recalled. "We wanted black, but Steve said, 'We've tried other colors than white, and they don't work.' "A few days later Jobs relented and accepted the idea, tentatively.

The commercial interspersed high-voltage shots of the band in partial silhouette with the usual silhouette of a dancing woman listening to an iPod. But even as it was being shot in London, the agreement with Apple was unraveling. Jobs began having second thoughts about the idea of a special black iPod, and the royalty rates were not fully pinned down. He called James Vincent, at Apple's ad agency, and told him to call London and put things on hold. "I don't think it's going to happen," Jobs said. "They don't realize how much value we are giving them, it's going south. Let's think of some other ad to do." Vincent, a lifelong U2 fan, knew how big the ad would be, both for the band and Apple, and begged for the chance to call Bono to try to get things on track. Jobs gave him Bono's mobile number, and he reached the singer in his kitchen in Dublin.

Bono was also having a few second thoughts. "I don't think this is going to work," he told Vincent. "The band is reluctant." Vincent asked what the problem was. "When we were teenagers in Dublin, we said we would never do naff stuff," Bono replied. Vincent, despite being British and familiar with rock slang, said he didn't know what that meant. "Doing rubbishy things for money," Bono explained. "We are all about our fans. We feel like we'd be letting them down if we went in an ad. It doesn't feel right. I'm sorry we wasted your time."

Vincent asked what more Apple could do to make it work. "We are giving you the most important thing we have to give, and that's our music," said Bono. "And what are you giving us back? Advertising, and our fans will think it's for you. We need something more." Vincent replied that the offer of the special U2 edition of the iPod and the royalty arrangement was a huge deal. "That's the most prized thing we have to give," he told Bono.

The singer said he was ready to try to put the deal back together, so Vincent immediately called Jony Ive, another big U2 fan (he had first seen them in concert in Newcastle in 1983), and described the situation. Then he called Jobs and suggested he send Ive to Dublin to show what the black iPod would look like. Jobs agreed. Vincent called Bono back, and asked if he knew Jony Ive, unaware that they had met before and admired each other. "Know Jony Ive?" Bono laughed. "I love that guy.

I drink his bathwater."

"That's a bit strong," Vincent replied, "but how about letting him come visit and show how cool your iPod would be?"

"I'm going to pick him up myself in my Maserati," Bono answered. "He's going to stay at my house, I'm going to take him out, and I will get him really drunk."

The next day, as Ive headed toward Dublin, Vincent had to fend off Jobs, who was still having second thoughts. "I don't know if we're doing the right thing," he said. "We don't want to do this for anyone else." He was worried about setting the precedent of artists getting a royalty from each iPod sold. Vincent assured him that the U2 deal would be special.

"Jony arrived in Dublin and I put him up at my guest house, a serene place over a railway track with a view of the sea," Bono recalled. "He shows me this beautiful black iPod with a deep red click wheel, and I say okay, we'll do it." They went to a local pub, hashed out some of the details, and then called Jobs in Cupertino to see if he would agree. Jobs haggled for a while over each detail of the finances, and over the design, before he finally embraced the deal. That impressed Bono. "It's actually amazing that a CEO cares that much about detail," he said. When it was resolved, Ive and Bono settled into some serious drinking. Both are comfortable in pubs. After a few pints, they decided to call Vincent back in California. He was not home, so Bono left a message on his answering machine, which Vincent made sure never to erase. "I'm sitting here in bubbling Dublin with your friend Jony," it said. "We're both a bit drunk, and we're happy with this wonderful iPod and I can't even believe it exists and I'm holding it in my hand. Thank you!"

Jobs rented a theater in San Jose for the unveiling of the TV commercial and special iPod. Bono and The Edge joined him onstage. The album sold 840,000 copies in its first week and debuted at number one on the *Billboard* chart. Bono told the press afterward that he had done the commercial without charge because "U2 will get as much value out of the commercial as Apple will." Jimmy Iovine added that it would allow the band to "reach a younger audience."

What was remarkable was that associating with a computer and electronics company was the best way for a rock band to seem hip and appeal to young people. Bono later explained that not all corporate sponsorships were deals with the devil. "Let's have a look," he told Greg Kot, the *Chicago Tribune* music critic. "The 'devil' here is a bunch of creative minds, more creative than a lot of people in rock bands. The lead singer is Steve Jobs. These men have helped design the most beautiful art object in music culture since the electric guitar. That's the iPod. The job of art is to chase ugliness away."

Bono got Jobs to do another deal with him in 2006, this one for his Product Red campaign that raised money and awareness to fight AIDS in Africa. Jobs was never much interested in philanthropy, but he agreed to do a special red iPod as part of Bono's campaign. It was not a whole-hearted commitment. He balked, for example, at using the campaign's signature treatment of putting the name of the company in parentheses with the word "red" in superscript after it, as in (APPLE)^{RED}. "I don't want Apple in parentheses," Jobs insisted. Bono replied, "But Steve, that's how we show unity for our cause." The conversation got heated — to the F-you stage — before they agreed to sleep on it. Finally Jobs compromised, sort of. Bono could do what he wanted in his ads, but Jobs would never put Apple in parentheses on any of his products or in any of his stores. The iPod was labeled (PRODUCT)^{RED}, not (APPLE)^{RED}.

"Steve can be sparky," Bono recalled, "but those moments have made us closer friends, because there are not many people in your life where you can have those robust discussions. He's very opinionated. After our shows, I talk to him and he's always got an opinion." Jobs and his family occasionally visited Bono and his wife and four kids at their home near Nice on the French Riviera. On one vacation, in 2008, Jobs chartered a boat and moored it near Bono's home. They ate meals together, and Bono played tapes of the songs U2 was preparing for what became the *No Line on the Horizon* album. But despite the friendship, Jobs was still a tough negotiator. They tried to make a deal for another ad and special release of the song "Get On Your Boots," but they could not come to terms. When Bono hurt his back in 2010 and had to cancel a tour, Powell sent him a gift basket with a DVD of the comedy duo Flight of the Conchords, the book *Mozart's Brain and the Fighter Pilot,* honey from her beehives, and pain cream.

Jobs wrote a note and attached it to the last item, saying, "Pain Cream — I love this stuff."

YO-YO MA

There was one classical musician Jobs revered both as a person and as a performer: Yo-Yo Ma, the versatile virtuoso who is as sweet and profound as the tones he creates on his cello. They had met in 1981, when Jobs was at the Aspen Design Conference and Ma was at the Aspen Music Festival. Jobs tended to be deeply moved by artists who displayed purity, and he became a fan. He invited Ma to play at his wedding, but he was out of the country on tour. He came by the Jobs house a few years later, sat in the living room, pulled out his 1733 Stradivarius cello, and played Bach. "This is what I would have played for your wedding," he told them. Jobs teared up and told him, "You playing is the best argument I've ever heard for the existence of God, because I don't really believe a human alone can do this." On a subsequent visit Ma allowed Jobs's daughter Erin to hold the cello while they sat around the kitchen. By that time Jobs had been struck by cancer, and he made Ma promise to play at his funeral.

CHAPTER THIRTY-THREE: PIXAR'S FRIENDS
. . . AND FOES

A BUG'S LIFE

When Apple developed the iMac, Jobs drove with Jony Ive to show it to the folks at Pixar. He felt that the machine had the spunky personality that would appeal to the creators of Buzz Lightyear and Woody, and he loved the fact that Ive and John Lasseter shared the talent to connect art with technology in a playful way.

Pixar was a haven where Jobs could escape the intensity in Cupertino. At Apple, the managers were often excitable and exhausted, Jobs tended to be volatile, and people felt nervous about where they stood with him. At Pixar, the storytellers and illustrators seemed more serene and behaved more gently, both with each other and even with Jobs. In other words, the tone at each place was set at the top, by Jobs at Apple, but by Lasseter at Pixar.

Jobs reveled in the earnest playfulness of moviemaking and got passionate about the algorithms that enabled such magic as allowing computer-generated raindrops to refract sunbeams or blades of grass to wave in the wind. But he was able to restrain himself from trying to control the creative process. It was at Pixar that he learned to let other creative people flourish and take the lead. Largely it was because he loved Lasseter, a gentle artist who, like Ive, brought out the best in Jobs.

Jobs's main role at Pixar was deal making, in which his natural intensity was an asset. Soon after the release of *Toy Story,* he clashed with Jeffrey Katzenberg, who had left Disney in the summer of 1994 and joined with Steven Spielberg and David Geffen to start DreamWorks SKG. Jobs believed that his Pixar team had told Katzenberg, while he was still at Disney, about its proposed second movie, *A Bug's Life,* and that he had then stolen the idea of an animated insect movie when he decided to produce *Antz* at DreamWorks. "When Jeffrey was still running Disney animation, we pitched him on *A Bug's Life,*" Jobs said. "In sixty years of animation history, nobody had thought of doing an animated movie about insects, until Lasseter. It was one of his brilliant creative sparks. And Jeffrey left and went to DreamWorks and all of a sudden had this idea for an animated movie about — Oh! — insects. And he pretended he'd never heard the pitch. He lied. He lied through his teeth."

Actually, not. The real story is a bit more interesting. Katzenberg never heard the *Bug's Life* pitch while at Disney. But after he left for DreamWorks, he stayed in touch with Lasseter, occasionally pinging him with one of his typical "Hey buddy, how you doing just checking in" quick phone calls. So when Lasseter happened to be at the Technicolor facility on the Universal lot, where DreamWorks was also located, he called Katzenberg and dropped by with a couple of colleagues. When Katzenberg asked what they were doing next, Lasseter told him. "We described to him *A Bug's Life,* with an ant as the main character, and told him the whole story of him organizing the other ants and enlisting a group of circus performer insects to fight off the grasshoppers," Lasseter recalled. "I should have been wary. Jeffrey kept asking questions about when it would be released."

Lasseter began to get worried when, in early 1996, he heard rumors that DreamWorks might be making its own computer-animated movie about ants. He called Katzenberg and asked him point-blank. Katzenberg hemmed, hawed, and asked where Lasseter had heard that. Lasseter asked again, and Katzenberg admitted it was true. "How could you?" yelled Lasseter, who very rarely raised his voice.

"We had the idea long ago," said Katzenberg, who explained that it had been pitched to him by a development director at DreamWorks.

"I don't believe you," Lasseter replied.

Katzenberg conceded that he had sped up *Antz* as a way to counter his former colleagues at Disney. DreamWorks' first major picture was to be *Prince of Egypt,* which was scheduled to be released for Thanksgiving 1998, and he was appalled when he heard that Disney was planning to release Pixar's *A Bug's Life* that same weekend. So he had rushed *Antz* into production to force Disney to change the release date of *A Bug's Life.*

"Fuck you," replied Lasseter, who did not normally use such language. He didn't speak to Katzenberg for another thirteen years.

Jobs was furious, and he was far more practiced than Lasseter at giving vent to his emotions. He called Katzenberg and started yelling. Katzenberg made an offer: He would delay production of *Antz* if Jobs and Disney would move *A Bug's Life* so that it didn't compete with *Prince of Egypt*. "It was a blatant extortion attempt, and I didn't go for it," Jobs recalled. He told Katzenberg there was nothing he could do to make Disney change the release date.

"Of course you can," Katzenberg replied. "You can move mountains. You taught me how!" He said that when Pixar was almost bankrupt, he had come to its rescue by giving it the deal to do *Toy Story*. "I was the one guy there for you back then, and now you're allowing them to use you to screw me." He suggested that if Jobs wanted to, he could simply slow down production on *A Bug's Life* without telling Disney. If he did, Katzenberg said, he would put *Antz* on hold. "Don't even go there," Jobs replied.

Katzenberg had a valid gripe. It was clear that Eisner and Disney were using the Pixar movie to get back at him for leaving Disney and starting a rival animation studio. "*Prince of Egypt* was the first thing we were making, and they scheduled something for our announced release date just to be hostile," he said. "My view was like that of the Lion King, that if you stick your hand in my cage and paw me, watch out."

No one backed down, and the rival ant movies provoked a press frenzy. Disney tried to keep Jobs quiet, on the theory that playing up the rivalry would serve to help *Antz,* but he was a man not easily muzzled. "The bad guys rarely win," he told the *Los Angeles Times*. In response, DreamWorks' savvy marketing maven, Terry Press, suggested, "Steve Jobs should take a pill."

Antz was released at the beginning of October 1998. It was not a bad movie. Woody Allen voiced the part of a neurotic ant living in a conformist society who yearns to express his individualism. "This is the kind of Woody Allen comedy Woody Allen no longer makes," *Time* wrote. It grossed a respectable $91 million domestically and $172 million worldwide.

A Bug's Life came out six weeks later, as planned. It had a more epic plot, which reversed Aesop's tale of "The Ant and the Grasshopper," plus a greater technical virtuosity, which allowed such startling details as the view of grass from a bug's vantage point. *Time* was much more effusive about it. "Its design work is so stellar — a wide-screen Eden of leaves and labyrinths populated by dozens of ugly, buggy, cuddly cutups — that it makes the DreamWorks film seem, by comparison, like radio," wrote Richard Corliss. It did twice as well as *Antz* at the box office, grossing $163 million domestically and $363 million worldwide. (It also beat *Prince of Egypt*.)

A few years later Katzenberg ran into Jobs and tried to smooth things over. He insisted that he had never heard the pitch for *A Bug's Life* while at Disney; if he had, his settlement with Disney would have given him a share of the profits, so it's not something he would lie about. Jobs laughed, and accepted as much. "I asked you to move your release date, and you wouldn't, so you can't be mad at me for protecting my child," Katzenberg told him. He recalled that Jobs "got really calm and Zen-like" and said he understood. But Jobs later said that he never really forgave Katzenberg:

Our film toasted his at the box office. Did that feel good? No, it still felt awful, because people started saying how everyone in Hollywood was doing insect movies. He took the brilliant originality away from John, and that can never be replaced. That's unconscionable, so I've never trusted him, even after he tried to make amends. He came up to me after he was successful with *Shrek* and said, "I'm a changed man, I'm finally at peace with myself," and all this crap. And it was like, give me a break, Jeffrey.

For his part, Katzenberg was much more gracious. He considered Jobs one of the "true geniuses in the world," and he learned to respect him despite their volatile dealings.

More important than beating *Antz* was showing that Pixar was not a one-hit wonder. *A Bug's Life* grossed as much as *Toy Story* had, proving that the first success was not a fluke. "There's a classic thing in business, which is the second-product syndrome," Jobs later said. It comes from not understanding what made your first product so successful. "I lived through that at Apple. My feeling was, if we got through our second film, we'd make it."

STEVE'S OWN MOVIE

Toy Story 2, which came out in November 1999, was even bigger, with a $485 million gross worldwide. Given that Pixar's success was now assured, it was time to start building a showcase headquarters. Jobs and the Pixar facilities team found an abandoned Del Monte fruit cannery in Emeryville, an industrial neighborhood between Berkeley and Oakland, just across the Bay Bridge from San Francisco. They tore it down, and Jobs commissioned Peter Bohlin, the architect of the Apple stores, to design a new building for the sixteen-acre plot.

Jobs obsessed over every aspect of the new building, from the overall concept to the tiniest detail regarding materials and construction. "Steve had this firm belief that the right kind of building can do great things for a culture," said Pixar's president Ed Catmull. Jobs controlled the creation of the building as if he were a director sweating each scene of a film. "The Pixar building was Steve's own movie," Lasseter said.

Lasseter had originally wanted a traditional Hollywood studio, with separate buildings for various projects and bungalows for development teams. But the Disney folks said they didn't like their new campus because the teams felt isolated, and Jobs agreed. In fact he decided they should go to the other extreme: one huge building around a central atrium designed to encourage random encounters.

Despite being a denizen of the digital world, or maybe because he knew all too well its isolating potential, Jobs was a strong believer in face-to-face meetings. "There's a temptation in our networked age to think that ideas can be developed by email and iChat," he said. "That's crazy. Creativity comes from spontaneous meetings, from random discussions. You run into someone, you ask what they're doing, you say 'Wow,' and soon you're cooking up all sorts of ideas."

So he had the Pixar building designed to promote encounters and unplanned collaborations. "If a building doesn't encourage that, you'll lose a lot of innovation and the magic that's sparked by serendipity," he said. "So we designed the building to make people get out of their offices and mingle in the central atrium with people they might not otherwise see." The front doors and main stairs and corridors all led to the atrium, the café and the mailboxes were there, the conference rooms had windows that looked out onto it, and the six-hundred-seat theater and two smaller screening rooms all spilled into it. "Steve's theory worked from day one," Lasseter recalled. "I kept running into people I hadn't seen for months. I've never seen a building that promoted collaboration and creativity as well as this one."

Jobs even went so far as to decree that there be only two huge bathrooms in the building, one for each gender, connected to the atrium. "He felt that very, very strongly," recalled Pam Kerwin, Pixar's general manager. "Some of us felt that was going too far. One pregnant woman said she shouldn't be forced to walk for ten minutes just to go to the bathroom, and that led to a big fight." It was one of the few times that Lasseter disagreed with Jobs. They reached a compromise: there would be two sets of bathrooms on either side of the atrium on both of the two floors.

Because the building's steel beams were going to be visible, Jobs pored over samples from manufacturers across the country to see which had the best color and texture. He chose a mill in Arkansas, told it to blast the steel to a pure color, and made sure the truckers used caution not to nick any of it. He also insisted that all the beams be bolted together, not welded. "We sandblasted the steel and clear-coated it, so you can actually see what it's like," he recalled. "When the steelworkers were putting up the beams, they would bring their families on the weekend to show them."

The wackiest piece of serendipity was "The Love Lounge." One of the animators found a small door on the back wall when he moved into his office. It opened to a low corridor that you could crawl through to a room clad in sheet metal that provided access to the air-conditioning valves. He and his colleagues commandeered the secret room, festooned it with Christmas lights and lava lamps, and furnished it with benches upholstered in animal prints, tasseled pillows, a fold-up cocktail table, liquor bottles, bar equipment, and napkins that read "The Love Lounge." A video camera installed in the corridor allowed occupants to monitor who might be approaching.

Lasseter and Jobs brought important visitors there and had them sign the wall. The signatures include Michael Eisner, Roy Disney, Tim Allen, and Randy Newman. Jobs loved it, but since he wasn't a drinker he sometimes referred to it as the Meditation Room. It reminded him, he said, of the one that he and Daniel Kottke had at Reed, but without the acid.

The Divorce

In testimony before a Senate committee in February 2002, Michael Eisner blasted the ads that Jobs had created for Apple's iTunes. "There are computer companies that have full-page ads and billboards that say: Rip, mix, burn," he declared. "In other words, they can create a theft and distribute it to all their friends if they buy this particular computer."

This was not a smart comment. It misunderstood the meaning of "rip" and assumed it involved ripping someone off, rather than importing files from a CD to a computer. More significantly, it truly pissed off Jobs, as Eisner should have known. That too was not smart. Pixar had recently released the fourth movie in its Disney deal, *Monsters, Inc.*, which turned out to be the most successful of them all, with $525 million in worldwide gross. Disney's Pixar deal was again coming up for renewal, and Eisner had not made it easier by publicly poking a stick at his partner's eye. Jobs was so incredulous he called a Disney executive to vent: "Do you know what Michael just did to me?"

Eisner and Jobs came from different backgrounds and opposite coasts, but they were similar in being strong-willed and without much inclination to find compromises. They both had a passion for making good products, which often meant micromanaging details and not sugarcoating their criticisms. Watching Eisner take repeated rides on the Wildlife Express train through Disney World's Animal Kingdom and coming up with smart ways to improve the customer experience was like watching Jobs play with the interface of an iPod and find ways it could be simplified. Watching them manage people was a less edifying experience.

Both were better at pushing people than being

pushed, which led to an unpleasant atmosphere when they started trying to do it to each other. In a disagreement, they tended to assert that the other party was lying. In addition, neither Eisner nor Jobs seemed to believe that he could learn anything from the other; nor would it have occurred to either even to fake a bit of deference by pretending to have anything to learn. Jobs put the onus on Eisner:

> The worst thing, to my mind, was that Pixar had successfully reinvented Disney's business, turning out great films one after the other while Disney turned out flop after flop. You would think the CEO of Disney would be curious how Pixar was doing that. But during the twenty-year relationship, he visited Pixar for a total of about two and a half hours, only to give little congratulatory speeches. He was never curious. I was amazed. Curiosity is very important.

That was overly harsh. Eisner had been up to Pixar a bit more than that, including visits when Jobs wasn't with him. But it was true that he showed little curiosity about the artistry or technology at the studio. Jobs likewise didn't spend much time trying to learn from Disney's management.

The open sniping between Jobs and Eisner began in the summer of 2002. Jobs had always admired the creative spirit of the great Walt Disney, especially because he had nurtured a company to last for generations. He viewed Walt's nephew Roy as an embodiment of this historic legacy and spirit. Roy was still on the Disney board, despite his own growing estrangement from Eisner, and Jobs let him know that he would not renew the Pixar-Disney deal as long as Eisner was still the CEO.

Roy Disney and Stanley Gold, his close associate on the Disney board, began warning other directors about the Pixar problem. That prompted Eisner to send the board an intemperate email in late August 2002. He was confident that Pixar would eventually renew its deal, he said, partly because Disney had rights to the Pixar movies and characters that had been made thus far. Plus, he said, Disney would be in a better negotiating position in a year, after Pixar finished *Finding Nemo.* "Yesterday we saw for the second time the new Pixar movie, *Finding Nemo,* that comes out next May," he wrote. "This will be a reality check for those guys. It's okay, but nowhere near as good as their previous films. Of course they think it is great." There were two major problems with this email: It leaked to the *Los Angeles Times,* provoking Jobs to go ballistic, and Eisner's assessment of the movie was wrong, very wrong.

Finding Nemo became Pixar's (and Disney's) biggest hit thus far. It easily beat out *The Lion King* to become, for the time being, the most successful animated movie in history. It grossed $340 million domestically and $868 million worldwide. Until 2010 it was also the most popular DVD of all time, with forty million copies sold, and spawned some of the most popular rides at Disney theme parks. In addition, it was a richly textured, subtle, and deeply beautiful artistic achievement that won the Oscar for best animated feature. "I liked the film because it was about taking risks and learning to let those you love take risks," Jobs said. Its success added $183 million to Pixar's cash reserves, giving it a hefty war chest of $521 million for the final showdown with Disney.

Shortly after *Finding Nemo* was finished, Jobs made Eisner an offer that was so one-sided it was clearly meant to be rejected. Instead of a fifty-fifty split on revenues, as in the existing deal, Jobs proposed a new arrangement in which Pixar would own outright the films it made and the characters in them, and it would merely pay Disney a 7.5% fee to distribute the movies. Plus, the last two films under the existing deal — *The Incredibles* and *Cars* were the ones in the works — would shift to the new distribution deal.

Eisner, however, held one powerful trump card. Even if Pixar didn't renew, Disney had the right to make sequels of *Toy Story* and the other movies that Pixar had made, and it owned all the characters, from Woody to Nemo, just as it owned Mickey Mouse and Donald Duck. Eisner was already planning — or threatening — to have Disney's own animation studio do a *Toy Story 3,* which Pixar had declined to do. "When you see what that company did putting out *Cinderella II,* you shudder at what would have happened," Jobs said.

Eisner was able to force Roy Disney off the board in November 2003, but that didn't end the turmoil. Disney released a scathing open letter. "The company has lost its focus, its creative energy, and its heritage," he wrote. His litany of Eisner's alleged failings included not building a constructive relationship with Pixar. By this point Jobs had decided that he no longer wanted to work with

Eisner. So in January 2004 he publicly announced that he was cutting off negotiations with Disney.

Jobs was usually disciplined in not making public the strong opinions that he shared with friends around his Palo Alto kitchen table. But this time he did not hold back. In a conference call with reporters, he said that while Pixar was producing hits, Disney animation was making "embarrassing duds." He scoffed at Eisner's notion that Disney made any creative contribution to the Pixar films: "The truth is there has been little creative collaboration with Disney for years. You can compare the creative quality of our films with the creative quality of Disney's last three films and judge each company's creative ability yourselves." In addition to building a better creative team, Jobs had pulled off the remarkable feat of building a brand that was now as big a draw for moviegoers as Disney's. "We think the Pixar brand is now the most powerful and trusted brand in animation." When Jobs called to give him a heads-up, Roy Disney replied, "When the wicked witch is dead, we'll be together again."

John Lasseter was aghast at the prospect of breaking up with Disney. "I was worried about my children, what they would do with the characters we'd created," he recalled. "It was like a dagger to my heart." When he told his top staff in the Pixar conference room, he started crying, and he did so again when he addressed the eight hundred or so Pixar employees gathered in the studio's atrium. "It's like you have these dear children and you have to give them up to be adopted by convicted child molesters." Jobs came to the atrium stage next and tried to calm things down. He explained why it might be necessary to break with Disney, and he assured them that Pixar as an institution had to keep looking forward to be successful. "He has the absolute ability to make you believe," said Oren Jacob, a longtime technologist at the studio. "Suddenly, we all had the confidence that, whatever happened, Pixar would flourish."

Bob Iger, Disney's chief operating officer, had to step in and do damage control. He was as sensible and solid as those around him were volatile. His background was in television; he had been president of the ABC Network, which was acquired in 1996 by Disney. His reputation was as a corporate suit, and he excelled at deft management, but he also had a sharp eye for talent, a good-humored ability to understand people, and a quiet flair that he was secure enough to keep muted. Unlike Eisner and Jobs, he had a disciplined calm, which helped him deal with large egos. "Steve did some grandstanding by announcing that he was ending talks with us," Iger later recalled. "We went into crisis mode, and I developed some talking points to settle things down."

Eisner had presided over ten great years at Disney, when Frank Wells served as his president. Wells freed Eisner from many management duties so he could make his suggestions, usually valuable and often brilliant, on ways to improve each movie project, theme park ride, television pilot, and countless other products. But after Wells was killed in a helicopter crash in 1994, Eisner never found the right manager. Katzenberg had demanded Wells's job, which is why Eisner ousted him. Michael Ovitz became president in 1995; it was not a pretty sight, and he was gone in less than two years. Jobs later offered his assessment:

For his first ten years as CEO, Eisner did a really good job. For the last ten years, he really did a bad job. And the change came when Frank Wells died. Eisner is a really good creative guy. He gives really good notes. So when Frank was running operations, Eisner could be like a bumblebee going from project to project trying to make them better. But when Eisner had to run things, he was a terrible manager. Nobody liked working for him. They felt they had no authority. He had this strategic planning group that was like the Gestapo, in that you couldn't spend any money, not even a dime, without them approving it. Even though I broke with him, I had to respect his achievements in the first ten years. And there was a part of him I actually liked. He's a fun guy to be around at times — smart, witty. But he had a dark side to him. His ego got the better of him. Eisner was reasonable and fair to me at first, but eventually, over the course of dealing with him for a decade, I came to see a dark side to him.

Eisner's biggest problem in 2004 was that he did not fully fathom how messed up his animation division was. Its two most recent movies, *Treasure Planet* and *Brother Bear,* did no honor to the Disney legacy, or to its balance sheets. Hit animation movies were the lifeblood of the company; they spawned theme park rides, toys, and television shows. *Toy Story* had led to a movie sequel, a *Dis-*

ney on Ice show, a *Toy Story Musical* performed on Disney cruise ships, a direct-to-video film featuring Buzz Lightyear, a computer storybook, two video games, a dozen action toys that sold twenty-five million units, a clothing line, and nine different attractions at Disney theme parks. This was not the case for *Treasure Planet*.

"Michael didn't understand that Disney's problems in animation were as acute as they were," Iger later explained. "That manifested itself in the way he dealt with Pixar. He never felt he needed Pixar as much as he really did." In addition, Eisner loved to negotiate and hated to compromise, which was not always the best combination when dealing with Jobs, who was the same way. "Every negotiation needs to be resolved by compromises," Iger said. "Neither one of them is a master of compromise."

The impasse was ended on a Saturday night in March 2005, when Iger got a phone call from former senator George Mitchell and other Disney board members. They told him that, starting in a few months, he would replace Eisner as Disney's CEO. When Iger got up the next morning, he called his daughters and then Steve Jobs and John Lasseter. He said, very simply and clearly, that he valued Pixar and wanted to make a deal. Jobs was thrilled. He liked Iger and even marveled at a small connection they had: his former girlfriend Jennifer Egan and Iger's wife, Willow Bay, had been roommates at Penn.

That summer, before Iger officially took over, he and Jobs got to have a trial run at making a deal. Apple was coming out with an iPod that would play video as well as music. It needed television shows to sell, and Jobs did not want to be too public in negotiating for them because, as usual, he wanted the product to be secret until he unveiled it onstage. Iger, who had multiple iPods and used them throughout the day, from his 5 a.m. workouts to late at night, had already been envisioning what it could do for television shows. So he immediately offered ABC's most popular shows, *Desperate Housewives* and *Lost*. "We negotiated that deal in a week, and it was complicated," Iger said. "It was important because Steve got to see how I worked, and because it showed everyone that Disney could in fact work with Steve."

For the announcement of the video iPod, Jobs rented a theater in San Jose, and he invited Iger to be his surprise guest onstage. "I had never been to one of his announcements, so I had no idea what a big deal it was," Iger recalled. "It was a real breakthrough for our relationship. He saw I was pro-technology and willing to take risks." Jobs did his usual virtuoso performance, running through all the features of the new iPod, how it was "one of the best things we've ever done," and how the iTunes Store would now be selling music videos and short films. Then, as was his habit, he ended with "And yes, there is one more thing:" The iPod would be selling TV shows. There was huge applause. He mentioned that the two most popular shows were on ABC. "And who owns ABC? Disney! I know these guys," he exulted.

When Iger then came onstage, he looked as relaxed and as comfortable as Jobs. "One of the things that Steve and I are incredibly excited about is the intersection between great content and great technology," he said. "It's great to be here to announce an extension of our relation with Apple," he added. Then, after the proper pause, he said, "Not with Pixar, but with Apple."

But it was clear from their warm embrace that a new Pixar-Disney deal was once again possible. "It signaled my way of operating, which was 'Make love not war,' " Iger recalled. "We had been at war with Roy Disney, Comcast, Apple, and Pixar. I wanted to fix all that, Pixar most of all."

Iger had just come back from opening the new Disneyland in Hong Kong, with Eisner at his side in his last big act as CEO. The ceremonies included the usual Disney parade down Main Street. Iger realized that the only characters in the parade that had been created in the past decade were Pixar's. "A lightbulb went off," he recalled. "I'm standing next to Michael, but I kept it completely to myself, because it was such an indictment of his stewardship of animation during that period. After ten years of *The Lion King, Beauty and the Beast,* and *Aladdin,* there were then ten years of nothing."

Iger went back to Burbank and had some financial analysis done. He discovered that they had actually lost money on animation in the past decade and had produced little that helped ancillary products. At his first meeting as the new CEO, he presented the analysis to the board, whose members expressed some anger that they had never been told this. "As animation goes, so goes our company," he told the board. "A hit animated film is a big wave, and the ripples go down to every part of our business — from characters in a parade, to music, to parks, to video games, TV, Internet, consumer products. If I don't have wave makers, the company is not going to succeed." He presented them with some choices. They could stick with the current animation management, which he didn't think would work. They could get rid of management and find someone else, but he said he didn't know who that would be. Or they could buy Pixar. "The problem is, I don't know if it's for sale, and if it is, it's going to be a huge amount of money," he said. The board authorized him to explore a deal.

Iger went about it in an unusual way. When he first talked to Jobs, he admitted the revelation that had occurred to him in Hong Kong and how it convinced him that Disney badly needed Pixar. "That's why I just loved Bob Iger," recalled Jobs. "He just blurted it out. Now that's the dumbest thing you can do as you enter a negotiation, at least according to the traditional rule book. He just put his cards out on the table and said, 'We're screwed.' I immediately liked the guy, because that's how I worked too. Let's just immediately put all the cards on the table and see where they fall." (In fact that was not usually Jobs's mode of operation. He often began negotiations by proclaiming that the other company's products or services sucked.)

Jobs and Iger took a lot of walks — around the Apple campus, in Palo Alto, at the Allen and Co. retreat in Sun Valley. At first they came up with a plan for a new distribution deal: Pixar would get back all the rights to the movies and characters it had already produced in return for Disney's getting an equity stake in Pixar, and it would pay Disney a simple fee to distribute its future movies. But Iger worried that such a deal would simply set

Pixar up as a competitor to Disney, which would be bad even if Disney had an equity stake in it. So he began to hint that maybe they should actually do something bigger. "I want you to know that I am really thinking out of the box on this," he said. Jobs seemed to encourage the advances. "It wasn't too long before it was clear to both of us that this discussion might lead to an acquisition discussion," Jobs recalled.

But first Jobs needed the blessing of John Lasseter and Ed Catmull, so he asked them to come over to his house. He got right to the point. "We need to get to know Bob Iger," he told them. "We may want to throw in with him and to help him remake Disney. He's a great guy." They were skeptical at first. "He could tell we were pretty shocked," Lasseter recalled.

"If you guys don't want to do it, that's fine, but I want you to get to know Iger before you decide," Jobs continued. "I was feeling the same as you, but I've really grown to like the guy." He explained how easy it had been to make the deal to put ABC shows on the iPod, and added, "It's night and day different from Eisner's Disney. He's straightforward, and there's no drama with him." Lasseter remembers that he and Catmull just sat there with their mouths slightly open.

Iger went to work. He flew from Los Angeles to Lasseter's house for dinner, and stayed up well past midnight talking. He also took Catmull out to dinner, and then he visited Pixar Studios, alone, with no entourage and without Jobs. "I went out and met all the directors one on one, and they each pitched me their movie," he said. Lasseter was proud of how much his team impressed Iger, which of course made him warm up to Iger. "I never had more pride in Pixar than that day," he said. "All the teams and pitches were amazing, and Bob was blown away."

Indeed after seeing what was coming up over the next few years — *Cars, Ratatouille, WALL-E* — Iger told his chief financial officer at Disney, "Oh my God, they've got great stuff. We've got to get this deal done. It's the future of the company." He admitted that he had no faith in the movies that Disney animation had in the works.

The deal they proposed was that Disney would purchase Pixar for $7.4 billion in stock. Jobs would thus become Disney's largest shareholder, with approximately 7% of the company's stock compared to 1.7% owned by Eisner and 1% by Roy Disney. Disney Animation would be put under Pixar, with Lasseter and Catmull running the combined unit. Pixar would retain its independent identity, its studio and headquarters would remain in Emeryville, and it would even keep its own email addresses.

Iger asked Jobs to bring Lasseter and Catmull to a secret meeting of the Disney board in Century City, Los Angeles, on a Sunday morning. The goal was to make them feel comfortable with what would be a radical and expensive deal. As they prepared to take the elevator from the parking garage, Lasseter said to Jobs, "If I start getting too excited or go on too long, just touch my leg." Jobs ended up having to do it once, but otherwise Lasseter made the perfect sales pitch. "I talked about how we make films, what our philosophies are, the honesty we have with each other, and how we nurture the creative talent," he recalled. The board asked a lot of questions, and Jobs let Lasseter answer most. But Jobs did talk about how exciting it was to connect art with technology. "That's what our culture is all about, just like at Apple," he said.

Before the Disney board got a chance to approve the merger, however, Michael Eisner arose from the departed to try to derail it. He called Iger and said it was far too expensive. "You can fix animation yourself," Eisner told him. "How?" asked Iger. "I know you can," said Eisner. Iger got a bit annoyed. "Michael, how come you say I can fix it, when you couldn't fix it yourself?" he asked.

Eisner said he wanted to come to a board meeting, even though he was no longer a member or an officer, and speak against the acquisition. Iger resisted, but Eisner called Warren Buffett, a big shareholder, and George Mitchell, who was the lead director. The former senator convinced Iger to let Eisner have his say. "I told the board that they didn't need to buy Pixar because they already owned 85% of the movies Pixar had already made," Eisner recounted. He was referring to the fact that for the movies already made, Disney was getting that percentage of the gross, plus it had the rights to make all the sequels and exploit the characters. "I made a presentation that said, here's the 15% of Pixar that Disney does not already own. So that's what you're getting. The rest is a bet on future Pixar films." Eisner admitted that Pixar had been enjoying a good run, but he said it could not continue. "I showed

633

the history of producers and directors who had X number of hits in a row and then failed. It happened to Spielberg, Walt Disney, all of them." To make the deal worth it, he calculated, each new Pixar movie would have to gross $1.3 billion. "It drove Steve crazy that I knew that," Eisner later said.

After he left the room, Iger refuted his argument point by point. "Let me tell you what was wrong with that presentation," he began. When the board had finished hearing them both, it approved the deal Iger proposed.

Iger flew up to Emeryville to meet Jobs and jointly announce the deal to the Pixar workers. But before they did, Jobs sat down alone with Lasseter and Catmull. "If either of you have doubts," he said, "I will just tell them no thanks and blow off this deal." He wasn't totally sincere. It would have been almost impossible to do so at that point. But it was a welcome gesture. "I'm good," said Lasseter. "Let's do it." Catmull agreed. They all hugged, and Jobs wept.

Everyone then gathered in the atrium. "Disney is buying Pixar," Jobs announced. There were a few tears, but as he explained the deal, the staffers began to realize that in some ways it was a reverse acquisition. Catmull would be the head of Disney animation, Lasseter its chief creative officer. By the end they were cheering. Iger had been standing on the side, and Jobs invited him to center stage. As he talked about the special culture of Pixar and how badly Disney needed to nurture it and learn from it, the crowd broke into applause.

■ ■ ■ ■

"My goal has always been not only to make great products, but to build great companies," Jobs later said. "Walt Disney did that. And the way we did the merger, we kept Pixar as a great company and helped Disney remain one as well."

CHAPTER THIRTY-FOUR:
TWENTY-FIRST-CENTURY MACS
SETTING APPLE APART

CLAMS, ICE CUBES, AND SUNFLOWERS

Ever since the introduction of the iMac in 1998, Jobs and Jony Ive had made beguiling design a signature of Apple's computers. There was a consumer laptop that looked like a tangerine clam, and a professional desktop computer that suggested a Zen ice cube. Like bell-bottoms that turn up in the back of a closet, some of these models looked better at the time than they do in retrospect, and they show a love of design that was, on occasion, a bit too exuberant. But they set Apple apart and provided the publicity bursts it needed to survive in a Windows world.

The Power Mac G4 Cube, released in 2000, was so alluring that one ended up on display in New York's Museum of Modern Art. An eight-inch perfect cube the size of a Kleenex box, it was the pure expression of Jobs's aesthetic. The sophistication came from minimalism. No buttons marred the surface. There was no CD tray, just a subtle slot. And as with the original Macintosh, there was no fan. Pure Zen. "When you see something that's so thoughtful on the outside you say, 'Oh, wow, it must be really thoughtful on the inside,' " he told *Newsweek*. "We make progress by eliminating things, by removing the superfluous."

The G4 Cube was almost ostentatious in its lack of ostentation, and it was powerful. But it was not a success. It had been designed as a high-end desktop, but Jobs wanted to turn it, as he did almost every product, into something that could be mass-marketed to consumers. The Cube ended up not serving either market well. Workaday professionals weren't seeking a jewel-like sculpture for their desks, and mass-market consumers were not eager to spend twice what they'd pay for a plain vanilla desktop. Jobs predicted that Apple would sell 200,000 Cubes per quarter. In its first quarter it sold half that. The next quarter it sold fewer than thirty thousand units. Jobs later admitted that he had overdesigned and overpriced the Cube, just as he had the NeXT computer. But gradually he was learning his lesson. In building devices like the iPod, he would control costs and make the tradeoffs necessary to get them launched on time and on budget.

Partly because of the poor sales of the Cube, Apple produced disappointing revenue numbers in September 2000. That was just when the tech bubble was deflating and Apple's education market was declining. The company's stock price, which had been above $60, fell 50% in one day, and by early December it was below $15.

None of this deterred Jobs from continuing to push for distinctive, even distracting, new design. When flat-screen displays became commercially viable, he decided it was time to replace the iMac, the translucent consumer desktop computer that looked as if it were from a *Jetsons* cartoon. Ive came up with a model that was somewhat conventional, with the guts of the computer attached to the back of the flat screen. Jobs didn't like it. As he often did, both at Pixar and at Apple, he slammed on the brakes to rethink things. There was something about the design that lacked purity, he felt. "Why have this flat display if you're going to glom all this stuff on its back?" he asked Ive. "We should let each element be true to itself."

Jobs went home early that day to mull over the problem, then called Ive to come by. They wandered into the garden, which Jobs's wife had planted with a profusion of sunflowers. "Every year I do something wild with the garden, and that time it involved masses of sunflowers, with a sunflower house for the kids," she recalled. "Jony and Steve were riffing on their design problem, then Jony asked, 'What if the screen was separated from the base like a sunflower?' He got excited and started sketching." Ive liked his designs to suggest a narrative, and he realized that a sunflower shape would convey that the flat screen was so fluid and responsive that it could reach for the sun.

In Ive's new design, the Mac's screen was attached to a movable chrome neck, so that it looked not only like a sunflower but also like a cheeky Luxo lamp. Indeed it evoked the playful personality of Luxo Jr. in the first short film that John Lasseter had made at Pixar. Apple took out many patents for the design, most crediting Ive, but on one of them, for "a computer system having a movable assembly attached to a flat panel display," Jobs listed himself as the primary inventor.

In hindsight, some of Apple's Macintosh designs may seem a bit too cute. But other computer makers were at the other extreme. It was an industry that you'd expect to be innovative, but instead it was dominated by cheaply designed generic boxes. After a few ill-conceived stabs at painting on blue colors and trying new shapes, companies such as Dell, Compaq, and HP commoditized computers by outsourcing manufacturing and competing on price. With its spunky designs and its pathbreaking applications like iTunes and iMovie, Apple was about the only place innovating.

INTEL INSIDE

Apple's innovations were more than skin-deep. Since 1994 it had been using a microprocessor, called the PowerPC, that was made by a partnership of IBM and Motorola. For a few years it was faster than Intel's chips, an advantage that Apple touted in humorous commercials. By the time of Jobs's return, however, Motorola had fallen behind in producing new versions of the chip. This provoked a fight between Jobs and Motorola's CEO Chris Galvin. When Jobs decided to stop licensing the Macintosh operating system to clone makers, right after his return to Apple in 1997, he suggested to Galvin that he might consider making an exception for Motorola's clone, the StarMax Mac, but only if Motorola sped up development of new PowerPC chips for laptops. The call got heated. Jobs offered his opinion that Motorola chips sucked. Galvin, who also had a temper, pushed back. Jobs hung up on him. The Motorola StarMax was canceled, and Jobs secretly began planning to move

Apple off the Motorola-IBM PowerPC chip and to adopt, instead, Intel's. This would not be a simple task. It was akin to writing a new operating system.

Jobs did not cede any real power to his board, but he did use its meetings to kick around ideas and think through strategies in confidence, while he stood at a whiteboard and led freewheeling discussions. For eighteen months the directors discussed whether to move to an Intel architecture. "We debated it, we asked a lot of questions, and finally we all decided it needed to be done," board member Art Levinson recalled.

Paul Otellini, who was then president and later became CEO of Intel, began huddling with Jobs. They had gotten to know each other when Jobs was struggling to keep NeXT alive and, as Otellini later put it, "his arrogance had been temporarily tempered." Otellini has a calm and wry take on people, and he was amused rather than put off when he discovered, upon dealing with Jobs at Apple in the early 2000s, "that his juices were going again, and he wasn't nearly as humble anymore." Intel had deals with other computer makers, and Jobs wanted a better price than they had. "We had to find creative ways to bridge the numbers," said Otellini. Most of the negotiating was done, as Jobs preferred, on long walks, sometimes on the trails up to the radio telescope known as the Dish above the Stanford campus. Jobs would start the walk by telling a story and explaining how he saw the history of computers evolving. By the end he would be haggling over price.

"Intel had a reputation for being a tough partner, coming out of the days when it was run by Andy

Grove and Craig Barrett," Otellini said. "I wanted to show that Intel was a company you could work with." So a crack team from Intel worked with Apple, and they were able to beat the conversion deadline by six months. Jobs invited Otellini to Apple's Top 100 management retreat, where he donned one of the famous Intel lab coats that looked like a bunny suit and gave Jobs a big hug. At the public announcement in 2005, the usually reserved Otellini repeated the act. "Apple and Intel, together at last," flashed on the big screen.

Bill Gates was amazed. Designing crazy-colored cases did not impress him, but a secret program to switch the CPU in a computer, completed seamlessly and on time, was a feat he truly admired. "If you'd said, 'Okay, we're going to change our microprocessor chip, and we're not going to lose a beat,' that sounds impossible," he told me years later, when I asked him about Jobs's accomplishments. "They basically did that."

OPTIONS

Among Jobs's quirks was his attitude toward money. When he returned to Apple in 1997, he portrayed himself as a person working for $1 a year, doing it for the benefit of the company rather than himself. Nevertheless he embraced the idea of option megagrants — granting huge bundles of options to buy Apple stock at a preset price — that were not subject to the usual good compensation practices of board committee reviews and performance criteria.

When he dropped the "interim" in his title and officially became CEO, he was offered (in addition to the airplane) a megagrant by Ed Woolard and the board at the beginning of 2000; defying the image he cultivated of not being interested in money, he had stunned Woolard by asking for even more options than the board had proposed. But soon after he got them, it turned out that it was for naught. Apple stock cratered in September 2000 — due to disappointing sales of the Cube plus the bursting of the Internet bubble — which made the options worthless.

Making matters worse was a June 2001 cover story in *Fortune* about overcompensated CEOs, "The Great CEO Pay Heist." A mug of Jobs, smiling smugly, filled the cover. Even though his options were underwater at the time, the technical method of valuing them when granted (known as a Black-Scholes valuation) set their worth at $872 million. *Fortune* proclaimed it "by far" the largest compensation package ever granted a CEO. It was the worst of all worlds: Jobs had almost no money that he could put in his pocket for his four years of hard and successful turnaround work at Apple, yet he had become the poster child of greedy CEOs, making him look hypocritical and undermining his self-image. He wrote a scathing letter to the editor, declaring that his options actually "are worth zero" and offering to sell them to *Fortune* for half of the supposed $872 million the magazine had reported.

In the meantime Jobs wanted the board to give him another big grant of options, since his old ones seemed worthless. He insisted, both to the board and probably to himself, that it was more about getting proper recognition than getting rich. "It wasn't so much about the money," he later said in a deposition in an SEC lawsuit over the options. "Everybody likes to be recognized by his peers. . . . I felt that the board wasn't really doing the same with me." He felt that the board should have come to him offering a new grant, without his having to suggest it. "I thought I was doing a pretty good job. It would have made me feel better at the time."

His handpicked board in fact doted on him. So they decided to give him another huge grant in August 2001, when the stock price was just under $18. The problem was that he worried about his image, especially after the *Fortune* article. He did not want to accept the new grant unless the board canceled his old options at the same time. But to do so would have adverse accounting implications, because it would be effectively repricing the old options. That would require taking a charge against current earnings. The only way to avoid this "variable accounting" problem was to cancel his old options at least six months after his new options were granted. In addition, Jobs started haggling with the board over how quickly the new options would vest.

It was not until mid-December 2001 that Jobs finally agreed to take the new options and, braving the optics, wait six months before his old ones were canceled. But by then the stock price (adjusting for a split) had gone up $3, to about $21. If the strike price of the new options was set at that new level, each would have thus been $3 less valuable. So Apple's legal counsel, Nancy Heinen, looked over the recent stock prices and helped to choose an October date, when the stock was $18.30. She also approved a set of minutes that purported to show that the board had approved the grant on that date. The backdating was potentially worth $20 million to Jobs.

Once again Jobs would end up suffering bad publicity without making a penny. Apple's stock price kept dropping, and by March 2003 even the new options were so low that Jobs traded in all of them for an outright grant of $75 million worth of shares, which amounted to about $8.3 million for each year he had worked since coming back in 1997 through the end of the vesting in 2006.

None of this would have mattered much if the *Wall Street Journal* had not run a powerful series in 2006 about backdated stock options. Apple wasn't mentioned, but its board appointed a committee of three members — Al Gore, Eric Schmidt of Google, and Jerry York, formerly of IBM and Chrysler — to investigate its own practices. "We decided at the outset that if Steve was at fault we would let the chips fall where they may," Gore recalled. The committee uncovered some irregularities with Jobs's grants and those of other top officers, and it immediately turned the findings over to the SEC. Jobs was aware of the backdating, the report said, but he ended up not benefiting financially. (A board committee at Disney also found that similar backdating had occurred at Pixar when Jobs was in charge.)

The laws governing such backdating practices were murky, especially since no one at Apple ended up benefiting from the dubiously dated grants. The SEC took eight months to do its own investigation, and in April 2007 it announced that it would not bring action against Apple "based in part on its swift, extensive, and extraordinary co-operation in the Commission's investigation [and its] prompt self-reporting." Although the SEC found that Jobs had been aware of the backdating, it cleared him of any misconduct because he "was unaware of the accounting implications."

The SEC did file complaints against Apple's former chief financial officer Fred Anderson, who was on the board, and general counsel Nancy Heinen. Anderson, a retired Air Force captain with a square jaw and deep integrity, had been a wise and calming influence at Apple, where he was known for his ability to control Jobs's tantrums. He was cited by the SEC only for "negligence" regarding the paperwork for one set of the grants (not the ones that went to Jobs), and the SEC allowed him to continue to serve on corporate boards. Nevertheless he ended up resigning from the Apple board.

Anderson thought he had been made a scape-goat. When he settled with the SEC, his lawyer issued a statement that cast some of the blame on Jobs. It said that Anderson had "cautioned Mr. Jobs that the executive team grant would have to be priced on the date of the actual board agreement or there could be an accounting charge," and that Jobs replied "that the board had given its prior approval."

Heinen, who initially fought the charges against her, ended up settling and paying a $2.2 million fine, without admitting or denying any wrongdoing. Likewise the company itself settled a shareholders' lawsuit by agreeing to pay $14 million in damages.

"Rarely have so many avoidable problems been created by one man's obsession with his own image," Joe Nocera wrote in the *New York Times*. "Then again, this is Steve Jobs we're talking about." Contemptuous of rules and regulations, he created a climate that made it hard for someone like Heinen to buck his wishes. At times, great creativity occurred. But people around him could pay a price. On compensation issues in particular, the difficulty of defying his whims drove some good people to make some bad mistakes.

The compensation issue in some ways echoed Jobs's parking quirk. He refused such trappings as having a "Reserved for CEO" spot, but he assumed for himself the right to park in the handicapped spaces. He wanted to be seen (both by himself and by others) as someone willing to work for $1 a year, but he also wanted to have huge stock grants bestowed upon him. Jangling inside him were the contradictions of a counterculture rebel turned business entrepreneur, someone who wanted to believe that he had turned on and tuned in without having sold out and cashed in.

CHAPTER THIRTY-FIVE: ROUND ONE
MEMENTO MORI

CANCER

Jobs would later speculate that his cancer was caused by the grueling year that he spent, starting in 1997, running both Apple and Pixar. As he drove back and forth, he had developed kidney stones and other ailments, and he would come home so exhausted that he could barely speak. "That's probably when this cancer started growing, because my immune system was pretty weak at that time," he said.

There is no evidence that exhaustion or a weak immune system causes cancer. However, his kidney problems did indirectly lead to the detection of his cancer. In October 2003 he happened to run into the urologist who had treated him, and she asked him to get a CAT scan of his kidneys and ureter. It had been five years since his last scan. The new scan revealed nothing wrong with his kidneys, but it did show a shadow on his pancreas, so she asked him to schedule a pancreatic study. He didn't. As usual, he was good at willfully ignoring inputs that he did not want to process. But she persisted. "Steve, this is really important," she said a few days later. "You need to do this."

Her tone of voice was urgent enough that he complied. He went in early one morning, and after studying the scan, the doctors met with him to deliver the bad news that it was a tumor. One of them even suggested that he should make sure his affairs were in order, a polite way of saying that he might have only months to live. That evening they performed a biopsy by sticking an endoscope down his throat and into his intestines so they could put a needle into his pancreas and get a few cells from the tumor. Powell remembers her husband's doctors tearing up with joy. It turned out to be an islet cell or pancreatic neuroendocrine tumor, which is rare but slower growing and thus more likely to be treated successfully. He was lucky that it was detected so early — as the by-product of a routine kidney screening — and thus could be surgically removed before it had definitely spread.

One of his first calls was to Larry Brilliant, whom he first met at the ashram in India. "Do you still believe in God?" Jobs asked him. Brilliant said that he did, and they discussed the many paths to God that had been taught by the Hindu guru Neem Karoli Baba. Then Brilliant asked Jobs what was wrong. "I have cancer," Jobs replied.

Art Levinson, who was on Apple's board, was chairing the board meeting of his own company, Genentech, when his cell phone rang and Jobs's name appeared on the screen. As soon as there was a break, Levinson called him back and heard the news of the tumor. He had a background in cancer biology, and his firm made cancer treatment drugs, so he became an advisor. So did Andy Grove of Intel, who had fought and beaten prostate cancer. Jobs called him that Sunday, and he drove right over to Jobs's house and stayed for two hours.

To the horror of his friends and wife, Jobs decided not to have surgery to remove the tumor, which was the only accepted medical approach. "I really didn't want them to open up my body, so I tried to see if a few other things would work," he told me years later with a hint of regret. Specifically, he kept to a strict vegan diet, with large quantities of fresh carrot and fruit juices. To that regimen he added acupuncture, a variety of herbal remedies, and occasionally a few other treatments he found on the Internet or by consulting people around the country, including a psychic. For a while he was under the sway of a doctor who operated a natural healing clinic in southern California that stressed the use of organic herbs, juice fasts, frequent bowel cleansings, hydrotherapy, and the expression of all negative feelings.

"The big thing was that he really was not ready to open his body," Powell recalled. "It's hard to push someone to do that." She did try, however. "The body exists to serve the spirit," she argued. His friends repeatedly urged him to have surgery and chemotherapy. "Steve talked to me when he was trying to cure himself by eating horseshit and horseshit roots, and I told him he was crazy," Grove recalled. Levinson said that he "pleaded every day" with Jobs and found it "enormously frustrating that I just couldn't connect with him." The fights almost ruined their friendship. "That's not how cancer works," Levinson insisted when Jobs discussed his diet treatments. "You cannot solve this without surgery and blasting it with toxic chemicals." Even the diet doctor Dean Ornish, a pioneer in alternative and nutritional methods of treating diseases, took a long walk with Jobs and insisted that sometimes traditional methods were the right option. "You really need surgery," Ornish told him.

Jobs's obstinacy lasted for nine months after his October 2003 diagnosis. Part of it was the product of the dark side of his reality distortion field. "I think Steve has such a strong desire for the world to be a certain way that he wills it to be that way," Levinson speculated. "Sometimes it doesn't work. Reality is unforgiving." The flip side of his wondrous ability to focus was his fearsome willingness to filter out things he did not wish to deal with. This led to many of his great breakthroughs, but it could also backfire. "He has that ability to ignore stuff he doesn't want to confront," Powell explained. "It's just the way he's wired." Whether it involved personal topics relating to his family and marriage, or professional issues relating to engineering or business challenges, or health and cancer issues, Jobs sometimes simply didn't engage.

In the past he had been rewarded for what his wife called his "magical thinking" — his assumption that he could will things to be as he wanted. But cancer does not work that way. Powell enlisted everyone close to him, including his sister Mona Simpson, to try to bring him around. In July 2004 a CAT scan showed that the tumor had grown and possibly spread. It forced him to face reality.

Jobs underwent surgery on Saturday, July 31, 2004, at Stanford University Medical Center. He did not have a full "Whipple procedure," which removes a large part of the stomach and intestine as well as the pancreas. The doctors considered it, but decided instead on a less radical approach, a modified Whipple that removed only part of the pancreas.

Jobs sent employees an email the next day, using his PowerBook hooked up to an AirPort Express in his hospital room, announcing his surgery. He assured them that the type of pancreatic cancer he had "represents about 1% of the total cases of pancreatic cancer diagnosed each year, and can be cured by surgical removal if diagnosed in time (mine was)." He said he would not require chemotherapy or radiation treatment, and he planned to return to work in September. "While I'm out, I've asked Tim Cook to be responsible for Apple's day to day operations, so we shouldn't miss a beat. I'm sure I'll be calling some of you way too much in August, and I look forward to seeing you in September."

One side effect of the operation would become a problem for Jobs because of his obsessive diets and the weird routines of purging and fasting that he had practiced since he was a teenager. Because the pancreas provides the enzymes that allow the stomach to digest food and absorb nutrients, removing part of the organ makes it hard to get enough protein. Patients are advised to make sure that they eat frequent meals and maintain a nutritious diet, with a wide variety of meat and fish proteins as well as full-fat milk products. Jobs had never done this, and he never would.

He stayed in the hospital for two weeks and then struggled to regain his strength. "I remember coming back and sitting in that rocking chair," he told me, pointing to one in his living room. "I didn't have the energy to walk. It took me a week before I could walk around the block. I pushed myself to walk to the gardens a few blocks away, then further, and within six months I had my energy almost back."

Unfortunately the cancer had spread. During the operation the doctors found three liver metastases. Had they operated nine months earlier, they might have caught it before it spread, though they would never know for sure. Jobs began chemotherapy treatments, which further complicated his eating challenges.

THE STANFORD COMMENCEMENT

Jobs kept his continuing battle with the cancer secret — he told everyone that he had been "cured" — just as he had kept quiet about his diagnosis in October 2003. Such secrecy was not surprising; it was part of his nature. What was more surprising was his decision to speak very personally and publicly about his cancer diagnosis. Although he rarely gave speeches other than his staged product demonstrations, he accepted Stanford's invitation to give its June 2005 commencement address. He was in a reflective mood after his health scare and turning fifty.

For help with the speech, he called the brilliant scriptwriter Aaron Sorkin (*A Few Good Men, The West Wing*). Jobs sent him some thoughts. "That was in February, and I heard nothing, so I ping him again in April, and he says, 'Oh, yeah,' and I send him a few more thoughts," Jobs recounted. "I

652

finally get him on the phone, and he keeps saying 'Yeah,' but finally it's the beginning of June, and he never sent me anything."

Jobs got panicky. He had always written his own presentations, but he had never done a commencement address. One night he sat down and wrote the speech himself, with no help other than bouncing ideas off his wife. As a result, it turned out to be a very intimate and simple talk, with the unadorned and personal feel of a perfect Steve Jobs product.

Alex Haley once said that the best way to begin a speech is "Let me tell you a story." Nobody is eager for a lecture, but everybody loves a story. And that was the approach Jobs chose. "Today, I want to tell you three stories from my life," he began. "That's it. No big deal. Just three stories."

The first was about dropping out of Reed College. "I could stop taking the required classes that didn't interest me, and begin dropping in on the ones that looked far more interesting." The second was about how getting fired from Apple turned out to be good for him. "The heaviness of being successful was replaced by the lightness of being a beginner again, less sure about everything." The students were unusually attentive, despite a plane circling overhead with a banner that exhorted "recycle all e-waste," and it was his third tale that enthralled them. It was about being diagnosed with cancer and the awareness it brought:

Remembering that I'll be dead soon is the most important tool I've ever encountered to help me make the big choices in life. Because almost everything — all external expectations, all pride, all fear of embarrassment or failure — these things just fall away in the face of death, leaving only what is truly important. Remembering that you are going to die is the best way I know to avoid the trap of thinking you have something to lose. You are already naked. There is no reason not to follow your heart.

The artful minimalism of the speech gave it simplicity, purity, and charm. Search where you will, from anthologies to YouTube, and you won't find a better commencement address. Others may have been more important, such as George Marshall's at Harvard in 1947 announcing a plan to rebuild Europe, but none has had more grace.

A LION AT FIFTY

For his thirtieth and fortieth birthdays, Jobs had celebrated with the stars of Silicon Valley and other assorted celebrities. But when he turned fifty in 2005, after coming back from his cancer surgery, the surprise party that his wife arranged featured mainly his closest friends and professional colleagues. It was at the comfortable San Francisco home of some friends, and the great chef Alice Waters prepared salmon from Scotland along with couscous and a variety of garden-raised vegetables. "It was beautifully warm and intimate, with everyone and the kids all able to sit in one room," Waters recalled. The entertainment was comedy improvisation done by the cast of *Whose Line Is It Anyway?* Jobs's close friend Mike Slade was there, along with colleagues from Apple and Pixar, including Lasseter, Cook, Schiller, Clow,

Rubinstein, and Tevanian.

Cook had done a good job running the company during Jobs's absence. He kept Apple's temperamental actors performing well, and he avoided stepping into the limelight. Jobs liked strong personalities, up to a point, but he had never truly empowered a deputy or shared the stage. It was hard to be his understudy. You were damned if you shone, and damned if you didn't. Cook had managed to navigate those shoals. He was calm and decisive when in command, but he didn't seek any notice or acclaim for himself. "Some people resent the fact that Steve gets credit for everything, but I've never given a rat's ass about that," said Cook. "Frankly speaking, I'd prefer my name never be in the paper."

When Jobs returned from his medical leave, Cook resumed his role as the person who kept the moving parts at Apple tightly meshed and remained unfazed by Jobs's tantrums. "What I learned about Steve was that people mistook some of his comments as ranting or negativism, but it was really just the way he showed passion. So that's how I processed it, and I never took issues personally." In many ways he was Jobs's mirror image: unflappable, steady in his moods, and (as the thesaurus in the NeXT would have noted) saturnine rather than mercurial. "I'm a good negotiator, but he's probably better than me because he's a cool customer," Jobs later said. After adding a bit more praise, he quietly added a reservation, one that was serious but rarely spoken: "But Tim's not a product person, per se."

In the fall of 2005, after returning from his medical leave, Jobs tapped Cook to become Apple's chief operating officer. They were flying together to Japan. Jobs didn't really *ask* Cook; he simply turned to him and said, "I've decided to make you COO."

Around that time, Jobs's old friends Jon Rubinstein and Avie Tevanian, the hardware and software lieutenants who had been recruited during the 1997 restoration, decided to leave. In Tevanian's case, he had made a lot of money and was ready to quit working. "Avie is a brilliant guy and a nice guy, much more grounded than Ruby and doesn't carry the big ego," said Jobs. "It was a huge loss for us when Avie left. He's a one-of-a-kind person — a genius."

Rubinstein's case was a little more contentious. He was upset by Cook's ascendency and frazzled after working for nine years under Jobs. Their shouting matches became more frequent. There was also a substantive issue: Rubinstein was repeatedly clashing with Jony Ive, who used to work for him and now reported directly to Jobs. Ive was always pushing the envelope with designs that dazzled but were difficult to engineer. It was Rubinstein's job to get the hardware built in a practical way, so he often balked. He was by nature cautious. "In the end, Ruby's from HP," said Jobs. "And he never delved deep, he wasn't aggressive."

There was, for example, the case of the screws that held the handles on the Power Mac G4. Ive decided that they should have a certain polish and shape. But Rubinstein thought that would be "astronomically" costly and delay the project for weeks, so he vetoed the idea. His job was to deliver products, which meant making trade-offs. Ive viewed that approach as inimical to innovation, so he would go both above him to Jobs and also around him to the midlevel engineers. "Ruby would say, 'You can't do this, it will delay,' and I would say, 'I think we can,' " Ive recalled. "And I would know, because I had worked behind his back with the product teams." In this and other cases, Jobs came down on Ive's side.

At times Ive and Rubinstein got into arguments that almost led to blows. Finally Ive told Jobs, "It's him or me." Jobs chose Ive. By that point Rubinstein was ready to leave. He and his wife had bought property in Mexico, and he wanted time off to build a home there. He eventually went to work for Palm, which was trying to match Apple's iPhone. Jobs was so furious that Palm was hiring some of his former employees that he complained to Bono, who was a cofounder of a private equity group, led by the former Apple CFO Fred Anderson, that had bought a controlling stake in Palm. Bono sent Jobs a note back saying, "You should chill out about this. This is like the Beatles ringing up because Herman and the Hermits have taken one of their road crew." Jobs later admitted that he had overreacted. "The fact that they completely failed salves that wound," he said.

Jobs was able to build a new management team that was less contentious and a bit more subdued. Its main players, in addition to Cook and Ive, were Scott Forstall running iPhone software, Phil Schiller in charge of marketing, Bob Mansfield doing Mac hardware, Eddy Cue handling Internet services, and Peter Oppenheimer as the chief financial officer. Even though there was a surface sameness to his top team — all were middle-aged white males — there was a range of styles. Ive was emotional and expressive; Cook was as cool as steel. They all knew they were expected to be deferential to Jobs while also pushing back on his ideas and being willing to argue — a tricky balance to maintain, but each did it well. "I realized very early that if you didn't voice your opinion, he would mow you down," said Cook. "He takes contrary positions to create more discussion, because it may lead to a better result. So if you don't feel comfortable disagreeing, then you'll never survive."

The key venue for freewheeling discourse was the Monday morning executive team gathering, which started at 9 and went for three or four hours. The focus was always on the future: What should each product do next? What new things should be developed? Jobs used the meeting to enforce a sense of shared mission at Apple. This served to centralize control, which made the company seem as tightly integrated as a good Apple product, and prevented the struggles between divisions that plagued decentralized companies.

Jobs also used the meetings to enforce focus. At Robert Friedland's farm, his job had been to prune the apple trees so that they would stay strong, and that became a metaphor for his pruning at Apple. Instead of encouraging each group to let product lines proliferate based on marketing considerations, or permitting a thousand ideas to bloom, Jobs insisted that Apple focus on just two or three priorities at a time. "There is no one better at turning off the noise that is going on around him," Cook said. "That allows him to focus on a few things and say no to many things. Few people are really good at that."

In order to institutionalize the lessons that he and his team were learning, Jobs started an in-house center called Apple University. He hired Joel Podolny, who was dean of the Yale School of Management, to compile a series of case studies analyzing important decisions the company had made, including the switch to the Intel micropro-cessor and the decision to open the Apple Stores. Top executives spent time teaching the cases to new employees, so that the Apple style of decision making would be embedded in the culture.

In ancient Rome, when a victorious general pa-raded through the streets, legend has it that he was sometimes trailed by a servant whose job it was to repeat to him, "Memento mori": Remember you will die. A reminder of mortality would help the hero keep things in perspective, instill some humility. Jobs's memento mori had been delivered by his doctors, but it did not instill humility. In-stead he roared back after his recovery with even more passion. The illness reminded him that he had nothing to lose, so he should forge ahead full speed. "He came back on a mission," said Cook. "Even though he was now running a large com-pany, he kept making bold moves that I don't think anybody else would have done."

For a while there was some evidence, or at least hope, that he had tempered his personal style, that facing cancer and turning fifty had caused him to be a bit less brutish when he was upset. "Right after he came back from his operation, he didn't do the humiliation bit as much," Tevanian recalled. "If he was displeased, he might scream and get hopping mad and use expletives, but he wouldn't do it in a way that would totally destroy the person he was talking to. It was just his way to get the person to do a better job." Tevanian reflected for a moment as he said this, then added a caveat: "Unless he thought someone was really bad and had to go, which happened every once in a while."

Eventually, however, the rough edges returned. Because most of his colleagues were used to it by then and had learned to cope, what upset them most was when his ire turned on strangers. "Once we went to a Whole Foods market to get a smoothie," Ive recalled. "And this older woman was making it, and he really got on her about how she was doing it. Then later, he sympathized. 'She's an older woman and doesn't want to be doing this job.' He didn't connect the two. He was being a purist in both cases."

On a trip to London with Jobs, Ive had the thankless task of choosing the hotel. He picked the Hempel, a tranquil five-star boutique hotel with a sophisticated minimalism that he thought Jobs would love. But as soon as they checked in, he braced himself, and sure enough his phone rang a minute later. "I hate my room," Jobs declared. "It's a piece of shit, let's go." So Ive gathered his luggage and went to the front desk, where Jobs bluntly told the shocked clerk what he thought. Ive realized that most people, himself among them, tend not to be direct when they feel something is shoddy

660

because they want to be liked, "which is actually a vain trait." That was an overly kind explanation. In any case, it was not a trait Jobs had.

Because Ive was so instinctively nice, he puzzled over why Jobs, whom he deeply liked, behaved as he did. One evening, in a San Francisco bar, he leaned forward with an earnest intensity and tried to analyze it:

> He's a very, very sensitive guy. That's one of the things that makes his antisocial behavior, his rudeness, so unconscionable. I can understand why people who are thick-skinned and unfeeling can be rude, but not sensitive people. I once asked him why he gets so mad about stuff. He said, "But I don't stay mad." He has this very childish ability to get really worked up about something, and it doesn't stay with him at all. But there are other times, I think honestly, when he's very frustrated, and his way to achieve catharsis is to hurt somebody. And I think he feels he has a liberty and a license to do that. The normal rules of social engagement, he feels, don't apply to him. Because of how very sensitive he is, he knows exactly how to efficiently and effectively hurt someone. And he does do that.

Every now and then a wise colleague would pull Jobs aside to try to get him to settle down. Lee Clow was a master. "Steve, can I talk to you?" he would quietly say when Jobs had belittled someone publicly. He would go into Jobs's office and explain how hard everyone was working. "When you humiliate them, it's more debilitating than stimulating," he said in one such session. Jobs would apologize and say he understood. But then he would lapse again. "It's simply who I am," he would say.

One thing that did mellow was his attitude toward Bill Gates. Microsoft had kept its end of the bargain it made in 1997, when it agreed to continue developing great software for the Macintosh. Also, it was becoming less relevant as a competitor, having failed thus far to replicate Apple's digital hub strategy. Gates and Jobs had very different approaches to products and innovation, but their rivalry had produced in each a surprising self-awareness.

For their All Things Digital conference in May 2007, the *Wall Street Journal* columnists Walt Mossberg and Kara Swisher worked to get them together for a joint interview. Mossberg first invited Jobs, who didn't go to many such conferences, and was surprised when he said he would do it if Gates would. On hearing that, Gates accepted as well.

Mossberg wanted the evening joint appearance to be a cordial discussion, not a debate, but that seemed less likely when Jobs unleashed a swipe at Microsoft during a solo interview earlier that day. Asked about the fact that Apple's iTunes software for Windows computers was extremely popular, Jobs joked, "It's like giving a glass of ice water to somebody in hell."

So when it was time for Gates and Jobs to meet in the green room before their joint session that evening, Mossberg was worried. Gates got there first, with his aide Larry Cohen, who had briefed him about Jobs's remark earlier that day. When Jobs ambled in a few minutes later, he grabbed a bottle of water from the ice bucket and sat down. After a moment or two of silence, Gates said, "So I guess I'm the representative from hell." He wasn't smiling. Jobs paused, gave him one of his impish grins, and handed him the ice water. Gates relaxed, and the tension dissipated.

The result was a fascinating duet, in which each wunderkind of the digital age spoke warily, and then warmly, about the other. Most memorably they gave candid answers when the technology strategist Lise Buyer, who was in the audience, asked what each had learned from observing the other. "Well, I'd give a lot to have Steve's taste," Gates answered. There was a bit of nervous laughter; Jobs had famously said, ten years earlier, that his problem with Microsoft was that it had absolutely no taste. But Gates insisted he was serious. Jobs was a "natural in terms of intuitive taste." He recalled how he and Jobs used to sit together reviewing the software that Microsoft was making for the Macintosh. "I'd see Steve make the decision based on a sense of people and product that, you know, is hard for me to explain. The way he does things is just different and I think it's magical. And in that case, wow."

Jobs stared at the floor. Later he told me that he was blown away by how honest and gracious Gates had just been. Jobs was equally honest, though not quite as gracious, when his turn came. He described the great divide between the Apple theology of building end-to-end integrated products and Microsoft's openness to licensing its software to competing hardware makers. In the music market, the integrated approach, as manifested in his iTunes-iPod package, was proving to be the better, he noted, but Microsoft's decoupled approach was faring better in the personal computer market. One question he raised in an offhand way was: Which approach might work better for mobile phones?

Then he went on to make an insightful point: This difference in design philosophy, he said, led him and Apple to be less good at collaborating with other companies. "Because Woz and I started the company based on doing the whole banana, we weren't so good at partnering with people," he said. "And I think if Apple could have had a little more of that in its DNA, it would have served it extremely well."

CHAPTER THIRTY-SIX:
THE iPHONE
THREE REVOLUTIONARY PRODUCTS IN ONE

AN iPOD THAT MAKES CALLS

By 2005 iPod sales were skyrocketing. An astonishing twenty million were sold that year, quadruple the number of the year before. The product was becoming more important to the company's bottom line, accounting for 45% of the revenue that year, and it was also burnishing the hipness of the company's image in a way that drove sales of Macs.

That is why Jobs was worried. "He was always obsessing about what could mess us up," board member Art Levinson recalled. The conclusion he had come to: "The device that can eat our lunch is the cell phone." As he explained to the board, the digital camera market was being decimated now that phones were equipped with cameras. The same could happen to the iPod, if phone manufacturers started to build music players into them. "Everyone carries a phone, so that could render the iPod unnecessary."

His first strategy was to do something that he had admitted in front of Bill Gates was not in his DNA: to partner with another company. He began talking to Ed Zander, the new CEO of Motorola, about making a companion to Motorola's popular RAZR, which was a cell phone and digital camera, that would have an iPod built in. Thus was born the ROKR. It ended up having neither the enticing minimalism of an iPod nor the convenient slimness of a RAZR. Ugly, difficult to load, and with an arbitrary hundred-song limit, it had all the hallmarks of a product that had been negotiated by a committee, which was counter to the way Jobs liked to work. Instead of hardware, software, and content all being controlled by one company, they were cobbled together by Motorola, Apple, and the wireless carrier Cingular. "You call this the phone of the future?" *Wired* scoffed on its November 2005 cover.

Jobs was furious. "I'm sick of dealing with these stupid companies like Motorola," he told Tony Fadell and others at one of the iPod product review meetings. "Let's do it ourselves." He had noticed something odd about the cell phones on the market: They all stank, just like portable music players used to. "We would sit around talking about how much we hated our phones," he recalled. "They were way too complicated. They had features nobody could figure out, including the address book. It was just Byzantine." George Riley, an outside lawyer for Apple, remembers sitting at meetings to go over legal issues, and Jobs would get bored, grab Riley's mobile phone, and start pointing out all the ways it was "brain-dead." So Jobs and his team became excited about the prospect of building a phone that they would want to use. "That's the best motivator of all," Jobs later said.

Another motivator was the potential market.

666

More than 825 million mobile phones were sold in 2005, to everyone from grammar schoolers to grandmothers. Since most were junky, there was room for a premium and hip product, just as there had been in the portable music-player market. At first he gave the project to the Apple group that was making the AirPort wireless base station, on the theory that it was a wireless product. But he soon realized that it was basically a consumer device, like the iPod, so he reassigned it to Fadell and his teammates.

Their initial approach was to modify the iPod. They tried to use the trackwheel as a way for a user to scroll through phone options and, without a keyboard, try to enter numbers. It was not a natural fit. "We were having a lot of problems using the wheel, especially in getting it to dial phone numbers," Fadell recalled. "It was cumbersome." It was fine for scrolling through an address book, but horrible at inputting anything. The team kept trying to convince themselves that users would mainly be calling people who were already in their address book, but they knew that it wouldn't really work.

At that time there was a second project under way at Apple: a secret effort to build a tablet computer. In 2005 these narratives intersected, and the ideas for the tablet flowed into the planning for the phone. In other words, the idea for the iPad actually came before, and helped to shape, the birth of the iPhone.

MULTI-TOUCH

One of the engineers developing a tablet PC at Microsoft was married to a friend of Laurene and

Steve Jobs, and for his fiftieth birthday he wanted to have a dinner party that included them along with Bill and Melinda Gates. Jobs went, a bit reluctantly. "Steve was actually quite friendly to me at the dinner," Gates recalled, but he "wasn't particularly friendly" to the birthday guy.

Gates was annoyed that the guy kept revealing information about the tablet PC he had developed for Microsoft. "He's our employee and he's revealing our intellectual property," Gates recounted. Jobs was also annoyed, and it had just the consequence that Gates feared. As Jobs recalled:

> This guy badgered me about how Microsoft was going to completely change the world with this tablet PC software and eliminate all notebook computers, and Apple ought to license his Microsoft software. But he was doing the device all wrong. It had a stylus. As soon as you have a stylus, you're dead. This dinner was like the tenth time he talked to me about it, and I was so sick of it that I came home and said, "Fuck this, let's show him what a tablet can really be."

Jobs went into the office the next day, gathered his team, and said, "I want to make a tablet, and it can't have a keyboard or a stylus." Users would be able to type by touching the screen with their fingers. That meant the screen needed to have a feature that became known as multi-touch, the ability to process multiple inputs at the same time. "So could you guys come up with a multi-touch, touch-sensitive display for me?" he asked. It took them about six months, but they came up with a crude but workable prototype.

Jony Ive had a different memory of how multi-touch was developed. He said his design team had already been working on a multi-touch input that was developed for the trackpads of Apple's Mac-Book Pro, and they were experimenting with ways to transfer that capability to a computer screen. They used a projector to show on a wall what it would look like. "This is going to change every-thing," Ive told his team. But he was careful not to show it to Jobs right away, especially since his people were working on it in their spare time and he didn't want to quash their enthusiasm. "Be-cause Steve is so quick to give an opinion, I don't show him stuff in front of other people," Ive re-called. "He might say, 'This is shit,' and snuff the idea. I feel that ideas are very fragile, so you have to be tender when they are in development. I real-ized that if he pissed on this, it would be so sad, because I knew it was so important."

Ive set up the demonstration in his conference room and showed it to Jobs privately, knowing that he was less likely to make a snap judgment if there was no audience. Fortunately he loved it. "This is the future," he exulted.

It was in fact such a good idea that Jobs realized that it could solve the problem they were having creating an interface for the proposed cell phone. That project was far more important, so he put the tablet development on hold while the multi-touch interface was adopted for a phone-size screen. "If it worked on a phone," he recalled, "I knew we could go back and use it on a tablet."

Jobs called Fadell, Rubinstein, and Schiller to

a secret meeting in the design studio conference room, where Ive gave a demonstration of multi-touch. "Wow!" said Fadell. Everyone liked it, but they were not sure that they would be able to make it work on a mobile phone. They decided to proceed on two paths: P1 was the code name for the phone being developed using an iPod trackwheel, and P2 was the new alternative using a multi-touch screen.

A small company in Delaware called Finger-Works was already making a line of multi-touch trackpads. Founded by two academics at the University of Delaware, John Elias and Wayne Westerman, FingerWorks had developed some tablets with multi-touch sensing capabilities and taken out patents on ways to translate various finger gestures, such as pinches and swipes, into useful functions. In early 2005 Apple quietly acquired the company, all of its patents, and the services of its two founders. FingerWorks quit selling its products to others, and it began filing its new patents in Apple's name.

After six months of work on the trackwheel P1 and the multi-touch P2 phone options, Jobs called his inner circle into his conference room to make a decision. Fadell had been trying hard to develop the trackwheel model, but he admitted they had not cracked the problem of figuring out a simple way to dial calls. The multi-touch approach was riskier, because they were unsure whether they could execute the engineering, but it was also more exciting and promising. "We all know this is the one we want to do," said Jobs, pointing to the touchscreen. "So let's make it work." It was what he liked to call a bet-the-company moment, high risk and high reward if it succeeded.

A couple of members of the team argued for having a keyboard as well, given the popularity of the BlackBerry, but Jobs vetoed the idea. A physical keyboard would take away space from the screen, and it would not be as flexible and adaptable as a touchscreen keyboard. "A hardware keyboard seems like an easy solution, but it's constraining," he said. "Think of all the innovations we'd be able to adapt if we did the keyboard onscreen with software. Let's bet on it, and then we'll find a way to make it work." The result was a device that displays a numerical pad when you want to dial a phone number, a typewriter keyboard when you want to write, and whatever buttons you might need for each particular activity. And then they all disappear when you're watching a video. By having software replace hardware, the interface became fluid and flexible.

Jobs spent part of every day for six months helping to refine the display. "It was the most complex fun I've ever had," he recalled. "It was like being the one evolving the variations on 'Sgt. Pepper.'" A lot of features that seem simple now were the result of creative brainstorms. For example, the team worried about how to prevent the device from playing music or making a call accidentally when it was jangling in your pocket. Jobs was congenitally averse to having on-off switches, which he deemed "inelegant." The solution was "Swipe to Open," the simple and fun on-screen slider that activated the device when it had gone dormant. Another breakthrough was the sensor that figured out when you put the phone to your ear, so that your lobes didn't accidentally activate some function. And of course the icons came in his favorite shape, the primitive he made Bill Atkinson design into the software of the first Macintosh: rounded rectangles. In session after session, with Jobs immersed in every detail, the team members figured out ways to simplify what other phones made complicated. They added a big bar to guide you in putting calls on hold or making conference calls, found easy ways to navigate through email, and created icons you could scroll through horizontally to get to different apps — all of which were easier because they could be used visually on the screen rather than by using a keyboard built into the hardware.

GORILLA GLASS

Jobs became infatuated with different materials the way he did with certain foods. When he went back to Apple in 1997 and started work on the iMac, he had embraced what could be done with translucent and colored plastic. The next phase was metal. He and Ive replaced the curvy plastic PowerBook G3 with the sleek titanium PowerBook G4, which they redesigned two years later in aluminum, as if just to demonstrate how much they liked different metals. Then they did an iMac and an iPod Nano in anodized aluminum, which meant that the metal had been put in an acid bath and electrified so that its surface oxidized. Jobs was told it could not be done in the quantities they needed, so he had a factory built in China to handle it. Ive went there, during the SARS epidemic, to oversee the process. "I stayed for three months in a dormitory to work

672

on the process," he recalled. "Ruby and others said it would be impossible, but I wanted to do it because Steve and I felt that the anodized aluminum had a real integrity to it."

Next was glass. "After we did metal, I looked at Jony and said that we had to master glass," said Jobs. For the Apple stores, they had created huge windowpanes and glass stairs. For the iPhone, the original plan was for it to have a plastic screen, like the iPod. But Jobs decided it would feel much more elegant and substantive if the screens were glass. So he set about finding a glass that would be strong and resistant to scratches.

The natural place to look was Asia, where the glass for the stores was being made. But Jobs's friend John Seeley Brown, who was on the board of Corning Glass in Upstate New York, told him that he should talk to that company's young and dynamic CEO, Wendell Weeks. So he dialed the main Corning switchboard number and asked to be put through to Weeks. He got an assistant, who offered to pass along the message. "No, I'm Steve Jobs," he replied. "Put me through." The assistant refused. Jobs called Brown and complained that he had been subjected to "typical East Coast bullshit." When Weeks heard that, he called the main Apple switchboard and asked to speak to Jobs. He was told to put his request in writing and send it in by fax. When Jobs was told what happened, he took a liking to Weeks and invited him to Cupertino.

Jobs described the type of glass Apple wanted for the iPhone, and Weeks told him that Corning had developed a chemical exchange process in the 1960s that led to what they dubbed "gorilla glass."

It was incredibly strong, but it had never found a market, so Corning quit making it. Jobs said he doubted it was good enough, and he started explaining to Weeks how glass was made. This amused Weeks, who of course knew more than Jobs about that topic. "Can you shut up," Weeks interjected, "and let me teach you some science?" Jobs was taken aback and fell silent. Weeks went to the whiteboard and gave a tutorial on the chemistry, which involved an ion-exchange process that produced a compression layer on the surface of the glass. This turned Jobs around, and he said he wanted as much gorilla glass as Corning could make within six months. "We don't have the capacity," Weeks replied. "None of our plants make the glass now."

"Don't be afraid," Jobs replied. This stunned Weeks, who was good-humored and confident but not used to Jobs's reality distortion field. He tried to explain that a false sense of confidence would not overcome engineering challenges, but that was a premise that Jobs had repeatedly shown he didn't accept. He stared at Weeks unblinking. "Yes, you can do it," he said. "Get your mind around it. You can do it."

As Weeks retold this story, he shook his head in astonishment. "We did it in under six months," he said. "We produced a glass that had never been made." Corning's facility in Harrisburg, Kentucky, which had been making LCD displays, was converted almost overnight to make gorilla glass full-time. "We put our best scientists and engineers on it, and we just made it work." In his airy office, Weeks has just one framed memento on display. It's a message Jobs sent the day the iPhone came out: "We couldn't have done it without you."

THE DESIGN

On many of his major projects, such as the first *Toy Story* and the Apple store, Jobs pressed "pause" as they neared completion and decided to make major revisions. That happened with the design of the iPhone as well. The initial design had the glass screen set into an aluminum case. One Monday morning Jobs went over to see Ive. "I didn't sleep last night," he said, "because I realized that I just don't love it." It was the most important product he had made since the first Macintosh, and it just didn't look right to him. Ive, to his dismay, instantly realized that Jobs was right. "I remember feeling absolutely embarrassed that he had to make the observation."

The problem was that the iPhone should have been all about the display, but in their current design the case competed with the display instead of getting out of the way. The whole device felt too masculine, task-driven, efficient. "Guys, you've killed yourselves over this design for the last nine months, but we're going to change it," Jobs told Ive's team. "We're all going to have to work nights and weekends, and if you want we can hand out some guns so you can kill us now." Instead of balking, the team agreed. "It was one of my proudest moments at Apple," Jobs recalled.

The new design ended up with just a thin stainless steel bezel that allowed the gorilla glass display to go right to the edge. Every part of the device seemed to defer to the screen. The new look was austere, yet also friendly. You could fondle it. It meant they had to redo the circuit boards, antenna, and processor placement inside, but Jobs ordered the change. "Other companies may have shipped," said Fadell, "but we pressed the reset button and started over."

One aspect of the design, which reflected not only Jobs's perfectionism but also his desire to control, was that the device was tightly sealed. The case could not be opened, even to change the battery. As with the original Macintosh in 1984, Jobs did not want people fiddling inside. In fact when Apple discovered in 2011 that third-party repair shops were opening up the iPhone 4, it replaced the tiny screws with a tamper-resistant Pentalobe screw that was impossible to open with a commercially available screwdriver. By not having a replaceable battery, it was possible to make the iPhone much thinner. For Jobs, thinner was always better. "He's always believed that thin is beautiful," said Tim Cook. "You can see that in all of the work. We have the thinnest notebook, the thinnest smartphone, and we made the iPad thin and then even thinner."

THE LAUNCH

When it came time to launch the iPhone, Jobs decided, as usual, to grant a magazine a special sneak preview. He called John Huey, the editor in chief of Time Inc., and began with his typical superlative: "This is the best thing we've ever done." He wanted to give *Time* the exclusive, "but there's nobody smart enough at *Time* to write it, so I'm going to give it to someone else." Huey introduced him to Lev Grossman, a savvy technology writer (and novelist) at *Time.* In his piece Grossman correctly noted that the iPhone did not really invent many new features, it just made these features a lot more usable. "But that's important. When our tools don't work, we tend to blame ourselves, for being too stupid or not reading the manual or having too-fat fingers. . . . When our tools are broken, we feel broken. And when somebody fixes one, we feel a tiny bit more whole."

For the unveiling at the January 2007 Macworld in San Francisco, Jobs invited back Andy Hertzfeld, Bill Atkinson, Steve Wozniak, and the 1984 Macintosh team, as he had done when he launched the iMac. In a career of dazzling product presentations, this may have been his best. "Every once in a while a revolutionary product comes along that changes everything," he began. He referred to two earlier examples: the original Macintosh, which "changed the whole computer industry," and the first iPod, which "changed the entire music industry." Then he carefully built up to the product he was about to launch: "Today, we're introducing three revolutionary products of this class. The first one is a widescreen iPod with touch controls. The second is a revolutionary mobile phone. And the third is a breakthrough Internet communications device." He repeated the list for emphasis, then asked, "Are you getting it? These are not three separate devices, this is one device, and we are calling it iPhone."

When the iPhone went on sale five months later, at the end of June 2007, Jobs and his wife walked to the Apple store in Palo Alto to take in the excitement. Since he often did that on the day new products went on sale, there were some fans hanging out in anticipation, and they greeted him as they would have Moses if he had walked in to buy the Bible. Among the faithful were Hertzfeld and Atkinson. "Bill stayed in line all night," Hertzfeld said. Jobs waved his arms and started laughing. "I sent him one," he said. Hertzfeld replied, "He needs six."

The iPhone was immediately dubbed "the Jesus Phone" by bloggers. But Apple's competitors emphasized that, at $500, it cost too much to be successful. "It's the most expensive phone in the world," Microsoft's Steve Ballmer said in a CNBC interview. "And it doesn't appeal to business customers because it doesn't have a keyboard." Once again Microsoft had underestimated Jobs's product. By the end of 2010, Apple had sold ninety million iPhones, and it reaped more than half of the total profits generated in the global cell phone market.

"Steve understands desire," said Alan Kay, the Xerox PARC pioneer who had envisioned a "Dynabook" tablet computer forty years earlier. Kay was good at making prophetic assessments, so Jobs asked him what he thought of the iPhone. "Make the screen five inches by eight inches, and you'll rule the world," Kay said. He did not know that the design of the iPhone had started with, and would someday lead to, ideas for a tablet computer that would fulfill — indeed exceed — his vision for the Dynabook.

CHAPTER THIRTY-SEVEN: ROUND TWO

THE CANCER RECURS

THE BATTLES OF 2008

By the beginning of 2008 it was clear to Jobs and his doctors that his cancer was spreading. When they had taken out his pancreatic tumors in 2004, he had the cancer genome partially sequenced. That helped his doctors determine which pathways were broken, and they were treating him with targeted therapies that they thought were most likely to work.

He was also being treated for pain, usually with morphine-based analgesics. One day in February 2008 when Powell's close friend Kathryn Smith was staying with them in Palo Alto, she and Jobs took a walk. "He told me that when he feels really bad, he just concentrates on the pain, goes into the pain, and that seems to dissipate it," she recalled. That wasn't exactly true, however. When Jobs was in pain, he let everyone around him know it.

There was another health issue that became increasingly problematic, one that medical researchers didn't focus on as rigorously as they did cancer or pain. He was having eating problems and losing weight. Partly this was because he had lost much of his pancreas, which produces the enzymes needed to digest protein and other nutrients. It was also because both the cancer and the morphine reduced his appetite. And then there was the psychological component, which the doctors barely knew how to address: Since his early teens, he had indulged his weird obsession with extremely restrictive diets and fasts.

Even after he married and had children, he retained his dubious eating habits. He would spend weeks eating the same thing — carrot salad with lemon, or just apples — and then suddenly spurn that food and declare that he had stopped eating it. He would go on fasts, just as he did as a teenager, and he became sanctimonious as he lectured others at the table on the virtues of whatever eating regimen he was following. Powell had been a vegan when they were first married, but after her husband's operation she began to diversify their family meals with fish and other proteins. Their son, Reed, who had been a vegetarian, became a "hearty omnivore." They knew it was important for his father to get diverse sources of protein.

The family hired a gentle and versatile cook, Bryar Brown, who once worked for Alice Waters at Chez Panisse. He came each afternoon and made a panoply of healthy offerings for dinner, which used the herbs and vegetables that Powell grew in their garden. When Jobs expressed any whim — carrot salad, pasta with basil, lemongrass soup — Brown would quietly and patiently find a way to make it. Jobs had always been an extremely opinionated eater, with a tendency to instantly judge any food as either fantastic or terrible. He could taste two avocados that most mortals would find indistinguishable, and declare that one was the best avocado ever grown and the other inedible.

Beginning in early 2008 Jobs's eating disorders got worse. On some nights he would stare at the floor and ignore all of the dishes set out on the long kitchen table. When others were halfway through their meal, he would abruptly get up and leave, saying nothing. It was stressful for his family. They watched him lose forty pounds during the spring of 2008.

His health problems became public again in March 2008, when *Fortune* published a piece called "The Trouble with Steve Jobs." It revealed that he had tried to treat his cancer with diets for nine months and also investigated his involvement in the backdating of Apple stock options. As the story was being prepared, Jobs invited — summoned — *Fortune*'s managing editor Andy Serwer to Cupertino to pressure him to spike it. He leaned into Serwer's face and asked, "So, you've uncovered the fact that I'm an asshole. Why is that news?" Jobs made the same rather self-aware argument when he called Serwer's boss at Time Inc., John Huey, from a satellite phone he brought to Hawaii's Kona Village. He offered to convene a panel of fellow CEOs and be part of a discussion about what health issues are proper to disclose, but only if *Fortune* killed its piece. The magazine didn't.

When Jobs introduced the iPhone 3G in June 2008, he was so thin that it overshadowed the product announcement. In *Esquire* Tom Junod described the "withered" figure onstage as being "gaunt as a pirate, dressed in what had heretofore been the vestments of his invulnerability." Apple released a statement saying, untruthfully, that his weight loss was the result of "a common bug." The following month, as questions persisted, the company released another statement saying that Jobs's health was "a private matter."

Joe Nocera of the *New York Times* wrote a column denouncing the handling of Jobs's health issues. "Apple simply can't be trusted to tell the truth about its chief executive," he wrote in late July. "Under Mr. Jobs, Apple has created a culture of secrecy that has served it well in many ways — the speculation over which products Apple will unveil at the annual Macworld conference has been one of the company's best marketing tools. But that same culture poisons its corporate governance." As he was writing the column and getting the standard "a private matter" comment from all at Apple, he got an unexpected call from Jobs himself. "This is Steve Jobs," he began. "You think I'm an arrogant asshole who thinks he's above the law, and I think you're a slime bucket who gets most of his facts wrong." After that rather arresting opening, Jobs offered up some information about his health, but only if Nocera would keep it off the record. Nocera honored the request, but he was able to report that, while Jobs's health problems amounted to more than a common bug, "they weren't life-threatening and he doesn't have a recurrence of cancer." Jobs had given Nocera more information than he was willing to give his own board and shareholders, but it was not the full truth.

Partly due to concern about Jobs's weight loss, Apple's stock price drifted from $188 at the beginning of June 2008 down to $156 at the end of

July. Matters were not helped in late August when *Bloomberg News* mistakenly released its prepackaged obituary of Jobs, which ended up on Gawker. Jobs was able to roll out Mark Twain's famous quip a few days later at his annual music event. "Reports of my death are greatly exaggerated," he said, as he launched a line of new iPods. But his gaunt appearance was not reassuring. By early October the stock price had sunk to $97.

That month Doug Morris of Universal Music was scheduled to meet with Jobs at Apple. Instead Jobs invited him to his house. Morris was surprised to see him so ill and in pain. Morris was about to be honored at a gala in Los Angeles for City of Hope, which raised money to fight cancer, and he wanted Jobs to be there. Charitable events were something Jobs avoided, but he decided to do it, both for Morris and for the cause. At the event, held in a big tent on Santa Monica beach, Morris told the two thousand guests that Jobs was giving the music industry a new lease on life. The performances — by Stevie Nicks, Lionel Richie, Erykah Badu, and Akon — went on past midnight, and Jobs had severe chills. Jimmy Iovine gave him a hooded sweatshirt to wear, and he kept the hood over his head all evening. "He was so sick, so cold, so thin," Morris recalled.

Fortune's veteran technology writer Brent Schlender was leaving the magazine that December, and his swan song was to be a joint interview with Jobs, Bill Gates, Andy Grove, and Michael Dell. It had been hard to organize, and just a few days before it was to happen, Jobs called to back out. "If they ask why, just tell them I'm an asshole," he said. Gates was annoyed, then discovered what the health situation was. "Of course, he had a very, very good reason," said Gates. "He just didn't want to say." That became more apparent when Apple announced on December 16 that Jobs was canceling his scheduled appearance at the January Macworld, the forum he had used for big product launches for the past eleven years.

The blogosphere erupted with speculation about his health, much of which had the odious smell of truth. Jobs was furious and felt violated. He was also annoyed that Apple wasn't being more active in pushing back. So on January 5, 2009, he wrote and released a misleading open letter. He claimed that he was skipping Macworld because he wanted to spend more time with his family. "As many of you know, I have been losing weight throughout 2008," he added. "My doctors think they have found the cause — a hormone imbalance that has been robbing me of the proteins my body needs to be healthy. Sophisticated blood tests have confirmed this diagnosis. The remedy for this nutritional problem is relatively simple."

There was a kernel of truth to this, albeit a small one. One of the hormones created by the pancreas is glucagon, which is the flip side of insulin. Glucagon causes your liver to release blood sugar. Jobs's tumor had metastasized into his liver and was wreaking havoc. In effect, his body was devouring itself, so his doctors gave him drugs to try to lower the glucagon level. He did have a hormone imbalance, but it was because his cancer had spread into his liver. He was in personal denial about this, and he also wanted to be in public denial. Unfortunately that was legally problematic, because he ran a publicly traded company. But Jobs was furious about the way the blogosphere was treating him, and he wanted to strike back.

He was very sick at this point, despite his upbeat statement, and also in excruciating pain. He had undertaken another round of cancer drug therapy, and it had grueling side effects. His skin started drying out and cracking. In his quest for alternative approaches, he flew to Basel, Switzerland, to try an experimental hormone-delivered radiotherapy. He also underwent an experimental treatment developed in Rotterdam known as peptide receptor radionuclide therapy.

After a week filled with increasingly insistent legal advice, Jobs finally agreed to go on medical leave. He made the announcement on January 14, 2009, in another open letter to the Apple staff. At first he blamed the decision on the prying of bloggers and the press. "Unfortunately, the curiosity over my personal health continues to be a distraction not only for me and my family, but everyone else at Apple," he said. But then he admitted that the remedy for his "hormone imbalance" was not as simple as he had claimed. "During the past week I have learned that my health-related issues are more complex than I originally thought." Tim Cook would again take over daily operations, but Jobs said that he would remain CEO, continue to be involved in major decisions, and be back by June.

Jobs had been consulting with Bill Campbell and Art Levinson, who were juggling the dual roles of being his personal health advisors and also the

685

co-lead directors of the company. But the rest of the board had not been as fully informed, and the shareholders had initially been misinformed. That raised some legal issues, and the SEC opened an investigation into whether the company had withheld "material information" from shareholders. It would constitute security fraud, a felony, if the company had allowed the dissemination of false information or withheld true information that was relevant to the company's financial prospects. Because Jobs and his magic were so closely identified with Apple's comeback, his health seemed to meet this standard. But it was a murky area of the law; the privacy rights of the CEO had to be weighed. This balance was particularly difficult in the case of Jobs, who both valued his privacy and embodied his company more than most CEOs. He did not make the task easier. He became very emotional, both ranting and crying at times, when railing against anyone who suggested that he should be less secretive.

Campbell treasured his friendship with Jobs, and he didn't want to have any fiduciary duty to violate his privacy, so he offered to step down as a director. "The privacy side is so important to me," he later said. "He's been my friend for about a million years." The lawyers eventually determined that Campbell didn't need to resign from the board but that he should step aside as co-lead director. He was replaced in that role by Andrea Jung of Avon. The SEC investigation ended up going nowhere, and the board circled the wagons to protect Jobs from calls that he release more information. "The press wanted us to blurt out more personal details," recalled Al Gore. "It was really up to Steve to go beyond what the law requires, but he was adamant that he didn't want his privacy invaded. His wishes should be respected." When I asked Gore whether the board should have been more forthcoming at the beginning of 2009, when Jobs's health issues were far worse than shareholders were led to believe, he replied, "We hired outside counsel to do a review of what the law required and what the best practices were, and we handled it all by the book. I sound defensive, but the criticism really pissed me off."

One board member disagreed. Jerry York, the former CFO at Chrysler and IBM, did not say anything publicly, but he confided to a reporter at the *Wall Street Journal,* off the record, that he was "disgusted" when he learned that the company had concealed Jobs's health problems in late 2008. "Frankly, I wish I had resigned then." When York died in 2010, the *Journal* put his comments on the record. York had also provided off-the-record information to *Fortune,* which the magazine used when Jobs went on his third health leave, in 2011.

Some at Apple didn't believe the quotes attributed to York were accurate, since he had not officially raised objections at the time. But Bill Campbell knew that the reports rang true; York had complained to him in early 2009. "Jerry had a little more white wine than he should have late at night, and he would call at two or three in the morning and say, 'What the fuck, I'm not buying that shit about his health, we've got to make sure.' And then I'd call him the next morning and he'd say, 'Oh fine, no problem.' So on some of those evenings, I'm sure he got raggy and talked to reporters."

MEMPHIS

The head of Jobs's oncology team was Stanford University's George Fisher, a leading researcher on gastrointestinal and colorectal cancers. He had been warning Jobs for months that he might have to consider a liver transplant, but that was the type of information that Jobs resisted processing. Powell was glad that Fisher kept raising the possibility, because she knew it would take repeated proddings to get her husband to consider the idea.

He finally became convinced in January 2009, just after he claimed his "hormonal imbalance" could be treated easily. But there was a problem. He was put on the wait list for a liver transplant in California, but it became clear he would never get one there in time. The number of available donors with his blood type was small. Also, the metrics used by the United Network for Organ Sharing, which establishes policies in the United States, favored those suffering from cirrhosis and hepatitis over cancer patients.

There is no legal way for a patient, even one as wealthy as Jobs, to jump the queue, and he didn't. Recipients are chosen based on their MELD score (Model for End-Stage Liver Disease), which uses lab tests of hormone levels to determine how urgently a transplant is needed, and on the length of time they have been waiting. Every donation is closely audited, data are available on public websites (optn.transplant.hrsa.gov/), and you can monitor your status on the wait list at any time.

Powell became the troller of the organ-donation websites, checking in every night to see how many were on the wait lists, what their MELD scores were, and how long they had been on. "You can do the math, which I did, and it would have been way past June before he got a liver in California, and the doctors felt that his liver would give out in about April," she recalled. So she started asking questions and discovered that it was permissible to be on the list in two different states at the same time, which is something that about 3% of potential recipients do. Such multiple listing is not discouraged by policy, even though critics say it favors the rich, but it is difficult. There were two major requirements: The potential recipient had to be able to get to the chosen hospital within eight hours, which Jobs could do thanks to his plane, and the doctors from that hospital had to evaluate the patient in person before adding him or her to the list.

George Riley, the San Francisco lawyer who often served as Apple's outside counsel, was a caring Tennessee gentleman, and he had become close to Jobs. His parents had both been doctors at Methodist University Hospital in Memphis, he was born there, and he was a friend of James Eason, who ran the transplant institute there. Eason's unit was one of the best and busiest in the nation; in 2008 he and his team did 121 liver transplants. He had no problem allowing people from elsewhere to multiple-list in Memphis. "It's not gaming the system," he said. "It's people choosing where they want their health care. Some people would leave Tennessee to go to California or somewhere else to seek treatment. Now we have people coming from California to Tennessee." Riley arranged for Eason to fly to Palo Alto and conduct the required evaluation there.

By late February 2009 Jobs had secured a place on the Tennessee list (as well as the one in California), and the nervous waiting began. He was declining rapidly by the first week in March, and the waiting time was projected to be twenty-one days. "It was dreadful," Powell recalled. "It didn't look like we would make it in time." Every day became more excruciating. He moved up to third on the list by mid-March, then second, and finally first. But then days went by. The awful reality was that upcoming events like St. Patrick's Day and March Madness (Memphis was in the 2009 tournament and was a regional site) offered a greater likelihood of getting a donor because the drinking causes a spike in car accidents.

Indeed, on the weekend of March 21, 2009, a young man in his midtwenties was killed in a car crash, and his organs were made available. Jobs and his wife flew to Memphis, where they landed just before 4 a.m. and were met by Eason. A car was waiting on the tarmac, and everything was staged so that the admitting paperwork was done as they rushed to the hospital.

The transplant was a success, but not reassuring. When the doctors took out his liver, they found spots on the peritoneum, the thin membrane that surrounds internal organs. In addition, there were tumors throughout the liver, which meant it was likely that the cancer had migrated elsewhere as well. It had apparently mutated and grown quickly.

They took samples and did more genetic mapping.

A few days later they needed to perform another procedure. Jobs insisted against all advice they not pump out his stomach, and when they sedated him, he aspirated some of the contents into his lungs and developed pneumonia. At that point they thought he might die. As he described it later:

> I almost died because in this routine procedure they blew it. Laurene was there and they flew my children in, because they did not think I would make it through the night. Reed was looking at colleges with one of Laurene's brothers. We had a private plane pick him up near Dartmouth and tell them what was going on. A plane also picked up the girls. They thought it might be the last chance they had to see me conscious. But I made it.

Powell took charge of overseeing the treatment, staying in the hospital room all day and watching each of the monitors vigilantly. "Laurene was a beautiful tiger protecting him," recalled Jony Ive, who came as soon as Jobs could receive visitors. Her mother and three brothers came down at various times to keep her company. Jobs's sister Mona Simpson also hovered protectively. She and George Riley were the only people Jobs would allow to fill in for Powell at his bedside. "Laurene's family helped us take care of the kids — her mom and brothers were great," Jobs later said. "I was very fragile and not cooperative. But an experience like that binds you together in a deep way."

Powell came every day at 7 a.m. and gathered

the relevant data, which she put on a spreadsheet. "It was very complicated because there were a lot of different things going on," she recalled. When James Eason and his team of doctors arrived at 9 a.m., she would have a meeting with them to coordinate all aspects of Jobs's treatment. At 9 p.m., before she left, she would prepare a report on how each of the vital signs and other measurements were trending, along with a set of questions she wanted answered the next day. "It allowed me to engage my brain and stay focused," she recalled.

Eason did what no one at Stanford had fully done: take charge of all aspects of the medical care. Since he ran the facility, he could coordinate the transplant recovery, cancer tests, pain treatments, nutrition, rehabilitation, and nursing. He would even stop at the convenience store to get the energy drinks Jobs liked.

Two of the nurses were from tiny towns in Mississippi, and they became Jobs's favorites. They were solid family women and not intimidated by him. Eason arranged for them to be assigned only to Jobs. "To manage Steve, you have to be persistent," recalled Tim Cook. "Eason managed Steve and forced him to do things that no one else could, things that were good for him that may not have been pleasant."

Despite all the coddling, Jobs at times almost went crazy. He chafed at not being in control, and he sometimes hallucinated or became angry. Even when he was barely conscious, his strong personality came through. At one point the pulmonologist tried to put a mask over his face when he was deeply sedated. Jobs ripped it off and mumbled

692

that he hated the design and refused to wear it. Though barely able to speak, he ordered them to bring five different options for the mask and he would pick a design he liked. The doctors looked at Powell, puzzled. She was finally able to distract him so they could put on the mask. He also hated the oxygen monitor they put on his finger. He told them it was ugly and too complex. He suggested ways it could be designed more simply. "He was very attuned to every nuance of the environment and objects around him, and that drained him," Powell recalled.

One day, when he was still floating in and out of consciousness, Powell's close friend Kathryn Smith came to visit. Her relationship with Jobs had not always been the best, but Powell insisted that she come by the bedside. He motioned her over, signaled for a pad and pen, and wrote, "I want my iPhone." Smith took it off the dresser and brought it to him. Taking her hand, he showed her the "swipe to open" function and made her play with the menus.

Jobs's relationship with Lisa Brennan-Jobs, his daughter with Chrisann, had frayed. She had graduated from Harvard, moved to New York City, and rarely communicated with her father. But she flew down to Memphis twice, and he appreciated it. "It meant a lot to me that she would do that," he recalled. Unfortunately he didn't tell her at the time. Many of the people around Jobs found Lisa could be as demanding as her father, but Powell welcomed her and tried to get her involved. It was a relationship she wanted to restore.

As Jobs got better, much of his feisty personality returned. He still had his bile ducts. "When he started to recover, he passed quickly through the phase of gratitude, and went right back into the mode of being grumpy and in charge," Kat Smith recalled. "We were all wondering if he was going to come out of this with a kinder perspective, but he didn't."

He also remained a finicky eater, which was more of a problem than ever. He would eat only fruit smoothies, and he would demand that seven or eight of them be lined up so he could find an option that might satisfy him. He would touch the spoon to his mouth for a tiny taste and pronounce, "That's no good. That one's no good either." Finally Eason pushed back. "You know, this isn't a matter of taste," he lectured. "Stop thinking of this as food. Start thinking of it as medicine."

Jobs's mood buoyed when he was able to have visitors from Apple. Tim Cook came down regularly and filled him in on the progress of new products. "You could see him brighten every time the talk turned to Apple," Cook said. "It was like the light turned on." He loved the company deeply, and he seemed to live for the prospect of returning. Details would energize him. When Cook described a new model of the iPhone, Jobs spent the next hour discussing not only what to call it — they agreed on iPhone 3GS — but also the size and font of the "GS," including whether the letters should be capitalized (yes) and italicized (no).

One day Riley arranged a surprise after-hours visit to Sun Studio, the redbrick shrine where Elvis, Johnny Cash, B.B. King, and many other rock-and-roll pioneers recorded. They were given

694

a private tour and a history lecture by one of the young staffers, who sat with Jobs on the cigarette-scarred bench that Jerry Lee Lewis used. Jobs was arguably the most influential person in the music industry at the time, but the kid didn't recognize him in his emaciated state. As they were leaving, Jobs told Riley, "That kid was really smart. We should hire him for iTunes." So Riley called Eddy Cue, who flew the boy out to California for an interview and ended up hiring him to help build the early R&B and rock-and-roll sections of iTunes. When Riley went back to see his friends at Sun Studio later, they said that it proved, as their slogan said, that your dreams can still come true at Sun Studio.

RETURN

At the end of May 2009 Jobs flew back from Memphis on his jet with his wife and sister. They were met at the San Jose airfield by Tim Cook and Jony Ive, who came aboard as soon as the plane landed. "You could see in his eyes his excitement at being back," Cook recalled. "He had fight in him and was raring to go." Powell pulled out a bottle of sparkling apple cider and toasted her husband, and everyone embraced.

Ive was emotionally drained. He drove to Jobs's house from the airport and told him how hard it had been to keep things going while he was away. He also complained about the stories saying that Apple's innovation depended on Jobs and would disappear if he didn't return. "I'm really hurt," Ive told him. He felt "devastated," he said, and under-appreciated.

Jobs was likewise in a dark mental state after his return to Palo Alto. He was coming to grips with the thought that he might *not* be indispensable to the company. Apple stock had fared well while he was away, going from $82 when he announced his leave in January 2009 to $140 when he returned at the end of May. On one conference call with analysts shortly after Jobs went on leave, Cook departed from his unemotional style to give a rousing declaration of why Apple would continue to soar even with Jobs absent:

> We believe that we are on the face of the earth to make great products, and that's not changing. We are constantly focusing on innovating. We believe in the simple not the complex. We believe that we need to own and control the primary technologies behind the products that we make, and participate only in markets where we can make a significant contribution. We believe in saying no to thousands of projects, so that we can really focus on the few that are truly important and meaningful to us. We believe in deep collaboration and cross-pollination of our groups, which allow us to innovate in a way that others cannot. And frankly, we don't settle for anything less than excellence in every group in the company, and we have the self-honesty to admit when we're wrong and the courage to change. And I think, regardless of who is in what job, those values are so embedded in this company that Apple will do extremely well.

It sounded like something Jobs would say (and had said), but the press dubbed it "the Cook doctrine."

Jobs was rankled and deeply depressed, especially about the last line. He didn't know whether to be proud or hurt that it might be true. There was talk that he might step aside and become chairman rather than CEO. That made him all the more motivated to get out of his bed, overcome the pain, and start taking his restorative long walks again.

A board meeting was scheduled a few days after he returned, and Jobs surprised everyone by making an appearance. He ambled in and was able to stay for most of the meeting. By early June he was holding daily meetings at his house, and by the end of the month he was back at work.

Would he now, after facing death, be more mellow? His colleagues quickly got an answer. On his first day back, he startled his top team by throwing a series of tantrums. He ripped apart people he had not seen for six months, tore up some marketing plans, and chewed out a couple of people whose work he found shoddy. But what was truly telling was the pronouncement he made to a couple of friends late that afternoon. "I had the greatest time being back today," he said. "I can't believe how creative I'm feeling, and how the whole team is." Tim Cook took it in stride. "I've never seen Steve hold back from expressing his view or passion," he later said. "But that was good."

Friends noted that Jobs had retained his feistiness. During his recuperation he signed up for Comcast's high-definition cable service, and one day he called Brian Roberts, who ran the company. "I thought he was calling to say something nice about it," Roberts recalled. "Instead, he told me 'It sucks.' " But Andy Hertzfeld noticed that, beneath the gruffness, Jobs had become more honest. "Before, if you asked Steve for a favor, he might do the exact opposite," Hertzfeld said. "That was the perversity in his nature. Now he actually tries to be helpful."

His public return came on September 9, when he took the stage at the company's regular fall music event. He got a standing ovation that lasted almost a minute, then he opened on an unusually personal note by mentioning that he was the recipient of a liver donation. "I wouldn't be here without such generosity," he said, "so I hope all of us can be as generous and elect to become organ donors." After a moment of exultation — "I'm vertical, I'm back at Apple, and I'm loving every day of it" — he unveiled the new line of iPod Nanos, with video cameras, in nine different colors of anodized aluminum.

By the beginning of 2010 he had recovered most of his strength, and he threw himself back into work for what would be one of his, and Apple's, most productive years. He had hit two consecutive home runs since launching Apple's digital hub strategy: the iPod and the iPhone. Now he was going to swing for another.

CHAPTER THIRTY-EIGHT: THE iPAD

INTO THE POST-PC ERA

YOU SAY YOU WANT A REVOLUTION

Back in 2002, Jobs had been annoyed by the Microsoft engineer who kept proselytizing about the tablet computer software he had developed, which allowed users to input information on the screen with a stylus or pen. A few manufacturers released tablet PCs that year using the software, but none made a dent in the universe. Jobs had been eager to show how it should be done right — no stylus! — but when he saw the multi-touch technology that Apple was developing, he had decided to use it first to make an iPhone.

In the meantime, the tablet idea was percolating within the Macintosh hardware group. "We have no plans to make a tablet," Jobs declared in an interview with Walt Mossberg in May 2003. "It turns out people want keyboards. Tablets appeal to rich guys with plenty of other PCs and devices already." Like his statement about having a "hormone imbalance," that was misleading; at most of his annual Top 100 retreats, the tablet was among the future projects discussed. "We showed the idea off at many of these retreats, because Steve never lost his desire to do a tablet," Phil Schiller recalled.

The tablet project got a boost in 2007 when Jobs was considering ideas for a low-cost netbook computer. At an executive team brainstorming session one Monday, Ive asked why it needed a keyboard hinged to the screen; that was expensive and bulky. Put the keyboard on the screen using a multi-touch interface, he suggested. Jobs agreed. So the resources were directed to revving up the tablet project rather than designing a netbook.

The process began with Jobs and Ive figuring out the right screen size. They had twenty models made — all rounded rectangles, of course — in slightly varying sizes and aspect ratios. Ive laid them out on a table in the design studio, and in the afternoon they would lift the velvet cloth hiding them and play with them. "That's how we nailed what the screen size was," Ive said.

As usual Jobs pushed for the purest possible simplicity. That required determining what was the core essence of the device. The answer: the display screen. So the guiding principle was that everything they did had to defer to the screen. "How do we get out of the way so there aren't a ton of features and buttons that distract from the display?" Ive asked. At every step, Jobs pushed to remove and simplify.

At one point Jobs looked at the model and was slightly dissatisfied. It didn't feel casual and friendly enough, so that you would naturally scoop it up and whisk it away. Ive put his finger, so to speak, on the problem: They needed to signal that you could grab it with one hand, on impulse. The bottom of the edge needed to be slightly rounded, so that you'd feel comfortable just scooping it up rather than lifting it carefully. That meant engineering had to design the necessary connection ports and buttons in a simple lip that was thin enough to wash away gently underneath.

If you had been paying attention to patent filings, you would have noticed the one numbered D504889 that Apple applied for in March 2004 and was issued fourteen months later. Among the inventors listed were Jobs and Ive. The application carried sketches of a rectangular electronic tablet with rounded edges, which looked just the way the iPad turned out, including one of a man holding it casually in his left hand while using his right index finger to touch the screen.

Since the Macintosh computers were now using Intel chips, Jobs initially planned to use in the iPad the low-voltage Atom chip that Intel was developing. Paul Otellini, Intel's CEO, was pushing hard to work together on a design, and Jobs's inclination was to trust him. His company was making the fastest processors in the world. But Intel was used to making processors for machines that plugged into a wall, not ones that had to preserve battery life. So Tony Fadell argued strongly for something based on the ARM architecture, which was simpler and used less power. Apple had been an early partner with ARM, and chips using its architecture were in the original iPhone. Fadell gathered support from other engineers and proved that it was possible to confront Jobs and turn him around. "Wrong, wrong, wrong!" Fadell shouted at one meeting when Jobs insisted it was best to trust Intel to make a good mobile chip. Fadell even put his Apple badge on the table, threatening to resign.

Eventually Jobs relented. "I hear you," he said. "I'm not going to go against my best guys." In fact he went to the other extreme. Apple licensed the ARM architecture, but it also bought a 150-person microprocessor design firm in Palo Alto, called P.A. Semi, and had it create a custom system-on-a-chip, called the A4, which was based on the ARM architecture and manufactured in South Korea by Samsung. As Jobs recalled:

At the high-performance end, Intel is the best. They build the fastest chip, if you don't care about power and cost. But they build just the processor on one chip, so it takes a lot of other parts. Our A4 has the processor and the graphics, mobile operating system, and memory control all in the chip. We tried to help Intel, but they don't listen much. We've been telling them for years that their graphics suck. Every quarter we schedule a meeting with me and our top three guys and Paul Otellini. At the beginning, we were doing wonderful things together. They wanted this big joint project to do chips for future iPhones. There were two reasons we didn't go with them. One was that they are just really slow. They're like a steamship, not very flexible. We're used to going pretty fast. Second is that we just didn't want to teach them everything, which they could go and sell to our competitors.

According to Otellini, it would have made sense for the iPad to use Intel chips. The problem, he said, was that Apple and Intel couldn't agree on price. Also, they disagreed on who would control the design. It was another example of Jobs's desire, indeed compulsion, to control every aspect of a product, from the silicon to the flesh.

THE LAUNCH, JANUARY 2010

The usual excitement that Jobs was able to gin up for a product launch paled in comparison to the frenzy that built for the iPad unveiling on January 27, 2010, in San Francisco. The *Economist* put him on its cover robed, haloed, and holding what was dubbed "the Jesus Tablet." The *Wall Street Journal* struck a similarly exalted note: "The last time there was this much excitement about a tablet, it had some commandments written on it."

As if to underscore the historic nature of the launch, Jobs invited back many of the old-timers from his early Apple days. More poignantly, James Eason, who had performed his liver transplant the year before, and Jeffrey Norton, who had operated on his pancreas in 2004, were in the audience, sitting with his wife, his son, and Mona Simpson.

Jobs did his usual masterly job of putting a new device into context, as he had done for the iPhone three years earlier. This time he put up a screen that showed an iPhone and a laptop with a question mark in between. "The question is, is there room for something in the middle?" he asked. That "something" would have to be good at web browsing, email, photos, video, music, games, and ebooks. He drove a stake through the heart of the netbook concept. "Netbooks aren't better at anything!" he said. The invited guests and employees

703

cheered. "But we have something that is. We call it the iPad."

To underscore the casual nature of the iPad, Jobs ambled over to a comfortable leather chair and side table (actually, given his taste, it was a Le Corbusier chair and an Eero Saarinen table) and scooped one up. "It's so much more intimate than a laptop," he enthused. He proceeded to surf to the *New York Times* website, send an email to Scott Forstall and Phil Schiller ("Wow, we really are announcing the iPad"), flip through a photo album, use a calendar, zoom in on the Eiffel Tower on Google Maps, watch some video clips (*Star Trek* and Pixar's *Up*), show off the iBook shelf, and play a song (Bob Dylan's "Like a Rolling Stone," which he had played at the iPhone launch). "Isn't that awesome?" he asked.

With his final slide, Jobs emphasized one of the themes of his life, which was embodied by the iPad: a sign showing the corner of Technology Street and Liberal Arts Street. "The reason Apple can create products like the iPad is that we've always tried to be at the intersection of technology and liberal arts," he concluded. The iPad was the digital reincarnation of the *Whole Earth Catalog,* the place where creativity met tools for living.

For once, the initial reaction was not a Hallelujah Chorus. The iPad was not yet available (it would go on sale in April), and some who watched Jobs's demo were not quite sure what it was. An iPhone on steroids? "I haven't been this let down since Snooki hooked up with The Situation," wrote *Newsweek*'s Daniel Lyons (who moonlighted as "The Fake

Steve Jobs" in an online parody). Gizmodo ran a contributor's piece headlined "Eight Things That Suck about the iPad" (no multitasking, no cameras, no Flash . . .). Even the name came in for ridicule in the blogosphere, with snarky comments about feminine hygiene products and maxi pads. The hashtag "#iTampon" was the number-three trending topic on Twitter that day.

There was also the requisite dismissal from Bill Gates. "I still think that some mixture of voice, the pen and a real keyboard — in other words a netbook — will be the mainstream," he told Brent Schlender. "So, it's not like I sit there and feel the same way I did with the iPhone where I say, 'Oh my God, Microsoft didn't aim high enough.' It's a nice reader, but there's nothing on the iPad I look at and say, 'Oh, I wish Microsoft had done it.' " He continued to insist that the Microsoft approach of using a stylus for input would prevail. "I've been predicting a tablet with a stylus for many years," he told me. "I will eventually turn out to be right or be dead."

The night after his announcement, Jobs was annoyed and depressed. As we gathered in his kitchen for dinner, he paced around the table calling up emails and web pages on his iPhone.

I got about eight hundred email messages in the last twenty-four hours. Most of them are complaining. There's no USB cord! There's no this, no that. Some of them are like, "Fuck you, how can you do that?" I don't usually write people back, but I replied, "Your parents would be so proud of how you turned out." And some don't

705

like the iPad name, and on and on. I kind of got depressed today. It knocks you back a bit.

He did get one congratulatory call that day that he appreciated, from President Obama's chief of staff, Rahm Emanuel. But he noted at dinner that the president had not called him since taking office.

The public carping subsided when the iPad went on sale in April and people got their hands on it. Both *Time* and *Newsweek* put it on the cover. "The tough thing about writing about Apple products is that they come with a lot of hype wrapped around them," Lev Grossman wrote in *Time.* "The other tough thing about writing about Apple products is that sometimes the hype is true." His main reservation, a substantive one, was "that while it's a lovely device for consuming content, it doesn't do much to facilitate its creation." Computers, especially the Macintosh, had become tools that allowed people to make music, videos, websites, and blogs, which could be posted for the world to see. "The iPad shifts the emphasis from creating content to merely absorbing and manipulating it. It mutes you, turns you back into a passive consumer of other people's masterpieces." It was a criticism Jobs took to heart. He set about making sure that the next version of the iPad would emphasize ways to facilitate artistic creation by the user.

Newsweek's cover line was "What's So Great about the iPad? Everything." Daniel Lyons, who had zapped it with his "Snooki" comment at the launch, revised his opinion. "My first thought, as

I watched Jobs run through his demo, was that it seemed like no big deal," he wrote. "It's a bigger version of the iPod Touch, right? Then I got a chance to use an iPad, and it hit me: I want one." Lyons, like others, realized that this was Jobs's pet project, and it embodied all that he stood for. "He has an uncanny ability to cook up gadgets that we didn't know we needed, but then suddenly can't live without," he wrote. "A closed system may be the only way to deliver the kind of techno-Zen experience that Apple has become known for."

Most of the debate over the iPad centered on the issue of whether its closed end-to-end integration was brilliant or doomed. Google was starting to play a role similar to the one Microsoft had played in the 1980s, offering a mobile platform, Android, that was open and could be used by all hardware makers. *Fortune* staged a debate on this issue in its pages. "There's no excuse to be closed," wrote Michael Copeland. But his colleague Jon Fortt rebutted, "Closed systems get a bad rap, but they work beautifully and users benefit. Probably no one in tech has proved this more convincingly than Steve Jobs. By bundling hardware, software, and services, and controlling them tightly, Apple is consistently able to get the jump on its rivals and roll out polished products." They agreed that the iPad would be the clearest test of this question since the original Macintosh. "Apple has taken its control-freak rep to a whole new level with the A4 chip that powers the thing," wrote Fortt. "Cupertino now has absolute say over the silicon, device, operating system, App Store, and payment system."

Jobs went to the Apple store in Palo Alto shortly

before noon on April 5, the day the iPad went on sale. Daniel Kottke — his acid-dropping soul mate from Reed and the early days at Apple, who no longer harbored a grudge for not getting founders' stock options — made a point of being there. "It had been fifteen years, and I wanted to see him again," Kottke recounted. "I grabbed him and told him I was going to use the iPad for my song lyrics. He was in a great mood and we had a nice chat after all these years." Powell and their youngest child, Eve, watched from a corner of the store.

Wozniak, who had once been a proponent of making hardware and software as open as possible, continued to revise that opinion. As he often did, he stayed up all night with the enthusiasts waiting in line for the store to open. This time he was at San Jose's Valley Fair Mall, riding a Segway. A reporter asked him about the closed nature of Apple's ecosystem. "Apple gets you into their playpen and keeps you there, but there are some advantages to that," he replied. "I like open systems, but I'm a hacker. But most people want things that are easy to use. Steve's genius is that he knows how to make things simple, and that sometimes requires controlling everything."

The question "What's on your iPad?" replaced "What's on your iPod?" Even President Obama's staffers, who embraced the iPad as a mark of their tech hipness, played the game. Economic Advisor Larry Summers had the Bloomberg financial information app, Scrabble, and *The Federalist Papers.* Chief of Staff Rahm Emanuel had a slew of newspapers, Communications Advisor Bill Burton had *Vanity Fair* and one entire season of the television series *Lost,* and Political Director David Axelrod had Major League Baseball and NPR.

Jobs was stirred by a story, which he forwarded to me, by Michael Noer on *Forbes.com.* Noer was reading a science fiction novel on his iPad while staying at a dairy farm in a rural area north of Bogotá, Colombia, when a poor six-year-old boy who cleaned the stables came up to him. Curious, Noer handed him the device. With no instruction, and never having seen a computer before, the boy started using it intuitively. He began swiping the screen, launching apps, playing a pinball game. "Steve Jobs has designed a powerful computer that an illiterate six-year-old can use without instruction," Noer wrote. "If that isn't magical, I don't know what is."

In less than a month Apple sold one million iPads. That was twice as fast as it took the iPhone to reach that mark. By March 2011, nine months after its release, fifteen million had been sold. By some measures it became the most successful consumer product launch in history.

ADVERTISING

Jobs was not happy with the original ads for the iPad. As usual, he threw himself into the marketing, working with James Vincent and Duncan Milner at the ad agency (now called TBWA/Media Arts Lab), with Lee Clow advising from a semiretired perch. The commercial they first produced was a gentle scene of a guy in faded jeans and sweatshirt reclining in a chair, looking at email, a photo album, the *New York Times,* books, and video on an iPad propped on his lap. There were no words, just the background beat of "There Goes My Love" by the Blue Van. "After he approved it, Steve decided he hated it," Vincent recalled. "He thought it looked like a Pottery Barn commercial." Jobs later told me:

> It had been easy to explain what the iPod was — a thousand songs in your pocket — which allowed us to move quickly to the iconic silhouette ads. But it was hard to explain what an iPad was. We didn't want to show it as a computer, and yet we didn't want to make it so soft that it looked like a cute TV. The first set of ads showed we didn't know what we were doing. They had a cashmere and Hush Puppies feel to them.

James Vincent had not taken a break in months. So when the iPad finally went on sale and the ads started airing, he drove with his family to the Coachella Music Festival in Palm Springs, which featured some of his favorite bands, including Muse, Faith No More, and Devo. Soon after he arrived, Jobs called. "Your commercials suck," he said. "The iPad is revolutionizing the world, and we need something big. You've given me small shit."

"Well, what do you want?" Vincent shot back. "You've not been able to tell me what you want."

"I don't know," Jobs said. "You have to bring me something new. Nothing you've shown me is even close."

Vincent argued back and suddenly Jobs went ballistic. "He just started screaming at me," Vincent recalled. Vincent could be volatile himself, and the volleys escalated.

When Vincent shouted, "You've got to tell me what you want," Jobs shot back, "You've got to show me some stuff, and I'll know it when I see it."

"Oh, great, let me write that on my brief for my creative people: I'll know it when I see it."

Vincent got so frustrated that he slammed his fist into the wall of the house he was renting and put a large dent in it. When he finally went outside to his family, sitting by the pool, they looked at him nervously. "Are you okay?" his wife finally asked.

It took Vincent and his team two weeks to come up with an array of new options, and he asked to present them at Jobs's house rather than the office, hoping that it would be a more relaxed environment. Laying storyboards on the coffee table, he and Milner offered twelve approaches. One was inspirational and stirring. Another tried humor, with Michael Cera, the comic actor, wandering through a fake house making funny comments about the way people could use iPads. Others featured the iPad with celebrities, or set starkly on a white background, or starring in a little sitcom, or in a straightforward product demonstration.

After mulling over the options, Jobs realized what he wanted. Not humor, nor a celebrity, nor a demo. "It's got to make a statement," he said. "It needs to be a manifesto. This is big." He had announced that the iPad would change the world, and he wanted a campaign that reinforced that declaration. Other companies would come out with copycat tablets in a year or so, he said, and he wanted people to remember that the iPad was the real thing. "We need ads that stand up and declare

711

what we have done."

He abruptly got out of his chair, looking a bit weak but smiling. "I've got to go have a massage now," he said. "Get to work."

So Vincent and Milner, along with the copywriter Eric Grunbaum, began crafting what they dubbed "The Manifesto." It would be fast-paced, with vibrant pictures and a thumping beat, and it would proclaim that the iPad was revolutionary. The music they chose was Karen O's pounding refrain from the Yeah Yeah Yeahs' "Gold Lion." As the iPad was shown doing magical things, a strong voice declared, "iPad is thin. iPad is beautiful. . . . It's crazy powerful. It's magical. . . . It's video, photos. More books than you could read in a lifetime. It's already a revolution, and it's only just begun."

Once the Manifesto ads had run their course, the team again tried something softer, shot as day-in-the-life documentaries by the young filmmaker Jessica Sanders. Jobs liked them — for a little while. Then he turned against them for the same reason he had reacted against the original Pottery Barn–style ads. "Dammit," he shouted, "they look like a Visa commercial, typical ad agency stuff."

He had been asking for ads that were different and new, but eventually he realized he did not want to stray from what he considered the Apple voice. For him, that voice had a distinctive set of qualities: simple, declarative, clean. "We went down that lifestyle path, and it seemed to be growing on Steve, and suddenly he said, 'I hate that stuff, it's not Apple,'" recalled Lee Clow. "He told us to get back to the Apple voice. It's a very simple, honest voice." And so they went back to a clean white background, with just a close-up showing off all the things that "iPad is . . ." and could do.

APPS

The iPad commercials were not about the device, but about what you could do with it. Indeed its success came not just from the beauty of the hardware but from the applications, known as apps, that allowed you to indulge in all sorts of delightful activities. There were thousands — and soon hundreds of thousands — of apps that you could download for free or for a few dollars. You could sling angry birds with the swipe of your finger, track your stocks, watch movies, read books and magazines, catch up on the news, play games, and waste glorious amounts of time. Once again the integration of the hardware, software, and store made it easy. But the apps also allowed the platform to be sort of open, in a very controlled way, to outside developers who wanted to create software and content for it — open, that is, like a carefully curated and gated community garden.

The apps phenomenon began with the iPhone. When it first came out in early 2007, there were no apps you could buy from outside developers, and Jobs initially resisted allowing them. He didn't want outsiders to create applications for the iPhone that could mess it up, infect it with viruses, or pollute its integrity.

Board member Art Levinson was among those pushing to allow iPhone apps. "I called him a half dozen times to lobby for the potential of the apps," he recalled. If Apple didn't allow them, indeed encourage them, another smartphone maker would, giving itself a competitive advantage. Apple's marketing chief Phil Schiller agreed. "I couldn't imagine that we would create something as powerful as the iPhone and not empower developers to make lots of apps," he recalled. "I knew customers would love them." From the outside, the venture capitalist John Doerr argued that permitting apps would spawn a profusion of new entrepreneurs who would create new services.

Jobs at first quashed the discussion, partly because he felt his team did not have the bandwidth to figure out all of the complexities that would be involved in policing third-party app developers. He wanted focus. "So he didn't want to talk about it," said Schiller. But as soon as the iPhone was launched, he was willing to hear the debate. "Every time the conversation happened, Steve seemed a little more open," said Levinson. There were freewheeling discussions at four board meetings.

Jobs soon figured out that there was a way to have the best of both worlds. He would permit outsiders to write apps, but they would have to meet strict standards, be tested and approved by Apple, and be sold only through the iTunes Store. It was a way to reap the advantage of empowering thousands of software developers while retaining enough control to protect the integrity of the iPhone and the simplicity of the customer experience. "It was an absolutely magical solution that hit the sweet spot," said Levinson. "It gave us the benefits of openness while retaining end-to-end control."

The App Store for the iPhone opened on iTunes

in July 2008; the billionth download came nine months later. By the time the iPad went on sale in April 2010, there were 185,000 available iPhone apps. Most could also be used on the iPad, although they didn't take advantage of the bigger screen size. But in less than five months, developers had written twenty-five thousand new apps that were specifically configured for the iPad. By July 2011 there were 500,000 apps for both devices, and there had been more than fifteen billion downloads of them.

The App Store created a new industry overnight. In dorm rooms and garages and at major media companies, entrepreneurs invented new apps. John Doerr's venture capital firm created an iFund of $200 million to offer equity financing for the best ideas. Magazines and newspapers that had been giving away their content for free saw one last chance to put the genie of that dubious business model back into the bottle. Innovative publishers created new magazines, books, and learning materials just for the iPad. For example, the high-end publishing house Callaway, which had produced books ranging from Madonna's *Sex* to *Miss Spider's Tea Party,* decided to "burn the boats" and give up print altogether to focus on publishing books as interactive apps. By June 2011 Apple had paid out $2.5 billion to app developers.

The iPad and other app-based digital devices heralded a fundamental shift in the digital world. Back in the 1980s, going online usually meant dialing into a service like AOL, CompuServe, or Prodigy that charged fees for access to a carefully curated walled garden filled with content plus some exit gates that allowed braver users access to the Internet at large. The second phase, beginning in the early 1990s, was the advent of browsers that allowed everyone to freely surf the Internet using the hypertext transfer protocols of the World Wide Web, which linked billions of sites. Search engines arose so that people could easily find the websites they wanted. The release of the iPad portended a new model. Apps resembled the walled gardens of old. The creators could charge fees and offer more functions to the users who downloaded them. But the rise of apps also meant that the openness and linked nature of the web were sacrificed. Apps were not as easily linked or searchable. Because the iPad allowed the use of both apps and web browsing, it was not at war with the web model. But it did offer an alternative, for both the consumers and the creators of content.

Publishing and Journalism

With the iPod, Jobs had transformed the music business. With the iPad and its App Store, he began to transform all media, from publishing to journalism to television and movies.

Books were an obvious target, since Amazon's Kindle had shown there was an appetite for electronic books. So Apple created an iBooks Store, which sold electronic books the way the iTunes Store sold songs. There was, however, a slight difference in the business model. For the iTunes Store, Jobs had insisted that all songs be sold at one inexpensive price, initially 99 cents. Amazon's Jeff Bezos had tried to take a similar approach with ebooks, insisting on selling them for at most $9.99.

Jobs came in and offered publishers what he had refused to offer record companies: They could set any price they wanted for their wares in the iBooks Store, and Apple would take 30%. Initially that meant prices were higher than on Amazon. Why would people pay Apple more? "That won't be the case," Jobs answered, when Walt Mossberg asked him that question at the iPad launch event. "The price will be the same." He was right.

The day after the iPad launch, Jobs described to me his thinking on books:

> Amazon screwed it up. It paid the wholesale price for some books, but started selling them below cost at $9.99. The publishers hated that — they thought it would trash their ability to sell hardcover books at $28. So before Apple even got on the scene, some booksellers were starting to withhold books from Amazon. So we told the publishers, "We'll go to the agency model, where you set the price, and we get our 30%, and yes, the customer pays a little more, but that's what you want anyway." But we also asked for a guarantee that if anybody else is selling the books cheaper than we are, then we can sell them at the lower price too. So they went to Amazon and said, "You're going to sign an agency contract or we're not going to give you the books."

Jobs acknowledged that he was trying to have it both ways when it came to music and books. He had refused to offer the music companies the agency model and allow them to set their own prices. Why? Because he didn't have to. But with books he did. "We were not the first people in the books business," he said. "Given the situation that existed, what was best for us was to do this akido move and end up with the agency model. And we pulled it off."

Right after the iPad launch event, Jobs traveled to New York in February 2010 to meet with executives in the journalism business. In two days he saw Rupert Murdoch, his son James, and the management of their *Wall Street Journal;* Arthur Sulzberger Jr. and the top executives at the *New York Times;* and executives at *Time, Fortune,* and other Time Inc. magazines. "I would love to help quality journalism," he later said. "We can't depend on bloggers for our news. We need real reporting and editorial oversight more than ever. So I'd love to find a way to help people create digital products where they actually can make money." Since he had gotten people to pay for music, he hoped he could do the same for journalism.

Publishers, however, turned out to be leery of his lifeline. It meant that they would have to give 30% of their revenue to Apple, but that wasn't the biggest problem. More important, the publishers feared that, under his system, they would no longer have a direct relationship with their subscribers; they wouldn't have their email address and credit card number so they could bill them, communicate with them, and market new products to them. Instead Apple would own the customers, bill them, and have their information in its own database. And because of its privacy policy, Apple would not share this information unless a customer gave explicit permission to do so.

Jobs was particularly interested in striking a deal with the *New York Times,* which he felt was a great newspaper in danger of declining because it had not figured out how to charge for digital content. "One of my personal projects this year, I've decided, is to try to help — whether they want it or not — the *Times,*" he told me early in 2010. "I think it's important to the country for them to figure it out."

During his New York trip, he went to dinner with fifty top *Times* executives in the cellar private dining room at Pranna, an Asian restaurant. (He ordered a mango smoothie and a plain vegan pasta, neither of which was on the menu.) There he showed off the iPad and explained how important it was to find a modest price point for digital content that consumers would accept. He drew a chart of possible prices and volume. How many readers would they have if the *Times* were free? They already knew the answer to that extreme on the chart, because they were giving it away for free on the web already and had about twenty million regular visitors. And if they made it really expensive? They had data on that too; they charged print subscribers more than $300 a year and had about a million of them. "You should go after the midpoint, which is about ten million digital subscribers," he told them. "And that means your digital subs should be very cheap and simple, one click and $5 a month at most."

When one of the *Times* circulation executives insisted that the paper needed the email and credit card information for all of its subscribers, even if they subscribed through the App Store, Jobs said that Apple would not give it out. That angered the executive. It was unthinkable, he said, for the *Times* not to have that information. "Well, you can ask them for it, but if they won't voluntarily give it to you, don't blame me," Jobs said. "If you don't like it, don't use us. I'm not the one who got you in this jam. You're the ones who've spent the past five years giving away your paper online and not collecting anyone's credit card information."

Jobs also met privately with Arthur Sulzberger Jr. "He's a nice guy, and he's really proud of his new building, as he should be," Jobs said later. "I talked to him about what I thought he ought to do, but then nothing happened." It took a year, but in April 2011 the *Times* started charging for its digital edition and selling some subscriptions through Apple, abiding by the policies that Jobs established. It did, however, decide to charge approximately four times the $5 monthly charge that Jobs had suggested.

At the Time-Life Building, *Time*'s editor Rick Stengel played host. Jobs liked Stengel, who had assigned a talented team led by Josh Quittner to make a robust iPad version of the magazine each week. But he was upset to see Andy Serwer of *Fortune* there. Tearing up, he told Serwer how angry he still was about *Fortune*'s story two years earlier revealing details of his health and the stock options problems. "You kicked me when I was down," he said.

The bigger problem at Time Inc. was the same as the one at the *Times:* The magazine company did not want Apple to own its subscribers and prevent it from having a direct billing relation-ship. Time Inc. wanted to create apps that would direct readers to its own website in order to buy a subscription. Apple refused. When *Time* and other magazines submitted apps that did this, they were denied the right to be in the App Store.

Jobs tried to negotiate personally with the CEO of Time Warner, Jeff Bewkes, a savvy pragmatist with a no-bullshit charm to him. They had dealt with each other a few years earlier over video rights for the iPod Touch; even though Jobs had not been able to convince him to do a deal involving HBO's exclusive rights to show movies soon after their release, he admired Bewkes's straight and decisive style. For his part, Bewkes respected Jobs's ability to be both a strategic thinker and a master of the tiniest details. "Steve can go readily from the over-arching principals into the details," he said.

When Jobs called Bewkes about making a deal for Time Inc. magazines on the iPad, he started off by warning that the print business "sucks," that "nobody really wants your magazines," and that Apple was offering a great opportunity to sell digital subscriptions, but "your guys don't get it." Bewkes didn't agree with any of those premises. He said he was happy for Apple to sell digital subscriptions for Time Inc. Apple's 30% take was not the problem. "I'm telling you right now, if you sell a sub for us, you can have 30%," Bewkes told him.

"Well, that's more progress than I've made with anybody," Jobs replied.

"I have only one question," Bewkes continued. "If you sell a subscription to my magazine, and I

721

give you the 30%, who has the subscription — you or me?"

"I can't give away all the subscriber info because of Apple's privacy policy," Jobs replied.

"Well, then, we have to figure something else out, because I don't want my whole subscription base to become subscribers of yours, for you to then aggregate at the Apple store," said Bewkes. "And the next thing you'll do, once you have a monopoly, is come back and tell me that my magazine shouldn't be $4 a copy but instead should be $1. If someone subscribes to our magazine, we need to know who it is, we need to be able to create online communities of those people, and we need the right to pitch them directly about renewing."

Jobs had an easier time with Rupert Murdoch, whose News Corp. owned the *Wall Street Journal, New York Post,* newspapers around the world, Fox Studios, and the Fox News Channel. When Jobs met with Murdoch and his team, they also pressed the case that they should share ownership of the subscribers that came in through the App Store. But when Jobs refused, something interesting happened. Murdoch is not known as a pushover, but he knew that he did not have the leverage on this issue, so he accepted Jobs's terms. "We would prefer to own the subscribers, and we pushed for that," recalled Murdoch. "But Steve wouldn't do a deal on those terms, so I said, 'Okay, let's get on with it.' We didn't see any reason to mess around. He wasn't going to bend — and I wouldn't have bent if I were in his position — so I just said yes."

Murdoch even launched a digital-only daily newspaper, *The Daily,* tailored specifically for the iPad. It would be sold in the App Store, on the terms dictated by Jobs, at 99 cents a week. Murdoch himself took a team to Cupertino to show the proposed design. Not surprisingly, Jobs hated it. "Would you allow our designers to help?" he asked. Murdoch accepted. "The Apple designers had a crack at it," Murdoch recalled, "and our folks went back and had another crack, and ten days later we went back and showed them both, and he actually liked our team's version better. It stunned us."

The Daily, which was neither tabloidy nor serious, but instead a rather midmarket product like *USA Today,* was not very successful. But it did help create an odd-couple bonding between Jobs and Murdoch. When Murdoch asked him to speak at his June 2010 News Corp. annual management retreat, Jobs made an exception to his rule of never doing such appearances. James Murdoch led him in an after-dinner interview that lasted almost two hours. "He was very blunt and critical of what newspapers were doing in technology," Murdoch recalled. "He told us we were going to find it hard to get things right, because you're in New York, and anyone who's any good at tech works in Silicon Valley." This did not go down very well with the president of the Wall Street Journal Digital Network, Gordon McLeod, who pushed back a bit. At the end, McLeod came up to Jobs and said, "Thanks, it was a wonderful evening, but you probably just cost me my job." Murdoch chuckled a bit when he described the scene to me. "It ended up being true," he said. McLeod was out within three months.

In return for speaking at the retreat, Jobs got Murdoch to hear him out on Fox News, which he believed was destructive, harmful to the nation, and a blot on Murdoch's reputation. "You're blowing it with Fox News," Jobs told him over dinner. "The axis today is not liberal and conservative, the axis is constructive-destructive, and you've cast your lot with the destructive people. Fox has become an incredibly destructive force in our society. You can be better, and this is going to be your legacy if you're not careful." Jobs said he thought Murdoch did not really like how far Fox had gone. "Rupert's a builder, not a tearer-downer," he said. "I've had some meetings with James, and I think he agrees with me. I can just tell."

Murdoch later said he was used to people like Jobs complaining about Fox. "He's got sort of a left-wing view on this," he said. Jobs asked him to have his folks make a reel of a week of Sean Hannity and Glenn Beck shows — he thought that they were more destructive than Bill O'Reilly — and Murdoch agreed to do so. Jobs later told me that he was going to ask Jon Stewart's team to put together a similar reel for Murdoch to watch. "I'd be happy to see it," Murdoch said, "but he hasn't sent it to me."

Murdoch and Jobs hit it off well enough that Murdoch went to his Palo Alto house for dinner twice more during the next year. Jobs joked that he had to hide the dinner knives on such occasions, because he was afraid that his liberal wife was going to eviscerate Murdoch when he walked in. For his part, Murdoch was reported to have uttered a great line about the organic vegan dishes typically served: "Eating dinner at Steve's is a great experience, as long as you get out before the local restaurants close." Alas, when I asked Murdoch if he had ever said that, he didn't recall it.

One visit came early in 2011. Murdoch was due to pass through Palo Alto on February 24, and he texted Jobs to tell him so. He didn't know it was Jobs's fifty-sixth birthday, and Jobs didn't mention it when he texted back inviting him to dinner. "It was my way of making sure Laurene didn't veto the plan," Jobs joked. "It was my birthday, so she had to let me have Rupert over." Erin and Eve were there, and Reed jogged over from Stanford near the end of the dinner. Jobs showed off the designs for his planned boat, which Murdoch thought looked beautiful on the inside but "a bit plain" on the outside. "It certainly shows great optimism about his health that he was talking so much about building it," Murdoch later said.

At dinner they talked about the importance of infusing an entrepreneurial and nimble culture into a company. Sony failed to do that, Murdoch said. Jobs agreed. "I used to believe that a really big company couldn't have a clear corporate culture," Jobs said. "But I now believe it can be done. Murdoch's done it. I think I've done it at Apple."

Most of the dinner conversation was about education. Murdoch had just hired Joel Klein, the former chancellor of the New York City Department of Education, to start a digital curriculum division. Murdoch recalled that Jobs was somewhat dismissive of the idea that technology could transform education. But Jobs agreed with Murdoch that the paper textbook business would be blown away by digital learning materials.

In fact Jobs had his sights set on textbooks as the next business he wanted to transform. He believed it was an $8 billion a year industry ripe for digital destruction. He was also struck by the fact that many schools, for security reasons, don't have lockers, so kids have to lug a heavy backpack around. "The iPad would solve that," he said. His idea was to hire great textbook writers to create digital versions, and make them a feature of the iPad. In addition, he held meetings with the major publishers, such as Pearson Education, about partnering with Apple. "The process by which states certify textbooks is corrupt," he said. "But if we can make the textbooks free, and they come with the iPad, then they don't have to be certified. The crappy economy at the state level will last for a decade, and we can give them an opportunity to circumvent that whole process and save money."

CHAPTER THIRTY-NINE: NEW BATTLES

AND ECHOES OF OLD ONES

GOOGLE: OPEN VERSUS CLOSED

A few days after he unveiled the iPad in January 2010, Jobs held a "town hall" meeting with employees at Apple's campus. Instead of exulting about their transformative new product, however, he went into a rant against Google for producing the rival Android operating system. Jobs was furious that Google had decided to compete with Apple in the phone business. "We did not enter the search business," he said. "They entered the phone business. Make no mistake. They want to kill the iPhone. We won't let them." A few minutes later, after the meeting moved on to another topic, Jobs returned to his tirade to attack Google's famous values slogan. "I want to go back to that other question first and say one more thing. This 'Don't be evil' mantra, it's bullshit."

Jobs felt personally betrayed. Google's CEO Eric Schmidt had been on the Apple board during the development of the iPhone and iPad, and Google's founders, Larry Page and Sergey Brin, had treated him as a mentor. He felt ripped off. Android's touchscreen interface was adopting more and more of the features — multi-touch, swiping, a grid of app icons — that Apple had created.

Jobs had tried to dissuade Google from developing Android. He had gone to Google's headquarters near Palo Alto in 2008 and gotten into a shouting match with Page, Brin, and the head of the Android development team, Andy Rubin. (Because Schmidt was then on the Apple board, he recused himself from discussions involving the iPhone.) "I said we would, if we had good relations, guarantee Google access to the iPhone and guarantee it one or two icons on the home screen," he recalled. But he also threatened that if Google continued to develop Android and used any iPhone features, such as multi-touch, he would sue. At first Google avoided copying certain features, but in January 2010 HTC introduced an Android phone that boasted multi-touch and many other aspects of the iPhone's look and feel. That was the context for Jobs's pronouncement that Google's "Don't be evil" slogan was "bullshit."

So Apple filed suit against HTC (and, by extension, Android), alleging infringement of twenty of its patents. Among them were patents covering various multi-touch gestures, swipe to open, double-tap to zoom, pinch and expand, and the sensors that determined how a device was being held. As he sat in his house in Palo Alto the week the lawsuit was filed, he became angrier than I had ever seen him:

Our lawsuit is saying, "Google, you fucking ripped off the iPhone, wholesale ripped us off." Grand theft. I will spend my last dying breath if I need to, and I will spend every penny of Apple's $40 billion in the bank, to right this wrong. I'm going to destroy Android, because it's a stolen product. I'm willing to go to thermonuclear war on this. They are scared to death, because they know they are guilty. Outside of Search, Google's products — Android, Google Docs — are shit.

A few days after this rant, Jobs got a call from Schmidt, who had resigned from the Apple board the previous summer. He suggested they get together for coffee, and they met at a café in a Palo Alto shopping center. "We spent half the time talking about personal matters, then half the time on his perception that Google had stolen Apple's user interface designs," recalled Schmidt. When it came to the latter subject, Jobs did most of the talking. Google had ripped him off, he said in colorful language. "We've got you red-handed," he told Schmidt. "I'm not interested in settling. I don't want your money. If you offer me $5 billion, I won't want it. I've got plenty of money. I want you to stop using our ideas in Android, that's all I want." They resolved nothing.

Underlying the dispute was an even more fundamental issue, one that had unnerving historical resonance. Google presented Android as an "open" platform; its open-source code was freely available for multiple hardware makers to use on whatever phones or tablets they built. Jobs, of course, had a dogmatic belief that Apple should closely integrate its operating systems with its hardware. In the 1980s Apple had not licensed out its Macintosh operating system, and Microsoft eventually gained dominant market share by licensing its system to multiple hardware makers and, in Jobs's mind, ripping off Apple's interface.

The comparison between what Microsoft wrought in the 1980s and what Google was trying to do in 2010 was not exact, but it was close enough to be unsettling — and infuriating. It exemplified the great debate of the digital age: closed versus open, or as Jobs framed it, integrated versus fragmented. Was it better, as Apple believed and as Jobs's own controlling perfectionism almost compelled, to tie the hardware and software and content handling into one tidy system that assured a simple user experience? Or was it better to give users and manufacturers more choice and free up avenues for more innovation, by creating software systems that could be modified and used on different devices? "Steve has a particular way that he wants to run Apple, and it's the same as it was twenty years ago, which is that Apple is a brilliant innovator of closed systems," Schmidt later told me. "They don't want people to be on their platform without permission. The benefits of a closed platform is control. But Google has a specific belief that open is the better approach, because it leads to more options and competition and consumer choice."

So what did Bill Gates think as he watched Jobs, with his closed strategy, go into battle against Google, as he had done against Microsoft twenty-five years earlier? "There are some benefits to being more closed, in terms of how much you control the experience, and certainly at times he's had the benefit of that," Gates told me. But refusing to license the Apple iOS, he added, gave competitors like Android the chance to gain greater volume.

In addition, he argued, competition among a variety of devices and manufacturers leads to greater consumer choice and more innovation. "These companies are not all building pyramids next to Central Park," he said, poking fun at Apple's Fifth Avenue store, "but they are coming up with innovations based on competing for consumers." Most of the improvements in PCs, Gates pointed out, came because consumers had a lot of choices, and that would someday be the case in the world of mobile devices. "Eventually, I think, open will succeed, but that's where I come from. In the long run, the coherence thing, you can't stay with that."

Jobs believed in "the coherence thing." His faith in a controlled and closed environment remained unwavering, even as Android gained market share. "Google says we exert more control than they do, that we are closed and they are open," he railed when I told him what Schmidt had said. "Well, look at the results — Android's a mess. It has different screen sizes and versions, over a hundred permutations." Even if Google's approach might eventually win in the marketplace, Jobs found it repellent. "I like being responsible for the whole user experience. We do it not to make money. We do it because we want to make great products, not crap like Android."

FLASH, THE APP STORE, AND CONTROL

Jobs's insistence on end-to-end control was manifested in other battles as well. At the town hall meeting where he attacked Google, he also assailed Adobe's multimedia platform for websites, Flash, as a "buggy" battery hog made by "lazy" people. The iPod and iPhone, he said, would never run Flash. "Flash is a spaghetti-ball piece of technology that has lousy performance and really bad security problems," he said to me later that week.

He even banned apps that made use of a compiler created by Adobe that translated Flash code so that it would be compatible with Apple's iOS. Jobs disdained the use of compilers that allowed developers to write their products once and have them ported to multiple operating systems. "Allowing Flash to be ported across platforms means things get dumbed down to the lowest common denominator," he said. "We spend lots of effort to make our platform better, and the developer doesn't get any benefit if Adobe only works with functions that every platform has. So we said that we want developers to take advantage of our better features, so that their apps work better on our platform than they work on anybody else's." On that he was right. Losing the ability to differentiate Apple's platforms — allowing them to become commoditized like HP and Dell machines — would have meant death for the company.

There was, in addition, a more personal reason. Apple had invested in Adobe in 1985, and together the two companies had launched the desktop publishing revolution. "I helped put Adobe on the map," Jobs claimed. In 1999, after he returned to Apple, he had asked Adobe to start making its video editing software and other products for the iMac and its new operating system, but Adobe refused. It focused on making its products for Windows. Soon after, its founder, John Warnock, retired. "The soul of Adobe disappeared when

Warnock left," Jobs said. "He was the inventor, the person I related to. It's been a bunch of suits since then, and the company has turned out crap."

When Adobe evangelists and various Flash supporters in the blogosphere attacked Jobs for being too controlling, he decided to write and post an open letter. Bill Campbell, his friend and board member, came by his house to go over it. "Does it sound like I'm just trying to stick it to Adobe?" he asked Campbell. "No, it's facts, just put it out there," the coach said. Most of the letter focused on the technical drawbacks of Flash. But despite Campbell's coaching, Jobs couldn't resist venting at the end about the problematic history between the two companies. "Adobe was the last major third party developer to fully adopt Mac OS X," he noted.

Apple ended up lifting some of its restrictions on cross-platform compilers later in the year, and Adobe was able to come out with a Flash authoring tool that took advantage of the key features of Apple's iOS. It was a bitter war, but one in which Jobs had the better argument. In the end it pushed Adobe and other developers of compilers to make better use of the iPhone and iPad interface and its special features.

Jobs had a tougher time navigating the controversies over Apple's desire to keep tight control over which apps could be downloaded onto the iPhone and iPad. Guarding against apps that contained viruses or violated the user's privacy made sense; preventing apps that took users to other websites to buy subscriptions, rather than doing it through the iTunes Store, at least had a business rationale. But Jobs and his team went further: They decided to ban any app that defamed people, might be politically explosive, or was deemed by Apple's censors to be pornographic.

The problem of playing nanny became apparent when Apple rejected an app featuring the animated political cartoons of Mark Fiore, on the rationale that his attacks on the Bush administration's policy on torture violated the restriction against defamation. Its decision became public, and was subjected to ridicule, when Fiore won the 2010 Pulitzer Prize for editorial cartooning in April. Apple had to reverse itself, and Jobs made a public apology. "We're guilty of making mistakes," he said. "We're doing the best we can, we're learning as fast as we can — but we thought this rule made sense."

It was more than a mistake. It raised the specter of Apple's controlling what apps we got to see and read, at least if we wanted to use an iPad or iPhone. Jobs seemed in danger of becoming the Orwellian Big Brother he had gleefully destroyed in Apple's "1984" Macintosh ad. He took the issue seriously. One day he called the *New York Times* columnist Tom Friedman to discuss how to draw lines without looking like a censor. He asked Friedman to head an advisory group to help come up with guidelines, but the columnist's publisher said it would be a conflict of interest, and no such committee was formed.

The pornography ban also caused problems. "We believe we have a moral responsibility to keep porn off the iPhone," Jobs declared in an email to a customer. "Folks who want porn can buy an Android."

This prompted an email exchange with Ryan Tate, the editor of the tech gossip site Valleywag. Sipping a stinger cocktail one evening, Tate shot off an email to Jobs decrying Apple's heavy-handed control over which apps passed muster. "If Dylan was 20 today, how would he feel about your company?" Tate asked. "Would he think the iPad had the faintest thing to do with 'revolution'? Revolutions are about freedom."

To Tate's surprise, Jobs responded a few hours later, after midnight. "Yep," he said, "freedom from programs that steal your private data. Freedom from programs that trash your battery. Freedom from porn. Yep, freedom. The times they are a changin', and some traditional PC folks feel like their world is slipping away. It is."

In his reply, Tate offered some thoughts on Flash and other topics, then returned to the censorship issue. "And you know what? I don't want 'freedom from porn.' Porn is just fine! And I think my wife would agree."

"You might care more about porn when you have kids," replied Jobs. "It's not about freedom, it's about Apple trying to do the right thing for its users." At the end he added a zinger: "By the way, what have you done that's so great? Do you create anything, or just criticize others' work and belittle their motivations?"

Tate admitted to being impressed. "Rare is the CEO who will spar one-on-one with customers and bloggers like this," he wrote. "Jobs deserves big credit for breaking the mold of the typical

American executive, and not just because his company makes such hugely superior products: Jobs not only built and then rebuilt his company around some very strong opinions about digital life, but he's willing to defend them in public. Vigorously. Bluntly. At two in the morning on a weekend." Many in the blogosphere agreed, and they sent Jobs emails praising his feistiness. Jobs was proud as well; he forwarded his exchange with Tate and some of the kudos to me.

Still, there was something unnerving about Apple's decreeing that those who bought their products shouldn't look at controversial political cartoons or, for that matter, porn. The humor site eSarcasm.com launched a "Yes, Steve, I want porn" web campaign. "We are dirty, sex-obsessed miscreants who need access to smut 24 hours a day," the site declared. "Either that, or we just enjoy the idea of an uncensored, open society where a techno-dictator doesn't decide what we can and cannot see."

At the time Jobs and Apple were engaged in a battle with Valleywag's affiliated website, Gizmodo, which had gotten hold of a test version of the unreleased iPhone 4 that a hapless Apple engineer had left in a bar. When the police, responding to Apple's complaint, raided the house of the reporter, it raised the question of whether control freakiness had combined with arrogance.

Jon Stewart was a friend of Jobs and an Apple fan. Jobs had visited him privately in February when he took his trip to New York to meet with media executives. But that didn't stop Stewart from going after him on *The Daily Show.* "It wasn't supposed to be this way! Microsoft was supposed to be the evil one!" Stewart said, only half-jokingly. Behind him, the word "appholes" appeared on the screen. "You guys were the rebels, man, the underdogs. But now, are you becoming The Man? Remember back in 1984, you had those awesome ads about overthrowing Big Brother? Look in the mirror, man!"

By late spring the issue was being discussed among board members. "There is an arrogance," Art Levinson told me over lunch just after he had raised it at a meeting. "It ties into Steve's personality. He can react viscerally and lay out his convictions in a forceful manner." Such arrogance was fine when Apple was the feisty underdog. But now Apple was dominant in the mobile market. "We need to make the transition to being a big company and dealing with the hubris issue," said Levinson. Al Gore also talked about the problem at board meetings. "The context for Apple is changing dramatically," he recounted. "It's not hammer-thrower against Big Brother. Now Apple's big, and people see it as arrogant." Jobs became defensive when the topic was raised. "He's still adjusting to it," said Gore. "He's better at being the underdog than being a humble giant."

Jobs had little patience for such talk. The reason Apple was being criticized, he told me then, was that "companies like Google and Adobe are lying about us and trying to tear us down." What did he think of the suggestion that Apple sometimes acted arrogantly? "I'm not worried about that," he said, "because we're not arrogant."

ANTENNAGATE: DESIGN VERSUS ENGINEERING

In many consumer product companies, there's tension between the designers, who want to make a product look beautiful, and the engineers, who need to make sure it fulfills its functional requirements. At Apple, where Jobs pushed both design and engineering to the edge, that tension was even greater.

When he and design director Jony Ive became creative coconspirators back in 1997, they tended to view the qualms expressed by engineers as evidence of a can't-do attitude that needed to be overcome. Their faith that awesome design could force superhuman feats of engineering was reinforced by the success of the iMac and iPod. When engineers said something couldn't be done, Ive and Jobs pushed them to try, and usually they succeeded. There were occasional small problems. The iPod Nano, for example, was prone to getting scratched because Ive believed that a clear coating would lessen the purity of his design. But that was not a crisis.

When it came to designing the iPhone, Ive's design desires bumped into a fundamental law of physics that could not be changed even by a reality distortion field. Metal is not a great material to put near an antenna. As Michael Faraday showed, electromagnetic waves flow around the surface of metal, not through it. So a metal enclosure around a phone can create what is known as a Faraday cage, diminishing the signals that get in or out. The original iPhone started with a plastic band at the bottom, but Ive thought that would wreck the design integrity and asked that there be an aluminum rim all around. After that ended up working out, Ive designed the iPhone 4 with a steel rim. The steel would be the structural support, look really sleek, and serve as part of the phone's antenna.

There were significant challenges. In order to serve as an antenna, the steel rim had to have a tiny gap. But if a person covered that gap with a finger or sweaty palm, there could be some signal loss. The engineers suggested a clear coating over the metal to help prevent this, but again Ive felt that this would detract from the brushed-metal look. The issue was presented to Jobs at various meetings, but he thought the engineers were crying wolf. You can make this work, he said. And so they did.

And it worked, almost perfectly. But not totally perfectly. When the iPhone 4 was released in June 2010, it looked awesome, but a problem soon became evident: If you held the phone a certain way, especially using your left hand so your palm covered the tiny gap, you could lose your connection. It occurred with perhaps one in a hundred calls. Because Jobs insisted on keeping his unreleased products secret (even the phone that Gizmodo scored in a bar had a fake case around it), the iPhone 4 did not go through the live testing that most electronic devices get. So the flaw was not caught before the massive rush to buy it began. "The question is whether the twin policies of putting design in front of engineering and having a policy of supersecrecy surrounding unreleased products helped Apple," Tony Fadell said later. "On the whole, yes, but unchecked power is a bad thing, and that's what happened."

Had it not been the Apple iPhone 4, a product that had everyone transfixed, the issue of a few extra dropped calls would not have made news. But it became known as "Antennagate," and it boiled to a head in early July, when *Consumer Reports* did some rigorous tests and said that it could not recommend the iPhone 4 because of the antenna problem.

Jobs was in Kona Village, Hawaii, with his family when the issue arose. At first he was defensive. Art Levinson was in constant contact by phone, and Jobs insisted that the problem stemmed from Google and Motorola making mischief. "They want to shoot Apple down," he said.

Levinson urged a little humility. "Let's try to figure out if there's something wrong," he said. When he again mentioned the perception that Apple was arrogant, Jobs didn't like it. It went against his black-white, right-wrong way of viewing the world. Apple was a company of principle, he felt. If others failed to see that, it was their fault, not a reason for Apple to play humble.

Jobs's second reaction was to be hurt. He took the criticism personally and became emotionally anguished. "At his core, he doesn't do things that he thinks are blatantly wrong, like some pure pragmatists in our business," Levinson said. "So if he feels he's right, he will just charge ahead rather than question himself." Levinson urged him not to get depressed. But Jobs did. "Fuck this, it's not worth it," he told Levinson. Finally Tim Cook was able to shake him out of his lethargy. He quoted someone as saying that Apple was becoming the new Microsoft, complacent and arrogant. The next day Jobs changed his attitude. "Let's get to the bottom of this," he said.

When the data about dropped calls were assembled from AT&T, Jobs realized there was a problem, even if it was more minor than people were making it seem. So he flew back from Hawaii. But before he left, he made some phone calls. It was time to gather a couple of trusted old hands, wise men who had been with him during the original Macintosh days thirty years earlier.

His first call was to Regis McKenna, the public relations guru. "I'm coming back from Hawaii to deal with this antenna thing, and I need to bounce some stuff off of you," Jobs told him. They agreed to meet at the Cupertino boardroom at 1:30 the next afternoon. The second call was to the adman Lee Clow. He had tried to retire from the Apple account, but Jobs liked having him around. His colleague James Vincent was summoned as well.

Jobs also decided to bring his son Reed, then a high school senior, back with him from Hawaii. "I'm going to be in meetings 24/7 for probably two days and I want you to be in every single one because you'll learn more in those two days than you would in two years at business school," he told him. "You're going to be in the room with the best people in the world making really tough decisions and get to see how the sausage is made." Jobs got a little misty-eyed when he recalled the experience. "I would go through that all again just for that opportunity to have him see me at work," he said. "He got to see what his dad does."

They were joined by Katie Cotton, the steady public relations chief at Apple, and seven other top executives. The meeting lasted all afternoon. "It was one of the greatest meetings of my life," Jobs later said. He began by laying out all the data they had gathered. "Here are the facts. So what should we do about it?"

McKenna was the most calm and straightforward. "Just lay out the truth, the data," he said. "Don't appear arrogant, but appear firm and confident." Others, including Vincent, pushed Jobs to be more apologetic, but McKenna said no. "Don't go into the press conference with your tail between your legs," he advised. "You should just say: 'Phones aren't perfect, and we're not perfect. We're human and doing the best we can, and here's the data.' " That became the strategy. When the topic turned to the perception of arrogance, McKenna urged him not to worry too much. "I don't think it would work to try to make Steve look humble," McKenna explained later. "As Steve says about himself, 'What you see is what you get.' "

At the press event that Friday, held in Apple's auditorium, Jobs followed McKenna's advice. He did not grovel or apologize, yet he was able to defuse the problem by showing that Apple understood it and would try to make it right. Then he changed the framework of the discussion, saying that all cell phones had some problems. Later he told me that he had sounded a bit "too annoyed" at the event, but in fact he was able to strike a tone that was unemotional and straightforward. He captured it in four short, declarative sentences: "We're not perfect. Phones are not perfect. We all know that. But we want to make our users happy."

If anyone was unhappy, he said, they could return the phone (the return rate turned out to be 1.7%, less than a third of the return rate for the iPhone 3GS or most other phones) or get a free bumper case from Apple. He went on to report data showing that other mobile phones had similar problems. That was not totally true. Apple's antenna design made it slightly worse than most other phones, including earlier versions of the iPhone. But it was true that the media frenzy over the iPhone 4's dropped calls was overblown. "This is blown so out of proportion that it's incredible," he said. Instead of being appalled that he didn't grovel or order a recall, most customers realized that he was right.

The wait list for the phone, which was already sold out, went from two weeks to three. It remained the company's fastest-selling product ever. The media debate shifted to the issue of whether Jobs was right to assert that other smartphones had the same antenna problems. Even if the answer was no, that was a better story to face than one about whether the iPhone 4 was a defective dud.

Some media observers were incredulous. "In a bravura demonstration of stonewalling, righteousness, and hurt sincerity, Steve Jobs successfully took to the stage the other day to deny the problem, dismiss the criticism, and spread the blame among other smartphone makers," Michael Wolff of newser.com wrote. "This is a level of modern marketing, corporate spin, and crisis management about which you can only ask with stupefied incredulity and awe: How do they get away with it? Or, more accurately, how does he get away with it?" Wolff attributed it to Jobs's mesmerizing effect as "the last charismatic individual." Other CEOs would be offering abject apologies and swallowing massive recalls, but Jobs didn't have to. "The grim, skeletal appearance, the absolutism, the ecclesiastical bearing, the sense of his relationship with the sacred, really works, and, in this instance, allows him the privilege of magisterially deciding what is meaningful and what is trivial."

Scott Adams, the creator of the cartoon strip *Dilbert,* was also incredulous, but far more admiring. He wrote a blog entry a few days later (which Jobs proudly emailed around) that marveled at how Jobs's "high ground maneuver" was destined to be studied as a new public relations standard. "Apple's response to the iPhone 4 problem didn't follow the public relations playbook, because Jobs decided to rewrite the playbook," Adams wrote. "If you want to know what genius looks like, study Jobs' words." By proclaiming up front that phones are not perfect, Jobs changed the context of the argument with an indisputable assertion. "If Jobs had not changed the context from the iPhone 4 to all smartphones in general, I could make you a hilarious comic strip about a product so poorly made that it won't work if it comes in contact with a human hand. But as soon as the context is changed to 'all smartphones have problems,' the humor opportunity is gone. Nothing kills humor like a general and boring truth."

HERE COMES THE SUN

There were a few things that needed to be resolved for the career of Steve Jobs to be complete. Among them was an end to the Thirty Years' War with the band he loved, the Beatles. In 2007 Apple had settled its trademark battle with Apple Corps, the holding company of the Beatles, which had first sued the fledgling computer company over use of the name in 1978. But that still did not get the Beatles into the iTunes Store. The band was the last major holdout, primarily because it had not resolved with EMI music, which owned most of its songs, how to handle the digital rights.

By the summer of 2010 the Beatles and EMI had sorted things out, and a four-person summit was held in the boardroom in Cupertino. Jobs and his vice president for the iTunes Store, Eddy Cue, played host to Jeff Jones, who managed the Beatles' interests, and Roger Faxon, the chief of EMI music. Now that the Beatles were ready to go digital, what could Apple offer to make that milestone special? Jobs had been anticipating this day for a long time. In fact he and his advertising team, Lee Clow and James Vincent, had mocked up some ads and commercials three years earlier when strategizing on how to lure the Beatles on board.

"Steve and I thought about all the things that we could possibly do," Cue recalled. That included taking over the front page of the iTunes Store, buying billboards featuring the best photographs of the band, and running a series of television ads in classic Apple style. The topper was offering a $149 box set that included all thirteen Beatles studio albums, the two-volume "Past Masters" collection, and a nostalgia-inducing video of the 1964 Washington Coliseum concert.

Once they reached an agreement in principle,

Jobs personally helped choose the photographs for the ads. Each commercial ended with a still black-and-white shot of Paul McCartney and John Lennon, young and smiling, in a recording studio looking down at a piece of music. It evoked the old photographs of Jobs and Wozniak looking at an Apple circuit board. "Getting the Beatles on iTunes was the culmination of why we got into the music business," said Cue.

Chapter Forty: To Infinity
THE CLOUD, THE SPACESHIP, AND BEYOND

The iPad 2

Even before the iPad went on sale, Jobs was thinking about what should be in the iPad 2. It needed front and back cameras — everyone knew that was coming — and he definitely wanted it to be thinner. But there was a peripheral issue that he focused on that most people hadn't thought about: The cases that people used covered the beautiful lines of the iPad and detracted from the screen. They made fatter what should be thinner. They put a pedestrian cloak on a device that should be magical in all of its aspects.

Around that time he read an article about magnets, cut it out, and handed it to Jony Ive. The magnets had a cone of attraction that could be precisely focused. Perhaps they could be used to align a detachable cover. That way, it could snap onto the front of an iPad but not have to engulf the entire device. One of the guys in Ive's group worked out how to make a detachable cover that could connect with a magnetic hinge. When you began to open it, the screen would pop to life like the face of a tickled baby, and then the cover could fold into a stand.

It was not high-tech; it was purely mechanical. But it was enchanting. It also was another example of Jobs's desire for end-to-end integration: The cover and the iPad had been designed together so that the magnets and hinge all connected seamlessly. The iPad 2 would have many improvements, but this cheeky little cover, which most other CEOs would never have bothered with, was the one that would elicit the most smiles.

Because Jobs was on another medical leave, he was not expected to be at the launch of the iPad 2, scheduled for March 2, 2011, in San Francisco. But when the invitations were sent out, he told me that I should try to be there. It was the usual scene: top Apple executives in the front row, Tim Cook eating energy bars, and the sound system blaring the appropriate Beatles songs, building up to "You Say You Want a Revolution" and "Here Comes the Sun." Reed Jobs arrived at the last minute with two rather wide-eyed freshman dorm mates.

"We've been working on this product for a while, and I just didn't want to miss today," Jobs said as he ambled onstage looking scarily gaunt but with a jaunty smile. The crowd erupted in whoops, hollers, and a standing ovation.

He began his demo of the iPad 2 by showing off the new cover. "This time, the case and the product were designed together," he explained. Then he moved on to address a criticism that had been rankling him because it had some merit: The original iPad had been better at consuming content than at creating it. So Apple had adapted its two best creative applications for the Macintosh, GarageBand and iMovie, and made powerful versions available for the iPad. Jobs showed how easy it was to compose and orchestrate a song, or put music and special effects into your home videos, and post or share such creations using the new iPad.

Once again he ended his presentation with the slide showing the intersection of Liberal Arts Street and Technology Street. And this time he gave one of the clearest expressions of his credo, that true creativity and simplicity come from integrating the whole widget — hardware and software, and for that matter content and covers and salesclerks — rather than allowing things to be open and fragmented, as happened in the world of Windows PCs and was now happening with Android devices:

It's in Apple's DNA that technology alone is not enough. We believe that it's technology married with the humanities that yields us the result that makes our heart sing. Nowhere is that more true than in these post-PC devices. Folks are rushing into this tablet market, and they're looking at it as the next PC, in which the hardware and the software are done by different companies. Our experience, and every bone in our body, says that is not the right approach. These are post-PC devices that need to be even more intuitive and easier to use than a PC, and where the software and the hardware and the applications need to be intertwined in an even more seamless way than they are on a PC. We think we have the right architecture not just in silicon, but in our organization, to build these kinds of products.

It was an architecture that was bred not just into the organization he had built, but into his own soul.

After the launch event, Jobs was energized. He came to the Four Seasons hotel to join me, his wife, and Reed, plus Reed's two Stanford pals, for lunch. For a change he was eating, though still with some pickiness. He ordered fresh-squeezed juice, which he sent back three times, declaring that each new offering was from a bottle, and a pasta primavera, which he shoved away as inedible after one taste. But then he ate half of my crab Louie salad and ordered a full one for himself, followed by a bowl of ice cream. The indulgent hotel was even able to produce a glass of juice that finally met his standards.

At his house the following day he was still on a high. He was planning to fly to Kona Village the next day, alone, and I asked to see what he had put on his iPad 2 for the trip. There were three movies: *Chinatown, The Bourne Ultimatum,* and *Toy Story 3.* More revealingly, there was just one book that he had downloaded: *The Autobiography of a Yogi,* the guide to meditation and spirituality that he had first read as a teenager, then reread in India, and had read once a year ever since.

Midway through the morning he decided he wanted to eat something. He was still too weak to drive, so I drove him to a café in a shopping mall. It was closed, but the owner was used to Jobs knocking on the door at off-hours, and he happily let us in. "He's taken on a mission to try to fatten me up," Jobs joked. His doctors had pushed him to eat eggs as a source of high-quality protein, and he ordered an omelet. "Living with a disease like this, and all the pain, constantly reminds you of your own mortality, and that can do strange things to your brain if you're not careful," he said. "You don't make plans more than a year out, and that's bad. You need to force yourself to plan as if you will live for many years."

An example of this magical thinking was his plan to build a luxurious yacht. Before his liver transplant, he and his family used to rent a boat for vacations, traveling to Mexico, the South Pacific, or the Mediterranean. On many of these cruises, Jobs got bored or began to hate the design of the boat, so they would cut the trip short and fly to Kona Village. But sometimes the cruise worked well. "The best vacation I've ever been on was when we went down the coast of Italy, then to Athens — which is a pit, but the Parthenon is mind-blowing — and then to Ephesus in Turkey, where they have these ancient public lavatories in marble with a place in the middle for musicians to serenade." When they got to Istanbul, he hired a history professor to give his family a tour. At the end they went to a Turkish bath, where the professor's lecture gave Jobs an insight about the globalization of youth:

I had a real revelation. We were all in robes, and they made some Turkish coffee for us. The professor explained how the coffee was made very different from anywhere else, and I realized, "So fucking what?" Which kids even in Turkey give a shit about Turkish coffee? All day I had looked at young people in Istanbul. They were all drinking what every other kid in the world drinks, and they were wearing clothes that look like they were bought at the Gap, and they are all using cell phones. They were like kids everywhere else. It hit me that, for young people, this whole world is the same now. When we're making products, there is no such thing as a Turkish phone, or a music player that young people in Turkey would want that's different from one young people elsewhere would want. We're just one world now.

After the joy of that cruise, Jobs had amused himself by beginning to design, and then repeatedly redesigning, a boat he said he wanted to build someday. When he got sick again in 2009, he almost canceled the project. "I didn't think I would be alive when it got done," he recalled. "But that made me so sad, and I decided that working on the design was fun to do, and maybe I have a shot at being alive when it's done. If I stop work on the boat and then I make it alive for another two years, I would be really pissed. So I've kept going."

After our omelets at the café, we went back to his house and he showed me all of the models and architectural drawings. As expected, the planned yacht was sleek and minimalist. The teak decks were perfectly flat and unblemished by any accoutrements. As at an Apple store, the cabin windows were large panes, almost floor to ceiling, and the main living area was designed to have walls of glass that were forty feet long and ten feet high. He had gotten the chief engineer of the Apple stores to design a special glass that was able to provide structural support.

By then the boat was under construction by the Dutch custom yacht builders Feadship, but Jobs was still fiddling with the design. "I know that it's possible I will die and leave Laurene with a half-built boat," he said. "But I have to keep going on it. If I don't, it's an admission that I'm about to die."

He and Powell would be celebrating their twentieth wedding anniversary a few days later, and he admitted that at times he had not been as appreciative of her as she deserved. "I'm very lucky, because you just don't know what you're getting into when you get married," he said. "You have an intuitive feeling about things. I couldn't have done better, because not only is Laurene smart and beautiful, she's turned out to be a really good person." For a moment he teared up. He talked about his other girlfriends, particularly Tina Redse, but said he ended up in the right place. He also reflected on how selfish and demanding he could be. "Laurene had to deal with that, and also with me being sick," he said. "I know that living with me is not a bowl of cherries."

Among his selfish traits was that he tended not to remember anniversaries or birthdays. But in this case, he decided to plan a surprise. They had gotten married at the Ahwahnee Hotel in Yosemite, and he decided to take Powell back there on their anniversary. But when Jobs called, the place was fully booked. So he had the hotel approach the people who had reserved the suite where he and Powell had stayed and ask if they would relinquish it. "I offered to pay for another weekend," Jobs recalled, "and the man was very nice and said, 'Twenty years, please take it, it's yours.'"

He found the photographs of the wedding, taken by a friend, and had large prints made on thick paper boards and placed in an elegant box. Scrolling through his iPhone, he found the note that he had composed to be included in the box and read it aloud:

> We didn't know much about each other twenty years ago. We were guided by our intuition; you swept me off my feet. It was snowing when we got married at the Ahwahnee. Years passed, kids came, good times, hard times, but never bad times. Our love and respect has endured and grown. We've been through so much together and here we are right back where we started 20 years ago — older, wiser — with wrinkles on our faces and hearts. We now know many of life's joys, sufferings, secrets and wonders and we're still here together. My feet have never returned to the ground.

By the end of the recitation he was crying uncontrollably. When he composed himself, he noted that he had also made a set of the pictures for each of his kids. "I thought they might like to see that I was young once."

iCloud

In 2001 Jobs had a vision: Your personal computer would serve as a "digital hub" for a variety of lifestyle devices, such as music players, video recorders, phones, and tablets. This played to Apple's

754

strength of creating end-to-end products that were simple to use. The company was thus transformed from a high-end niche computer company to the most valuable technology company in the world.

By 2008 Jobs had developed a vision for the next wave of the digital era. In the future, he believed, your desktop computer would no longer serve as the hub for your content. Instead the hub would move to "the cloud." In other words, your content would be stored on remote servers managed by a company you trusted, and it would be available for you to use on any device, anywhere. It would take him three years to get it right.

He began with a false step. In the summer of 2008 he launched a product called MobileMe, an expensive ($99 per year) subscription service that allowed you to store your address book, documents, pictures, videos, email, and calendar remotely in the cloud and to sync them with any device. In theory, you could go to your iPhone or any computer and access all aspects of your digital life. There was, however, a big problem: The service, to use Jobs's terminology, sucked. It was complex, devices didn't sync well, and email and other data got lost randomly in the ether. "Apple's MobileMe Is Far Too Flawed to Be Reliable," was the headline on Walt Mossberg's review in the *Wall Street Journal.*

Jobs was furious. He gathered the MobileMe team in the auditorium on the Apple campus, stood onstage, and asked, "Can anyone tell me what MobileMe is supposed to do?" After the team members offered their answers, Jobs shot back: "So why the fuck doesn't it do that?" Over the next half hour he continued to berate them. "You've tarnished Apple's reputation," he said. "You should hate each other for having let each other down. Mossberg, our friend, is no longer writing good things about us." In front of the whole audience, he got rid of the leader of the MobileMe team and replaced him with Eddy Cue, who oversaw all Internet content at Apple. As *Fortune*'s Adam Lashinsky reported in a dissection of the Apple corporate culture, "Accountability is strictly enforced."

By 2010 it was clear that Google, Amazon, Microsoft, and others were aiming to be the company that could best store all of your content and data in the cloud and sync it on your various devices. So Jobs redoubled his efforts. As he explained it to me that fall:

> We need to be the company that manages your relationship with the cloud — streams your music and videos from the cloud, stores your pictures and information, and maybe even your medical data. Apple was the first to have the insight about your computer becoming a digital hub. So we wrote all of these apps — iPhoto, iMovie, iTunes — and tied in our devices, like the iPod and iPhone and iPad, and it's worked brilliantly. But over the next few years, the hub is going to move from your computer into the cloud. So it's the same digital hub strategy, but the hub's in a different place. It means you will always have access to your content and you won't have to sync.
>
> It's important that we make this transformation, because of what Clayton Christensen calls "the

innovator's dilemma," where people who invent something are usually the last ones to see past it, and we certainly don't want to be left behind. I'm going to take MobileMe and make it free, and we're going to make syncing content simple. We are building a server farm in North Carolina. We can provide all the syncing you need, and that way we can lock in the customer.

Jobs discussed this vision at his Monday morning meetings, and gradually it was refined to a new strategy. "I sent emails to groups of people at 2 a.m. and batted things around," he recalled. "We think about this a lot because it's not a job, it's our life." Although some board members, including Al Gore, questioned the idea of making MobileMe free, they supported it. It would be their strategy for attracting customers into Apple's orbit for the next decade.

The new service was named iCloud, and Jobs unveiled it in his keynote address to Apple's Worldwide Developers Conference in June 2011. He was still on medical leave and, for some days in May, had been hospitalized with infections and pain. Some close friends urged him not to make the presentation, which would involve lots of preparation and rehearsals. But the prospect of ushering in another tectonic shift in the digital age seemed to energize him.

When he came onstage at the San Francisco Convention Center, he was wearing a VON-ROSEN black cashmere sweater on top of his usual Issey Miyake black turtleneck, and he had thermal underwear beneath his blue jeans. But he looked more gaunt than ever. The crowd gave him a prolonged standing ovation — "That always helps, and I appreciate it," he said — but within minutes Apple's stock dropped more than $4, to $340. He was making a heroic effort, but he looked weak.

He handed the stage over to Phil Schiller and Scott Forstall to demo the new operating systems for Macs and mobile devices, then came back on to show off iCloud himself. "About ten years ago, we had one of our most important insights," he said. "The PC was going to become the hub for your digital life. Your videos, your photos, your music. But it has broken down in the last few years. Why?" He riffed about how hard it was to get all of your content synced to each of your devices. If you have a song you've downloaded on your iPad, a picture you've taken on your iPhone, and a video you've stored on your computer, you can end up feeling like an old-fashioned switchboard operator as you plug USB cables into and out of things to get the content shared. "Keeping these devices in sync is driving us crazy," he said to great laughter. "We have a solution. It's our next big insight. We are going to demote the PC and the Mac to be just a device, and we are going to move the digital hub into the cloud."

Jobs was well aware that this "big insight" was in fact not really new. Indeed he joked about Apple's previous attempt: "You may think, Why should I believe them? They're the ones who brought me MobileMe." The audience laughed nervously. "Let me just say it wasn't our finest hour." But as he demonstrated iCloud, it was clear that it would be better. Mail, contacts, and calendar entries synced instantly. So did apps, photos, books, and documents. Most impressively, Jobs and Eddy Cue had made deals with the music companies (unlike the folks at Google and Amazon). Apple would have eighteen *million* songs on its cloud servers. If you had any of these on any of your devices or computers — whether you had bought it legally or pirated it — Apple would let you access a high-quality version of it on all of your devices without having to go through the time and effort to upload it to the cloud. "It all just works," he said.

That simple concept — that everything would just work seamlessly — was, as always, Apple's competitive advantage. Microsoft had been advertising "Cloud Power" for more than a year, and three years earlier its chief software architect, the legendary Ray Ozzie, had issued a rallying cry to the company: "Our aspiration is that individuals will only need to license their media once, and use any of their . . . devices to access and enjoy their media." But Ozzie had quit Microsoft at the end of 2010, and the company's cloud computing push was never manifested in consumer devices. Amazon and Google both offered cloud services in 2011, but neither company had the ability to integrate the hardware and software and content of a variety of devices. Apple controlled every link in the chain and designed them all to work together: the devices, computers, operating systems, and application software, along with the sale and storage of the content.

Of course, it worked seamlessly only if you were

using an Apple device and stayed within Apple's gated garden. That produced another benefit for Apple: customer stickiness. Once you began using iCloud, it would be difficult to switch to a Kindle or Android device. Your music and other content would not sync to them; in fact they might not even work. It was the culmination of three decades spent eschewing open systems. "We thought about whether we should do a music client for Android," Jobs told me over breakfast the next morning. "We put iTunes on Windows in order to sell more iPods. But I don't see an advantage of putting our music app on Android, except to make Android users happy. And I don't want to make Android users happy."

A New Campus

When Jobs was thirteen, he had looked up Bill Hewlett in the phone book, called him to score a part he needed for a frequency counter he was trying to build, and ended up getting a summer job at the instruments division of Hewlett-Packard. That same year HP bought some land in Cupertino to expand its calculator division. Wozniak went to work there, and it was on this site that he designed the Apple I and Apple II during his moonlighting hours.

When HP decided in 2010 to abandon its Cupertino campus, which was just about a mile east of Apple's One Infinite Loop headquarters, Jobs quietly arranged to buy it and the adjoining property. He admired the way that Hewlett and Packard had built a lasting company, and he prided himself on having done the same at Apple. Now he wanted a showcase headquarters, something that no West Coast technology company had. He eventually accumulated 150 acres, much of which had been apricot orchards when he was a boy, and threw himself into what would become a legacy project that combined his passion for design with his passion for creating an enduring company. "I want to leave a signature campus that expresses the values of the company for generations," he said.

He hired what he considered to be the best architectural firm in the world, that of Sir Norman Foster, which had done smartly engineered buildings such as the restored Reichstag in Berlin and 30 St. Mary Axe in London. Not surprisingly, Jobs got so involved in the planning, both the vision and the details, that it became almost impossible to settle on a final design. This was to be his lasting edifice, and he wanted to get it right. Foster's firm assigned fifty architects to the team, and every three weeks throughout 2010 they showed Jobs revised models and options. Over and over he would come up with new concepts, sometimes entirely new shapes, and make them restart and provide more alternatives.

When he first showed me the models and plans in his living room, the building was shaped like a huge winding racetrack made of three joined semi-circles around a large central courtyard. The walls were floor-to-ceiling glass, and the interior had rows of office pods that allowed the sunlight to stream down the aisles. "It permits serendipitous and fluid meeting spaces," he said, "and everybody gets to participate in the sunlight."

761

The next time he showed me the plans, a month later, we were in Apple's large conference room across from his office, where a model of the proposed building covered the table. He had made a major change. The pods would all be set back from the windows so that long corridors would be bathed in sun. These would also serve as the common spaces. There was a debate with some of the architects, who wanted to allow the windows to be opened. Jobs had never liked the idea of people being able to open things. "That would just allow people to screw things up," he declared. On that, as on other details, he prevailed.

When he got home that evening, Jobs showed off the drawings at dinner, and Reed joked that the aerial view reminded him of male genitalia. His father dismissed the comment as reflecting the mind-set of a teenager. But the next day he mentioned the comment to the architects. "Unfortunately, once I've told you that, you're never going to be able to erase that image from your mind," he said. By the next time I visited, the shape had been changed to a simple circle.

The new design meant that there would not be a straight piece of glass in the building. All would be curved and seamlessly joined. Jobs had long been fascinated with glass, and his experience demanding huge custom panes for Apple's retail stores made him confident that it would be possible to make massive curved pieces in quantity. The planned center courtyard was eight hundred feet across (more than three typical city blocks, or almost the length of three football fields), and he showed it to me with overlays indicating how it could surround St. Peter's Square in Rome. One of his lingering memories was of the orchards that had once dominated the area, so he hired a senior arborist from Stanford and decreed that 80% of the property would be landscaped in a natural manner, with six thousand trees. "I asked him to make sure to include a new set of apricot orchards," Jobs recalled. "You used to see them everywhere, even on the corners, and they're part of the legacy of this valley."

By June 2011 the plans for the four-story, three-million-square-foot building, which would hold more than twelve thousand employees, were ready to unveil. He decided to do so in a quiet and unpublicized appearance before the Cupertino City Council on the day after he had announced iCloud at the Worldwide Developers Conference.

Even though he had little energy, he had a full schedule that day. Ron Johnson, who had developed Apple's stores and run them for more than a decade, had decided to accept an offer to be the CEO of J.C. Penney, and he came by Jobs's house in the morning to discuss his departure. Then Jobs and I went into Palo Alto to a small yogurt and oatmeal café called Fraiche, where he talked animatedly about possible future Apple products. Later that day he was driven to Santa Clara for the quarterly meeting that Apple had with top Intel executives, where they discussed the possibility of using Intel chips in future mobile devices. That night U2 was playing at the Oakland Coliseum, and Jobs had considered going. Instead he decided to use that evening to show his plans to the Cupertino Council.

Arriving without an entourage or any fanfare, and looking relaxed in the same black sweater he had worn for his developers conference speech, he stood on a podium with clicker in hand and spent twenty minutes showing slides of the design to council members. When a rendering of the sleek, futuristic, perfectly circular building appeared on the screen, he paused and smiled. "It's like a spaceship has landed," he said. A few moments later he added, "I think we have a shot at building the best office building in the world."

The following Friday, Jobs sent an email to a colleague from the distant past, Ann Bowers, the widow of Intel's cofounder Bob Noyce. She had been Apple's human resources director and den mother in the early 1980s, in charge of reprimanding Jobs after his tantrums and tending to the wounds of his coworkers. Jobs asked if she would come see him the next day. Bowers happened to be in New York, but she came by his house that Sunday when she returned. By then he was sick again, in pain and without much energy, but he was eager to show her the renderings of the new headquarters. "You should be proud of Apple," he said. "You should be proud of what we built."

Then he looked at her and asked, intently, a question that almost floored her: "Tell me, what was I like when I was young?"

Bowers tried to give him an honest answer. "You were very impetuous and very difficult," she replied. "But your vision was compelling. You told us, 'The journey is the reward.' That turned out to be true."

"Yes," Jobs answered. "I did learn some things along the way." Then, a few minutes later, he repeated it, as if to reassure Bowers and himself. "I did learn some things. I really did."

CHAPTER FORTY-ONE:
ROUND THREE

THE TWILIGHT STRUGGLE

FAMILY TIES

Jobs had an aching desire to make it to his son's graduation from high school in June 2010. "When I was diagnosed with cancer, I made my deal with God or whatever, which was that I really wanted to see Reed graduate, and that got me through 2009," he said. As a senior, Reed looked eerily like his father at eighteen, with a knowing and slightly rebellious smile, intense eyes, and a shock of dark hair. But from his mother he had inherited a sweetness and painfully sensitive empathy that his father lacked. He was demonstrably affectionate and eager to please. Whenever his father was sitting sullenly at the kitchen table and staring at the floor, which happened often when he was ailing, the only thing sure to cause his eyes to brighten was Reed walking in.

Reed adored his father. Soon after I started working on this book, he dropped in to where I was staying and, as his father often did, suggested we take a walk. He told me, with an intensely earnest look, that his father was not a cold profit-seeking businessman but was motivated by a love of what he did and a pride in the products he was making.

After Jobs was diagnosed with cancer, Reed began spending his summers working in a Stanford oncology lab doing DNA sequencing to find genetic markers for colon cancer. In one experiment, he traced how mutations go through families. "One of the very few silver linings about me getting sick is that Reed's gotten to spend a lot of time studying with some very good doctors," Jobs said. "His enthusiasm for it is exactly how I felt about computers when I was his age. I think the biggest innovations of the twenty-first century will be the intersection of biology and technology. A new era is beginning, just like the digital one was when I was his age."

Reed used his cancer study as the basis for the senior report he presented to his class at Crystal Springs Uplands School. As he described how he used centrifuges and dyes to sequence the DNA of tumors, his father sat in the audience beaming, along with the rest of his family. "I fantasize about Reed getting a house here in Palo Alto with his family and riding his bike to work as a doctor at Stanford," Jobs said afterward.

Reed had grown up fast in 2009, when it looked as if his father was going to die. He took care of his younger sisters while his parents were in Memphis, and he developed a protective paternalism. But when his father's health stabilized in the spring of 2010, he regained his playful, teasing personality. One day during dinner he was discussing with his family where to take his girlfriend for dinner. His father suggested Il Fornaio, an elegant standard in Palo Alto, but Reed said he had been unable to get reservations. "Do you want me to try?" his father asked. Reed resisted; he wanted to handle it himself. Erin, the somewhat shy middle child, suggested that she could outfit a tepee in their garden and she and Eve, the younger sister, would serve them a romantic meal there. Reed stood up and hugged her. He would take her up on that some other time, he promised.

One Saturday Reed was one of the four contestants on his school's Quiz Kids team competing on a local TV station. The family — minus Eve, who was in a horse show — came to cheer him on. As the television crew bumbled around getting ready, his father tried to keep his impatience in check and remain inconspicuous among the parents sitting in the rows of folding chairs. But he was clearly recognizable in his trademark jeans and black turtleneck, and one woman pulled up a chair right next to him and started to take his picture. Without looking at her, he stood up and moved to the other end of the row. When Reed came on the set, his nameplate identified him as "Reed Powell." The host asked the students what they wanted to be when they grew up. "A cancer researcher," Reed answered.

Jobs drove his two-seat Mercedes SL55, taking Reed, while his wife followed in her own car with Erin. On the way home, she asked Erin why she thought her father refused to have a license plate on his car. "To be a rebel," she answered. I later put the question to Jobs. "Because people follow me sometimes, and if I have a license plate, they can track down where I live," he replied. "But that's kind of getting obsolete now with Google Maps. So I guess, really, it's just because I don't."

During Reed's graduation ceremony, his father sent me an email from his iPhone that simply exulted, "Today is one of my happiest days. Reed is graduating from High School. Right now. And, against all odds, I am here." That night there was a party at their house with close friends and family. Reed danced with every member of his family, including his father. Later Jobs took his son out to the barnlike storage shed to offer him one of his two bicycles, which he wouldn't be riding again. Reed joked that the Italian one looked a bit too gay, so Jobs told him to take the solid eight-speed next to it. When Reed said he would be indebted, Jobs answered, "You don't need to be indebted, because you have my DNA." A few days later *Toy Story 3* opened. Jobs had nurtured this Pixar trilogy from the beginning, and the final installment was about the emotions surrounding the departure of Andy for college. "I wish I could always be with you," Andy's mother says. "You always will be," he replies.

Jobs's relationship with his two younger daughters was somewhat more distant. He paid less attention to Erin, who was quiet, introspective, and seemed not to know exactly how to handle him, especially when he was emitting wounding barbs. She was a poised and attractive young woman, with a personal sensitivity more mature than her father's. She thought that she might want to be an architect, perhaps because of her father's interest in the field, and she had a good sense of design. But when her father was showing Reed the drawings for the new Apple campus, she sat on the other side of the kitchen, and it seemed not to occur to him to call her over as well. Her big hope that spring of 2010 was that her father would take her to the Oscars. She loved the movies. Even more, she wanted to fly with her father on his private plane and walk up the red carpet with him. Powell was quite willing to forgo the trip and tried to talk her husband into taking Erin. But he dismissed the idea.

At one point as I was finishing this book, Powell told me that Erin wanted to give me an interview. It's not something that I would have requested, since she was then just turning sixteen, but I agreed. The point Erin emphasized was that she understood why her father was not always attentive, and she accepted that. "He does his best to be both a father and the CEO of Apple, and he juggles those pretty well," she said. "Sometimes I wish I had more of his attention, but I know the work he's doing is very important and I think it's really cool, so I'm fine. I don't really need more attention."

Jobs had promised to take each of his children on a trip of their choice when they became teenagers. Reed chose to go to Kyoto, knowing how much his father was entranced by the Zen calm of that beautiful city. Not surprisingly, when Erin turned thirteen, in 2008, she chose Kyoto as well. Her father's illness caused him to cancel the trip, so he promised to take her in 2010, when he was better. But that June he decided he didn't want to go. Erin was crestfallen but didn't protest. Instead her mother took her to France with family friends, and they rescheduled the Kyoto trip for July.

Powell worried that her husband would again cancel, so she was thrilled when the whole family took off in early July for Kona Village, Hawaii, which was the first leg of the trip. But in Hawaii Jobs developed a bad toothache, which he ignored, as if he could will the cavity away. The tooth collapsed and had to be fixed. Then the iPhone 4 antenna crisis hit, and he decided to rush back to Cupertino, taking Reed with him. Powell and Erin stayed in Hawaii, hoping that Jobs would return and continue with the plans to take them to Kyoto.

To their relief, and mild surprise, Jobs actually did return to Hawaii after his press conference to pick them up and take them to Japan. "It's a miracle," Powell told a friend. While Reed took care of Eve back in Palo Alto, Erin and her parents stayed at the Tawaraya Ryokan, an inn of sublime simplicity that Jobs loved. "It was fantastic," Erin recalled.

Twenty years earlier Jobs had taken Erin's half-sister, Lisa Brennan-Jobs, to Japan when she was about the same age. Among her strongest memories was sharing with him delightful meals and watching him, usually such a picky eater, savor unagi sushi and other delicacies. Seeing him take joy in eating made Lisa feel relaxed with him for the first time. Erin recalled a similar experience: "Dad knew where he wanted to go to lunch every day. He told me he knew an incredible soba shop, and he took me there, and it was so good that it's been hard to ever eat soba again because nothing comes close." They also found a tiny neighborhood sushi restaurant, and Jobs tagged it on his iPhone as "best sushi I've ever had." Erin agreed.

They also visited Kyoto's famous Zen Buddhist temples; the one Erin loved most was Saihō-ji,

771

known as the "moss temple" because of its Golden Pond surrounded by gardens featuring more than a hundred varieties of moss. "Erin was really really happy, which was deeply gratifying and helped improve her relationship with her father," Powell recalled. "She deserved that."

Their younger daughter, Eve, was quite a different story. She was spunky, self-assured, and in no way intimidated by her father. Her passion was horseback riding, and she became determined to make it to the Olympics. When a coach told her how much work it would require, she replied, "Tell me exactly what I need to do. I will do it." He did, and she began diligently following the program.

Eve was an expert at the difficult task of pinning her father down; she often called his assistant at work directly to make sure something got put on his calendar. She was also pretty good as a negotiator. One weekend in 2010, when the family was planning a trip, Erin wanted to delay the departure by half a day, but she was afraid to ask her father. Eve, then twelve, volunteered to take on the task, and at dinner she laid out the case to her father as if she were a lawyer before the Supreme Court. Jobs cut her off — "No, I don't think I want to" — but it was clear that he was more amused than annoyed. Later that evening Eve sat down with her mother and deconstructed the various ways that she could have made her case better.

Jobs came to appreciate her spirit — and see a lot of himself in her. "She's a pistol and has the strongest will of any kid I've ever met," he said. "It's like payback." He had a deep understanding of her personality, perhaps because it bore some resemblance to his. "Eve is more sensitive than a lot of people think," he explained. "She's so smart that she can roll over people a bit, so that means she can alienate people, and she finds herself alone. She's in the process of learning how to be who she is, but tempers it around the edges so that she can have the friends that she needs."

Jobs's relationship with his wife was sometimes complicated but always loyal. Savvy and compassionate, Laurene Powell was a stabilizing influence and an example of his ability to compensate for some of his selfish impulses by surrounding himself with strong-willed and sensible people. She weighed in quietly on business issues, firmly on family concerns, and fiercely on medical matters. Early in their marriage, she cofounded and launched College Track, a national after-school program that helps disadvantaged kids graduate from high school and get into college. Since then she had become a leading force in the education reform movement. Jobs professed an admiration for his wife's work: "What she's done with College Track really impresses me." But he tended to be generally dismissive of philanthropic endeavors and never visited her after-school centers.

In February 2010 Jobs celebrated his fifty-fifth birthday with just his family. The kitchen was decorated with streamers and balloons, and his kids gave him a red-velvet toy crown, which he wore. Now that he had recovered from a grueling year of health problems, Powell hoped that he would become more attentive to his family. But for the most part he resumed his focus on his work. "I think it was hard on the family, especially the girls," she told me. "After two years of him being ill, he finally gets a little better, and they expected he would focus a bit on them, but he didn't." She wanted to make sure, she said, that both sides of his personality were reflected in this book and put into context. "Like many great men whose gifts are extraordinary, he's not extraordinary in every realm," she said. "He doesn't have social graces, such as putting himself in other people's shoes, but he cares deeply about empowering humankind, the advancement of humankind, and putting the right tools in their hands."

PRESIDENT OBAMA

On a trip to Washington in the early fall of 2010, Powell had met with some of her friends at the White House who told her that President Obama was going to Silicon Valley that October. She suggested that he might want to meet with her husband. Obama's aides liked the idea; it fit into his new emphasis on competitiveness. In addition, John Doerr, the venture capitalist who had become one of Jobs's close friends, had told a meeting of the President's Economic Recovery Advisory Board about Jobs's views on why the United States was losing its edge. He too suggested that Obama should meet with Jobs. So a half hour was put on the president's schedule for a session at the Westin San Francisco Airport.

There was one problem: When Powell told her husband, he said he didn't want to do it. He was annoyed that she had arranged it behind his back. "I'm not going to get slotted in for a token meeting so that he can check off that he met with a CEO," he told her. She insisted that Obama was "really psyched to meet with you." Jobs replied that if that were the case, then Obama should call and personally ask for the meeting. The standoff went on for five days. She called in Reed, who was at Stanford, to come home for dinner and try to persuade his father. Jobs finally relented.

The meeting actually lasted forty-five minutes, and Jobs did not hold back. "You're headed for a one-term presidency," Jobs told Obama at the outset. To prevent that, he said, the administration needed to be a lot more business-friendly. He described how easy it was to build a factory in China, and said that it was almost impossible to do so these days in America, largely because of regulations and unnecessary costs.

Jobs also attacked America's education system, saying that it was hopelessly antiquated and crippled by union work rules. Until the teachers' unions were broken, there was almost no hope for education reform. Teachers should be treated as professionals, he said, not as industrial assembly-line workers. Principals should be able to hire and fire them based on how good they were. Schools should be staying open until at least 6 p.m. and be in session eleven months of the year. It was absurd, he added, that American classrooms were still based on teachers standing at a board and using textbooks. All books, learning materials, and assessments should be digital and interactive, tailored to each student and providing feedback in real time.

Jobs offered to put together a group of six or seven CEOs who could really explain the innovation challenges facing America, and the president accepted. So Jobs made a list of people for a Washington meeting to be held in December. Unfortunately, after Valerie Jarrett and other presidential aides had added names, the list had expanded to more than twenty, with GE's Jeffrey Immelt in the lead. Jobs sent Jarrett an email saying it was a bloated list and he had no intention of coming. In fact his health problems had flared anew by then, so he would not have been able to go in any case, as Doerr privately explained to the president.

In February 2011, Doerr began making plans to host a small dinner for President Obama in Silicon Valley. He and Jobs, along with their wives, went to dinner at Evvia, a Greek restaurant in Palo Alto, to draw up a tight guest list. The dozen chosen tech titans included Google's Eric Schmidt, Yahoo's Carol Bartz, Facebook's Mark Zuckerberg, Cisco's John Chambers, Oracle's Larry Ellison, Genentech's Art Levinson, and Netflix's Reed Hastings. Jobs's attention to the details of the dinner extended to the food. Doerr sent him the proposed menu, and he responded that some of the dishes proposed by the caterer — shrimp, cod, lentil salad — were far too fancy "and not who you are, John." He particularly objected to the dessert that was planned, a cream pie tricked out with chocolate truffles, but the White House advance staff overruled him by telling the caterer that the president liked cream pie. Because Jobs had lost so much weight that he was easily chilled, Doerr kept the house so warm that Zuckerberg found himself sweating profusely.

Jobs, sitting next to the president, kicked off

the dinner by saying, "Regardless of our political persuasions, I want you to know that we're here to do whatever you ask to help our country." Despite that, the dinner initially became a litany of suggestions of what the president could do for the businesses there. Chambers, for example, pushed a proposal for a repatriation tax holiday that would allow major corporations to avoid tax payments on overseas profits if they brought them back to the United States for investment during a certain period. The president was annoyed, and so was Zuckerberg, who turned to Valerie Jarrett, sitting to his right, and whispered, "We should be talking about what's important to the country. Why is he just talking about what's good for him?"

Doerr was able to refocus the discussion by calling on everyone to suggest a list of action items. When Jobs's turn came, he stressed the need for more trained engineers and suggested that any foreign students who earned an engineering degree in the United States should be given a visa to stay in the country. Obama said that could be done only in the context of the "Dream Act," which would allow illegal aliens who arrived as minors and finished high school to become legal residents — something that the Republicans had blocked. Jobs found this an annoying example of how politics can lead to paralysis. "The president is very smart, but he kept explaining to us reasons why things can't get done," he recalled. "It infuriates me."

Jobs went on to urge that a way be found to train more American engineers. Apple had 700,000 factory workers employed in China, he said, and that was because it needed 30,000 engineers on-site to support those workers. "You can't find that many in America to hire," he said. These factory engineers did not have to be PhDs or geniuses; they simply needed to have basic engineering skills for manufacturing. Tech schools, community colleges, or trade schools could train them. "If you could educate these engineers," he said, "we could move more manufacturing plants here." The argument made a strong impression on the president. Two or three times over the next month he told his aides, "We've got to find ways to train those 30,000 manufacturing engineers that Jobs told us about."

Jobs was pleased that Obama followed up, and they talked by telephone a few times after the meeting. He offered to help create Obama's political ads for the 2012 campaign. (He had made the same offer in 2008, but he'd become annoyed when Obama's strategist David Axelrod wasn't totally deferential.) "I think political advertising is terrible. I'd love to get Lee Clow out of retirement, and we can come up with great commercials for him," Jobs told me a few weeks after the dinner. Jobs had been fighting pain all week, but the talk of politics energized him. "Every once in a while, a real ad pro gets involved, the way Hal Riney did with 'It's morning in America' for Reagan's re-election in 1984. So that's what I'd like to do for Obama."

THIRD MEDICAL LEAVE, 2011

The cancer always sent signals as it reappeared. Jobs had learned that. He would lose his appetite and begin to feel pains throughout his body. His

778

doctors would do tests, detect nothing, and reassure him that he still seemed clear. But he knew better. The cancer had its signaling pathways, and a few months after he felt the signs the doctors would discover that it was indeed no longer in remission.

Another such downturn began in early November 2010. He was in pain, stopped eating, and had to be fed intravenously by a nurse who came to the house. The doctors found no sign of more tumors, and they assumed that this was just another of his periodic cycles of fighting infections and digestive maladies. He had never been one to suffer pain stoically, so his doctors and family had become somewhat inured to his complaints.

He and his family went to Kona Village for Thanksgiving, but his eating did not improve. The dining there was in a communal room, and the other guests pretended not to notice as Jobs, looking emaciated, rocked and moaned at meals, not touching his food. It was a testament to the resort and its guests that his condition never leaked out. When he returned to Palo Alto, Jobs became increasingly emotional and morose. He thought he was going to die, he told his kids, and he would get choked up about the possibility that he would never celebrate any more of their birthdays.

By Christmas he was down to 115 pounds, which was more than fifty pounds below his normal weight. Mona Simpson came to Palo Alto for the holiday, along with her ex-husband, the television comedy writer Richard Appel, and their children. The mood picked up a bit. The families played parlor games such as Novel, in which participants try to fool each other by seeing who can write the most convincing fake opening sentence to a book, and things seemed to be looking up for a while. He was even able to go out to dinner at a restaurant with Powell a few days after Christmas. The kids went off on a ski vacation for New Year's, with Powell and Mona Simpson taking turns staying at home with Jobs in Palo Alto.

By the beginning of 2011, however, it was clear that this was not merely one of his bad patches. His doctors detected evidence of new tumors, and the cancer-related signaling further exacerbated his loss of appetite. They were struggling to determine how much drug therapy his body, in its emaciated condition, would be able to take. Every inch of his body felt like it had been punched, he told friends, as he moaned and sometimes doubled over in pain.

It was a vicious cycle. The first signs of cancer caused pain. The morphine and other painkillers he took suppressed his appetite. His pancreas had been partly removed and his liver had been re-placed, so his digestive system was faulty and had trouble absorbing protein. Losing weight made it harder to embark on aggressive drug therapies. His emaciated condition also made him more suscep-tible to infections, as did the immunosuppressants he sometimes took to keep his body from rejecting his liver transplant. The weight loss reduced the lipid layers around his pain receptors, causing him to suffer more. And he was prone to extreme mood swings, marked by prolonged bouts of anger and depression, which further suppressed his appetite.

Jobs's eating problems were exacerbated over the years by his psychological attitude toward food.

When he was young, he learned that he could induce euphoria and ecstasy by fasting. So even though he knew that he should eat — his doctors were begging him to consume high-quality protein — lingering in the back of his subconscious, he admitted, was his instinct for fasting and for diets like Arnold Ehret's fruit regimen that he had embraced as a teenager. Powell kept telling him that it was crazy, even pointing out that Ehret had died at fifty-six when he stumbled and knocked his head, and she would get angry when he came to the table and just stared silently at his lap. "I wanted him to force himself to eat," she said, "and it was incredibly tense at home." Bryar Brown, their part-time cook, would still come in the afternoon and make an array of healthy dishes, but Jobs would touch his tongue to one or two dishes and then dismiss them all as inedible. One evening he announced, "I could probably eat a little pumpkin pie," and the even-tempered Brown created a beautiful pie from scratch in an hour. Jobs ate only one bite, but Brown was thrilled.

Powell talked to eating disorder specialists and psychiatrists, but her husband tended to shun them. He refused to take any medications, or be treated in any way, for his depression. "When you have feelings," he said, "like sadness or anger about your cancer or your plight, to mask them is to lead an artificial life." In fact he swung to the other extreme. He became morose, tearful, and dramatic as he lamented to all around him that he was about to die. The depression became part of the vicious cycle by making him even less likely to eat.

Pictures and videos of Jobs looking emaciated began to appear online, and soon rumors were swirling about how sick he was. The problem, Powell realized, was that the rumors were true, and they were not going to go away. Jobs had agreed only reluctantly to go on medical leave two years earlier, when his liver was failing, and this time he also resisted the idea. It would be like leaving his homeland, unsure that he would ever return. When he finally bowed to the inevitable, in January 2011, the board members were expecting it; the telephone meeting in which he told them that he wanted another leave took only three minutes. He had often discussed with the board, in executive session, his thoughts about who could take over if anything happened to him, presenting both short-term and longer-term combinations of options. But there was no doubt that, in this current situation, Tim Cook would again take charge of day-to-day operations.

The following Saturday afternoon, Jobs allowed his wife to convene a meeting of his doctors. He realized that he was facing the type of problem that he never permitted at Apple. His treatment was fragmented rather than integrated. Each of his myriad maladies was being treated by different specialists — oncologists, pain specialists, nutritionists, hepatologists, and hematologists — but they were not being coordinated in a cohesive approach, the way James Eason had done in Memphis. "One of the big issues in the health care industry is the lack of caseworkers or advocates that are the quarterback of each team," Powell said. This was particularly true at Stanford, where nobody seemed in charge of figuring out how nutrition was related to pain care and to oncology. So Powell asked the various Stanford specialists to come to their house for a meeting that also included some outside doctors with a more aggressive and integrated approach, such as David Agus of USC. They agreed on a new regimen for dealing with the pain and for coordinating the other treatments.

Thanks to some pioneering science, the team of doctors had been able to keep Jobs one step ahead of the cancer. He had become one of the first twenty people in the world to have all of the genes of his cancer tumor as well as of his normal DNA sequenced. It was a process that, at the time, cost more than $100,000.

The gene sequencing and analysis were done collaboratively by teams at Stanford, Johns Hopkins, and the Broad Institute of MIT and Harvard. By knowing the unique genetic and molecular signature of Jobs's tumors, his doctors had been able to pick specific drugs that directly targeted the defective molecular pathways that caused his cancer cells to grow in an abnormal manner. This approach, known as molecular targeted therapy, was more effective than traditional chemotherapy, which attacks the process of division of all the body's cells, cancerous or not. This targeted therapy was not a silver bullet, but at times it seemed close to one: It allowed his doctors to look at a large number of drugs — common and uncommon, already available or only in development — to see which three or four might work best. Whenever his cancer mutated and repaved around one of these drugs, the doctors had another drug lined up to go next.

Although Powell was diligent in overseeing her husband's care, he was the one who made the final decision on each new treatment regimen. A typical example occurred in May 2011, when he held a meeting with George Fisher and other doctors from Stanford, the gene-sequencing analysts from the Broad Institute, and his outside consultant David Agus. They all gathered around a table at a suite in the Four Seasons hotel in Palo Alto. Powell did not come, but their son, Reed, did. For three hours there were presentations from the Stanford and Broad researchers on the new information they had learned about the genetic signatures of his cancer. Jobs was his usual feisty self. At one point he stopped a Broad Institute analyst who had made the mistake of using PowerPoint slides. Jobs chided him and explained why Apple's Keynote presentation software was better; he even offered to teach him how to use it. By the end of the meeting, Jobs and his team had gone through all of the molecular data, assessed the rationales for each of the potential therapies, and come up with a list of tests to help them better prioritize these.

One of his doctors told him that there was hope that his cancer, and others like it, would soon be considered a manageable chronic disease, which could be kept at bay until the patient died of something else. "I'm either going to be one of the first to be able to outrun a cancer like this, or I'm going to be one of the last to die from it," Jobs told me right after one of the meetings with his doctors. "Either among the first to make it to shore, or the last to get dumped."

When his 2011 medical leave was announced, the situation seemed so dire that Lisa Brennan-Jobs got back in touch after more than a year and arranged to fly from New York the following week. Her relationship with her father had been built on layers of resentment. She was understandably scarred by having been pretty much abandoned by him for her first ten years. Making matters worse, she had inherited some of his prickliness and, he felt, some of her mother's sense of grievance. "I told her many times that I wished I'd been a better dad when she was five, but now she should let things go rather than be angry the rest of her life," he recalled just before Lisa arrived.

The visit went well. Jobs was beginning to feel a little better, and he was in a mood to mend fences and express his affection for those around him. At age thirty-two, Lisa was in a serious relationship for one of the first times in her life. Her boyfriend was a struggling young filmmaker from California, and Jobs went so far as to suggest she move back to Palo Alto if they got married. "Look, I don't know how long I am for this world," he told her. "The doctors can't really tell me. If you want to see more of me, you're going to have to move out here. Why don't you consider it?" Even though Lisa did not move west, Jobs was pleased at how the reconciliation had worked out. "I hadn't been sure I wanted her to visit, because I was sick and didn't want other complications. But I'm very glad she came. It helped settle a lot of things in me."

■ ■ ■ ■

Jobs had another visit that month from someone who wanted to repair fences. Google's cofounder Larry Page, who lived less than three blocks away, had just announced plans to retake the reins of the company from Eric Schmidt. He knew how to flatter Jobs: He asked if he could come by and get tips on how to be a good CEO. Jobs was still furious at Google. "My first thought was, 'Fuck you,' " he recounted. "But then I thought about it and realized that everybody helped me when I was young, from Bill Hewlett to the guy down the block who worked for HP. So I called him back and said sure." Page came over, sat in Jobs's living room, and listened to his ideas on building great products and durable companies. Jobs recalled:

We talked a lot about focus. And choosing people. How to know who to trust, and how to build a team of lieutenants he can count on. I described the blocking and tackling he would have to do to keep the company from getting flabby or being larded with B players. The main thing I stressed was focus. Figure out what Google wants to be when it grows up. It's now all over the map. What are the five products you want to focus on? Get rid of the rest, because they're dragging you down. They're turning you into Microsoft. They're causing you to turn out products that are adequate but not great. I tried to be as helpful as I could. I will continue to do that with people like Mark Zuckerberg too. That's how I'm going to spend part of the time I have left. I can help the next generation remember the lineage of great companies here and how to continue the tradition. The Valley has been very supportive of me. I should do my best to repay.

The announcement of Jobs's 2011 medical leave prompted others to make a pilgrimage to the house in Palo Alto. Bill Clinton, for example, came by and talked about everything from the Middle East to American politics. But the most poignant visit was from the other tech prodigy born in 1955, the guy who, for more than three decades, had been Jobs's rival and partner in defining the age of personal computers.

Bill Gates had never lost his fascination with Jobs. In the spring of 2011 I was at a dinner with him in Washington, where he had come to discuss his foundation's global health endeavors. He expressed amazement at the success of the iPad and how Jobs, even while sick, was focusing on ways to improve it. "Here I am, merely saving the world from malaria and that sort of thing, and Steve is still coming up with amazing new products," he said wistfully. "Maybe I should have stayed in that game." He smiled to make sure that I knew he was joking, or at least half joking.

Through their mutual friend Mike Slade, Gates made arrangements to visit Jobs in May. The day before it was supposed to happen, Jobs's assistant called to say he wasn't feeling well enough. But it was rescheduled, and early one afternoon Gates drove to Jobs's house, walked through the back gate to the open kitchen door, and saw Eve studying at the table. "Is Steve around?" he asked. Eve pointed him to the living room.

They spent more than three hours together, just the two of them, reminiscing. "We were like the old guys in the industry looking back," Jobs recalled. "He was happier than I've ever seen him, and I kept thinking how healthy he looked." Gates was similarly struck by how Jobs, though scarily gaunt, had more energy than he expected. He was open about his health problems and, at least that day, feeling optimistic. His sequential regimens of targeted drug treatments, he told Gates, were like "jumping from one lily pad to another," trying to stay a step ahead of the cancer.

Jobs asked some questions about education, and Gates sketched out his vision of what schools in the future would be like, with students watching lectures and video lessons on their own while using the classroom time for discussions and problem solving. They agreed that computers had, so far, made surprisingly little impact on schools — far less than on other realms of society such as media and medicine and law. For that to change, Gates said, computers and mobile devices would have to focus on delivering more personalized lessons and providing motivational feedback.

They also talked a lot about the joys of family, including how lucky they were to have good kids and be married to the right women. "We laughed about how fortunate it was that he met Laurene, and she's kept him semi-sane, and I met Melinda, and she's kept me semi-sane," Gates recalled. "We also discussed how it's challenging to be one of our children, and how do we mitigate that. It was pretty personal." At one point Eve, who in the past had been in horse shows with Gates's daughter Jennifer, wandered in from the kitchen, and Gates asked her what jumping routines she liked best.

As their hours together drew to a close, Gates complimented Jobs on "the incredible stuff" he had created and for being able to save Apple in the late 1990s from the bozos who were about to destroy it. He even made an interesting concession. Throughout their careers they had adhered to competing philosophies on one of the most fundamental of all digital issues: whether hardware and software should be tightly integrated or more open. "I used to believe that the open, horizontal model would prevail," Gates told him. "But you proved that the integrated, vertical model could also be great." Jobs responded with his own admission. "Your model worked too," he said.

They were both right. Each model had worked in the realm of personal computers, where Macintosh coexisted with a variety of Windows machines, and that was likely to be true in the realm of mobile devices as well. But after recounting their discussion, Gates added a caveat: "The integrated approach works well when Steve is at the helm. But it doesn't mean it will win many rounds in the future." Jobs similarly felt compelled to add a caveat about Gates after describing their meeting: "Of course, his fragmented model worked, but it didn't make really great products. It produced crappy products. That was the problem. The big problem. At least over time."

"THAT DAY HAS COME"

Jobs had many other ideas and projects that he

hoped to develop. He wanted to disrupt the text-book industry and save the spines of spavined students bearing backpacks by creating electronic texts and curriculum material for the iPad. He was also working with Bill Atkinson, his friend from the original Macintosh team, on devising new digital technologies that worked at the pixel level to allow people to take great photographs using their iPhones even in situations without much light. And he very much wanted to do for television sets what he had done for computers, music players, and phones: make them simple and elegant. "I'd like to create an integrated television set that is completely easy to use," he told me. "It would be seamlessly synced with all of your devices and with iCloud." No longer would users have to fiddle with complex remotes for DVD players and cable channels. "It will have the simplest user interface you could imagine. I finally cracked it."

But by July 2011, his cancer had spread to his bones and other parts of his body, and his doctors were having trouble finding targeted drugs that could beat it back. He was in pain, sleeping erratically, had little energy, and stopped going to work. He and Powell had reserved a sailboat for a family cruise scheduled for the end of that month, but those plans were scuttled. He was eating almost no solid food, and he spent most of his days in his bedroom watching television.

In August, I got a message that he wanted me to come visit. When I arrived at his house, at mid-morning on a Saturday, he was still asleep, so I sat with his wife and kids in the garden, filled with a profusion of yellow roses and various types of daisies, until he sent word that I should come in. I found him curled up on the bed, wearing khaki shorts and a white turtleneck. His legs were shockingly sticklike, but his smile was easy and his mind quick. "We better hurry, because I have very little energy," he said.

He wanted to show me some of his personal pictures and let me pick a few to use in the book. Because he was too weak to get out of bed, he pointed to various drawers in the room, and I carefully brought him the photographs in each. As I sat on the side of the bed, I held them up, one at a time, so he could see them. Some prompted stories; others merely elicited a grunt or a smile. I had never seen a picture of his father, Paul Jobs, and I was startled when I came across a snapshot of a handsome hardscrabble 1950s dad holding a toddler. "Yes, that's him," he said. "You can use it." He then pointed to a box near the window that contained a picture of his father looking at him lovingly at his wedding. "He was a great man," Jobs said quietly. I murmured something along the lines of "He would have been proud of you." Jobs corrected me: "He *was* proud of me."

For a while, the pictures seemed to energize him. We discussed what various people from his past, ranging from Tina Redse to Mike Markkula to Bill Gates, now thought of him. I recounted what Gates had said after he described his last visit with Jobs, which was that Apple had shown that the integrated approach could work, but only "when Steve is at the helm." Jobs thought that was silly. "Anyone could make better products that way, not just me," he said. So I asked him to name another company that made great products by insisting on end-to-end integration. He thought for a while, trying to come up with an example. "The car companies," he finally said, but then he added, "Or at least they used to."

When our discussion turned to the sorry state of the economy and politics, he offered a few sharp opinions about the lack of strong leadership around the world. "I'm disappointed in Obama," he said. "He's having trouble leading because he's reluctant to offend people or piss them off." He caught what I was thinking and assented with a little smile: "Yes, that's not a problem I ever had."

After two hours, he grew quiet, so I got off the bed and started to leave. "Wait," he said, as he waved to me to sit back down. It took a minute or two for him to regain enough energy to talk. "I had a lot of trepidation about this project," he finally said, referring to his decision to cooperate with this book. "I was really worried."

"Why did you do it?" I asked.

"I wanted my kids to know me," he said. "I wasn't always there for them, and I wanted them to know why and to understand what I did. Also, when I got sick, I realized other people would write about me if I died, and they wouldn't know anything. They'd get it all wrong. So I wanted to make sure someone heard what I had to say."

He had never, in two years, asked anything about what I was putting in the book or what conclusions I had drawn. But now he looked at me and said, "I know there will be a lot in your book I won't like." It was more a question than a statement, and when he stared at me for a response, I nodded, smiled, and said I was sure that would be true. "That's good," he said. "Then it won't seem like an in-house book. I won't read it for a while, because I don't want to get mad. Maybe I will read it in a year — if I'm still around." By then, his eyes were closed and his energy gone, so I quietly took my leave.

As his health deteriorated throughout the summer, Jobs slowly began to face the inevitable: He would not be returning to Apple as CEO. So it was time for him to resign. He wrestled with the decision for weeks, discussing it with his wife, Bill Campbell, Jony Ive, and George Riley. "One of the things I wanted to do for Apple was to set an example of how you do a transfer of power right," he told me. He joked about all the rough transitions that had occurred at the company over the past thirty-five years. "It's always been a drama, like a third-world country. Part of my goal has been to make Apple the world's best company, and having an orderly transition is key to that."

The best time and place to make the transition, he decided, was at the company's regularly scheduled August 24 board meeting. He was eager to do it in person, rather than merely send in a letter or attend by phone, so he had been pushing himself to eat and regain strength. The day before the meeting, he decided he could make it, but he needed the help of a wheelchair. Arrangements were made to have him driven to headquarters and wheeled to the boardroom as secretly as possible.

He arrived just before 11 a.m., when the board members were finishing committee reports and other routine business. Most knew what was about to happen. But instead of going right to the topic on everyone's mind, Tim Cook and Peter Oppenheimer, the chief financial officer, went through the results for the quarter and the projections for the year ahead. Then Jobs said quietly that he had something personal to say. Cook asked if he and the other top managers should leave, and Jobs paused for more than thirty seconds before he decided they should. Once the room was cleared of all but the six outside directors, he began to read aloud from a letter he had dictated and revised over the previous weeks. "I have always said if there ever came a day when I could no longer meet my duties and expectations as Apple's CEO, I would be the first to let you know," it began. "Unfortunately, that day has come."

The letter was simple, direct, and only eight sentences long. In it he suggested that Cook replace him, and he offered to serve as chairman of the board. "I believe Apple's brightest and most innovative days are ahead of it. And I look forward to watching and contributing to its success in a new role."

There was a long silence. Al Gore was the first to speak, and he listed Jobs's accomplishments during his tenure. Mickey Drexler added that watching Jobs transform Apple was "the most incredible thing I've ever seen in business," and Art Levinson praised Jobs's diligence in ensuring that there was a smooth transition. Campbell said nothing, but there were tears in his eyes as the formal resolutions transferring power were passed.

Over lunch, Scott Forstall and Phil Schiller

came in to display mockups of some products that Apple had in the pipeline. Jobs peppered them with questions and thoughts, especially about what capacities the fourth-generation cellular networks might have and what features needed to be in future phones. At one point Forstall showed off a voice recognition app. As he feared, Jobs grabbed the phone in the middle of the demo and proceeded to see if he could confuse it. "What's the weather in Palo Alto?" he asked. The app answered. After a few more questions, Jobs challenged it: "Are you a man or a woman?" Amazingly, the app answered in its robotic voice, "They did not assign me a gender." For a moment the mood lightened.

When the talk turned to tablet computing, some expressed a sense of triumph that HP had suddenly given up the field, unable to compete with the iPad. But Jobs turned somber and declared that it was actually a sad moment. "Hewlett and Packard built a great company, and they thought they had left it in good hands," he said. "But now it's being dismembered and destroyed. It's tragic. I hope I've left a stronger legacy so that will never happen at Apple." As he prepared to leave, the board members gathered around to give him a hug.

After meeting with his executive team to explain the news, Jobs rode home with George Riley. When they arrived at the house, Powell was in the backyard harvesting honey from her hives, with help from Eve. They took off their screen helmets and brought the honey pot to the kitchen, where Reed and Erin had gathered, so that they could all celebrate the graceful transition. Jobs took a spoonful of the honey and pronounced it wonder-fully sweet.

That evening, he stressed to me that his hope was to remain as active as his health allowed. "I'm going to work on new products and marketing and the things that I like," he said. But when I asked how it really felt to be relinquishing control of the company he had built, his tone turned wistful, and he shifted into the past tense. "I've had a very lucky career, a very lucky life," he replied. "I've done all that I can do."

Chapter Forty Two: Legacy
THE BRIGHTEST HEAVEN OF INVENTION

FireWire

His personality was reflected in the products he created. Just as the core of Apple's philosophy, from the original Macintosh in 1984 to the iPad a generation later, was the end-to-end integration of hardware and software, so too was it the case with Steve Jobs: His passions, perfectionism, demons, desires, artistry, devilry, and obsession for control were integrally connected to his approach to business and the products that resulted.

The unified field theory that ties together Jobs's personality and products begins with his most salient trait: his intensity. His silences could be as searing as his rants; he had taught himself to stare without blinking. Sometimes this intensity was charming, in a geeky way, such as when he was explaining the profundity of Bob Dylan's music or why whatever product he was unveiling at that moment was the most amazing thing that Apple had ever made. At other times it could be terrifying, such as when he was fulminating about Google or Microsoft ripping off Apple.

This intensity encouraged a binary view of the world. Colleagues referred to the hero/shithead dichotomy. You were either one or the other, sometimes on the same day. The same was true of products, ideas, even food: Something was either "the best thing ever," or it was shitty, brain-dead, inedible. As a result, any perceived flaw could set off a rant. The finish on a piece of metal, the curve of the head of a screw, the shade of blue on a box, the intuitiveness of a navigation screen — he would declare them to "completely suck" until that moment when he suddenly pronounced them "absolutely perfect." He thought of himself as an artist, which he was, and he indulged in the temperament of one.

His quest for perfection led to his compulsion for Apple to have end-to-end control of every product that it made. He got hives, or worse, when contemplating great Apple software running on another company's crappy hardware, and he likewise was allergic to the thought of unapproved apps or content polluting the perfection of an Apple device. This ability to integrate hardware and software and content into one unified system enabled him to impose simplicity. The astronomer Johannes Kepler declared that "nature loves simplicity and unity." So did Steve Jobs.

This instinct for integrated systems put him squarely on one side of the most fundamental divide in the digital world: open versus closed. The hacker ethos handed down from the Homebrew Computer Club favored the open approach, in which there was little centralized control and people were free to modify hardware and software, share code, write to open standards, shun proprietary systems, and have content and apps that were compatible with a variety of devices and operating systems. The young Wozniak was in that camp: The Apple II he designed was easily opened and sported plenty of slots and ports that people could jack into as they pleased. With the Macintosh Jobs became a founding father of the other camp. The Macintosh would be like an appliance, with the hardware and software tightly woven together and closed to modifications. The hacker ethos would be sacrificed in order to create a seamless and simple user experience.

This led Jobs to decree that the Macintosh operating system would not be available for any other company's hardware. Microsoft pursued the opposite strategy, allowing its Windows operating system to be promiscuously licensed. That did not produce the most elegant computers, but it did lead to Microsoft's dominating the world of operating systems. After Apple's market share shrank to less than 5%, Microsoft's approach was declared the winner in the personal computer realm.

In the longer run, however, there proved to be some advantages to Jobs's model. Even with a small market share, Apple was able to maintain a huge profit margin while other computer makers were commoditized. In 2010, for example, Apple had just 7% of the revenue in the personal computer market, but it grabbed 35% of the operating profit.

More significantly, in the early 2000s Jobs's insistence on end-to-end integration gave Apple an advantage in developing a digital hub strategy, which allowed your desktop computer to link seamlessly with a variety of portable devices. The iPod, for example, was part of a closed and tightly integrated system. To use it, you had to use Apple's iTunes software and download content from its iTunes Store. The result was that the iPod, like the iPhone and iPad that followed, was an elegant delight in contrast to the kludgy rival products that did not offer a seamless end-to-end experience.

The strategy worked. In May 2000 Apple's market value was one-twentieth that of Microsoft. In May 2010 Apple surpassed Microsoft as the world's most valuable technology company, and by September 2011 it was worth 70% more than Microsoft. In the first quarter of 2011 the market for Windows PCs shrank by 1%, while the market for Macs grew 28%.

By then the battle had begun anew in the world of mobile devices. Google took the more open approach, and it made its Android operating system available for use by any maker of tablets or cell phones. By 2011 its share of the mobile market matched Apple's. The drawback of Android's openness was the fragmentation that resulted. Various handset and tablet makers modified Android into dozens of variants and flavors, making it hard for apps to remain consistent or make full use if its features. There were merits to both approaches. Some people wanted the freedom to use more open systems and have more choices of hardware; others clearly preferred Apple's tight integration and control, which led to products that had simpler interfaces, longer battery life, greater user-friendliness, and easier handling of content.

The downside of Jobs's approach was that his

desire to delight the user led him to resist empowering the user. Among the most thoughtful proponents of an open environment is Jonathan Zittrain of Harvard. He begins his book *The Future of the Internet — And How to Stop It* with the scene of Jobs introducing the iPhone, and he warns of the consequences of replacing personal computers with "sterile appliances tethered to a network of control." Even more fervent is Cory Doctorow, who wrote a manifesto called "Why I Won't Buy an iPad" for Boing Boing. "There's a lot of thoughtfulness and smarts that went into the design. But there's also a palpable contempt for the owner," he wrote. "Buying an iPad for your kids isn't a means of jump-starting the realization that the world is yours to take apart and reassemble; it's a way of telling your offspring that even changing the batteries is something you have to leave to the professionals."

For Jobs, belief in an integrated approach was a matter of righteousness. "We do these things not because we are control freaks," he explained. "We do them because we want to make great products, because we care about the user, and because we like to take responsibility for the entire experience rather than turn out the crap that other people make." He also believed he was doing people a service: "They're busy doing whatever they do best, and they want us to do what we do best. Their lives are crowded; they have other things to do than think about how to integrate their computers and devices."

This approach sometimes went against Apple's short-term business interests. But in a world filled with junky devices, inscrutable error messages, and annoying interfaces, it led to astonishing products marked by beguiling user experiences. Using an Apple product could be as sublime as walking in one of the Zen gardens of Kyoto that Jobs loved, and neither experience was created by worshipping at the altar of openness or by letting a thousand flowers bloom. Sometimes it's nice to be in the hands of a control freak.

Jobs's intensity was also evident in his ability to focus. He would set priorities, aim his laser attention on them, and filter out distractions. If something engaged him — the user interface for the original Macintosh, the design of the iPod and iPhone, getting music companies into the iTunes Store — he was relentless. But if he did not want to deal with something — a legal annoyance, a business issue, his cancer diagnosis, a family tug — he would resolutely ignore it. That focus allowed him to say no. He got Apple back on track by cutting all except a few core products. He made devices simpler by eliminating buttons, software simpler by eliminating features, and interfaces simpler by eliminating options.

He attributed his ability to focus and his love of simplicity to his Zen training. It honed his appreciation for intuition, showed him how to filter out anything that was distracting or unnecessary, and nurtured in him an aesthetic based on minimalism.

Unfortunately his Zen training never quite produced in him a Zen-like calm or inner serenity, and that too is part of his legacy. He was often tightly coiled and impatient, traits he made no effort to hide. Most people have a regulator between their mind and mouth that modulates their brutish sentiments and spikiest impulses. Not Jobs. He made a point of being brutally honest. "My job is to say when something sucks rather than sugarcoat it," he said. This made him charismatic and inspiring, yet also, to use the technical term, an asshole at times.

Andy Hertzfeld once told me, "The one question I'd truly love Steve to answer is, 'Why are you sometimes so mean?' " Even his family members wondered whether he simply lacked the filter that restrains people from venting their wounding thoughts or willfully bypassed it. Jobs claimed it was the former. "This is who I am, and you can't expect me to be someone I'm not," he replied when I asked him the question. But I think he actually could have controlled himself, if he had wanted. When he hurt people, it was not because he was lacking in emotional awareness. Quite the contrary: He could size people up, understand their inner thoughts, and know how to relate to them, cajole them, or hurt them at will.

The nasty edge to his personality was not necessary. It hindered him more than it helped him. But it did, at times, serve a purpose. Polite and velvety leaders, who take care to avoid bruising others, are generally not as effective at forcing change. Dozens of the colleagues whom Jobs most abused ended their litany of horror stories by saying that he got them to do things they never dreamed possible. And he created a corporation crammed with A players.

■ ■ ■ ■

The saga of Steve Jobs is the Silicon Valley creation myth writ large: launching a startup in his parents' garage and building it into the world's most valuable company. He didn't invent many things outright, but he was a master at putting together ideas, art, and technology in ways that invented the future. He designed the Mac after appreciating the power of graphical interfaces in a way that Xerox was unable to do, and he created the iPod after grasping the joy of having a thousand songs in your pocket in a way that Sony, which had all the assets and heritage, never could accomplish. Some leaders push innovations by being good at the big picture. Others do so by mastering details. Jobs did both, relentlessly. As a result he launched a series of products over three decades that transformed whole industries:

- The Apple II, which took Wozniak's circuit board and turned it into the first personal computer that was not just for hobbyists.
- The Macintosh, which begat the home computer revolution and popularized graphical user interfaces.
- *Toy Story* and other Pixar blockbusters, which opened up the miracle of digital imagination.
- Apple stores, which reinvented the role of a store in defining a brand.
- The iPod, which changed the way we consume music.
- The iTunes Store, which saved the music industry.

- The iPhone, which turned mobile phones into music, photography, video, email, and web devices.
- The App Store, which spawned a new content-creation industry.
- The iPad, which launched tablet computing and offered a platform for digital newspapers, magazines, books, and videos.
- iCloud, which demoted the computer from its central role in managing our content and let all of our devices sync seamlessly.
- And Apple itself, which Jobs considered his greatest creation, a place where imagination was nurtured, applied, and executed in ways so creative that it became the most valuable company on earth.

Was he smart? No, not exceptionally. Instead, he was a genius. His imaginative leaps were instinctive, unexpected, and at times magical. He was, indeed, an example of what the mathematician Mark Kac called a magician genius, someone whose insights come out of the blue and require intuition more than mere mental processing power. Like a pathfinder, he could absorb information, sniff the winds, and sense what lay ahead.

Steve Jobs thus became the greatest business executive of our era, the one most certain to be remembered a century from now. History will place him in the pantheon right next to Edison and Ford. More than anyone else of his time, he made products that were completely innovative, combining the power of poetry and processors. With a ferocity that could make working with him as unsettling as it was inspiring, he also built the world's most creative company. And he was able to infuse into its DNA the design sensibilities, perfectionism, and imagination that make it likely to be, even decades from now, the company that thrives best at the intersection of artistry and technology.

AND ONE MORE THING . . .

Biographers are supposed to have the last word. But this is a biography of Steve Jobs. Even though he did not impose his legendary desire for control on this project, I suspect that I would not be conveying the right feel for him — the way he asserted himself in any situation — if I just shuffled him onto history's stage without letting him have some last words.

Over the course of our conversations, there were many times when he reflected on what he hoped his legacy would be. Here are those thoughts, in his own words:

My passion has been to build an enduring company where people were motivated to make great products. Everything else was secondary. Sure, it was great to make a profit, because that was what allowed you to make great products. But the products, not the profits, were the motivation. Sculley flipped these priorities to where the goal was to make money. It's a subtle difference, but it ends up meaning everything: the people you hire, who gets promoted, what you discuss in meetings.

Some people say, "Give the customers what they want." But that's not my approach. Our job is to figure out what they're going to want before they do. I think Henry Ford once said, "If I'd asked customers what they wanted, they would have told me, 'A faster horse!' " People don't know what they want until you show it to them. That's why I never rely on market research. Our task is to read things that are not yet on the page.

Edwin Land of Polaroid talked about the intersection of the humanities and science. I like that intersection. There's something magical about that place. There are a lot of people innovating, and that's not the main distinction of my career. The reason Apple resonates with people is that there's a deep current of humanity in our innovation. I think great artists and great engineers are similar, in that they both have a desire to express themselves. In fact some of the best people working on the original Mac were poets and musicians on the side. In the seventies computers became a way for people to express their creativity. Great artists like Leonardo da Vinci and Michelangelo were also great at science. Michelangelo knew a lot about how to quarry stone, not just how to be a sculptor.

People pay us to integrate things for them, because they don't have the time to think about this stuff 24/7. If you have an extreme passion for producing great products, it pushes you to be integrated, to connect your hardware and your software and content management. You want to break new ground, so you have to do it yourself. If you want to allow your products to be open to other hardware or software, you have to give up some of your vision.

At different times in the past, there were companies that exemplified Silicon Valley. It was Hewlett-Packard for a long time. Then, in the semiconductor era, it was Fairchild and Intel. I think that it was Apple for a while, and then that faded. And then today, I think it's Apple and Google — and a little more so Apple. I think Apple has stood the test of time. It's been around for a while, but it's still at the cutting edge of what's going on.

It's easy to throw stones at Microsoft. They've clearly fallen from their dominance. They've become mostly irrelevant. And yet I appreciate what they did and how hard it was. They were very good at the business side of things. They were never as ambitious product-wise as they should have been. Bill likes to portray himself as a man of the product, but he's really not. He's a businessperson. Winning business was more important than making great products. He ended up the wealthiest guy around, and if that was his goal, then he achieved it. But it's never been my goal, and I wonder, in the end, if it was his goal. I admire him for the company he built — it's impressive — and I enjoyed working with him. He's bright and actually has a good sense of humor. But Microsoft never had the humanities and liberal arts in its DNA. Even when they saw the Mac, they couldn't copy it well. They totally didn't get it.

I have my own theory about why decline happens at companies like IBM or Microsoft. The company does a great job, innovates and becomes a monopoly or close to it in some field, and then the quality of the product becomes less important. The company starts valuing the great salesmen, because they're the ones who can move the needle on revenues, not the product engineers and designers. So the salespeople end up running the company. John Akers at IBM was a smart, eloquent, fantastic salesperson, but he didn't know anything about product. The same thing happened at Xerox. When the sales guys run the company, the product guys don't matter so much, and a lot of them just turn off. It happened at Apple when Sculley came in, which was my fault, and it happened when Ballmer took over at Microsoft. Apple was lucky and it rebounded, but I don't think anything will change at Microsoft as long as Ballmer is running it.

I hate it when people call themselves "entrepreneurs" when what they're really trying to do is launch a startup and then sell or go public, so they can cash in and move on. They're unwilling to do the work it takes to build a real company, which is the hardest work in business. That's how you really make a contribution and add to the legacy of those who went before. You build a company that will still stand for something a generation or two from now. That's what Walt Disney did, and Hewlett and Packard, and the people who built Intel. They created a company to last, not just to make money. That's what I want Apple to be.

I don't think I run roughshod over people, but if something sucks, I tell people to their face. It's my job to be honest. I know what I'm talking about, and I usually turn out to be right. That's the culture I tried to create. We are brutally honest with each other, and anyone can tell me they think I am full of shit and I can tell them the same. And we've had some rip-roaring arguments, where we are yelling at each other, and it's some of the best times I've ever had. I feel totally comfortable saying "Ron, that store looks like shit" in front of everyone else. Or I might say "God, we really fucked up the engineering on this" in front of the person that's responsible. That's the ante for being in the room: You've got to be able to be super honest. Maybe there's a better way, a gentlemen's club where we all wear ties and speak in this Brahmin language and velvet code-words, but I don't know that way, because I am middle class from California.

I was hard on people sometimes, probably harder than I needed to be. I remember the time when Reed was six years old, coming home, and I had just fired somebody that day, and I imagined what it was like for that person to tell his family and his young son that he had lost his job. It was hard. But somebody's got to do it. I figured that it was always my job to make sure that the team was excellent, and if I didn't do it, nobody was going to do it.

You always have to keep pushing to innovate. Dylan could have sung protest songs forever and probably made a lot of money, but he didn't. He had to move on, and when he did, by going electric in 1965, he alienated a lot of people. His 1966 Europe tour was his greatest. He would come on and do a set of acoustic guitar, and the audiences loved him. Then he brought out what became The Band, and they would all do an electric set, and the audience sometimes booed. There was one

810

point where he was about to sing "Like a Roll-ing Stone" and someone from the audience yells "Judas!" And Dylan then says, "Play it fucking loud!" And they did. The Beatles were the same way. They kept evolving, moving, refining their art. That's what I've always tried to do — keep mov-ing. Otherwise, as Dylan says, if you're not busy being born, you're busy dying.

What drove me? I think most creative people want to express appreciation for being able to take advantage of the work that's been done by others before us. I didn't invent the language or mathematics I use. I make little of my own food, none of my own clothes. Everything I do depends on other members of our species and the shoul-ders that we stand on. And a lot of us want to contribute something back to our species and to add something to the flow. It's about trying to ex-press something in the only way that most of us know how — because we can't write Bob Dylan songs or Tom Stoppard plays. We try to use the talents we do have to express our deep feelings, to show our appreciation of all the contributions that came before us, and to add something to that flow. That's what has driven me.

CODA

One sunny afternoon, when he wasn't feeling well, Jobs sat in the garden behind his house and re-flected on death. He talked about his experiences in India almost four decades earlier, his study of Buddhism, and his views on reincarnation and spiritual transcendence. "I'm about fifty-fifty on believing in God," he said. "For most of my

811

life, I've felt that there must be more to our existence than meets the eye."

He admitted that, as he faced death, he might be overestimating the odds out of a desire to believe in an afterlife. "I like to think that something survives after you die," he said. "It's strange to think that you accumulate all this experience, and maybe a little wisdom, and it just goes away. So I really want to believe that something survives, that maybe your consciousness endures."

He fell silent for a very long time. "But on the other hand, perhaps it's like an on-off switch," he said. "*Click!* And you're gone."

Then he paused again and smiled slightly. "Maybe that's why I never liked to put on-off switches on Apple devices."

ACKNOWLEDGMENTS

I'm deeply grateful to John and Ann Doerr, Laurene Powell, Mona Simpson, and Ken Auletta, all of whom helped get this project launched and provided invaluable support along the way. Alice Mayhew, who has been my editor at Simon & Schuster for thirty years, and Jonathan Karp, the publisher, both were extraordinarily diligent and attentive in shepherding this book, as was Amanda Urban, my agent. Crary Pullen was dogged in tracking down photos, and my assistant, Pat Zindulka, calmly facilitated things. I also want to thank my father, Irwin, and my daughter, Betsy, for reading the book and offering advice. And as always, I am most deeply indebted to my wife, Cathy, for her editing, suggestions, wise counsel, and so very much more.

SOURCES

INTERVIEWS (CONDUCTED 2009–2011)

Al Alcorn, Roger Ames, Fred Anderson, Bill Atkinson, Joan Baez, Marjorie Powell Barden, Jeff Bewkes, Bono, Ann Bowers, Stewart Brand, Chrisann Brennan, Larry Brilliant, John Seeley Brown, Tim Brown, Nolan Bushnell, Greg Calhoun, Bill Campbell, Berry Cash, Ed Catmull, Ray Cave, Lee Clow, Debi Coleman, Tim Cook, Katie Cotton, Eddy Cue, Andrea Cunningham, John Doerr, Millard Drexler, Jennifer Egan, Al Eisenstat, Michael Eisner, Larry Ellison, Philip Elmer-DeWitt, Gerard Errera, Tony Fadell, Jean-Louis Gassée, Bill Gates, Adele Goldberg, Craig Good, Austan Goolsbee, Al Gore, Andy Grove, Bill Hambrecht, Michael Hawley, Andy Hertzfeld, Joanna Hoffman, Elizabeth Holmes, Bruce Horn, John Huey, Jimmy Iovine, Jony Ive, Oren Jacob, Erin Jobs, Reed Jobs, Steve Jobs, Ron Johnson, Mitch Kapor, Susan Kare (email), Jeffrey Katzenberg, Pam Kerwin, Kristina Kiehl, Joel Klein, Daniel Kottke, Andy Lack, John Lasseter, Art Levinson, Steven Levy, Dan'l Lewin, Maya Lin, Yo-Yo Ma, Mike Markkula, John Markoff, Wynton Marsalis, Regis McKenna, Mike Merin,

Bob Metcalfe, Doug Morris, Walt Mossberg, Rupert Murdoch, Mike Murray, Nicholas Negroponte, Dean Ornish, Paul Otellini, Norman Pearlstine, Laurene Powell, Josh Quittner, Tina Redse, George Riley, Brian Roberts, Arthur Rock, Jeff Rosen, Alain Rossmann, Jon Rubinstein, Phil Schiller, Eric Schmidt, Barry Schuler, Mike Scott, John Sculley, Andy Serwer, Mona Simpson, Mike Slade, Alvy Ray Smith, Gina Smith, Kathryn Smith, Rick Stengel, Larry Tesler, Avie Tevanian, Guy "Bud" Tribble, Don Valentine, Paul Vidich, James Vincent, Alice Waters, Ron Wayne, Wendell Weeks, Ed Woolard, Stephen Wozniak, Del Yocam, Jerry York.

BIBLIOGRAPHY

Amelio, Gil. *On the Firing Line.* HarperBusiness, 1998.

Berlin, Leslie. *The Man behind the Microchip.* Oxford, 2005.

Butcher, Lee. *The Accidental Millionaire.* Paragon House, 1988.

Carlton, Jim. *Apple.* Random House, 1997.

Cringely, Robert X. *Accidental Empires.* Addison Wesley, 1992.

Deutschman, Alan. *The Second Coming of Steve Jobs.* Broadway Books, 2000.

Elliot, Jay, with William Simon. *The Steve Jobs Way.* Vanguard, 2011.

Freiberger, Paul, and Michael Swaine. *Fire in the Valley.* McGraw-Hill, 1984.

Garr, Doug. *Woz.* Avon, 1984.

Hertzfeld, Andy. *Revolution in the Valley.* O'Reilly, 2005. (See also his website, folklore.org.)

Hiltzik, Michael. *Dealers of Lightning.* HarperBusiness, 1999.

Jobs, Steve. Smithsonian oral history interview with Daniel Morrow, April 20, 1995.

———. Stanford commencement address, June 12, 2005.

Kahney, Leander. *Inside Steve's Brain.* Portfolio, 2008. (See also his website, cultofmac.com.)

Kawasaki, Guy. *The Macintosh Way.* Scott, Foresman, 1989.

Knopper, Steve. *Appetite for Self-Destruction.* Free Press, 2009.

Kot, Greg. *Ripped.* Scribner, 2009.

Kunkel, Paul. *AppleDesign.* Graphis Inc., 1997.

Levy, Steven. *Hackers.* Doubleday, 1984.

———. *Insanely Great.* Viking Penguin, 1994.

———. *The Perfect Thing.* Simon & Schuster, 2006.

Linzmayer, Owen. *Apple Confidential 2.0.* No Starch Press, 2004.

Malone, Michael. *Infinite Loop.* Doubleday, 1999.

Markoff, John. *What the Dormouse Said.* Viking Penguin, 2005.

McNish, Jacquie. *The Big Score.* Doubleday Canada, 1998.

Moritz, Michael. *Return to the Little Kingdom.* Overlook Press, 2009. Originally published, without prologue and epilogue, as *The Little Kingdom* (Morrow, 1984).

Nocera, Joe. *Good Guys and Bad Guys.* Portfolio, 2008.

Paik, Karen. *To Infinity and Beyond!* Chronicle Books, 2007.

Price, David. *The Pixar Touch.* Knopf, 2008.

Rose, Frank. *West of Eden.* Viking, 1989.

Sculley, John. *Odyssey.* Harper & Row, 1987.

Sheff, David. "Playboy Interview: Steve Jobs." *Playboy,* February 1985.

Simpson, Mona. *Anywhere but Here.* Knopf, 1986.

———. *A Regular Guy.* Knopf, 1996.

Smith, Douglas, and Robert Alexander. *Fumbling the Future.* Morrow, 1988.

Stross, Randall. *Steve Jobs and the NeXT Big Thing.* Atheneum, 1993.

"Triumph of the Nerds," PBS Television, hosted by Robert X. Cringely, June 1996.

Wozniak, Steve, with Gina Smith. *iWoz.* Norton, 2006.

Young, Jeffrey. *Steve Jobs.* Scott, Foresman, 1988.

———, and William Simon. *iCon.* John Wiley, 2005.

NOTES

Chapter 1: Childhood

The Adoption: Interviews with Steve Jobs, Laurene Powell, Mona Simpson, Del Yocam, Greg Calhoun, Chrisann Brennan, Andy Hertzfeld. Moritz, 44–45; Young, 16–17; Jobs, Smithsonian oral history; Jobs, Stanford commencement address; Andy Behrendt, "Apple Computer Mogul's Roots Tied to Green Bay," (Green Bay) *Press Gazette,* Dec. 4, 2005; Georgina Dickinson, "Dad Waits for Jobs to iPhone," *New York Post* and *The Sun* (London), Aug. 27, 2011; Mohannad Al-Haj Ali, "Steve Jobs Has Roots in Syria," *Al Hayat,* Jan. 16, 2011; Ulf Froitzheim, "Porträt Steve Jobs," *Unternehmen,* Nov. 26, 2007.

Silicon Valley: Interviews with Steve Jobs, Laurene Powell. Jobs, Smithsonian oral history; Moritz, 46; Berlin, 155–177; Malone, 21–22.

School: Interview with Steve Jobs. Jobs, Smithsonian oral history; Sculley, 166; Malone, 11, 28, 72; Young, 25, 34–35; Young and Simon, 18; Moritz, 48, 73–74. Jobs's address was originally 11161 Crist Drive, before the subdivision was incorporated into the town from the county. Some sources mention that Jobs worked at both Haltek and another store

819

with a similar name, Halted. When asked, Jobs says he can remember working only at Haltek.

Chapter 2: Odd Couple

Woz: Interviews with Steve Wozniak, Steve Jobs. Wozniak, 12–16, 22, 50–61, 86–91; Levy, *Hackers,* 245; Moritz, 62–64; Young, 28; Jobs, Macworld address, Jan. 17, 2007.

The Blue Box: Interviews with Steve Jobs, Steve Wozniak. Ron Rosenbaum, "Secrets of the Little Blue Box," *Esquire,* Oct. 1971. Wozniak answer, woz.org/letters/general/03.html; Wozniak, 98–115. For slightly varying accounts, see Markoff, 272; Moritz, 78–86; Young, 42–45; Malone, 30–35.

Chapter 3: The Dropout

Chrisann Brennan: Interviews with Chrisann Brennan, Steve Jobs, Steve Wozniak, Tim Brown. Moritz, 75–77; Young, 41; Malone, 39.

Reed College: Interviews with Steve Jobs, Daniel Kottke, Elizabeth Holmes. Freiberger and Swaine, 208; Moritz, 94–100; Young, 55; "The Updated Book of Jobs," *Time,* Jan. 3, 1983.

Robert Friedland: Interviews with Steve Jobs, Daniel Kottke, Elizabeth Holmes. In September 2010 I met with Friedland in New York City to discuss his background and relationship with Jobs, but he did not want to be quoted on the record. McNish, 11–17; Jennifer Wells, "Canada's Next Billionaire," *Maclean's,* June 3, 1996; Richard Read, "Financier's Saga of Risk," *Mines and Communities* magazine, Oct. 16, 2005; Jennifer Hunter, "But What Would His Guru Say?" (Toronto) *Globe and Mail,* Mar. 18, 1988; Moritz, 96,

109; Young, 56.

. . . *Drop Out:* Interviews with Steve Jobs, Steve Wozniak; Jobs, Stanford commencement address; Moritz, 97.

Chapter 4: Atari and India

Atari: Interviews with Steve Jobs, Al Alcorn, Nolan Bushnell, Ron Wayne. Moritz, 103–104.

India: Interviews with Daniel Kottke, Steve Jobs, Al Alcorn, Larry Brilliant.

The Search: Interviews with Steve Jobs, Daniel Kottke, Elizabeth Holmes, Greg Calhoun. Young, 72; Young and Simon, 31–32; Moritz, 107.

Breakout: Interviews with Nolan Bushnell, Al Alcorn, Steve Wozniak, Ron Wayne, Andy Hertzfeld. Wozniak, 144–149; Young, 88; Linzmayer, 4.

Chapter 5: The Apple I

Machines of Loving Grace: Interviews with Steve Jobs, Bono, Stewart Brand. Markoff, xii; Stewart Brand, "We Owe It All to the Hippies," *Time,* Mar. 1, 1995; Jobs, Stanford commencement address; Fred Turner, *From Counterculture to Cyberculture* (Chicago, 2006).

The Homebrew Computer Club: Interviews with Steve Jobs, Steve Wozniak. Wozniak, 152–172; Freiberger and Swaine, 99; Linzmayer, 5; Moritz, 144; Steve Wozniak, "Homebrew and How Apple Came to Be," www.atariarchives.org; Bill Gates, "Open Letter to Hobbyists," Feb. 3, 1976.

Apple Is Born: Interviews with Steve Jobs, Steve Wozniak, Mike Markkula, Ron Wayne. Steve Jobs, address to the Aspen Design Conference, June 15, 1983, tape in Aspen Institute archives;

Apple Computer Partnership Agreement, County of Santa Clara, Apr. 1, 1976, and Amendment to Agreement, Apr. 12, 1976; Bruce Newman, "Apple's Lost Founder," *San Jose Mercury News,* June 2, 2010; Wozniak, 86, 176–177; Moritz, 149–151; Freiberger and Swaine, 212–213; Ashlee Vance, "A Haven for Spare Parts Lives on in Silicon Valley," *New York Times,* Feb. 4, 2009; Paul Terrell interview, Aug. 1, 2008, mac-history.net.

Garage Band: Interviews with Steve Wozniak, Elizabeth Holmes, Daniel Kottke, Steve Jobs. Wozniak, 179–189; Moritz, 152–163; Young, 95–111; R. S. Jones, "Comparing Apples and Oranges," *Interface,* July 1976.

Chapter 6: The Apple II

An Integrated Package: Interviews with Steve Jobs, Steve Wozniak, Al Alcorn, Ron Wayne. Wozniak, 165, 190–195; Young, 126; Moritz, 169–170, 194–197; Malone, v, 103.

Mike Markkula: Interviews with Regis McKenna, Don Valentine, Steve Jobs, Steve Wozniak, Mike Markkula, Arthur Rock. Nolan Bushnell, keynote address at the ScrewAttack Gaming Convention, Dallas, July 5, 2009; Steve Jobs, talk at the International Design Conference at Aspen, June 15, 1983; Mike Markkula, "The Apple Marketing Philosophy" (courtesy of Mike Markkula), Dec. 1979; Wozniak, 196–199. See also Moritz, 182–183; Malone, 110–111.

Regis McKenna: Interviews with Regis McKenna, John Doerr, Steve Jobs. Ivan Raszl, "Interview with Rob Janoff," Creativebits.org, Aug. 3, 2009.

The First Launch Event: Interviews with Steve

822

Wozniak, Steve Jobs. Wozniak, 201–206; Moritz, 199–201; Young, 139.

Mike Scott: Interviews with Mike Scott, Mike Markkula, Steve Jobs, Steve Wozniak, Arthur Rock. Young, 135; Freiberger and Swaine, 219, 222; Moritz, 213; Elliot, 4.

Chapter 7: Chrisann and Lisa

Interviews with Chrisann Brennan, Steve Jobs, Elizabeth Holmes, Greg Calhoun, Daniel Kottke, Arthur Rock. Moritz, 285; "The Updated Book of Jobs," *Time,* Jan. 3, 1983; "Striking It Rich," *Time,* Feb. 15, 1982.

Chapter 8: Xerox and Lisa

A New Baby: Interviews with Andrea Cunningham, Andy Hertzfeld, Steve Jobs, Bill Atkinson. Wozniak, 226; Levy, *Insanely Great,* 124; Young, 168–170; Bill Atkinson, oral history, Computer History Museum, Mountain View, CA; Jef Raskin, "Holes in the Histories," *Interactions,* July 1994; Jef Raskin, "Hubris of a Heavyweight," *IEEE Spectrum,* July 1994; Jef Raskin, oral history, April 13, 2000, Stanford Library Department of Special Collections; Linzmayer, 74, 85–89.

Xerox PARC: Interviews with Steve Jobs, John Seeley Brown, Adele Goldberg, Larry Tesler, Bill Atkinson. Freiberger and Swaine, 239; Levy, *Insanely Great,* 66–80; Hiltzik, 330–341; Linzmayer, 74–75; Young, 170–172; Rose, 45–47; *Triumph of the Nerds,* PBS, part 3.

"Great Artists Steal": Interviews with Steve Jobs, Larry Tesler, Bill Atkinson. Levy, *Insanely Great,* 77, 87–90; *Triumph of the Nerds,* PBS, part 3;

Bruce Horn, "Where It All Began" (1966), www. mackido.com; Hiltzik, 343, 367–370; Malcolm Gladwell, "Creation Myth," *New Yorker,* May 16, 2011; Young, 178–182.

Chapter 9: Going Public

Options: Interviews with Daniel Kottke, Steve Jobs, Steve Wozniak, Andy Hertzfeld, Mike Markkula, Bill Hambrecht. "Sale of Apple Stock Barred," *Boston Globe,* Dec. 11, 1980.

Baby You're a Rich Man: Interviews with Larry Brilliant, Steve Jobs. Steve Ditlea, "An Apple on Every Desk," *Inc.,* Oct. 1, 1981; "Striking It Rich," *Time,* Feb. 15, 1982; "The Seeds of Success," *Time,* Feb. 15, 1982; Moritz, 292–295; Sheff.

Chapter 10: The Mac Is Born

Jef Raskin's Baby: Interviews with Bill Atkinson, Steve Jobs, Andy Hertzfeld, Mike Markkula. Jef Raskin, "Recollections of the Macintosh Project," "Holes in the Histories," "The Genesis and History of the Macintosh Project," "Reply to Jobs, and Personal Motivation," "Design Considerations for an Anthropophilic Computer," and "Computers by the Millions," Raskin papers, Stanford University Library; Jef Raskin, "A Conversation," *Ubiquity,* June 23, 2003; Levy, *Insanely Great,* 107–121; Hertzfeld, 19; "Macintosh's Other Designers," *Byte,* Aug. 1984; Young, 202, 208–214; "Apple Launches a Mac Attack," *Time,* Jan. 30, 1984; Malone, 255–258.

Texaco Towers: Interviews with Andrea Cunningham, Bruce Horn, Andy Hertzfeld, Mike Scott, Mike Markkula. Hertzfeld, 19–20, 26–27; Wozniak, 241–242.

Chapter 11: The Reality Distortion Field

Interviews with Bill Atkinson, Steve Wozniak, Debi Coleman, Andy Hertzfeld, Bruce Horn, Joanna Hoffman, Al Eisenstat, Ann Bowers, Steve Jobs. Some of these tales have variations. See Hertzfeld, 24, 68, 161.

Chapter 12: The Design

A Bauhaus Aesthetic: Interviews with Dan'l Lewin, Steve Jobs, Maya Lin, Debi Coleman. Steve Jobs in conversation with Charles Hampden-Turner, International Design Conference in Aspen, June 15, 1983. (The design conference audiotapes are stored at the Aspen Institute. I want to thank Deborah Murphy for finding them.)

Like a Porsche: Interviews with Bill Atkinson, Alain Rossmann, Mike Markkula, Steve Jobs. "The Macintosh Design Team," *Byte,* Feb. 1984; Hertzfeld, 29–31, 41, 46, 63, 68; Sculley, 157; Jerry Manock, "Invasion of Texaco Towers," Folklore .org; Kunkel, 26–30; Jobs, Stanford commencement address; email from Susan Kare; Susan Kare, "World Class Cities," in Hertzfeld, 165; Laurence Zuckerman, "The Designer Who Made the Mac Smile," *New York Times,* Aug. 26, 1996; Susan Kare interview, Sept. 8, 2000, Stanford University Library, Special Collections; Levy, *Insanely Great,* 156; Hartmut Esslinger, *A Fine Line* (Jossey-Bass, 2009), 7–9; David Einstein, "Where Success Is by Design," *San Francisco Chronicle,* Oct. 6, 1995; Sheff.

Chapter 13: Building the Mac

Competition: Interview with Steve Jobs. Levy,

Insanely Great, 125; Sheff; Hertzfeld, 71–73; *Wall Street Journal* advertisement, Aug. 24, 1981.

End-to-End Control: Interview with Berry Cash. Kahney, 241; Dan Farber, "Steve Jobs, the iPhone and Open Platforms," ZDNet.com, Jan. 13, 2007; Tim Wu, *The Master Switch* (Knopf, 2010), 254–276; Mike Murray, "Mac Memo" to Steve Jobs, May 19, 1982 (courtesy of Mike Murray).

Machines of the Year: Interviews with Daniel Kottke, Steve Jobs, Ray Cave. "The Computer Moves In," *Time,* Jan. 3, 1983; "The Updated Book of Jobs," *Time,* Jan. 3, 1983; Moritz, 11; Young, 293; Rose, 9–11; Peter McNulty, "Apple's Bid to Stay in the Big Time," *Fortune,* Feb. 7, 1983; "The Year of the Mouse," *Time,* Jan. 31, 1983.

Let's Be Pirates! Interviews with Ann Bowers, Andy Hertzfeld, Bill Atkinson, Arthur Rock, Mike Markkula, Steve Jobs, Debi Coleman; email from Susan Kare. Hertzfeld, 76, 135–138, 158, 160, 166; Moritz, 21–28; Young, 295–297, 301–303; Susan Kare interview, Sept. 8, 2000, Stanford University Library; Jeff Goodell, "The Rise and Fall of Apple Computer," *Rolling Stone,* Apr. 4, 1996; Rose, 59–69, 93.

Chapter 14: Enter Sculley

The Courtship: Interviews with John Sculley, Andy Hertzfeld, Steve Jobs. Rose, 18, 74–75; Sculley, 58–90, 107; Elliot, 90–93; Mike Murray, "Special Mac Sneak" memo to staff, Mar. 3, 1983 (courtesy of Mike Murray); Hertzfeld, 149–150.

The Honeymoon: Interviews with Steve Jobs, John Sculley, Joanna Hoffman. Sculley, 127–130, 154–155, 168, 179; Hertzfeld, 195.

Chapter 15: The Launch

Real Artists Ship: Interviews with Andy Hertz-feld, Steve Jobs. Video of Apple sales conference, Oct. 1983; "Personal Computers: And the Winner Is . . . IBM," *Business Week,* Oct. 3, 1983; Hertzfeld, 208–210; Rose, 147–153; Levy, *Insanely Great,* 178–180; Young, 327–328.

The "1984" Ad: Interviews with Lee Clow, John Sculley, Mike Markkula, Bill Campbell, Steve Jobs. Steve Hayden interview, *Weekend Edition,* NPR, Feb. 1, 2004; Linzmayer, 109–114; Sculley, 176.

Publicity Blast: Hertzfeld, 226–227; Michael Rogers, "It's the Apple of His Eye," *Newsweek,* Jan. 30, 1984; Levy, *Insanely Great,* 17–27.

January 24, 1984: Interviews with John Sculley, Steve Jobs, Andy Hertzfeld. Video of Jan. 1984 Apple shareholders meeting; Hertzfeld, 213–223; Sculley, 179–181; William Hawkins, "Jobs' Revolutionary New Computer," *Popular Science,* Jan. 1989.

Chapter 16: Gates and Jobs

The Macintosh Partnership: Interviews with Bill Gates, Steve Jobs, Bruce Horn. Hertzfeld, 52–54; Steve Lohr, "Creating Jobs," *New York Times,* Jan. 12, 1997; *Triumph of the Nerds,* PBS, part 3; Rusty Weston, "Partners and Adversaries," *MacWeek,* Mar. 14, 1989; Walt Mossberg and Kara Swisher, interview with Bill Gates and Steve Jobs, *All Things Digital,* May 31, 2007; Young, 319–320; Carlton, 28; Brent Schlender, "How Steve Jobs Linked Up with IBM," *Fortune,* Oct. 9, 1989; Steven Levy, "A Big Brother?" *Newsweek,* Aug. 18, 1997.

The Battle of the GUI: Interviews with Bill Gates, Steve Jobs. Hertzfeld, 191–193; Michael Schrage, "IBM Compatibility Grows," *Washington Post,* Nov. 29, 1983; *Triumph of the Nerds,* PBS, part 3.

Chapter 17: Icarus

Flying High: Interviews with Steve Jobs, Debi Coleman, Bill Atkinson, Andy Hertzfeld, Alain Rossmann, Joanna Hoffman, Jean-Louis Gassée, Nicholas Negroponte, Arthur Rock, John Sculley. Sheff; Hertzfeld, 206–207, 230; Sculley, 197–199; Young, 308–309; George Gendron and Bo Burlingham, "Entrepreneur of the Decade," *Inc.,* Apr. 1, 1989.

Falling: Interviews with Joanna Hoffman, John Sculley, Lee Clow, Debi Coleman, Andrea Cunningham, Steve Jobs. Sculley, 201, 212–215; Levy, *Insanely Great,* 186–192; Michael Rogers, "It's the Apple of His Eye," *Newsweek,* Jan. 30, 1984; Rose, 207, 233; Felix Kessler, "Apple Pitch," *Fortune,* Apr. 15, 1985; Linzmayer, 145.

Thirty Years Old: Interviews with Mallory Walker, Andy Hertzfeld, Debi Coleman, Elizabeth Holmes, Steve Wozniak, Don Valentine. Sheff.

Exodus: Interviews with Andy Hertzfeld, Steve Wozniak, Bruce Horn. Hertzfeld, 253, 263–264; Young, 372–376; Wozniak, 265–266; Rose, 248–249; Bob Davis, "Apple's Head, Jobs, Denies Ex-Partner Use of Design Firm," *Wall Street Journal,* Mar. 22, 1985.

Showdown, Spring 1985: Interviews with Steve Jobs, Al Alcorn, John Sculley, Mike Murray. Elliot, 15; Sculley, 205–206, 227, 238–244; Young, 367–379; Rose, 238, 242, 254–255; Mike Murray,

"Let's Wake Up and Die Right," memo to undis-
closed recipients, Mar. 7, 1985 (courtesy of Mike
Murray).

Plotting a Coup: Interviews with Steve Jobs, John
Sculley. Rose, 266–275; Sculley, ix–x, 245–246;
Young, 388–396; Elliot, 112.

Seven Days in May: Interviews with Jean-Louis
Gassée, Steve Jobs, Bill Campbell, Al Eisenstat,
John Sculley, Mike Murray, Mike Markkula, Debi
Coleman. Bro Uttal, "Behind the Fall of Steve
Jobs," *Fortune,* Aug. 5, 1985; Sculley, 249–260;
Rose, 275–290; Young, 396–404.

Like a Rolling Stone: Interviews with Mike Mur-
ray, Mike Markkula, Steve Jobs, John Sculley, Bob
Metcalfe, George Riley, Andy Hertzfeld, Tina
Redse, Mike Merin, Al Eisenstat, Arthur Rock.
Tina Redse email to Steve Jobs, July 20, 2010; "No
Job for Jobs," AP, July 26, 1985; "Jobs Talks about
His Rise and Fall," *Newsweek,* Sept. 30, 1985;
Hertzfeld, 269–271; Young, 387, 403–405; Young
and Simon, 116; Rose, 288–292; Sculley, 242–
245, 286–287; letter from Al Eisenstat to Arthur
Hartman, July 23, 1985 (courtesy of Al Eisenstat).

Chapter 18: NeXT

The Pirates Abandon Ship: Interviews with Dan'l
Lewin, Steve Jobs, Bill Campbell, Arthur Rock,
Mike Markkula, John Sculley, Andrea Cunning-
ham, Joanna Hoffman. Patricia Bellew Gray and
Michael Miller, "Apple Chairman Jobs Resigns,"
Wall Street Journal, Sept. 18, 1985; Gerald Lube-
now and Michael Rogers, "Jobs Talks about His
Rise and Fall," *Newsweek,* Sept. 30, 1985; Bro
Uttal, "The Adventures of Steve Jobs," *Fortune,*

Oct. 14, 1985; Susan Kerr, "Jobs Resigns," *Computer Systems News*, Sept. 23, 1985; "Shaken to the Very Core," *Time*, Sept. 30, 1985; John Eckhouse, "Apple Board Fuming at Steve Jobs," *San Francisco Chronicle*, Sept. 17, 1985; Hertzfeld, 132–133; Sculley, 313–317; Young, 415–416; Young and Simon, 127; Rose, 307–319; Stross, 73; Deutschman, 36; Complaint for Breaches of Fiduciary Obligations, *Apple Computer v. Steven P. Jobs and Richard A. Page*, Superior Court of California, Santa Clara County, Sept. 23, 1985; Patricia Bellew Gray, "Jobs Asserts Apple Undermined Efforts to Settle Dispute," *Wall Street Journal*, Sept. 25, 1985.

To Be on Your Own: Interviews with Arthur Rock, Susan Kare, Steve Jobs, Al Eisenstat. "Logo for Jobs' New Firm," *San Francisco Chronicle*, June 19, 1986; Phil Patton, "Steve Jobs: Out for Revenge," *New York Times*, Aug. 6, 1989; Paul Rand, NeXT Logo presentation, 1985; Doug Evans and Allan Pottasch, video interview with Steve Jobs on Paul Rand, 1993; Steve Jobs to Al Eisenstat, Nov. 4, 1985; Eisenstat to Jobs, Nov. 8, 1985; Agreement between Apple Computer Inc. and Steven P. Jobs, and Request for Dismissal of Lawsuit without Prejudice, filed in the Superior Court of California, Santa Clara County, Jan. 17, 1986; Deutschman, 47, 43; Stross, 76, 118–120, 245; Kunkel, 58–63; "Can He Do It Again?" *Business Week*, Oct. 24, 1988; Joe Nocera, "The Second Coming of Steve Jobs," *Esquire*, Dec. 1986, reprinted in *Good Guys and Bad Guys* (Portfolio, 2008), 49; Brenton Schlender, "How Steve Jobs Linked Up with IBM," *Fortune*, Oct. 9, 1989.

The Computer: Interviews with Mitch Kapor, Michael Hawley, Steve Jobs. Peter Denning and Karen Frenkle, "A Conversation with Steve Jobs," *Communications of the Association for Computer Machinery,* Apr. 1, 1989; John Eckhouse, "Steve Jobs Shows Off Ultra-Robotic Assembly Line," *San Francisco Chronicle,* June 13, 1989; Stross, 122–125; Deutschman, 60–63; Young, 425; Katie Hafner, "Can He Do It Again?" *Business Week,* Oct. 24, 1988; *The Entrepreneurs,* PBS, Nov. 5, 1986, directed by John Nathan.

Perot to the Rescue: Stross, 102–112; "Perot and Jobs," *Newsweek,* Feb. 9, 1987; Andrew Pollack, "Can Steve Jobs Do It Again?" *New York Times,* Nov. 8, 1987; Katie Hafner, "Can He Do It Again?" *Business Week,* Oct. 24, 1988; Pat Steger, "A Gem of an Evening with King Juan Carlos," *San Francisco Chronicle,* Oct. 5, 1987; David Remnick, "How a Texas Playboy Became a Billionaire," *Washington Post,* May 20, 1987.

Gates and NeXT: Interviews with Bill Gates, Adele Goldberg, Steve Jobs. Brit Hume, "Steve Jobs Pulls Ahead," *Washington Post,* Oct. 31, 1988; Brent Schlender, "How Steve Jobs Linked Up with IBM," *Fortune,* Oct. 9, 1989; Stross, 14; Linzmayer, 209; "William Gates Talks," *Washington Post,* Dec. 30, 1990; Katie Hafner, "Can He Do It Again?" *Business Week,* Oct. 24, 1988; John Thompson, "Gates, Jobs Swap Barbs," *Computer System News,* Nov. 27, 1989.

IBM: Brent Schlender, "How Steve Jobs Linked Up with IBM," *Fortune,* Oct. 9, 1989; Phil Patton, "Out for Revenge," *New York Times,* Aug. 6, 1989; Stross, 140–142; Deutschman, 133.

The Launch, October 1988: Stross, 166–186; Wes Smith, "Jobs Has Returned," *Chicago Tribune,* Nov. 13, 1988; Andrew Pollack, "NeXT Produces a Gala," *New York Times,* Oct. 10, 1988; Brenton Schlender, "Next Project," *Wall Street Journal,* Oct. 13, 1988; Katie Hafner, "Can He Do It Again?" *Business Week,* Oct. 24, 1988; Deutschman, 128; "Steve Jobs Comes Back," *Newsweek,* Oct. 24, 1988; "The NeXT Generation," *San Jose Mercury News,* Oct. 10, 1988.

Chapter 19: Pixar

Lucasfilm's Computer Division: Interviews with Ed Catmull, Alvy Ray Smith, Steve Jobs, Pam Kerwin, Michael Eisner. Price, 71–74, 89–101; Paik, 53–57, 226; Young and Simon, 169; Deutschman, 115.

Animation: Interviews with John Lasseter, Steve Jobs. Paik, 28–44; Price, 45–56.

Tin Toy: Interviews with Pam Kerwin, Alvy Ray Smith, John Lasseter, Ed Catmull, Steve Jobs, Jeffrey Katzenberg, Michael Eisner, Andy Grove. Steve Jobs email to Albert Yu, Sept. 23, 1995; Albert Yu to Steve Jobs, Sept. 25, 1995; Steve Jobs to Andy Grove, Sept. 25, 1995; Andy Grove to Steve Jobs, Sept. 26, 1995; Steve Jobs to Andy Grove, Oct. 1, 1995; Price, 104–114; Young and Simon, 166.

Chapter 20: A Regular Guy

Joan Baez: Interviews with Joan Baez, Steve Jobs, Joanna Hoffman, Debi Coleman, Andy Hertzfeld. Joan Baez, *And a Voice to Sing With* (Summit, 1989), 144, 380.

Finding Joanne and Mona: Interviews with Steve Jobs, Mona Simpson.

The Lost Father: Interviews with Steve Jobs, Laurene Powell, Mona Simpson, Ken Auletta, Nick Pileggi.

Lisa: Interviews with Chrisann Brennan, Avie Tevanian, Joanna Hoffman, Andy Hertzfeld. Lisa Brennan-Jobs, "Confessions of a Lapsed Vegetarian," *Southwest Review,* 2008; Young, 224; Deutschman, 76.

The Romantic: Interviews with Jennifer Egan, Tina Redse, Steve Jobs, Andy Hertzfeld, Joanna Hoffman. Deutschman, 73, 138. Mona Simpson's *A Regular Guy* is a novel loosely based on the relationship between Jobs, Lisa and Chrisann Brennan, and Tina Redse, who is the basis for the character named Olivia.

Chapter 21: Family Man

Laurene Powell: Interviews with Laurene Powell, Steve Jobs, Kathryn Smith, Avie Tevanian, Andy Hertzfeld, Marjorie Powell Barden.

The Wedding, March 18, 1991: Interviews with Steve Jobs, Laurene Powell, Andy Hertzfeld, Joanna Hoffman, Avie Tevanian, Mona Simpson. Simpson, *A Regular Guy,* 357.

A Family Home: Interviews with Steve Jobs, Laurene Powell, Andy Hertzfeld. David Weinstein, "Taking Whimsy Seriously," *San Francisco Chronicle,* Sept. 13, 2003; Gary Wolfe, "Steve Jobs," *Wired,* Feb. 1996; "Former Apple Designer Charged with Harassing Steve Jobs," AP, June 8, 1993.

Lisa Moves In: Interviews with Steve Jobs, Laurene Powell, Mona Simpson, Andy Hertzfeld. Lisa Brennan-Jobs, "Driving Jane," *Harvard Advocate,*

Spring 1999; Simpson, *A Regular Guy,* 251; email from Chrisann Brennan, Jan. 19, 2011; Bill Workman, "Palo Alto High School's Student Scoop," *San Francisco Chronicle,* Mar. 16, 1996; Lisa Brennan-Jobs, "Waterloo," *Massachusetts Review,* Spring 2006; Deutschman, 258; Chrisann Brennan website, chrysanthemum.com; Steve Lohr, "Creating Jobs," *New York Times,* Jan. 12, 1997.

Children: Interviews with Steve Jobs, Laurene Powell.

Chapter 22: Toy Story

Jeffrey Katzenberg: Interviews with John Lasseter, Ed Catmull, Jeffrey Katzenberg, Alvy Ray Smith, Steve Jobs. Price, 84–85, 119–124; Paik, 71, 90; Robert Murphy, "John Cooley Looks at Pixar's Creative Process," *Silicon Prairie News,* Oct. 6, 2010.

Cut! Interviews with Steve Jobs, Jeffrey Katzenberg, Ed Catmull, Larry Ellison. Paik, 90; Deutschman, 194–198; "Toy Story: The Inside Buzz," *Entertainment Weekly,* Dec. 8, 1995.

To Infinity! Interviews with Steve Jobs, Michael Eisner. Janet Maslin, "There's a New Toy in the House. Uh-Oh," *New York Times,* Nov. 22, 1995; "A Conversation with Steve Jobs and John Lasseter," *Charlie Rose,* PBS, Oct. 30, 1996; John Markoff, "Apple Computer Co-Founder Strikes Gold," *New York Times,* Nov. 30, 1995.

Chapter 23: The Second Coming

Things Fall Apart: Interview with Jean-Louis Gassée. Bart Ziegler, "Industry Has Next to No Patience with Jobs' NeXT," AP, Aug. 19, 1990;

Stross, 226–228; Gary Wolf, "The Next Insanely Great Thing," *Wired,* Feb. 1996; Anthony Perkins, "Jobs' Story," *Red Herring,* Jan. 1, 1996.

Apple Falling: Interviews with Steve Jobs, John Sculley, Larry Ellison. Sculley, 248, 273; Deutschman, 236; Steve Lohr, "Creating Jobs," *New York Times,* Jan. 12, 1997; Amelio, 190 and preface to the hardback edition; Young and Simon, 213–214; Linzmayer, 273–279; Guy Kawasaki, "Steve Jobs to Return as Apple CEO," *Macworld,* Nov. 1, 1994.

Slouching toward Cupertino: Interviews with Jon Rubinstein, Steve Jobs, Larry Ellison, Avie Tevanian, Fred Anderson, Larry Tesler, Bill Gates, John Lasseter. John Markoff, "Why Apple Sees Next as a Match Made in Heaven," *New York Times,* Dec. 23, 1996; Steve Lohr, "Creating Jobs," *New York Times,* Jan. 12, 1997; Rajiv Chandrasekaran, "Steve Jobs Returning to Apple," *Washington Post,* Dec. 21, 1996; Louise Kehoe, "Apple's Prodigal Son Returns," *Financial Times,* Dec. 23, 1996; Amelio, 189–201, 238; Carlton, 409; Linzmayer, 277; Deutschman, 240.

Chapter 24: The Restoration

Hovering Backstage: Interviews with Steve Jobs, Avie Tevanian, Jon Rubinstein, Ed Woolard, Larry Ellison, Fred Anderson, email from Gina Smith. Sheff; Brent Schlender, "Something's Rotten in Cupertino," *Fortune,* Mar. 3, 1997; Dan Gillmore, "Apple's Prospects Better Than Its CEO's Speech," *San Jose Mercury News,* Jan. 13, 1997; Carlton, 414–416, 425; Malone, 531; Deutschman, 241–245; Amelio, 219, 238–247, 261; Linzmayer,

201; Kaitlin Quistgaard, "Apple Spins Off New-ton," *Wired.com,* May 22, 1997; Louise Kehoe, "Doubts Grow about Leadership at Apple," *Financial Times,* Feb. 25, 1997; Dan Gillmore, "Ellison Mulls Apple Bid," *San Jose Mercury News,* Mar. 27, 1997; Lawrence Fischer, "Oracle Seeks Public Views on Possible Bid for Apple," *New York Times,* Mar. 28, 1997; Mike Barnicle, "Roadkill on the Info Highway," *Boston Globe,* Aug. 5, 1997.

Exit, Pursued by a Bear: Interviews with Ed Woolard, Steve Jobs, Mike Markkula, Steve Wozniak, Fred Anderson, Larry Ellison, Bill Campbell. Privately printed family memoir by Ed Woolard (courtesy of Woolard); Amelio, 247, 261, 267; Gary Wolf, "The World According to Woz," *Wired,* Sept. 1998; Peter Burrows and Ronald Grover, "Steve Jobs' Magic Kingdom," *Business Week,* Feb. 6, 2006; Peter Elkind, "The Trouble with Steve Jobs," *Fortune,* Mar. 5, 2008; Arthur Levitt, *Take on the Street* (Pantheon, 2002), 204–206.

Macworld Boston, August 1997: Steve Jobs, Macworld Boston speech, Aug. 6, 1997.

The Microsoft Pact: Interviews with Joel Klein, Bill Gates, Steve Jobs. Cathy Booth, "Steve's Job," *Time,* Aug. 18, 1997; Steven Levy, "A Big Brother?" *Newsweek,* Aug. 18, 1997. Jobs's cell phone call with Gates was reported by *Time* photographer Diana Walker, who shot the picture of him crouching onstage that appeared on the *Time* cover and in the photo section of this book.

Chapter 25: Think Different

Here's to the Crazy Ones: Interviews with Steve Jobs, Lee Clow, James Vincent, Norman Pearlstine. Cathy Booth, "Steve's Job," *Time,* Aug. 18, 1997; John Heilemann, "Steve Jobs in a Box," *New York,* June 17, 2007.

iCEO: Interviews with Steve Jobs, Fred Anderson. Video of Sept. 1997 staff meeting (courtesy of Lee Clow); "Jobs Hints That He May Want to Stay at Apple," *New York Times,* Oct. 10, 1997; Jon Swartz, "No CEO in Sight for Apple," *San Francisco Chronicle,* Dec. 12, 1997; Carlton, 437.

Killing the Clones: Interviews with Bill Gates, Steve Jobs, Ed Woolard. Steve Wozniak, "How We Failed Apple," *Newsweek,* Feb. 19, 1996; Linzmayer, 245–247, 255; Bill Gates, "Licensing of Mac Technology," a memo to John Sculley, June 25, 1985; Tom Abate, "How Jobs Killed Mac Clone Makers," *San Francisco Chronicle,* Sept. 6, 1997.

Product Line Review: Interviews with Phil Schiller, Ed Woolard, Steve Jobs. Deutschman, 248; Steve Jobs, speech at iMac launch event, May 6, 1998; video of Sept. 1997 staff meeting.

Chapter 26: Design Principles

Jony Ive: Interviews with Jony Ive, Steve Jobs, Phil Schiller. John Arlidge, "Father of Invention," *Observer* (London), Dec. 21, 2003; Peter Burrows, "Who Is Jonathan Ive?" *Business Week,* Sept. 25, 2006; "Apple's One-Dollar-a-Year Man," *Fortune,* Jan. 24, 2000; Rob Walker, "The Guts of a New Machine," *New York Times,* Nov. 30, 2003; Leander Kahney, "Design According to Ive," *Wired.com,* June 25, 2003.

Inside the Studio: Interview with Jony Ive. U.S. Patent and Trademark Office, online database, patft.uspto.gov; Leander Kahney, "Jobs Awarded Patent for iPhone Packaging," *Cult of Mac,* July 22, 2009; Harry McCracken, "Patents of Steve Jobs," *Technologizer.com,* May 28, 2009.

Chapter 27: The iMac

Back to the Future: Interviews with Phil Schiller, Avie Tevanian, Jon Rubinstein, Steve Jobs, Fred Anderson, Mike Markkula, Jony Ive, Lee Clow. Thomas Hormby, "Birth of the iMac," *Mac Observer,* May 25, 2007; Peter Burrows, "Who Is Jonathan Ive?" *Business Week,* Sept. 25, 2006; Lev Grossman, "How Apple Does It," *Time,* Oct. 16, 2005; Leander Kahney, "The Man Who Named the iMac and Wrote Think Different," *Cult of Mac,* Nov. 3, 2009; Levy, *The Perfect Thing,* 198; gawker.com/comment/21123257/; "Steve's Two Jobs," *Time,* Oct. 18, 1999.

The Launch, May 6, 1998: Interviews with Jony Ive, Steve Jobs, Phil Schiller, Jon Rubinstein. Steven Levy, "Hello Again," *Newsweek,* May 18, 1998; Jon Swartz, "Resurgence of an American Icon," *Forbes,* Apr. 14, 2000; Levy, *The Perfect Thing,* 95.

Chapter 28: CEO

Tim Cook: Interviews with Tim Cook, Steve Jobs, Jon Rubinstein. Peter Burrows, "Yes, Steve, You Fixed It. Congratulations. Now What?" *Business Week,* July 31, 2000; Tim Cook, Auburn commencement address, May 14, 2010; Adam Lashinsky, "The Genius behind Steve," *Fortune,* Nov. 10, 2008; Nick Wingfield, "Apple's No. 2 Has Low Profile," *Wall Street Journal,* Oct. 16, 2006.

Mock Turtlenecks and Teamwork: Interviews with Steve Jobs, James Vincent, Jony Ive, Lee Clow, Avie Tevanian, Jon Rubinstein. Lev Grossman, "How Apple Does It," *Time,* Oct. 16, 2005; Leander Kahney, "How Apple Got Everything Right by Doing Everything Wrong," *Wired,* Mar. 18, 2008.

From iCEO to CEO: Interviews with Ed Woolard, Larry Ellison, Steve Jobs. Apple proxy statement, Mar. 12, 2001.

Chapter 29: Apple Stores

The Customer Experience: Interviews with Steve Jobs, Ron Johnson. Jerry Useem, "America's Best Retailer," *Fortune,* Mar. 19, 2007; Gary Allen, "Apple Stores," ifoAppleStore.com.

The Prototype: Interviews with Art Levinson, Ed Woolard, Millard "Mickey" Drexler, Larry Ellison, Ron Johnson, Steve Jobs, Art Levinson. Cliff Edwards, "Sorry, Steve . . . ," *Business Week,* May 21, 2001.

Wood, Stone, Steel, Glass: Interviews with Ron Johnson, Steve Jobs. U.S. Patent Office, D478999, Aug. 26, 2003, US2004/0006939, Jan. 15, 2004; Gary Allen, "About Me," ifoapplestore.com.

Chapter 30: The Digital Hub

Connecting the Dots: Interviews with Lee Clow, Jony Ive, Steve Jobs. Sheff; Steve Jobs, Macworld keynote address, Jan. 9, 2001.

FireWire: Interviews with Steve Jobs, Phil Schiller, Jon Rubinstein. Steve Jobs, Macworld keynote address, Jan. 9, 2001; Joshua Quittner, "Apple's New Core," *Time,* Jan. 14, 2002; Mike Evangelist, "Steve Jobs, the Genuine Article," *Writer's Block*

Live, Oct. 7, 2005; Farhad Manjoo, "Invincible Apple," *Fast Company,* July 1, 2010; email from Phil Schiller.

iTunes: Interviews with Steve Jobs, Phil Schiller, Jon Rubinstein, Tony Fadell. Brent Schlender, "How Big Can Apple Get," *Fortune,* Feb. 21, 2005; Bill Kincaid, "The True Story of SoundJam," http://panic.com/extras/audionstory/popup-sjstory .html; Levy, *The Perfect Thing,* 49–60; Knopper, 167; Lev Grossman, "How Apple Does It," *Time,* Oct. 17, 2005; Markoff, xix.

The iPod: Interviews with Steve Jobs, Phil Schiller, Jon Rubinstein, Tony Fadell. Steve Jobs, iPod announcement, Oct. 23, 2001; Toshiba press releases, PR Newswire, May 10, 2000, and June 4, 2001; Tekla Perry, "From Podfather to Palm's Pilot," *IEEE Spectrum,* Sept. 2008; Leander Kahney, "Inside Look at Birth of the iPod," *Wired,* July 21, 2004; Tom Hormby and Dan Knight, "History of the iPod," *Low End Mac,* Oct. 14, 2005.

That's It! Interviews with Tony Fadell, Phil Schiller, Jon Rubinstein, Jony Ive, Steve Jobs. Levy, *The Perfect Thing,* 17, 59–60; Knopper, 169; Leander Kahney, "Straight Dope on the IPod's Birth," *Wired,* Oct. 17, 2006.

The Whiteness of the Whale: Interviews with James Vincent, Lee Clow, Steve Jobs. Wozniak, 298; Levy, *The Perfect Thing,* 73; Johnny Davis, "Ten Years of the iPod," *Guardian,* Mar. 18, 2011.

Chapter 31: The iTunes Store

Warner Music: Interviews with Paul Vidich, Steve Jobs, Doug Morris, Barry Schuler, Roger Ames, Eddy Cue. Paul Sloan, "What's Next for Apple,"

Business 2.0, Apr. 1, 2005; Knopper, 157–161,170; Devin Leonard, "Songs in the Key of Steve," *Fortune,* May 12, 2003; Tony Perkins, interview with Nobuyuki Idei and Sir Howard Stringer, World Economic Forum, Davos, Jan. 25, 2003; Dan Tynan, "The 25 Worst Tech Products of All Time," *PC World,* Mar. 26, 2006; Andy Langer, "The God of Music," *Esquire,* July 2003; Jeff Goodell, "Steve Jobs," *Rolling Stone,* Dec. 3, 2003.

Herding Cats: Interviews with Doug Morris, Roger Ames, Steve Jobs, Jimmy Iovine, Andy Lack, Eddy Cue, Wynton Marsalis. Knopper, 172; Devin Leonard, "Songs in the Key of Steve," *Fortune,* May 12, 2003; Peter Burrows, "Show Time!" *Business Week,* Feb. 2, 2004; Pui-Wing Tam, Bruce Orwall, and Anna Wilde Mathews, "Going Hollywood," *Wall Street Journal,* Apr. 25, 2003; Steve Jobs, keynote speech, Apr. 28, 2003; Andy Langer, "The God of Music," *Esquire,* July 2003; Steven Levy, "Not the Same Old Song," *Newsweek,* May 12, 2003.

Microsoft: Interviews with Steve Jobs, Phil Schiller, Tim Cook, Jon Rubinstein, Tony Fadell, Eddy Cue. Emails from Jim Allchin, David Cole, Bill Gates, Apr. 30, 2003 (these emails later became part of an Iowa court case and Steve Jobs sent me copies); Steve Jobs, presentation, Oct. 16, 2003; Walt Mossberg interview with Steve Jobs, All Things Digital conference, May 30, 2007; Bill Gates, "We're Early on the Video Thing," *Business Week,* Sept. 2, 2004.

Mr. Tambourine Man: Interviews with Andy Lack, Tim Cook, Steve Jobs, Tony Fadell, Jon Rubinstein. Ken Belson, "Infighting Left Sony behind

Apple in Digital Music," *New York Times,* Apr. 19, 2004; Frank Rose, "Battle for the Soul of the MP3 Phone," *Wired,* Nov. 2005; Saul Hansel, "Gates vs. Jobs: The Rematch," *New York Times,* Nov. 14, 2004; John Borland, "Can Glaser and Jobs Find Harmony?" *CNET News,* Aug. 17, 2004; Levy, *The Perfect Thing,* 169.

Chapter 32: Music Man

On His iPod: Interviews with Steve Jobs, James Vincent. Elisabeth Bumiller, "President Bush's iPod," *New York Times,* Apr. 11, 2005; Levy, *The Perfect Thing,* 26–29; Devin Leonard, "Songs in the Key of Steve," *Fortune,* May 12, 2003.

Bob Dylan: Interviews with Jeff Rosen, Andy Lack, Eddy Cue, Steve Jobs, James Vincent, Lee Clow. Matthew Creamer, "Bob Dylan Tops Music Chart Again — and Apple's a Big Reason Why," *Ad Age,* Oct. 8, 2006.

The Beatles; Bono; Yo-Yo Ma: Interviews with Bono, John Eastman, Steve Jobs, Yo-Yo Ma, George Riley.

Chapter 33: Pixar's Friends

A Bug's Life: Interviews with Jeffrey Katzenberg, John Lasseter, Steve Jobs. Price, 171–174; Paik, 116; Peter Burrows, "Antz vs. Bugs" and "Steve Jobs: Movie Mogul," *Business Week,* Nov. 23, 1998; Amy Wallace, "Ouch! That Stings," *Los Angeles Times,* Sept. 21, 1998; Kim Masters, "Battle of the Bugs," *Time,* Sept. 28, 1998; Richard Schickel, "Antz," *Time,* Oct. 12, 1998; Richard Corliss, "Bugs Funny," *Time,* Nov. 30, 1998.

Steve's Own Movie: Interviews with John Lasseter, Pam Kerwin, Ed Catmull, Steve Jobs. Paik, 168; Rick Lyman, "A Digital Dream Factory in Silicon Valley," *New York Times,* June 11, 2001.

The Divorce: Interviews with Mike Slade, Oren Jacob, Michael Eisner, Bob Iger, Steve Jobs, John Lasseter, Ed Catmull. James Stewart, *Disney War* (Simon & Schuster, 2005), 383; Price, 230–235; Benny Evangelista, "Parting Slam by Pixar's Jobs," *San Francisco Chronicle,* Feb. 5, 2004; John Markoff and Laura Holson, "New iPod Will Play TV Shows," *New York Times,* Oct. 13, 2005.

Chapter 34: Twenty-First-Century Macs

Clams, Ice Cubes, and Sunflowers: Interviews with Jon Rubinstein, Jony Ive, Laurene Powell, Steve Jobs, Fred Anderson, George Riley. Steven Levy, "Thinking inside the Box," *Newsweek,* July 31, 2000; Brent Schlender, "Steve Jobs," *Fortune,* May 14, 2001; Ian Fried, "Apple Slices Revenue Forecast Again," *CNET News,* Dec. 6, 2000; Linzmayer, 301; U.S. Design Patent D510577S, granted on Oct. 11, 2005.

Intel Inside: Interviews with Paul Otellini, Bill Gates, Art Levinson. Carlton, 436.

Options: Interviews with Ed Woolard, George Riley, Al Gore, Fred Anderson, Eric Schmidt. Geoff Colvin, "The Great CEO Heist," *Fortune,* June 25, 2001; Joe Nocera, "Weighing Jobs's Role in a Scandal," *New York Times,* Apr. 28, 2007; Deposition of Steven P. Jobs, Mar. 18, 2008, *SEC v. Nancy Heinen,* U.S. District Court, Northern District of California; William Barrett, "Nobody Loves Me," *Forbes,* May 11, 2009; Peter Elkind, "The Trouble with Steve Jobs," *Fortune,* Mar. 5, 2008.

Chapter 35: Round One

Cancer: Interviews with Steve Jobs, Laurene Powell, Art Levinson, Larry Brilliant, Dean Ornish, Bill Campbell, Andy Grove, Andy Hertzfeld.

The Stanford Commencement: Interviews with Steve Jobs, Laurene Powell. Steve Jobs, Stanford commencement address.

A Lion at Fifty: Interviews with Mike Slade, Alice Waters, Steve Jobs, Tim Cook, Avie Tevanian, Jony Ive, Jon Rubinstein, Tony Fadell, George Riley, Bono, Walt Mossberg, Steven Levy, Kara Swisher. Walt Mossberg and Kara Swisher interviews with Steve Jobs and Bill Gates, All Things Digital conference, May 30, 2007; Steven Levy, "Finally, Vista Makes Its Debut," *Newsweek,* Feb. 1, 2007.

Chapter 36: The iPhone

An iPod That Makes Calls: Interviews with Art Levinson, Steve Jobs, Tony Fadell, George Riley, Tim Cook. Frank Rose, "Battle for the Soul of the MP3 Phone," *Wired,* Nov. 2005.

Multi-touch: Interviews with Jony Ive, Steve Jobs, Tony Fadell, Tim Cook.

Gorilla Glass: Interviews with Wendell Weeks, John Seeley Brown, Steve Jobs.

The Design: Interviews with Jony Ive, Steve Jobs, Tony Fadell. Fred Vogelstein, "The Untold Story," *Wired,* Jan. 9, 2008.

The Launch: Interviews with John Huey, Nicholas Negroponte. Lev Grossman, "Apple's New Calling," *Time,* Jan. 22, 2007; Steve Jobs, speech, Macworld, Jan. 9, 2007; John Markoff, "Apple Introduces Innovative Cellphone," *New York Times,*

Jan. 10, 2007; John Heilemann, "Steve Jobs in a Box," *New York,* June 17, 2007; Janko Roettgers, "Alan Kay: With the Tablet, Apple Will Rule the World," *GigaOM,* Jan. 26, 2010.

Chapter 37: Round Two

The Battles of 2008: Interviews with Steve Jobs, Kathryn Smith, Bill Campbell, Art Levinson, Al Gore, John Huey, Andy Serwer, Laurene Powell, Doug Morris, Jimmy Iovine. Peter Elkind, "The Trouble with Steve Jobs," *Fortune,* Mar. 5, 2008; Joe Nocera, "Apple's Culture of Secrecy," *New York Times,* July 26, 2008; Steve Jobs, letter to the Apple community, Jan. 5 and Jan. 14, 2009; Doron Levin, "Steve Jobs Went to Switzerland in Search of Cancer Treatment," *Fortune.com,* Jan. 18, 2011; Yukari Kanea and Joann Lublin, "On Apple's Board, Fewer Independent Voices," *Wall Street Journal,* Mar. 24, 2010; Micki Maynard (Micheline Maynard), Twitter post, 2:45 p.m., Jan. 18, 2011; Ryan Chittum, "The Dead Source Who Keeps on Giving," *Columbia Journalism Review,* Jan. 18, 2011.

Memphis: Interviews with Steve Jobs, Laurene Powell, George Riley, Kristina Kiehl, Kathryn Smith. John Lauerman and Connie Guglielmo, "Jobs Liver Transplant," *Bloomberg,* Aug. 21, 2009.

Return: Interviews with Steve Jobs, George Riley, Tim Cook, Jony Ive, Brian Roberts, Andy Hertzfeld.

Chapter 38: The iPad

You Say You Want a Revolution: Interviews with

Steve Jobs, Phil Schiller, Tim Cook, Jony Ive, Tony Fadell, Paul Otellini. All Things Digital conference, May 30, 2003.

The Launch, January 2010: Interviews with Steve Jobs, Daniel Kottke. Brent Schlender, "Bill Gates Joins the iPad Army of Critics," *bnet.com,* Feb. 10, 2010; Steve Jobs, keynote address in San Francisco, Jan. 27, 2010; Nick Summers, "Instant Apple iPad Reaction," *Newsweek.com,* Jan. 27, 2010; Adam Frucci, "Eight Things That Suck about the iPad" Gizmodo, Jan. 27, 2010; Lev Grossman, "Do We Need the iPad?" *Time,* Apr. 1, 2010; Daniel Lyons, "Think Really Different," *Newsweek,* Mar. 26, 2010; Techmate debate, *Fortune,* Apr. 12, 2010; Eric Laningan, "Wozniak on the iPad" TwiT TV, Apr. 5, 2010; Michael Shear, "At White House, a New Question: What's on Your iPad?" *Washington Post,* June 7, 2010; Michael Noer, "The Stable Boy and the iPad," *Forbes.com,* Sept. 8, 2010.

Advertising: Interviews with Steve Jobs, James Vincent, Lee Clow.

Apps: Interviews with Art Levinson, Phil Schiller, Steve Jobs, John Doerr.

Publishing and Journalism: Interviews with Steve Jobs, Jeff Bewkes, Rick Stengel, Andy Serwer, Josh Quittner, Rupert Murdoch. Ken Auletta, "Publish or Perish," *New Yorker,* Apr. 26, 2010; Ryan Tate, "The Price of Crossing Steve Jobs," Gawker, Sept. 30, 2010.

Chapter 39: New Battles

Google: Open versus Closed: Interviews with Steve Jobs, Bill Campbell, Eric Schmidt, John Doerr, Tim Cook, Bill Gates. John Abell, "Google's

'Don't Be Evil' Mantra Is 'Bullshit,' " *Wired,* Jan. 30, 2010; Brad Stone and Miguel Helft, "A Battle for the Future Is Getting Personal," *New York Times,* March 14, 2010.

Flash, the App Store, and Control: Interviews with Steve Jobs, Bill Campbell, Tom Friedman, Art Levinson, Al Gore. Leander Kahney, "What Made Apple Freeze Out Adobe?" *Wired,* July 2010; Jean-Louis Gassée, "The Adobe-Apple Flame War," *Monday Note,* Apr. 11, 2010; Steve Jobs, "Thoughts on Flash," Apple.com, Apr. 29, 2010; Walt Mossberg and Kara Swisher, Steve Jobs interview, All Things Digital conference, June 1, 2010; Robert X. Cringely (pseudonym), "Steve Jobs: Savior or Tyrant?" *InfoWorld,* Apr. 21, 2010; Ryan Tate, "Steve Jobs Offers World 'Freedom from Porn,' " Valleywag, May 15, 2010; JR Raphael, "I Want Porn," esarcasm.com, Apr. 20, 2010; Jon Stewart, *The Daily Show,* Apr. 28, 2010.

Antennagate: Design versus Engineering: Interviews with Tony Fadell, Jony Ive, Steve Jobs, Art Levinson, Tim Cook, Regis McKenna, Bill Campbell, James Vincent. Mark Gikas, "Why Consumer Reports Can't Recommend the iPhone4," *Consumer Reports,* July 12, 2010; Michael Wolff, "Is There Anything That Can Trip Up Steve Jobs?" *newser.com* and *vanityfair.com,* July 19, 2010; Scott Adams, "High Ground Maneuver," dilbert.com, July 19, 2010.

Here Comes the Sun: Interviews with Steve Jobs, Eddy Cue, James Vincent.

Chapter 40: To Infinity

The iPad 2: Interviews with Larry Ellison, Steve

Jobs, Laurene Powell. Steve Jobs, speech, iPad 2 launch event, Mar. 2, 2011.

iCloud: Interviews with Steve Jobs, Eddy Cue. Steve Jobs, keynote address, Worldwide Developers Conference, June 6, 2011; Walt Mossberg, "Apple's Mobile Me Is Far Too Flawed to Be Reliable," *Wall Street Journal,* July 23, 2008; Adam Lashinsky, "Inside Apple," *Fortune,* May 23, 2011; Richard Waters, "Apple Races to Keep Users Firmly Wrapped in Its Cloud," *Financial Times,* June 9, 2011.

A New Campus: Interviews with Steve Jobs, Steve Wozniak, Ann Bowers. Steve Jobs, appearance before the Cupertino City Council, June 7, 2011.

Chapter 41: Round Three

Family Ties: Interviews with Laurene Powell, Erin Jobs, Steve Jobs, Kathryn Smith, Jennifer Egan. Email from Steve Jobs, June 8, 2010, 4:55 p.m.; Tina Redse to Steve Jobs, July 20, 2010, and Feb. 6, 2011.

President Obama: Interviews with David Axelrod, Steve Jobs, John Doerr, Laurene Powell, Valerie Jarrett, Eric Schmidt, Austan Goolsbee.

Third Medical Leave, 2011: Interviews with Kathryn Smith, Steve Jobs, Larry Brilliant.

Visitors: Interviews with Steve Jobs, Bill Gates, Mike Slade.

Chapter 42: Legacy

Jonathan Zittrain, *The Future of the Internet — And How to Stop It* (Yale, 2008), 2; Cory Doctorow, "Why I Won't Buy an iPad," Boing Boing, Apr. 2, 2010.

ILLUSTRATION CREDITS

Diana Walker — Contour by Getty Images: 1, 2, 3, 4, 5, 6, 7, 8, 9, 10, 11, 12, 13, 23
Courtesy of Steve Jobs: 14, 15, 17, 18, 19, 20, 21, 22
Courtesy of Kathryn Smith: 16

ABOUT THE AUTHOR

Walter Isaacson, the CEO of the Aspen Institute, has been the chairman of CNN and the managing editor of Time magazine. He is the author of *Einstein: His Life and Universe, Benjamin Franklin: An American Life,* and *Kissinger: A Biography,* and is the coauthor, with Evan Thomas, of *The Wise Men: Six Friends and the World They Made.* He and his wife live in Washington, D.C.